Nachhaltige Energieversorgung und Integration von Speichern

Detlef Schulz *(Hrsg.)*

Nachhaltige Energieversorgung und Integration von Speichern

Tagungsband zur NEIS 2015

Springer Vieweg

Herausgeber
Detlef Schulz
Helmut-Schmidt-Universität
Universität der Bundeswehr
Hamburg, Deutschland

ISBN 978-3-658-10957-8 ISBN 978-3-658-10958-5 (eBook)
DOI 10.1007/ 978-3-658-10958-5

Die Deutsche Nationalbibliothek verzeichnet diese Publikation in der Deutschen Nationalbibliographie; detaillierte bibliographische Daten sind im Internet über http://dnb.d-nb.de abrufbar.

Springer Vieweg

Gedruckt auf säurefreiem und chlorfrei gebleichtem Papier.

Springer Fachmedien Wiesbaden GmbH ist Teil der Fachverlagsgruppe
Springer Science+Business Media
(www.springer.com)

Mit freundlicher Unterstützung von
Cluster Erneuerbare Energien Hamburg

Vorwort

Der Umbau der elektrischen Energieversorgung zu einer Erzeugung ohne fossile und nukleare Brennstoffe, die sog. Dekarbonisierung, erfordert einen grundlegenden Systemwandel. Es besteht die Aufgabe, mit wetterabhängig fluktuierenden erneuerbaren Energien eine nachhaltige und sichere Versorgung zu gewährleisten. Dafür sind neue technische Lösungsansätze sowie geeignete politische und marktregulatorische Rahmenbedingungen notwendig.

In den Themenbereichen der Konferenz „Nachhaltige Energieversorgung und Integration von Energiespeichern" (NEIS 2015, www.neis-konferenz.de) werden die vielschichtigen Transformationsprozesse eines Systemwandels in der Energieversorgung abgebildet. Die NEIS 2015 fand am 10. und 11. September als dritte Veranstaltung dieser Konferenzreihe statt, die jährlich von der Professur für Elektrische Energiesysteme der Helmut-Schmidt-Universität/Universität der Bundeswehr Hamburg organisiert wird. Das hier bestehende Forschungscluster „Nachhaltige Energieversorgung" (http://www.hsu-hh.de/nev) bietet den Teilnehmern eine hervorragende Plattform für den interdisziplinären wissenschaftlichen Austausch.

Die Konferenzbeiträge der Keynote-Speaker Prof. PhD. Frede Blaabjerg von der Aaalborg University/Denmark zum Thema "Advanced Grid Integration of Renewables enabled by Power Electronics Technology" und Prof. Dr.-Ing. Christian Rehtanz von der Technischen Universität Dortmund zum Thema „Planung und Betrieb von Smart Grids" setzten wesentliche Impulse für nachfolgende Diskussionen. Hier wurden die Optionen für technische Entwicklungen gezeigt, die auch als „Schutz für zukünftige Investitionen" betrachtet werden können.

In den sechs folgenden Themenblöcken zeichneten sich teilweise eindeutige Trends ab. Im Themenblock „Rahmenbedingungen" wurde das Spannungsfeld von technischem und organisatorischem Umbruch und den dabei notwendigen und möglichen Anpassungen diskutiert. Im Block „Netzintegration" wurden technische Lösungsansätze bei zunehmender dezentraler Erzeugung präsentiert. Die Vorträge zum Thema „Regelleistung" beschäftigten sich mit dem Potenzial, der Simulation und der Wirtschaftlichkeitsberechnung von verschiedenen Speichertechnologien und Erzeugungsarten. Bei den „Mobilitätsanwendungen" wurden sowohl zukünftige Geschäftsmodelle als auch angepasste technische Lösungen präsentiert. Im Themenblock „Brennstoffzellen und thermische Systeme" stand die Simulation und Regelung verschiedener Technologien im Vordergrund. Abschließend wurden im Block „Gesamtsystem" der Energietransportbedarf sowie Energiespeichersysteme und ökonomische Bewertungsmethoden vorgestellt. Bei der Lektüre der einzelnen Tagungsbeiträge wünsche ich Ihnen viel Freude.

Mein Dank gilt an dieser Stelle den Keynote-Speakern, den Session-Leitern und den Vortragenden für ihre wissenschaftlichen Beiträge. Unserem sehr engagierten Team von Wissenschaftlern und fleißigen Helfern danke ich für die Organisation, Vorbereitung und Unterstützung bei der Durchführung der Konferenz. Hervorzuheben sind dabei die in diesem Jahr verantwortlichen Organisatoren Frau M. Sc. Gesa Kaatz und Herr M. Sc. Markus Dietmannsberger, bei denen ich mich besonders bedanke.

Für die Durchführung der interessanten Konferenz-Exkursion in die Netzleitwarte Hamburg bedanke ich mich bei Herrn Gero Boomgaarden (Leiter Netzbetrieb), Herrn Claus-Peter Siemens (Leiter Netzführung Mittelspannung) und Herrn Torsten Behrens (Netzführung) von der Stromnetz Hamburg GmbH.

Detlef Schulz
Hamburg, im Oktober 2015

Inhaltsverzeichnis

Keynotes

Advanced Grid Integration of Renewables Enabled by Power Electronics Technology

Frede Blaabjerg, Yongheng Yang, Ke Ma, Xiongfei Wang
Department of Energy Technology, Aalborg University
Pontoppidanstraede 101, Aalborg DK-9220, Denmark
fbl@et.aau.dk, yoy@et.aau.dk, kema@et.aau.dk, xwa@et.aau.dk

Abstract

In the past decades, many countries (e.g., Germany and Denmark) have experienced a significant change in their energy structures – from fossil-based resources to clean renewables. The scenario of highly penetrated renewables is going to be further enhanced in the future mixed energy paradigms. This requires that the production, distribution and use of the energy should be as technological efficient as possible and incentives to save energy at the end-user should also be strengthened. In order to realize the transition smoothly and effectively, energy conversion systems, currently based on power electronics technology, will again play an essential role in advancing the grid integration of renewables. In view of this issue, some of the most emerging renewable energies, e.g., wind energy and photovoltaic, which by means of power electronics are changing character as a major part in the electricity generation, are explored. Issues like demands to renewables, power converter technologies, control of the systems, and advanced grid integration are covered.

1 Introduction

Demands of reliable and environmental-friendly electricity generation from Renewable Energy Systems (RESs) have been the main driving force for the RES development [1]-[5]. Consequently, great efforts have been made by many countries (e.g., Germany, Spain, and Denmark) to alter their energy paradigms with more installations of renewables such as wind power, PhotoVoltaic (PV) power, hydropower, and biomass power. Among various renewable power systems, Wind Turbine System (WTS) and PV system technologies are still the most promising technologies, accounting for a large portion of renewable energy generation [4]-[14], and will expand more. However, the increasing adoption of RESs poses two major challenges. One is the energy structure transition – from the conventional and fossil-based energy to renewables. The other one is the wide-scale use of power electronics in the power generation, the grid integration, the power transmission/distribution and the end-user application.

On the other hand, the power electronics technology has become the enabling technology to advance the grid integration of various renewables. Hence, the power electronics systems should be highly efficient and exceedingly reliable. Basically, it should be able to transfer the renewable energies to the power grid. More important, it should be capable to exhibit advanced ancillary functions (e.g., Low Voltage Ride-Through, LVRT, grid support with reactive power injection). A wide-scale adoption of power electronics technology makes those completely weather-dependent energies more controllable, but increasingly intricate. Underpinned by intelligent control strategies, the power electronics technology can fulfill the requirements imposed by the distribution/transmission system operators as well as specific demands from the end-users, especially when more advanced power devices and more accurate knowledge of the mission profiles are available.

In this paper, the power electronics technology, enabling a clean and reliable power conversion from renewables, is discussed. In § II, the basic demands to RESs are presented, followed by the WTS and PV technologies including main power converter topologies. Then, typical control strategies for PV systems and wind turbines are presented considering the grid demands. Due to the increasing complexity of the future power systems integrated with a large amount of RESs, in § III, the advanced grid integration enabled by the power electronics technology is summarized. Finally, the conclusions and perspectives are given for the two main renewable energies.

2 Demands to Renewables

Fig. 1 demonstrates the architecture of a typical RES based power system, where the power electronics unit is the core of the system. Increasing penetration of RESs results in rigorous demands to the key part of the entire system, i.e., the power electronics. As shown in Fig. 1, the tasks of a power electronics based RES are as varied as they are demanded by the local operators or the end-users [4]-[6]. A very basic demand is to transfer the energy to the grid according to the renewable energy characteristics. Other specific demands to the RESs can be summarized as: a) reliable/secure power supply, b) high efficiency, low cost, small volume, and effective protection, c) control of active and reactive power injected into the grid, d) dynamic grid support (ride-through operation), and e) system monitoring and communication.

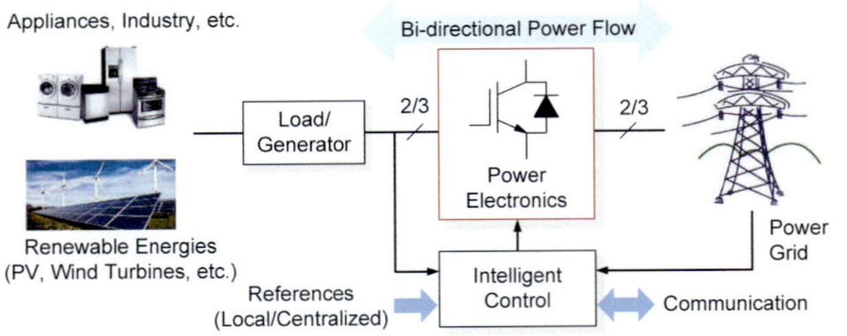

Fig. 1. Power electronics technologies and intelligent control techniques enabled renewable energy system.

2.1 General Requirements for Renewables

2.1.1 Demands to Wind Power Conversion Systems

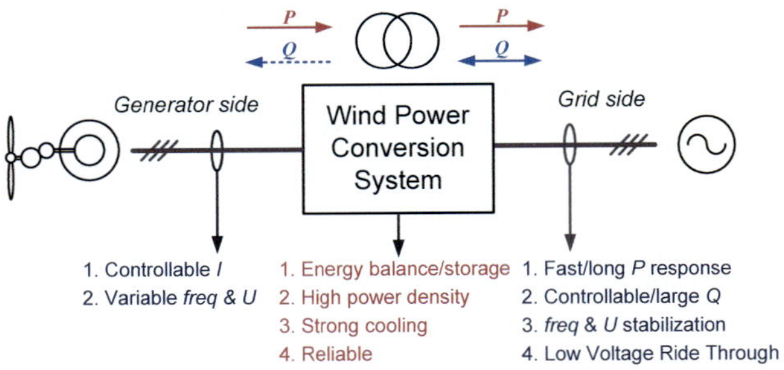

Fig. 2. Demands to wind turbine power systems.

Fig. 2 shows the demands to wind turbine power systems at different levels. Specifically, for the generator side, the current flowing in the generator rotor or stator should be regulated to control the electromagnetic torque, not only for maximizing the extracted power from the wind turbines, but also for the energy balancing in case of dynamics due to inertia mismatch between mechanical and electrical power. For the grid side, the converter must emulate the behaviors of conventional power plants regardless of the wind speed. This means it should help to maintain the frequency as well as voltage amplitude of the grid, and also withstand the grid faults or even contribute the faults recovery [4], [6], [11].

Due to the relative large power capacity, the failures of wind power conversion system will strongly impact the grid stability and result in high repairing cost; thereby the reliability performance is especially emphasized [14]-[17]. Also because of high power capacity, the voltage level of generator may need to be boosted up to facilitate the power transmission, and thus transformers are normally required. Furthermore, because of the limited space in the nacelle or tower of the WTS, the power density and cooling ability are crucial in wind power systems. Finally, due to the power mismatch between the wind turbine and the grid, and energy storage, balancing is an important issue and may result in extra cost of the system.

2.1.2 Demands to PV Power Conversion Systems

Due to the fast development, even tougher requirements have been released for PV systems [18]–[20]. These demands can be generally categorized into the three parts as shown in Fig. 3. However, the power capacity of a PV system is not as large as that of an individual WTS. Moreover, the power inertia of PV output is compatible with the grid behavior, and therefore, the demands to PV systems are less stringent than those to WTSs.

For the PV side, the current or voltage of PV panels should be controlled to extract the energy. In view of this, a DC-DC converter is commonly used in PV systems to flexibly track the maximum power, where the DC voltage should be maintained as a desirable value for the inverter. For the grid side, normally the Total Harmonic Distortion (THD) of the output current has to be restrained at a lower level (e.g., 5 %) [4], [18]. While for large PV power systems with higher power ratings (e.g., hundreds kW), the grid side also demands the PV inverter to stabilize the grid voltage by providing ancillary services. In response to grid faults, the PV inverters have to ride-through voltage faults, when a higher PV penetration level comes into reality [18]–[29], as what have been required for WTSs.

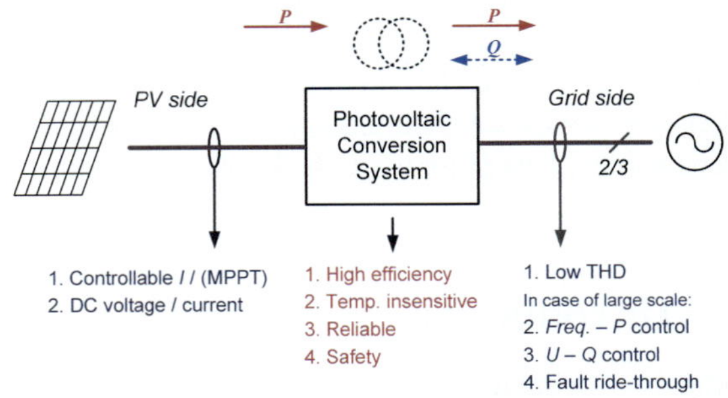

Fig. 3. Demands to PV power conversion systems.

Additionally, for the PV technology, the power capacity per generating unit is relative low but the cost of energy is relative high. Hence, there are very strong demands for high efficiency in PV systems. On the other hand, transformerless PV inverters have gained increasing popularity in the European market (e.g., Germany and Spain) [4], [30]–[32] in order to further extend the efficiency. However, in this case, the galvanic safety becomes a more crucial issue. Reduction of the potential leakage current is generally required in such applications. Furthermore, similar to the wind power conversion systems, reliability is also important for power electronics based PV systems, and motivated by extending the total energy production and reducing the cost [18], [33], [34]. Finally, because of exposure or smaller housing chamber, the PV converter system must be more temperature insensitive, which is also beneficial to the reliability performance.

2.2 Grid Integration Requirements

Fluctuation and unpredictability are the characteristics of renewable power, which is not preferable for grid operation. Hence, the grid integration focuses on the RES connection point in order to alleviate the potential harmfulness to the grid. In that case, the RESs should not only be passively and simply injecting available power to the grid, but also behave like an active generation unit, which can manage the power exchange with the grid according to demands, and provide frequency/voltage support for the power grid, as aforementioned.

Many grid integrated systems are demanded to control the active power at the Point-of-Common-Coupling (PCC). Normally, the active power has to be regulated based on the grid frequency. For example, in the Danish grid code, the active power should be decreased when the frequency rises above 48.7 Hz or 50.15 Hz depending on the power reserving strategy [35]. Similarly, the reactive power provided by the RESs has to be regulated in a certain range, as exemplified in Fig. 4, which gives a range of the reactive power delivered by the WTSs in respect to the active power [36]. Moreover, the Transmission System Operator (TSO) will normally specify the reactive power range in order to maintain the grid voltage levels. Notably, this reactive power control should be realized slowly under the time constant of minutes in steady state operation [37].

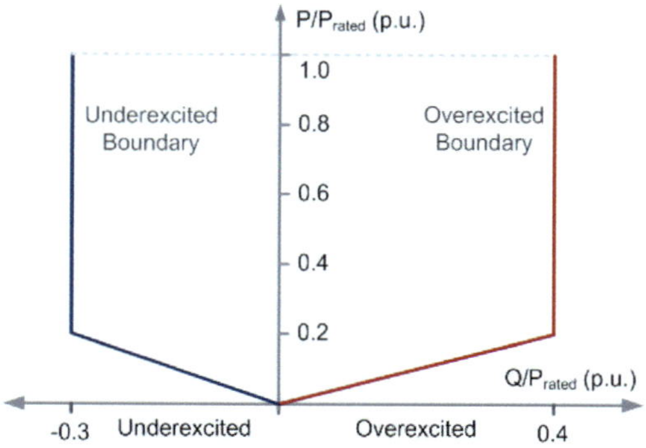

Fig. 4. Reactive power ranges for a wind farm specified in the German grid codes [36].

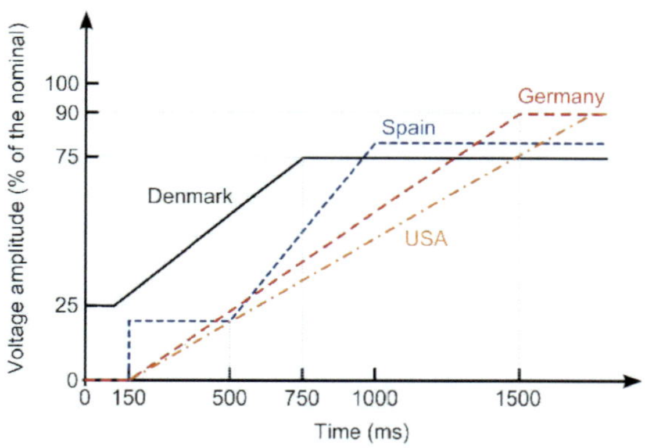

Fig. 5. Voltage profiles for the low voltage ride-through (LVRT) capability defined by different countries [37].

Besides the normal operation, the TSOs in different countries have released grid support requirements for the generating units under grid faults. As shown in Fig. 5 [36], [37], in which voltage profiles are defined for wind turbine systems in response to grid faults. Additionally, it has been a demand that the RES system should also provide reactive power (up to 100 % current capacity) to contribute to the voltage recovery. The requirements for more grid supports by the grid-connected renewables on one hand have increased the cost per produced kWh, but on the other hand made them more suitable to be largely utilized and integrated into the grid. It can be predicted that the stricter grid codes in the future will keep challenging the renewable systems and pushing forward the power electronic technologies.

3 Power Electronics Technology – advancing the grid integration

The design and operation of power electronics converters for both wind turbine and PV systems strongly rely on the grid requirements and the energy demand. It can be seen from the evolution of wind turbine power converters, which has changed from non-power-electronics-based topologies to full-scale power converters with increasing power ratings of individual wind turbine [4], [10]–[12]. As the demand of higher power ratings and efficiency increases for PV systems, the PV power converters also had experienced a clear change, and they are mostly transformerless nowadays [4], [8], [9], [31].

3.1 Power Converter Technology

3.1.1 Power Converters for WTSs

Depending on the types of generator, power electronics, speed controllability, and the way in which the aerodynamic power is limited, the wind turbine designs can generally be categorized into several concepts. In these applications, the power converters play different roles and have various power ratings. Up till now, the configuration of Doubly Fed Induction Generator (DFIG) with partial-scale power converters still dominates on the market, as shown in Fig. 6(a). In very near future, the configuration with Synchronous Generator (SG) with full-scale power converters is expected to take over the WTS market [10], [14], [17], as it is shown in Fig. 6(b). Each of the WTS concepts has its suitable converter topologies and some of them are illustrated in the following.

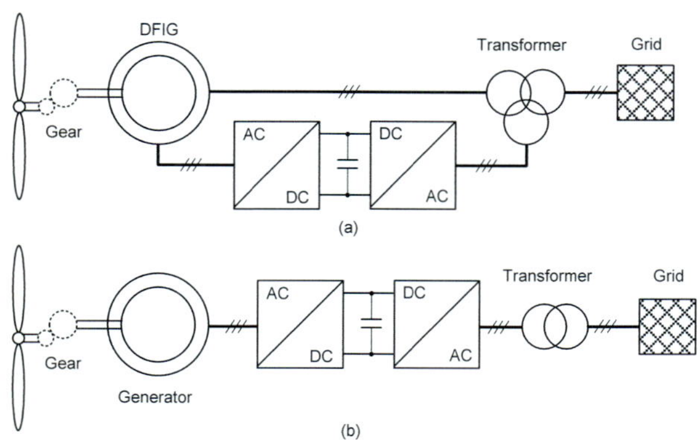

Fig. 6. Dominant WTS configurations: a) variable speed wind turbine with partial-scale power converter and a doubly fed induction generator and (b) variable speed wind turbine with full-scale power converter and an asynchronous/synchronous generator.

A. Two-level power converters

Due to simplicity, the two-Level Voltage Source Converter (2L-VSC) is the most common solution so far in the DIFG WTSs, since the power rating requirement is limited. In practice, two 2L-VSCs are configured in a Back-

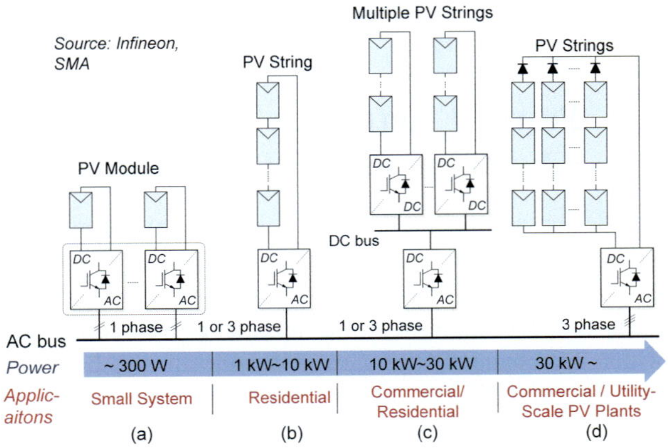

Fig. 7. Grid-connected PV systems with: (a) module inverter, (b) string inverter, (c) multi-string inverter, and (d) central inverter [8].

To-Back (BTB) way, as shown in Fig. 7. One advantage of the 2L-BTB solution lies in the full power controllability with a relatively simple structure and few components, which in return contribute to a well-proven, robust, and reliable performance as well as lower cost.

B. Multi-level power converters

With the abilities to achieve a higher voltage and a higher power level, multi-level power converters are getting more popularity in WTSs [13], [14]. The three-Level Neutral Point Clamped (3L-NPC) topology is one of the multi-level topologies on the market, as shown in Fig. 8. The 3L-NPC BTB solution achieves one more output voltage levels and less dv/dt stresses in contrast to the 2L-BTB, and thus it is possible to convert the power at MV level with lower currents, less paralleled devices, and smaller filter size. The main drawback of the 3L-NPC BTB is the mid-point voltage fluctuation of the DC-bus, which has been extensively investigated [14], [38].

In order to handle the fast growth in the power capacity, multi-cell converter configurations (i.e. parallel and/or series connection of converter units) are also being developed and becoming widely commercialized in the wind turbine industry [39].

3.1.2 Power Converters for PV Systems

For PV systems, a general classification of grid-connected PV inverters is shown in Fig. 7 [8], [31]. A common central inverter can be used in a PV plant larger than tens kWp with higher efficiency and lower cost [8]. Compared to central inverters, the string inverter can achieve MPPT separately, leading to better total energy yield. The module inverter acts on a single PV panel with a single MPPT. As shown in Fig. 9, another PV technology is an intermediate solution between the string inverter and the module inverter, being multi-string inverter. This configuration is flexible with a high overall efficiency because each PV string is controlled separately.

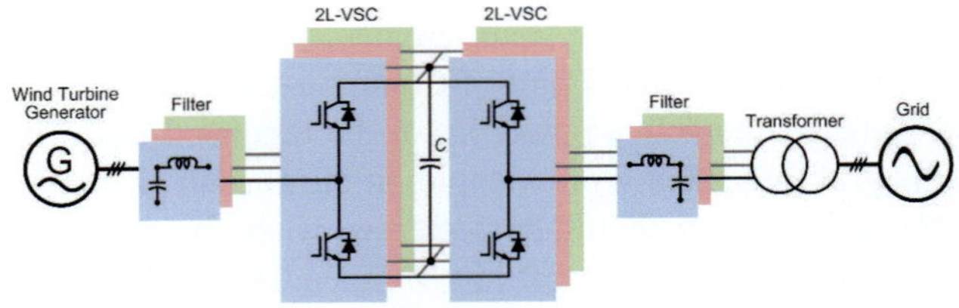

Fig. 8. 2L-VSC BTB voltage source converter for WTSs.

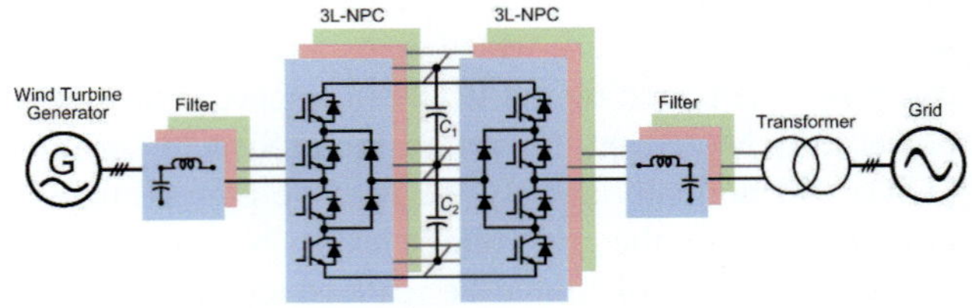

Fig. 9. 3L-NPC BTB topology for WTSs.

Additionally, the PV systems are still dominant in residential applications with much lower power ratings (e.g. several kW). Thus, single-phase topologies are more common. However, several PV power plants have come into service recently using central inverters (e.g., SMA Sunny Central CP XT inverter) and more are under construction. The power converter technology for this is similar to the grid-side converter technology in WTSs, where multi-level technologies can be employed.

In respect to the design of PV inverters, the efficiency and leakage current are two main considerations. As aforementioned, transformerless PV inverters are developed for higher efficiency [4], [8], [9], [40]–[45]. A widely adopted single-phase PV inverter is the Full-Bridge (FB) topology as shown in Fig. 10. In the light of safety issues, the FB with a bipolar modulation is feasible in transformerless PV applications. However, the conversion efficiency is not very satisfied. Hence, many other transformerless PV inverters are derived from the FB topology. For instance, the H6 inverter patented by Ingeteam [41] shown in Fig. 11 disconnects the PV panels/strings from the inverter using four extra devices to realize the "isolation"; while the Highly Efficient and Reliable Inverter Concept (HERIC inverter) by Sunways [42] provides an AC bypass. There have been other topologies reported in the literature [9], [43]–[45].

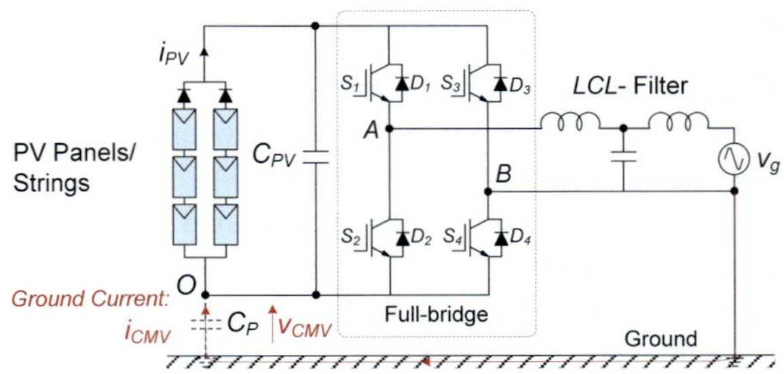

Fig. 10. Single-phase full-bridge PV inverter.

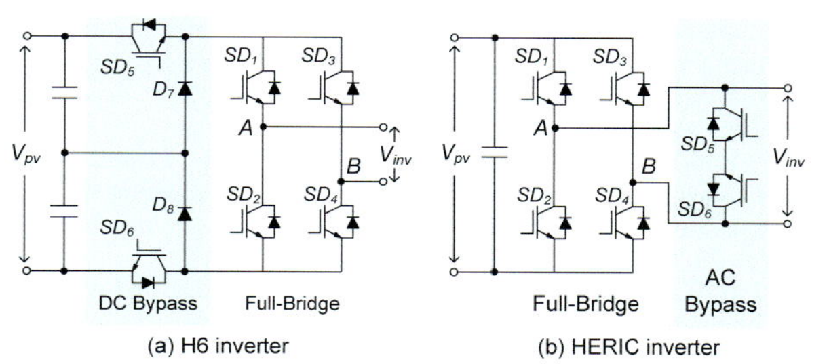

Fig. 11. Two transformerless PV inverters (H6 and HERIC) [41], [42]: (a) H6 inverter and (b) HERIC inverter.

3.2 General Control of Renewables

The DSO/TSO has given priority to finding a solution to guarantee stable operation of RESs and thus accept even more renewable energies. This consideration should be taken into account, which makes the control systems of RESs multi-functional, as shown in Fig. 12. The basic controls like current regulation, DC bus stabilization and grid synchronization have to be quickly performed by the power converter, where a Proportional-Integral (PI) controller and Proportional-Resonant (PR) controllers are typically used [6]. As the penetration level of RESs continues growing, it will be mandatory for the RESs to provide ancillary services, such as LVRT, reactive power control and frequency control through active power control, in order to ensure a reliable and efficient power conversion from such renewable energies. Hence, beyond the basic power extracting control, the RESs have also to perform advanced control functions, as shown in Fig. 12.

In addition, the injected current into the grid has to be synchronized well with the grid voltage, as standards require that in the field [4]–[8]. Therefore, the grid synchronization issue plays an important role for both WTSs and PV systems. To address this problem, Phase Locked Loop (PLL) based synchronization methods stand out of various reported solutions [6]. Evaluating criterions for synchronization methods are the dynamic response speed and the disturbance rejection capability. The Second Order Generalized Integrator based PLL (SOGI-PLL) presents a better performance compared to other methods, especially for single-phase systems [4], [6]. It can be a good candidate for the synchronization for RESs and used in industrial applications.

More important, in respect to the aforementioned control methods for WTSs and PV systems, a fast and accurate synchronization system will strongly contribute to the dynamic performance and the stability margin of the whole control systems. The knowledge of grid conditions significantly affects the control systems in different operation modes. For example, the detection of the grid faults and the extraction of positive and negative sequence currents are of importance for the control of RESs in LVRT operation modes.

3.3 Advanced Grid Integration

As the heart of every renewable energy generation system, the power electronics converter is responsible for the power generation from wind and solar PV energy efficiently and reliably. Thus, to realize a widespread adoption of such renewables, the power electronics technology will be more active into the grid in the future [46], [47]. Together with advanced control strategies, it can fulfill the upcoming stringent requirements regarding the efficiency, the controllability, the cost and the reliability.

It can be foreseen that more focuses will be devoted on:

1) Advancing the power electronics/devices technologies, as well as increasing the utilization of power electronics;
2) Developing proper and more stringent grid demands for a better integration of the renewables;
3) Lowering the cost of energy [48] by means of improving the conversion efficiency and reducing the system downtime.

Nevertheless, the power electronics technology can enable more advanced integration of renewables into the power grid.

4 Conclusions

In this paper, the power electronics technology as the enabling technology for advanced grid integration of renewables has been discussed. After the introduction of demands to renewables, an overview of the mainstream power converter topologies for PV systems and WTSs has also been given, together with their general control strategies. It can be concluded that the power electronics technology is playing an important role in the electricity generation, and is advancing the integration of renewables into the grid. In the future, it is expected that there will be more power electronics systems associated with intelligent control strategies for the renewable energy systems in order to further increase the power capacity, where the power electronics technology will contribute more to the advancement of grid integration.

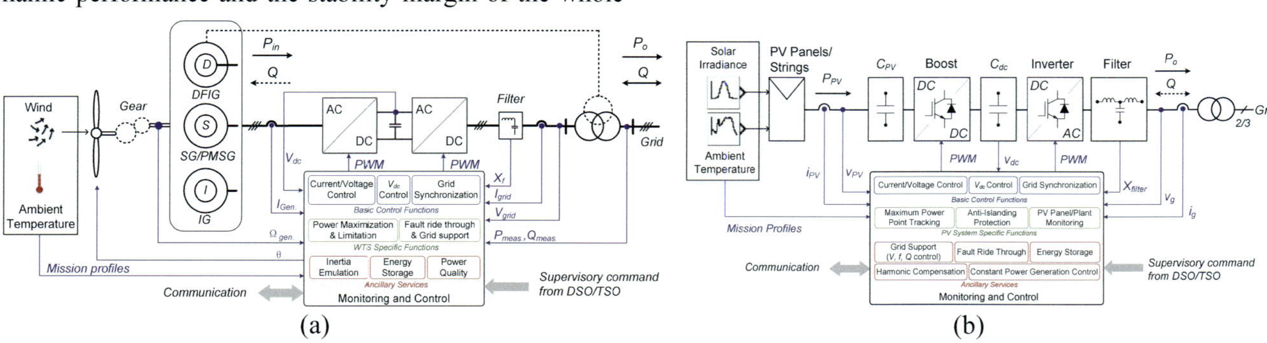

Fig. 12. General control function blocks for: (a) wind turbine power systems and (b) PV power systems.

5 References

[1] REN21, "Renewables 2015: Global Status Report (GSR)," [Online]. Available: www.ren21.net/, Jun. 2015.

[2] SolarPower Europe, "Global market outlook for solar power – 2015-2019," [Online]. Available: www.solarpowereurope.org/, Jun. 2015.

[3] E. J. Coster, J. M. A. Myrzik, B. Kruimer, and W. L. Kling, "Integration issues of distributed generation in distribution grids," *Proc. IEEE*, vol. 99, no. 1, pp. 28–39, Jan. 2011.

[4] R. Teodorescu, M. Liserre, and P. Rodriguez, *Grid Converters for Photovoltaic and Wind Power Systems*. Hoboken, NJ: Wiley, 2011.

[5] F. Blaabjerg and D. M. Ionel, "Renewable enegy devices and systems – state-of-the-art technology, research and development, challenges and future trends," *Electric Power Components and Systems*, vol. 43, no. 12, pp. 1319-1328, Jul. 2015.

[6] F. Blaabjerg, R. Teodorescu, M. Liserre, and A.V. Timbus, "Overview of control and grid synchronization for distributed power generation systems," *IEEE Trans. Ind. Electron.*, vol. 53, no. 5, pp. 1398-1409, Oct. 2006.

[7] Abu-Rub, H., Malinowski, M., and Al-Hadad, K., Eds., *Power Electronics for Renewable Energy Systems, Transportation and Industrial Applications*, Hoboken, NJ: Wiley, 2014.

[8] S.B. Kjaer, J.K. Pedersen, and F. Blaabjerg, "A review of single-phase grid-connected inverters for photovoltaic modules," *IEEE Trans. Ind. Appl.*, vol. 41, no. 5, pp. 1292-1306, Sept.-Oct. 2005.

[9] D. Meneses, F. Blaabjerg, O. García, and J.A. Cobos, "Review and comparison of step-up transformerless topologies for photovoltaic AC-module application," *IEEE Trans. Power Electron.*, vol. 28, no. 6, pp. 2649-2663, Jun. 2013.

[10] F. Blaabjerg and K. Ma, "Future on power electronics for wind turbine systems," *IEEE J. Emerg. Sel. Top. Power Electron.*, vol. 1, no. 3, pp. 139 - 152, Sept. 2013.

[11] J.M. Carrasco, L.G. Franquelo, J.T. Bialasiewicz, E. Galvan, R.C.P. Guisado, Ma.A.M. Prats, J.I. Leon, and N. Moreno-Alfonso, "Power-electronic systems for the grid integration of renewable energy sources: a survey," *IEEE Trans. Ind. Electron.*, vol. 53, no. 4, pp. 1002-1016, Jun. 2006.

[12] M. Liserre, T. Sauter, and J.Y. Hung, "Future energy systems: Integrating renewable energy sources into the smart power grid through industrial electronics," *IEEE Ind. Electron. Mag.*, vol. 4, no. 1, pp. 18-37, Mar. 2010.

[13] M. Liserre, R. Cardenas, M. Molinas, and J. Rodriguez, "Overview of multi-MW wind turbines and wind parks," *IEEE Trans. Ind. Electron.*, vol. 58, no. 4, pp. 1081-1095, Apr. 2011.

[14] K. Ma and F. Blaabjerg, "Multilevel converters for 10 MW wind turbines," in *Proc. of EPE'11*, pp. 1-10, Aug. 30 2011-Sept. 1 2011.

[15] F. Blaabjerg, K. Ma, and D. Zhou, "Power electronics and reliability in renewable energy systems," in *Proc. of ISIE*, pp. 19-30, 28-31 May 2012.

[16] H. Wang, M. Liserre, and F. Blaabjerg, "Toward reliable power electronics: challenges, design tools, and opportunities," *IEEE Ind. Electron. Mag.*, vol. 7, no. 2, pp. 17-26, Jun. 2013.

[17] K. Ma, L. Tutelea, I. Boldea, D. M. Ionel, and F. Blaabjerg, "Power electronic drives, controls, and electric generators for large wind turbines–an overview," *Electric Power Components and Systems*, vol. 43, no. 12, pp. 1406-1421, Jul. 2015.

[18] Y. Yang, P. Enjeti, H. Wang, and F. Blaabjerg, "Wide-scale adoption of photovoltaic energy: grid code modifications are explored in the distribution grid," *IEEE Ind. Appl. Mag.*, vol. 21, no. 5, pp. 21-31, Sept.-Oct. 2015.

[19] N.P. Papanikolaou, "Low-voltage ride-through concept in flyback inverterbased alternating current photovoltaic modules," *IET Power Electron.*, vol. 6, no. 7, pp. 1436-1448, Aug. 2013.

[20] Y. Bae, T.-K. Vu, and R.-Y. Kim, "Implemental control strategy for grid stabilization of grid-connected PV system based on german grid code in symmetrical low-to-medium voltage network," *IEEE Trans. Energy Conv.*, vol. 28, no. 3, pp. 619-631, Sept. 2013.

[21] E. ON GmbH, "Grid Code - High and extra high voltage." [Online]. Available: http://www.eon-netz.com/.

[22] Comitato Elettrotecnico Italiano, "CEI 0-21: Reference technical rules for connecting users to the active and passive LV distribution companies of electricity." [Online]. Available: http://www.ceiweb.it/.

[23] P. Rodriguez, A.V. Timbus, R. Teodorescu, M. Liserre, and F. Blaabjerg, "Flexible active power control of distributed power generation systems during grid faults," *IEEE Trans. Ind. Electron.*, vol. 54, no. 5, pp. 2583-2592, Oct. 2007.

[24] G.M.S. Azevedo, G. Vazquez, A. Luna, D. Aguilar, and A. Rolan, "Photovoltaic inverters with fault ride-through Capability," in *Proc. of ISIE'09*, pp. 549-553, 5-8 Jul. 2009.

[25] C.H. Benz, W.-T. Franke, and F.W. Fuchs, "Low voltage ride through capability of a 5 kW grid-tied solar inverter," in *Proc. of EPE/PEMC*, pp. T12-13-T12-20, 6-8 Sept. 2010.

[26] X. Bao, P. Tan, F. Zhuo, and X. Yue, "Low voltage ride through control strategy for high-power grid-connected photovoltaic inverter," in *Proc. of APEC'13*, pp. 97-100, 17-21 Mar. 2013.

[27] H.-C. Chen, C.-T. Lee, P.T. Cheng, R.Teodorescu, F. Blaabjerg, and S. Bhattacharya, "A flexible low-voltage ride-through operation for the distributed generation converters," in *Proc. of PEDS'13*, pp. 1354-1359, 22-25 Apr. 2013.

[28] Y. Yang, F. Blaabjerg, and Z. Zou, "Benchmarking of grid fault modes in single-phase grid-connected photovoltaic systems," *IEEE Trans. Ind. Appl.*, vol. 49, no. 5, pp. 2167-2176, Sept./Oct. 2013.

[29] Y. Yang, F. Blaabjerg, and H. Wang, "Low voltage ride-through of single-phase transformerless photovoltaic inverters," *IEEE Trans. Ind. Appl.*, vol. 50, no. 3, pp. 1942-1952, May/Jun. 2014.

[30] Meneses, D.; Blaabjerg, F.; García, O.; Cobos, J.A., "Review and Comparison of Step-Up Transformerless Topologies for Photovoltaic AC-Module Application," *IEEE Trans. Power Electron.*, vol. 28, no. 6, pp. 2649-2663, June 2013.

[31] Y. Yang and F. Blaabjerg, "Overview of single-phase grid-connected photovoltaic systems," *Electric Power Components and Systems*, vol. 43, no. 12, pp. 1352-1363, Jul. 2015.

[32] T. Kerekes, D. Sera, and L. Mathe, "Three-phase photovoltaic systems: structures, topologies, and control," *Electric Power Components and Systems*, vol. 43, no. 12, pp. 1364-1375, Jul. 2015.

[33] E. Koutroulis and F. Blaabjerg, "Design optimization of transformerless grid-connected PV inverters including reliability," *IEEE Trans. Power Electron.*, vol. 28, no. 1, pp. 325-335, Jan. 2013.

[34] Y. Yang, H. Wang, and F.Blaabjerg, "Reliability assessment of transformerless PV inverters considering mission profiles," *International Journal of Photoenergy*, vol. 2015, 10 pages, 2015.

[35] Energinet – *Technical regulation 3.2.5 for wind power plants with a power output greater than 11 kW*, Sep. 2010.

[36] E.ON-Netz – *Grid Code. Requirements for offshore grid connections in the E.ON Netz network*, April 2008.

[37] M. Altin, O. Goksu, R. Teodorescu, P. Rodriguez, B. Bak-Jensen, L. Helle, "Overview of recent grid codes for wind power integration," in *Proc. of OPTIM'2010*, pp.1152-1160, 2010.

[38] J. Rodriguez, S. Bernet, P.K. Steimer, and I.E. Lizama, "A survey on neutral-point-clamped inverters," *IEEE Trans. Ind. Electron.*, vol. 57, no. 7, pp. 2219-2230, Jul. 2010.

[39] B. Andresen and J. Birk, "A high power density converter system for the Gamesa G10x 4.5 MW Wind turbine," in *Proc. of EPE*, pp. 1-7, 2007.

[40] S.V. Araujo, P. Zacharias, and R. Mallwitz, "Highly efficient single-phase transformerless inverters for grid-connected PV systems," *IEEE Trans. Ind. Electron.*, vol. 57, no. 9, pp. 3118-3128, Sept. 2010.

[41] R. Gonzalez, J. Lopez, P. Sanchis, and L. Marroyo, "Transformerless inverter for single-phase photovoltaic systems," *IEEE Trans. Power Electron.*, vol. 22, no. 2, pp. 693-697, Mar. 2007.

[42] H. Schmidt, S. Christoph, and J. Ketterer, "Current inverter for direct/alternating currents, has direct and alternating connections with an intermediate power store, a bridge circuit, rectifier diodes and a inductive choke," German Patent DE10 221 592 A1, 4 Dec. 2003.

[43] I. Patrao, E. Figueres, F. Gonzalez-Espin, and G. Garcera, "Transformerless topologies for grid-connected single-phase photovoltaic inverters," *Renewable and Sustainable Energy Reviews*, vol. 15, no. 7, pp. 3423-3431, Sept. 2011.
[44] L. Zhang, K. Sun, L. Feng, H. Wu, and Y. Xing, "A family of neutral point clamped full-bridge topologies for transformerless photovoltaic grid-tied inverters," *IEEE Trans. Power Electron.*, vol. 28, no. 2, pp. 730-739, Feb. 2013.
[45] B. Gu, J. Dominic, J.-S. Lai, C.-L. Chen, T. LaBella, and B. Chen, "High reliability and efficiency single-phase transformerless inverter for grid-connected photovoltaic systems," *IEEE Trans. Power Electron.*, vol. 28, no. 5, pp. 2235-2245, May 2013.
[46] J.D., van Wyk and F.C. Lee, "On a future for power electronics," *IEEE Journal of Emerging and Selected Topics in Power Electronics,* , vol. 1, no. 2, pp. 59-72, Jun. 2013.
[47] J.G. Kassakian and T.M. Jahns, "Evolving and emerging applications of power electronics in systems," *IEEE Journal of Emerging and Selected Topics in Power Electronics*, vol. 1, no. 2, pp. 47-58, Jun. 2013.
[48] M. Campbell, J. Blunden, E. Smeloff, and P. Aschenbrenner, "Minimizing utility-scale PV power plant LCOE through the use of high capacity factor configurations," in *Proc. of IEEE PVSC*, pp. 421-426, 7-12 Jun. 2009.

This paper has been in part published in the *Proceedings of the 3rd International Conference on Electric Power and Energy Conversion Systems (EPECS)*, Oct. 2-4, 2013. DOI: 10.1109/EPECS.2013.6712980.

Planung und Betrieb von Smart Grids

Christian Rehtanz[1], André Seack[1], Sebastian Lehnhoff[2], Olav Krause[3]
[1] ie[3] - Institut für Energiesysteme, Energieeffizienz und Energiewirtschaft, Technische Universität Dortmund, christian.rehtanz@tu-dortmund.de
[2] OFFIS – Institut für Informatik, Oldenburg
[3] University of Queensland, Brisbane, Australien

Kurzfassung

Der Verteilnetzausbau ist ein wesentlicher Baustein für die Integration erneuerbarer Energien in das System der elektrischen Energieversorgung. Studien wie die dena-Verteilnetzstudie [1] haben gezeigt, dass in den nächsten Jahren ein signifikanter Netzausbau notwendig ist. Dieser kann jedoch durch Netzinnovationen reduziert werden. Im folgenden werden neue Ansätze zur Netzplanung und zum Netzbetrieb vorgestellt, die in der Lage sind, die Netzeffizienz zu erhöhen und den Ausbaubedarf zu verringern.

1 Smart Grids und Smart Markets

Unter Smart Grids werden neue Technologien im Netz verstanden, die die Effizienz des Netzbetriebs bzw. die Netzausnutzung verbessern. Durch eine bessere Überwachung können betriebliche Grenzen frühzeitig erkannt werden und Kapazitätsreserven reduziert werden. Zusätzliche Stellglieder zur Spannungs- oder Leistungsflussregelung können einen zusätzlichen Netzausbau vermeiden helfen. Insgesamt muss eine Koordination zwischen Überwachung und dem Einsatz dieser Stellglieder erfolgen.

Die Ergebnisse aktueller Studien zeigen, dass durch ein intelligentes Last- und Erzeugungsmanagement eine signifikante Kostenreduktionen beim Ausbau elektrischer Energiesysteme erzielt werden kann [1]. Ebenfalls wurde allerdings gezeigt, dass bei einer Implementierung von Smart-Market-Funktionen ohne Berücksichtigung von Netzrestriktionen – speziell in den unteren Spannungsebenen – diese Kostenreduktion nicht nur nicht erreicht werden wird, sondern dass es sogar zu einer Erhöhung des Netzausbaubedarfs kommen kann. Hieraus leitet sich der Bedarf für ein aktives Management von Einspeisern und Lasten unter Berücksichtigung der Netzrestriktionen und steuernden Elementen auf allen Spannungsebenen ab. Dieses macht vor allem in Mittel- und Niederspannungsnetzen, aufgrund ihrer Anzahl, ihrer speziellen Eigenschaften und im Hinblick auf Messdatenverfügbarkeit, neue Ansätze zur Zustandserfassung und Koordination erforderlich. Smart Grids und Smart Markets müssen also zukünftig Hand in Hand gehen.

Dieses Zusammenspiel muss sowohl in der Netzplanung als auch im Netzbetrieb geeignet berücksichtigt werden. Für die Planung müssen alle Mechanismen und Funktionen der Smart Markets und Smart Grids geeignete modelliert werden. Im Folgenden wird hierzu in Kapitel 2 ein agentenbasierter Ansatz zur flexiblen Generierung von Zeitreihen vorgestellt, der den genannten Anforderungen entspricht.

Im Betrieb von Verteilnetzen und hier insbesondere auf der Mittel- und Niederspannungsebene halten eine Reihe von neuen Funktionen Einzug. Die Schutz- und Regelungsfunktionen müssen mit den dezentralen Anlagen koordiniert werden. In Abschnitt 3 werden eine Reihe dieser Smart-Grids- und Smart-Markets-Funktionen vorgestellt. In den meisten Fällen muss hierzu der Netzzustand geeignet ermittelt werden. In Kapitel 4 wird daher ein neues Verfahren zur Adaptiven State Estimation vorgestellt, welches insbesondere auf den unteren Spannungsebenen bei geringer Dichte an Messinformationen flexibel angewendet werden kann.

2 Agentenbasierte Netzplanung

Die Auslegung der Transport- und Verteilnetze erfolgt nach konventionellen Methoden über die Definition von Extremszenarien. Die Netze werden dabei auf die maximal auftretende Einspeisung bei gleichzeitig geringer Last (Rückspeisefall) und auf die maximal im Netz auftretende Last bei sehr niedriger oder nicht vorhandener Einspeisung (Lastfall) ausgelegt. Da die Auftrittswahrscheinlichkeit der Extremszenarien nur schwer zu ermitteln ist, werden die Netze für selten oder nie auftretende Belastungssituationen dimensioniert [2]. Durch die zeitliche Überlagerung der Einspeisung und Last weicht die resultierende reale Versorgungsaufgabe im Netz jedoch oft von den Dimensionierungsszenarien ab (siehe Bild 1).

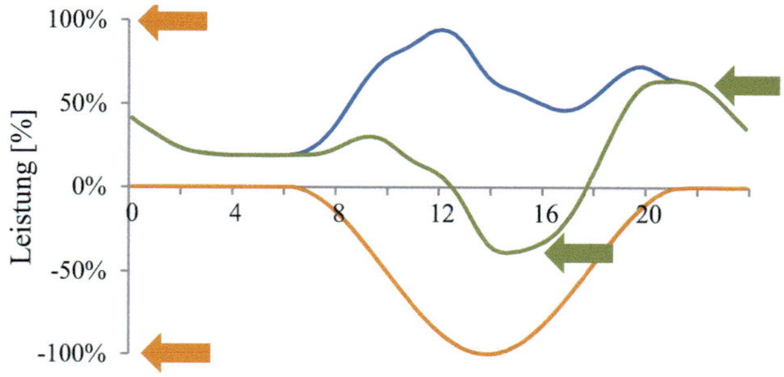

Bild 1: Dimensionierungsgrößen für die Verteilnetze mit detaillierte Zeitreihenbetrachtung (Grün) und in der konventionellen Planung (Orange)

Weiterhin kann das Verhalten von Netzanschlussnehmern bei der Teilnahme an Smart Markets oder die Interaktionen bei Smart Grid Konzepten im konventionellen Planungsprozess nicht berücksichtigt werden.

Besonders bei einer hohen Anzahl von Netzanschlussnehmern die Demand-Side-Management-Funktionen implementieren oder einer hohen Anzahl von Speichersystemen und Elektrofahrzeugen im Netz ist eine Bestimmung des Einflusses auf die Netzbelastungssituation essentiell. Eine detaillierte Modellierung der Netzbelastungssituation ist für die Bewertung zentraler und dezentraler Netzkontrollkonzepte („Smart Grid") erforderlich. Diese oft sehr komplexen Algorithmen können durch eine Extremwertbetrachtung nicht ausreichend evaluiert werden.

Um das zeitliche Verhalten aller Netzanschlussnehmer angemessen modellieren zu können, ist eine detaillierte Simulation aller Netznutzer auf Basis von Zeitreihen erforderlich. Für die Simulation des zeitlichen Verhaltens aller Netzanschlussnehmer wurde in [3][4][5] ein Multiagentensystem entwickelt, das jeden Netzteilnehmer im Verteilnetz durch einen einzelnen Agenten abbildet. Der Aufbau des Simulationssystems und eine Auswahl der berücksichtigten Netzanschlussnehmer ist in Bild 2 dargestellt. Externe Eingangsgrößen wie die Simulationszeit (1), die lokale Wettersituation (2) und der Marktpreis (3) beeinflussen das Verhalten von allen Agenten im System.

Regenerative Erzeugungsanlagen (4) können dabei ebenso berücksichtigt wie Haushaltskunden (5) und Industriekunden (5). Neuartige Netzteilnehmer wie Elektrofahrzeuge (6), Speichersysteme (7) und Wärmepumpen (8) werden ebenfalls auf der Ebene einzelner Anlagen berücksichtigt.

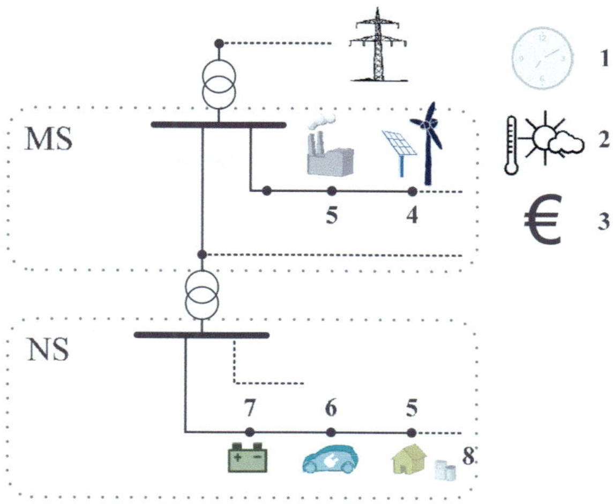

Bild 2: Schematisch Darstellung der Netzanschlussnehmer im agentenbasierten Simulationssystem

In Anhängigkeit der Parametrierung der Agenten können unterschiedliche Verhalten der einzelnen Netzteilnehmer simuliert werden. So kann die Reaktion eines Haushaltskunden auf ein zentrales Marktpreissignal oder das Mobilitätsverhalten unterschiedlicher Elektrofahrzeugtypen berücksichtigt werden. Zusätzlich sind verschiedene Interaktionskonzepte wie beispielsweise das BDEW-Ampelmodel [6] im System umgesetzt.

Über die Oberfläche der Simulationssoftware (Bild 3) können unterschiedliche Szenarien und Durchdringungsgrade für Smart-Market- oder Smart-Grid-Konzepte vorgegeben werden.

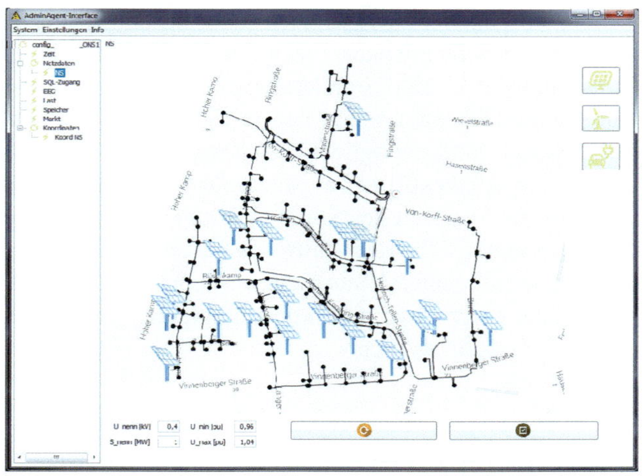

Bild 3: Oberfläche der Software zur agentenbasierten Verteilnetzplanung

Da im Simulationssystem eine Berechnung des Netzzustandes für jeden Zeitschritt erfolgt, generiert das System Zeitreihen der Auslastung und Knotenspannungen für jedes Betriebsmittel und Netzknoten.

Um die Ergebnisse zu verdeutlichen, ist in Bild 4 die Auslastungsdauerlinie einer Leitung für eine exemplarische Szenarienvariation dargestellt. Der Einfluss von unterschiedlichen Durchdringungsgraden für DSM bei Haushaltslasten auf die Transformatorauslastung kann unmittelbar aus den Zeitreihen abgeleitet werden.

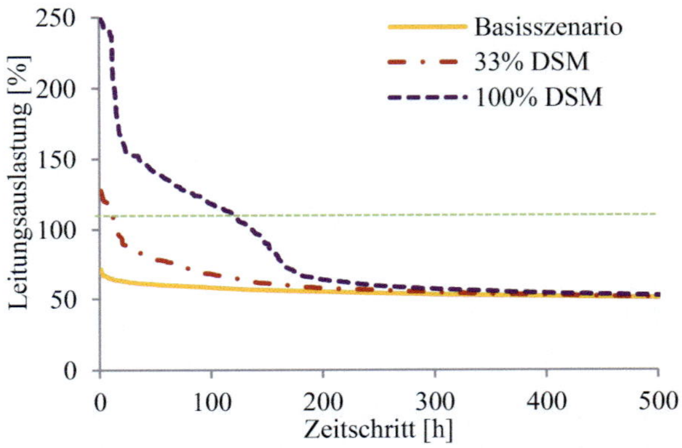

Bild 4: Exemplarische Auslastungsdauerlinie einer Leitung für drei Szenarien

Weiterhin liefert das entwickelte Simulationssystem Einspeise- und Lastzeitreihen für jeden konventionellen und innovativen Netzanschlussnehmer (Speicher, Wärmepumpen, Elektrofahrzeuge) im Verteilnetz. In den Simulationen kann zusätzlich das Regelverhalten von Netzbetriebsmitteln wie regelbaren Ortsnetzstationen und Smart Grid Interaktion berücksichtigt und der Einfluss auf die Netzbelastungssituation ermittelt werden.

Durch die generierten Zeitreihen kann anschließend die Auftrittswahrscheinlichkeit von Belastungssituationen ermittelt werden. Das Simulationssystem unterstützt damit einen effizienten sowie bedarfsorientierten Netzplanungsprozess und ermöglicht damit erstmalig einen Vergleich von konventionellen Netzausbaumaßnahmen mit innovativen Ansätzen des Smart Grid und Smart Market. Mit der Kenntnis über die Auftrittswahrscheinlichkeit von Belastungssituationen können diese Maßnahmen als wirtschaftliche Alternative dem konventionellen Netzausbau gegenübergestellt werden.

3 Komponenten und Funktionen in Smart Grids

Die primärtechnische Ausstattung von Smart Grids benötigt Stellglieder. Zunächst sind automatisch steuerbare Leistungsschalter in der Mittelspannungsebene zu ergänzen, die eine automatische Topologieanpassung erlauben.

Darüber hinaus sind innovative Komponenten beispielsweise Automatic Voltage Restorer (AVR) zur Längsspannungsregelung in der MS- und NS-Ebene oder auch regelbare Ortsnetztransformatoren im Netz denkbar, wie in Bild 5 dargestellt.

Des Weiteren sind aber auch die Komponenten des Smart Market, d. h. die umrichterbasierten Einspeiser oder Speicher zu betrachten. Diese können ebenfalls zur Spannungs-Blindleistungs-Steuerung verwendet werden.

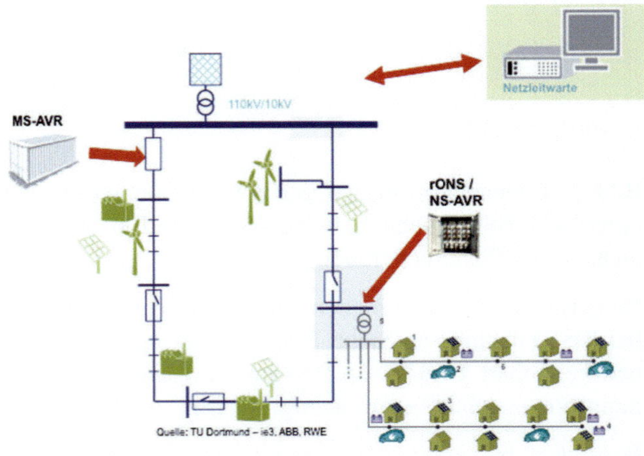

Bild 5: Primärtechnische Komponenten in Smart Grids auf Mittel- und Niederspannungsebene

Die Koordination aller dieser primärtechnischen Komponenten im Netz und des Smart Market sind über IKT zu steuern. Zur Basisausstattung zählen hier Intelligent Electronic Devices (IED) in der Schaltanlage sowie den Ortsnetzstationen. Diese Geräte sollten zukünftig eine standardisierte Hardwareplattform bilden, auf der die unterschiedlichen Smart-Grids-Funktionen modular ergänzt werden können.

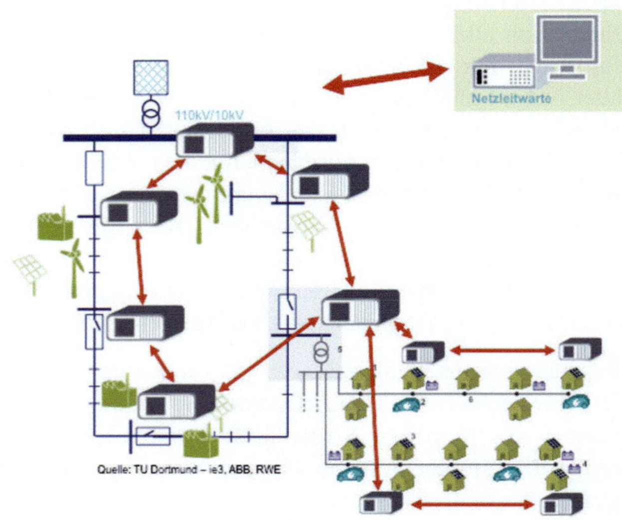

Bild 6: Sekundärtechnische Komponenten in Smart Grids

In der Niederspannungsebene sind Geräte zu ergänzen, die als Messwertaufnehmer oder zur Steuerung dezentraler Anlagen dienen. Alle diese Geräte müssen über eine geeignete IKT-Infrastruktur miteinander kommunizieren. Dedizierte zukünftige Funkstandards wären hierfür wünschenswert. Bis dahin wird zumeist der heutige Mobilfunk verwendet.

Die Funktionen, die diese Gerätearchitektur nutzen, wurden seitens des ie3 der TU Dortmund in unterschiedlichen Forschungsprojekten entworfen, die im Folgenden zusammen mit den Projektnamen und Fördermittelgebern genannt werden:

- Modulare Schutzfunktionen (BMWi, i-Protect)
- Fehlerortung (BMWi, i3S)
- Topologieoptimierung durch Trennstellenverlagerung (EU, Grid-4-EU)
- EE-Netz-Koordination gemäß Ampelkonzept (BMWi, Proaktives Verteilnetz)
- Weitbereichs-Spannungsregelung und Systemdienstleistungen aus dem Verteilnetz (Helmholtz Energie Allianz)
- Power Quality Monitoring

Für den Großteil dieser Funktionen ist es notwendig, eine Zustandsschätzung des aktuellen Verteilnetzzustandes vorzunehmen. Die Ermittlung muss über dieselbe Geräte- und Kommunikationsarchitektur erfolgen und sollte so weit wie möglich dezentral und automatisiert stattfinden. Der folgende Abschnitt erläutert ein neu entwickeltes Verfahren hierzu.

4 Adaptive Zustandsschätzung in Smart Grids

Die Koordination von innovativen Netzkomponenten, dezentralen Erzeugungsanlagen und beeinflussbaren Lasten untereinander und mit dem Netz stellt eine der wesentlichen Aufgaben eines zukünftigen Smart Grids dar.

Eine der zentralen Herausforderungen bei der Implementierung von Smart Grids und hiermit zusammenhängenden Smart-Market-Funktionalitäten auf Mittel- und Niederspannungsebene ist die geringe Messstellenabdeckung und die hierdurch resultierende begrenzte Beobachtbarkeit der Betriebszustände. Traditionelle Ansätze, die eine vollständige Beobachtbarkeit voraussetzen, stoßen hier an ihre Grenzen und können bestenfalls durch Verwendung von Modellreduktionen anwendbar gemacht werden. Die folgenden Ansätze wurden bereits in [7] vorgestellt. Hierin befinden sich auch detailliertere mathematische Ausführungen.

4.1 Klassische State Estimation

Als Standardverfahren der Netzzustandsabschätzung, der sogenannten State Estimation, kommt in den meisten Beobachtungssystemen elektrischer Energienetze das sogenannte „weighted least mean squares“ (WLMS) Estimationsverfahren nach Schweppe et. al. zum Einsatz [8]-[11].

Es handelt sich hierbei um ein Verfahren, das die Invertierung eines nicht-linearen Gleichungssystems durch eine iterative, auf lokaler Linearisierung und Invertierung basierende, schrittweise Verbesserung der Zustandsschätzung leistet. Im Zentrum dieses Verfahrens steht ein System von typischerweise nicht-linearen Gleichungen $\mathrm{m(x)}$, welche den funktionalen Zusammenhang zwischen dem internen, nicht direkt beobachtbaren Zustand x und den verfügbaren Beobachtungen m abbilden.

Die innere Invertierung basiert auf der Jacobimatrix $\mathrm{J(x_0)}$ der Systemgleichungen $\mathrm{m(x)}$, welche die Systemgleichungen im angenommenen Zustand $\mathrm{x_0}$ linear approximiert. Definiert werden kann die Jacobimatrix über den Nabla-Operator, wie in (1) dargestellt.

$$J(x_0) = m(x_0) \cdot \nabla \tag{1}$$

Da in überbestimmten Systemen die Anzahl der Messdaten – und damit die Anzahl der Systemgleichungen – größer ist, als die Anzahl der zu bestimmenden Zustandsvariablen, gilt für $\mathrm{J(x_0)}$, dass die Anzahl der Zeilen m die Anzahl der Spalten n übersteigt. Mit $\mathrm{m} > n$ existiert jedoch die Inverse $\mathrm{J^{-1}(x_0)}$ nicht mehr und eine direkte Invertierung des linearen Gleichungssystems scheitert.

Diesem Umstand geschuldet, basiert das WLMS-Verfahren auf der Bildung einer Pseudo-Inversen der Jacobimatrix $\mathrm{J(x_0)}$ (2).

$$J^I(x_0) = \underbrace{\left(J^{\mathrm{T}}(x_0) \cdot W \cdot J(x_0)\right)}_{G(x_0)}^{-1} \cdot J^{\mathrm{T}}(x_0) \cdot W \tag{2}$$

Dieses Verfahren setzt eine vollständige Beobachtbarkeit des überwachten Netzes und damit eine lückenlose Abdeckung des beobachteten Netzes mit Messeinrichtungen voraus, was speziell in Mittel- und Niederspannungsnetzen nicht zu erwarten ist. Gleichzeitig ist dieses Verfahren auch anfällig für Fehlfunktionen aufgrund von Kommunikationsstörungen, durch die es zu einer zumindest zeitweisen Nichtverfügbarkeit einzelner oder mehrerer Messdaten kommen kann. Reichen die weiter verfügbaren Messdaten nicht aus, alle Zustandsvariablen des beobachteten Netzes zu bestimmen, scheitert das Verfahren und es können keine Aussagen über den Betriebszustand mehr getroffen werden; auch nicht über solche Teile des beobachteten Netzes für die eventuell genügend Messdaten zur Verfügung gestanden hätten.

4.2 Adaptiver Ansatz

Wie auch das WLMS-Verfahren baut der hier beschriebene Ansatz auf der Verwendung der Pseudo-Inversen der Jacobimatrix der Systemgleichungen auf. Zur Berechnung der Pseudo-Inversen kommt hier jedoch nicht der in (2) dargestellte Ansatz, sondern das Verfahren der Singulärwertzerlegung zum Einsatz. Detaillierte Ausführungen befinden sich in [7].

Der Vorteil der Verwendung der Singulärwertzerlegung besteht darin, dass dieser Prozess auch für Matrizen mit Rang $\mathrm{r} < n$ anwendbar ist. In diesem Fall sind nicht mehr alle ersten n rechten Singulärvektoren funktional mit dem Messraum verbunden. Dies bedeutet wiederum, dass es Zustandskomponenten gibt, die nicht mehr aus den verfügbaren Messinformationen bestimmbar sind.

Das Verfahren ermöglicht eine individuelle Beobachtbarkeitsanalyse für – aus den Estimationsergebnissen berechnete – Parameter. Auf Basis der bestimmten ersten r Komponenten lassen sich alle Betriebsparameter des beobachteten Netzes berechnen (z. B. Leitungsströme, Wirkleistungsverluste, etc.). Ob diese beobachtbar sind – und damit der ermittelte Wert – lässt sich anhand der rechten Singulärvektoren r + 1 bis n entscheiden. Da diese gemeinsam den Teilraum aller nicht beobachtbaren Zustandskomponenten aufspannen – also derjenigen Zustandskomponenten, die auftreten können, ohne dass dies in den Messinformationen sichtbar werden würde – kann eine Sensitivitätsanalyse entscheiden, ob der berechnete Wert vollständig durch die verfügbare Messinformation bestimmt ist.

Wird die Pseudo-Inverse auf die beschriebene Weise berechnet und wird bei der Analyse des Estimationsergebnisses die genannte Sensitivitätsanalyse durchgeführt, ergibt sich ein System zur Netzzustandsabschätzung das zum einen keinerlei Anforderung an Art, Zusammenstellung und Verteilung von Messinformationen stellt und zum anderen im Fall eines nur teilweise bestimmten Netzzustandes für jede abgeleitete Größe individuell deren Beobachtbarkeit erkennen kann.

Dies erlaubt es ein Netzbeobachtungssystem zu entwickeln, das robust auf einen permanenten oder auch temporären Mangel an Messinformationen reagiert und darüber hinaus aus den verfügbaren Messinformationen dynamisch ermittelt, welche abgeleiteten Werte beobachtbar sind. Diese Funktionalität ist entscheidend für die Netzzustandsbeobachtung in Mittel- und Niederspannungsnetzen, in denen eine vollständige Abdeckung mit Messeinrichtungen technisch wie ökonomisch schwierig zu erreichen sein wird. Mit seinen Eigenschaften bildet der beschriebene Ansatz eine hervorragende Basis für die Entwicklung und Implementierung einer Reihe höherwertigerer Funktionen:

- Bestimmung minimaler Messstellenkonfiguration
- Analog zur Beobachtbarkeit auch optimale Steuerbarkeit durch Last-/EE-Management
- Modellparameter-Estimation (Werte der Netzelemente) aus überbestimmten Messgrößen

4.3 Anwendungsbeispiel

Anhand eines exemplarischen Niederspannungsnetzes, für das Smart-Meter-Messwerte vorlagen, konnte die Funktion der ASE getestet werden. Die Messwerte waren zu keinem Zeitpunkt vollständig und erhobene Messdaten wiesen eine hohe Fluktuation der Verfügbarkeit auf, da sich die Kommunikation mittels Power Line Carrier (PLC) als sehr instabil erwiesen hat. Diese realen und zu einem guten Teil repräsentativen Bedingungen erlauben es, die Flexibilität des neuen Verfahrens zu prüfen und zu demonstrieren. Messwerte wurden in allen drei Phasen erhoben und der Neutralleiter explizit modelliert. Bild 7 zeigt den Verteilnetzausschnitt.

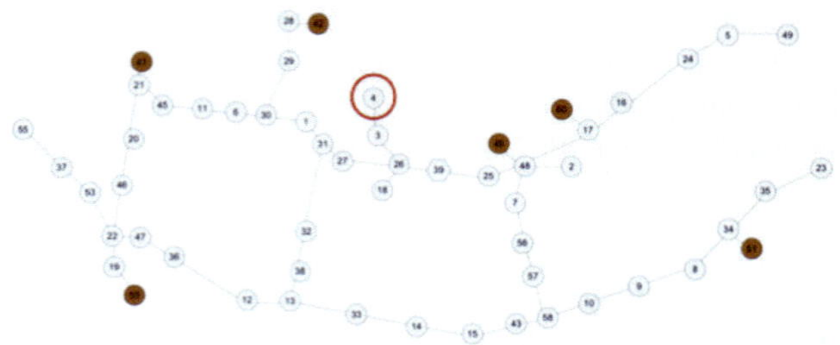

Bild 7: NS-Netzausschnitt mit gekennzeichnetem Knoten zur Ermittlung der Spannung

Bei dem dargestellten Verteilnetzausschnitt handelt es sich um ein reales Niederspannungsnetz, von dessen Knoten üblicherweise mehrere Endkunden versorgt und deren Verbrauch in Echtzeit erhoben wird. Dies führt zu einer hohen Heterogenität der Messgrößenzusammenstellung, da z. B. Spannungswerte mehrfach erhoben sein können und somit eine Ausgleichsrechnung erforderlich machen. Eine solche redundanten Messgrößenerhebung stellt eine teilweise Überbestimmtheit einzelner Zustandskomponenten dar und die entsprechende, fehlerminimierende Ausgleichsrechnung wird vom ASE automatisch geleistet. Als Beispiele abgeleiteter Betriebsparameter sind in Bild 8 die zugehörigen ermittelten Spannungswerte an dem gekennzeichneten Knoten und in Bild 9 die Wirkleistungsflüsse zwischen diesem Knoten und dem Nachbarknoten dargestellt.

Wie aus beiden Abbildungen deutlich wird, sind die entsprechenden Betriebsparameter nicht vollständig bestimmt und die Bestimmbarkeit variiert über der Zeit. Dennoch leistet der ASE ohne vorgelagerte Topologieanalyse, Modellreduktion oder Zuhilfenahme von geschätzten Werten für den Ersatz nicht verfügbarer Messgrößen (Pseudo-Messwerte) eine Zustandsestimation des gesamten Netzausschnittes und bestimmt die wahrscheinlichsten Werte der bestimmbaren Betriebsparameter.

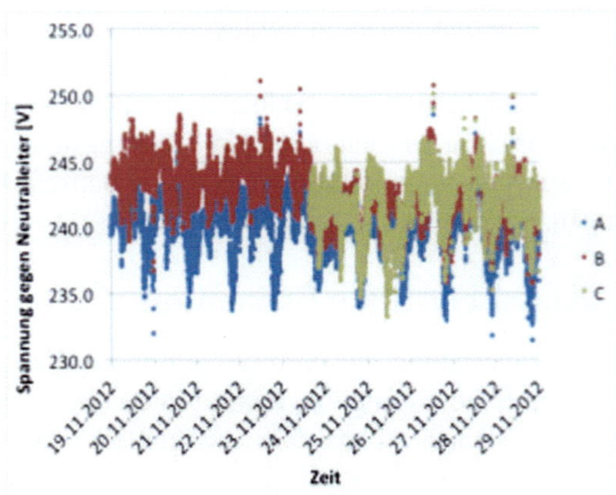

Bild 8: 3-phasige Spannungswerte am gekennzeichneten Knoten

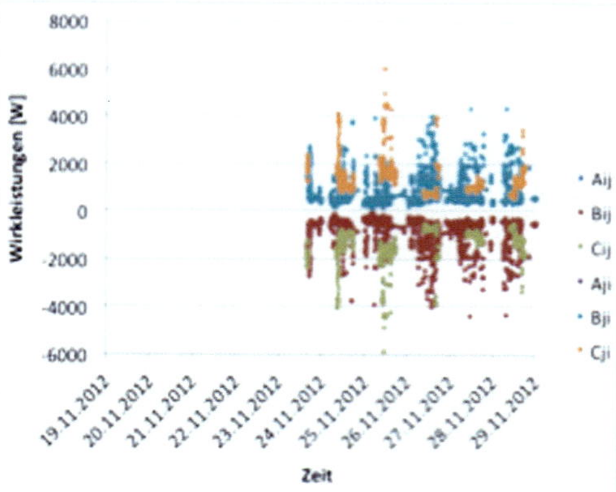

Bild 9: Leistungsflüsse zwischen gekennzeichnetem Knoten und Nachbarknoten

5 Zusammenfassung

Der Artikel stellt neue Ansätze für die Planung von Smart Grids in der Verteilnetzebene mittels eines agentenbasierten Systems zur Zeitreihensimulation vor. Smart Grids und Smart Markets Funktionen werden dargelegt und der Bedarf für spezielle Netzzustandsschätzungen wird hergeleitet. Ein neues Verfahren, die sogenannte Adaptive State Estimation, wird vorgestellt und erläutert.

Die Smart Grids Verfahren einschließlich der Adaptiven State Estimation werden derzeit auf einer standardisierten IKT-Infrastruktur implementiert. Systemtest dieser neuen Sekundärtechnikarchitektur und deren Funktionen werden in den Prüffeldern des ie3 der TU Dortmund, im SESA-Lab des OFFIS sowie im Advanced Power Engineering Lab der University of Queensland durchgeführt.

6 Literatur

[1] dena -Verteilnetzstudie: „Ausbau- und Innovationsbedarf in den Stromverteilnetzen in Deutschland bis 2030", 2012

[2] Schmidtner, T.: *Probabilistische Methoden in der Netzplanung „Niederspannung.* In VDE-Kongress 2012 – Smart grid, Berlin: VDE-Verlag, 2012. – ISBN 97838007344667

[3] Seack, A.; Kays, J.; Rehtanz, C.: *Time series based distribution grid planning approach with decentralised voltage regulation,* 18th Power Systems Computation Conference (PSCC), Wroclaw, Poland, 2014

[4] Seack A.; Kays J.; Jendernalik L.: *Agent based approach to model photovoltaic feed-in in distribution network planning.* Cigré 2013 Auckland Symposium, Auckland, New Zealand, September 2013

[5] Kays, J.: Agent-based simulation environment for Improving the Distribution Grid Planning, Dissertation, TU Dortmund, 2014, ISBN 978-3-86844-662-3

[6] Bundesverband der Energie und Wasserwirtschaft: *BDEW Roadmap – Realistische Schritte zur Umsetzung von Smart Grids in Deutschland.* Berlin, Februar 2013

[7] Krause, O.; Lehnhoff, S.; Rehtanz, C.: "Adaptive Netzzustandsschätzung in unterbestimmten Verteilnetzen", VDE-ETG-Fachtagung "Smart Grids und Smart Markets", 25.-26.03.2015, Kassel

[8] Schweppe, F. C.; Wildes, J. (1970). Power System Static-State Estimation, Part I: Exact Model. *Power Apparatus and Systems, IEEE Transactions on*, (1), 120–125. doi:10.1109/TPAS.1970.292678

[9] Schweppe, F. C.; Rom, D. B. (1970). Power System Static-State Estimation, Part II: Approximate Model. *Power Apparatus and Systems, IEEE Transactions on*, (1), 125–130. doi:10.1109/TPAS.1970.292679

[10] Schweppe, F. C. (1970). Power System Static-State Estimation, Part III: Implementation. *Power Apparatus and Systems, IEEE Transactions on*, (1), 130–135. doi:10.1109/TPAS.1970.292680

[11] Abur, A.; Exposito, A. G. (2004). *Power System State Estimation: Theory and Implementation.* Taylor & Francis.

Rahmenbedingungen

Vorteile der Netzanschlusszertifizierung in Deutschland

Dipl. Ing. Roman Bluhm, RB@netz-ing.de
NETZ-ING. AB GmbH, 24768 Rendsburg, Deutschland
NEIS Konferenz 2015 Hamburg

Kurzfassung

Der Ausbau der Regenerativen Energien in Deutschland schreitet im Rahmen der politischen Ausbauziele zügig voran. Diese Entwicklung stellt eine Herausforderung für die Netzbetreiber dar, denn sie müssen bei stark steigenden Anteilen von dezentral eingespeister erneuerbarer Energie weiterhin für die Sicherheit der Netze sorgen. Regenerative Erzeugungsanlagen sollen daher zunehmend Systemdienstleistungen erbringen, die vorher nur von konventionellen Kraftwerken eingefordert wurden. Die zahlreichen Netzanschlussvorhaben müssen auf die Einhaltung der geltenden Anforderungen überprüft werden. Diese Bewertungsverfahren sind im Sinne der Energiewende effizient, neutral und rechtssicher durchzuführen. In Deutschland hat sich dazu die Netzanschlussprüfung durch akkreditierte Zertifizierungsstellen etabliert, welche in diesem Beitrag detailliert erläutert wird.

1 Verfahren der Netzanschlussprüfung in Deutschland

Ursprünglich waren in Deutschland die Netzbetreiber für die Prüfung und Genehmigung der Netzanbindung von Erzeugungsanlagen erneuerbarer Energien zuständig. Die Anzahl der Verfahren nahm in den 2000er-Jahren stark zu und zum Teil wurden auch Interessenkonflikte zwischen Antragstellern und Netzbetreibern deutlich.

Aus diesem Grunde haben die Netzbetreiber mit der Technischen Richtlinie „Erzeugungsanlagen am Mittelspannungsnetz" [1] im Jahr 2008 ein Zertifizierungsverfahren durch neutrale Drittstellen eingeführt. Die neutralen Drittstellen werden zur Wahrung der Neutralität und zur einheitlichen technischen Vorgehensweise, konform mit den europäischen Regelungen, von der Deutschen Akkreditierungsstelle (DAkkS) nach DIN EN ISO/IEC 17065 [5] akkreditiert und überwacht.

Auch in der Systemdienstleistungsverordnung [2] (SDLWindV) wird vom Gesetzgeber dieses Zertifizierungsverfahren als Voraussetzung für den Erhalt der garantierten Einspeisevergütung für Windenergieanlagen gefordert.
Die beteiligten Netzbetreiber, Anlagenbetreiber, Anlagenhersteller und Gutachter haben daraufhin die genaue Vorgehensweise in der Technischen Richtlinie TR8 [3] der FGW e.V. definiert.

Die Prüfung umfasst insbesondere folgende Aspekte:

- Regenerative Erzeugungsanlagen werden ähnlich wie konventionelle Kraftwerke an der Frequenzhaltung im Netz beteiligt, so dass Leistungsschwankungen ausgeglichen werden können.
- Die statische Spannungshaltung im Netz wird durch regenerative Erzeugungsanlagen unterstützt indem Einrichtungen zur regelbaren Blindleistungsbereitstellung geschaffen werden.
- Die Aufrechterhaltung der Netzsicherheit wird unterstützt indem sich regenerative Erzeugungsanlagen während eines Netzfehlers nicht mehr abschalten.
- Zur dynamischen Netzstützung wird während eines Netzfehlers ein zusätzlicher Blindstrom in das Netz eingespeist.
- Schutzeinstellungen auf allen Ebenen werden geprüft und koordiniert eingestellt.
- Ungewünschte Rückwirkungen von Erzeugungsanlagen auf das Netz, wie Spannungsänderungen, Flicker, Oberschwingungen, werden minimiert.
- Das Regelungskonzept zur Fernsteuerbarkeit von Wirk- und Blindleistung wird überprüft und getestet.

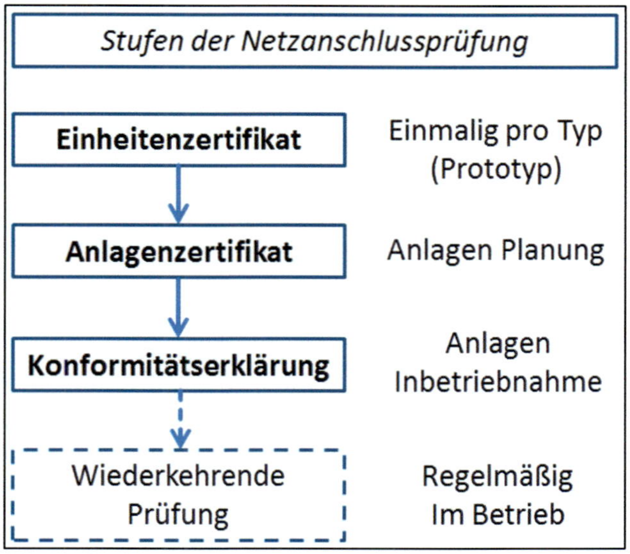

Bild 1: Stufen der Netzanschlussprüfung

Die Prüfung einer neuen Erzeugungsanlage erfolgt dabei in drei Stufen: Einheitenzertifizierung, Anlagenzertifizierung und Konformitätserklärung. Eine Übersicht der Stufen der Netzanschlussprüfung ist in **Bild 1** zu sehen. Die einzelnen Stufen und deren Vorteile werden im Folgenden genauer erklärt. Darüber hinaus wird das Konzept der wiederkehrenden Prüfung vorgestellt.

1.1 Einheitenzertifizierung

Die Einheitenzertifizierung ist die detaillierte Prüfung eines bestimmten Anlagentyps eines Herstellers, zum Beispiel ein Windenergieanlagen-Typ oder ein Photovoltaik-Wechselrichter-Typ. Die Richtlinie TR8 [3] nutzt dafür den Begriff Energieerzeugungseinheit (EZE). Hierzu sind umfangreiche Vermessungen im Feld durchzuführen, wie die Vermessung von Leistungskurve, Oberschwingungen, Flicker und das Verhalten bei Spannungseinbrüchen. Diese aufwendige und kostspielige Untersuchung muss nur einmal pro Anlagentyp erfolgen und spart somit Einzel-Messungen bei jeder weiteren installierten Einheit ein. Zusätzlich ist ein validiertes Simulationsmodell zu erstellen, welches das dynamische Verhalten des Anlagentyps bei Spannungseinbrüchen abbildet, so dass weitere Prüfungen auf Basis von Simulationen erfolgen können.

Die Ergebnisse im Einheitenzertifikat sind Grundlage für die Prüfung des Netzanschlusses aller Projekte mit Anlagen dieses Typs an einem ganz konkreten Anschlusspunkt in der Anlagenzertifizierung.

1.2 Anlagenzertifizierung

Die Anlagenzertifizierung überprüft die Planung einer Energieerzeugungsanlage (EZA), zum Beispiel eines Windparks oder einer Freiflächensolaranlage, an einem konkreten Netzanschlusspunkt (NAP). Dazu werden die Messergebnisse und das Simulationsmodell der verwendeten Anlagentypen aus der Einheitenzertifizierung herangezogen, sowie die vom zuständigen Netzbetreiber übermittelten technischen Daten des Netzanschlusspunktes, charakterisiert durch die Nennspannung, Netzkurzschlussleistung, Netzimpedanzwinkel, Sternpunktbehandlung und weitere Angaben. Es wird geprüft, ob das Verhalten der Summe aller Einzelanlagen am NAP die Anforderungen erfüllt. Dazu werden auch alle sonstigen Komponenten der Erzeugungsanlage, wie Kabel, Transformatoren oder Kompensationsanlagen in die Berechnung einbezogen. Die Berechnung der relevanten Lastfluss-Betriebspunkte und des dynamischen Verhaltens erfolgt normalerweise mit Hilfe der etablierten Software-Programme DigSilent PowerFactory oder Matlab-Simulink. Die Ansicht eines Simulationsmodells eines beispielhaften Windparks ist in **Bild 2** zu sehen.

In der Simulation wird z. B. das Verhalten eines kompletten Windparks bei einem Netzfehler mit kurzem Spannungseinbruch getestet, siehe **Bild 3**. Dabei wird geprüft, ob die Blindstromeinspeisung während und nach dem Fehler der Anforderung entspricht und, dass es zu keinen Instabilitäten kommt.

Teilweise ist auch eine anteilige Berechnung der Grenzwerte nötig, wenn entweder ein Park aus Altanlagen erweitert wird, oder wenn der Netzanschlusspunkt von mehreren Erzeugungsanlagen genutzt wird.

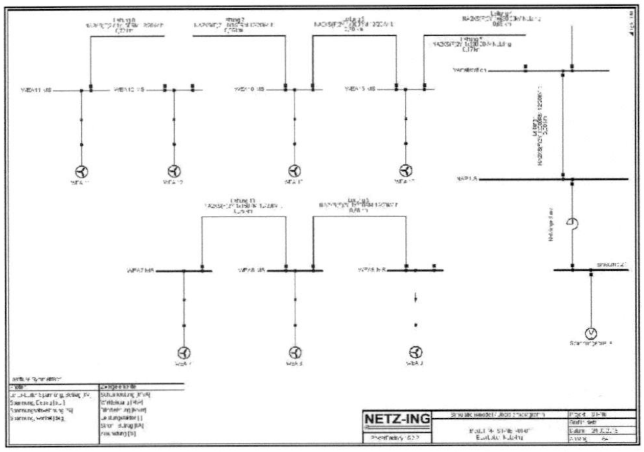

Bild 2 Beispiel Simulationsmodell Windpark in DIgSILENT PowerFactory

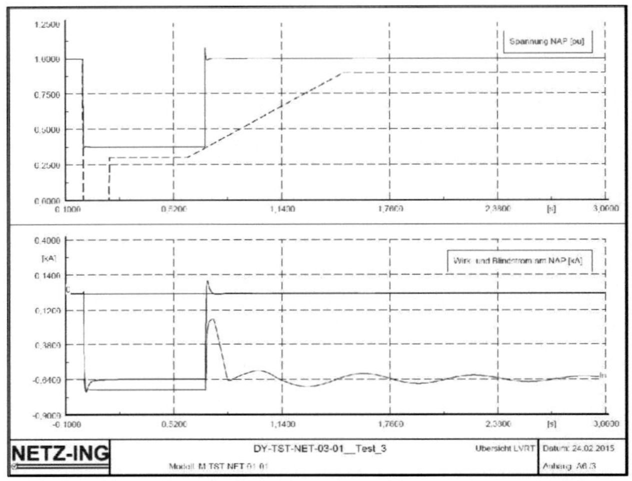

Bild 3 Beispielhafte Simulationsverläufe bei einem Spannungseinbruch. Oben: Spannung, unten: Wirk- und Blindstrom am NAP.

1.3 Konformitätserklärung

Nach der Inbetriebnahme einer Erzeugungsanlage ist in der Konformitätserklärung zu prüfen, ob die Anlage konform zu der im Anlagenzertifikat dokumentierten Planung errichtet wurde. Hierzu wird die Anlage normalerweise vom Gutachter vor Ort in Augenschein genommen um die Ausführung der Komponenten, Schutzeinstellungen und Reglerparameter zu kontrollieren. Die Dokumentation der tatsächlichen Ausführung erfolgt unter anderem durch Fotos der relevanten Komponenten, Typenschilder, Parametersets, wie hier in **Bild 4.**

Erfahrungsgemäß werden auch in diesem letzten Schritt oft noch relevante Fehleinstellungen identifiziert. Eine aktuelle Auswertung von in der FGW organisierten Zertifizierungsstellen hat gezeigt, dass von den untersuchten Projekten mit insgesamt 1.400 MW Anschlussleistung aus dem Jahr 2014 in über 90 % der Fälle fehlerhafte Punkte in der Konformitätserklärung ermittelt und korrigiert werden konnten. Eine noch größere Auswertung von Anlagenzertifikaten aus den Jahren 2013 und 2014 mit mehr als 10.000 MW Anschlussleistung hat gezeigt, dass bei ca. 80 % der Projekte Netz-relevante Fehler vorlagen, die so bereits in der Planungsphase behoben werden konnten.

Bild 4 Netzanschlusspunkt Windpark während einer Konformitätsprüfung

1.4 Wiederkehrende Prüfung

Die Wiederkehrende Prüfung ist bisher kein fester Bestandteil der Netzanschlussprüfung in Deutschland. Übertragen auf den Automobil-Sektor würde das bedeuten, dass ein KFZ zwar bei der Zulassung technisch überwacht wird, aber anschließend während der gesamten Lebensdauer nie wieder zur technischen Prüfung (z. B. TÜV) müsste. Diese Situation wäre in Deutschland undenkbar, obwohl die Lebensdauer von ca. 20 Jahren von Erzeugungsanlagen und KFZ durchaus vergleichbar ist und auch technische Veränderungen während der Betriebszeit in ähnlicher Weise auftreten können. Bei Windenergieanlagen erfolgen beispielsweise regelmäßig Software-Updates zur Optimierung der Betriebsführung oder zur Fehlerbehebung. Die Parameter der Software haben aber auch Einfluss auf das Verhalten der Anlage am Netz, wie zum Beispiel das Durchfahren von Netzfehlern (Low Voltage Ride Through, LVRT), die Art und Weise der Einspeisung von Blindstrom im Fehlerfall, oder auf die Schutzeinstellungen. Des Weiteren kann auch ein Komponententausch von Trafo, Generator oder Umrichter die Netzeigenschaften im Laufe des Betriebs verändern. Es kann daher aktuell niemand garantieren, dass eine Erzeugungsanlage nach 5, 10 oder 20 Jahren noch die Netzanforderungen erfüllt, auf die sie zur Inbetriebnahme geprüft worden ist.

Der Erfolg der Energiewende ist aber ganz entscheidend auch davon abhängig, dass eine möglichst große Anzahl der bestehenden Erzeugungsanlagen die Netzanforderungen dauerhaft erfüllt und somit die Systemstabilität sicher gewährleistet werden kann. Es gibt daher immer mehr Stimmen in der Branche die sich für die Einführung von Wiederkehrenden Prüfungen der Netzeigenschaften von Erzeugungsanlagen aussprechen. Da eine zusätzliche Prüfung auch zusätzliche Kosten verursachen würde, sind alle Beteiligten dazu angehalten eine effiziente Lösung mit vertretbarem Aufwand zu finden, insbesondere in Bezug auf den zeitlichen Turnus und die erforderlichen Prüfpunkte. Durchsetzen kann solch eine Prüfung am Ende nur eine gesetzliche Vorgabe oder eine ähnliche Anforderung der Netzbetreiber.

2 Vorteile des Verfahrens

Die Landschaft der Erzeugungsanlagen erneuerbarer Energien in Deutschland ist geprägt durch zahlreiche kleine und mittelgroße Akteure. Somit sind zahlreiche Beurteilungen von Netzanschlussvorhaben zu erstellen und zu bewerten. Diese Bewertungsverfahren müssen im Sinne der Energiewende effizient, neutral und rechtssicher durchgeführt werden.

In Zusammenarbeit aller Beteiligten wurde mit der Richtlinie TR8 [3] ein allgemein akzeptiertes, transparentes Verfahren zur Netzanschlussprüfung etabliert. Dieses Verfahren gewährleistet eine effektive, standardisierte und diskriminierungsfreie Beurteilung des Netzanschlussgesuches. Die gutachterliche Unabhängigkeit der verantwortlichen Mitarbeiter und deren Kompetenz auf technischen und netzphysikalischen Niveau wird regelmäßig durch die DAkkS nach DIN/ISO 17065 [5] überprüft.

Die Einführung des Verfahrens im Jahre 2009 führte zu einer sichtbaren Qualitätssteigerung bei der Planung und dem Betrieb von EZA. Das Verfahren stellt heute eine Grundvoraussetzung zur flächendeckenden tatsächlichen praktischen Umsetzung der geltenden Anforderungen dar und trägt somit wesentlich zum Erfolg der Energiewende bei.

Die Bereitstellung von Systemdienstleistungen (SDL) verursacht zweifellos Kosten. Die beteiligten Anlagenbetreiber und Netzbetreiber haben hier jedoch unterschiedliche wirtschaftliche Interessen. Die Systemstabilität liegt im Interesse der Allgemeinheit und kann nur über eine gesetzliche Regelung und deren Überprüfung durch unabhängige Prüfer sichergestellt werden.

Die Konzentration der Netzanschlussprüfung auf ca. 15 Zertifizierungsstellen spart volkswirtschaftlich erhebliche Kosten ein. Durch den freien Markt an Zertifizierungsstellen und die in den letzten Jahren gesammelte Erfahrung konnten die Kosten der Prüfung immer weiter reduziert werden. Mittlerweile haben die Zertifizierungsstellen ihre Ressourcen deutlich erweitert, so dass eine schnelle Bearbeitung der Prüfungen gewährleistet werden kann.

Die zahlreichen Netzbetreiber sind so nicht gezwungen eigene Experten vorzuhalten um personell und fachlich in der Lage zu sein, die notwendigen Prüfungen selbst durchzuführen und die gesetzlich vorgegebenen Bearbeitungsfristen einzuhalten.

Vertrauliche Unterlagen der Netzbetreiber und der Anlagenhersteller werden bei den Zertifizierungsstellen vertraulich behandelt, meist auf Basis von Geheimhaltungsvereinbarungen. Das Know-how bleibt so geschützt und muss nicht in die Breite gestreut werden.

Die akkreditierten Zertifizierungsstellen sind verpflichtet, sich ständig an der Weiterentwicklung von Richtlinien zu beteiligen. Diese Arbeit wird so sukzessive der Allgemeinheit zur Verfügung gestellt.

Das deutsche Verfahren kann aufgrund der europäischen Verankerung der Akkreditierung leicht in andere Länder übertragen werden, in denen ebenfalls der hohe Prüfaufwand zum Nachweis der Netzanschlusskriterien nicht mehr durch das bisherige Fachpersonal der Netzbetreiber geführt werden kann.

3 Ausblick

Der aktuelle Prozess der Netzanschlusszertifizierung in Deutschland und dessen Vorteile wurden oben eingehend erläutert. Um den Erfolg der Energiewende zu sichern, sollte dieser Nachweisprozess möglichst beibehalten werden.

Mit dem Auslaufen der gesetzlichen Verankerung in der SDLWindV [2] bis Ende 2016, sowie der geplanten Ablösung der BDEW Mittelspannungsrichtlinie [1] steht der Fortbestand dieses etablierten Prozesses und damit die Sicherstellung der Systemstabilität in der Elektroenergieversorgung auf dem Spiel. Der Gesetzgeber ist nun gefordert im Sinne einer sicheren Energiewende und zum Wohl der Allgemeinheit eine Weiterführung der Netzanschlussprüfung, zum Beispiel über eine neue Verankerung im EnWG [8], zu garantieren.

Des Weiteren könnte der Prozess in Zukunft um eine sinnvolle Lösung zur wiederkehrenden Prüfung erweitert werden, oder auch auf den Niederspannungsbereich ausgedehnt werden.

4 Autor

Roman Bluhm ist Geschäftsführer und stellvertretender Zertifizierungsstellenleiter der NETZ-ING. AB GmbH mit Sitz in Rendsburg und Obmann der AG EZA und Konformitätserklärung in der FGW e.V.

5 Literatur

[1] BDEW MSR 2008, „Technische Richtlinie Erzeugungsanlagen am Mittelspannungsnetz" Richtlinie für den Anschluss und Parallelbetrieb von Erzeugungsanlagen am Mittelspannungsnetz; Stand: Ausgabe Juni 2008; Hrsg.: BDEW, mit 4. Ergänzung, gültig ab 01.01.2013 (BDEW Mittelspannungsrichtlinie)

[2] SDLWindV, Verordnung zum EEG zu Systemdienstleistungen durch Windenergieanlagen (Systemdienstleistungsverordnung - SDLWindV) incl. Änderungsverordnung vom Juni 2010, 2. Änderungsverordnung vom April 2011 und Novellierung Juli 2011 und Juli 2014

[3] FGW TR8, Technische Richtlinien für Erzeugungseinheiten und -anlagen Teil 8: „Zertifizierung der elektrischen Eigenschaften von Erzeugungseinheiten und -anlagen am Mittel-, Hoch- und Höchstspannungsnetz"; Herausgeber: Fördergesellschaft Windenergie und anderer Erneuerbaren Energien, FGW e.V. Revision 6 vom 01.05.2013

[4] VDE-AR-N 4120, Technische Bedingungen für den Anschluss und Betrieb von Kundenanlagen an das Hochspannungsnetz (TAB Hochspannung), Januar 2015

[5] DIN EN ISO/IEC 17065, Konformitätsbewertung – Anforderungen an Stellen, die Produkte, Prozesse und Dienstleistungen zertifizieren, Januar 2013

[6] VDN Transmissioncode 2007-08: „Netz- und Systemregeln der deutschen Übertragungsnetzbetreiber"

[7] VDE-FNN 2010-02: „Lastenheft Blindleistungsrichtungs-Unterspannungsschutz (Q-U-Schutz)"

[8] EnWG, Gesetz über die Elektrizitäts- und Gasversorgung (Energiewirtschaftsgesetz — EnWG) von 2005 Zuletzt geändert durch Art. 6 G v. 21.7.2014 I 1066

[9] FGW e.V., „Auswertung der Anlagenzertifikate durch die AG Zertifizierungsstellen" und „Vorteile der Zertifizierung von Erzeugungsanlagen", 2015

[10] VDN-Leitfaden „EEG Erzeugungsanlagen am Hoch- und Höchstspannungsnetz", 2004

Rechtliche Rahmenbedingungen für Stromspeicher im „Strommarkt 2.0“ und im Zusammenspiel mit Netzausbau und EEG-Einspeisemanagement

Henning Thomas, LL.M.[1]

[1]Rechtsanwalt bei Becker Büttner Held, Kaiser-Wilhelm-Straße 93, 20355 Hamburg, henning.thomas@bbh-online.de

Kurzfassung

Der Einsatz von Stromspeichern zur Integration der erneuerbaren Energien und zum Umbau der Energiesysteme muss die Vorgaben des Energiewirtschaftsrechts beachten. Ein Sonderrecht für Stromspeicher besteht darin nicht, stattdessen ergibt sich der Rechtsrahmen aus den allgemeinen Bestimmungen des Energiewirtschaftsgesetzes und des Erneuerbare-Energien-Gesetzes sowie spezifischer für einzelne Fragen von Stromspeichern getroffener Einzelregelungen. In dem mit dem Weißbuch des Bundesministeriums für Wirtschaft und Energie (BMWi) vorgeschlagenen „Strommarkt 2.0“ werden sich Speicher als eine unter den Flexibilitätsoptionen unter den regulatorischen Rahmenbedingungen behaupten müssen. Für den Einsatz von Speichern besonders wichtige Rahmenbedingungen werden mit diesem Beitrag aufgezeigt und im Besonderen der Frage nachgegangen, wie die Speicher im Zusammenspiel mit Netzausbau und Einspeisemanagement aus rechtlicher Sicht zu verorten sind.

1 Stromspeicher unter den regulatorischen Rahmenbedingungen des „Strommarkts 2.0“

Mit dem Weißbuch „Ein Strommarkt für die Energiewende“ hat das BMWi für eine Grundsatzentscheidung zur Weiterentwicklung des bestehenden Strommarkts zu einem „Strommarkt 2.0“ ausgesprochen [1]. Dieser entwickelt den heutigen Strommarkt fort und ergänzt diesen durch eine Kapazitätsreserve. Die Einführung eines zusätzlichen Kapazitätsmarktes lehnt das BMWi ab.

Das Weißbuch des BMWi enthält wichtige Aussagen, die für die zukünftigen Rahmenbedingungen für Stromspeicher von zentraler Relevanz sind. Das BMWi möchte die bestehenden Marktmechanismen stärken und spricht sich für eine wettbewerbliche Preisbildung aus. Die freie Preisbildung am Strommarkt soll in einem Strommarktgesetz garantiert werden. Der Strommarkt 2.0 soll einen technologieoffenen Wettbewerb der Flexibilitätsoptionen eröffnen, zu denen auch die Stromspeicher gehören. Das Potenzial der Flexibilitätsoptionen sei „heute vielfältig und weit größer als der tatsächliche Bedarf“. Sehr viele Flexibilitätsoptionen seien bereits wirtschaftlich oder würden es bei veränderten Strompreisen, so dass sie keine gesonderte Förderung benötigten. Speicher würden sich voraussichtlich erst mittel- bis langfristig im Wettbewerb der Flexibilitätsoptionen durchsetzen können. Der Netzausbau sei in der Regel die kosteneffizienteste Flexibilitätsoption.

Das Weißbuch zeigt aber zugleich auch Anwendungsfelder für Stromspeicher auf. Zu den Anwendungsfeldern gehört neben der Teilnahme als Flexibilitätsoption am Strommarkt beispielsweise, dass Batteriespeicher in der Elektromobilität eingesetzt werden können. Vor allem betont das Weißbuch die zunehmend wichtige Verknüpfung der Sektoren Strom, Wärme und Verkehr („Power-to-X“). In einigen wenigen Fällen könnten Speicher zudem sogar den Netzausbau auf der Niederspannungsebene ersetzen, auch könnten Netzbetreiber Speicher für Systemdienstleistungen einsetzen.

- Bilanzkreistreue stärken (Maßnahme 3)
- Regelleistungsmärkte für neue Anbieter öffnen (Maßnahme 6)
- Zielmodell für staatlich veranlasste Preisbestandteile und Netzentgelte entwickeln (Maßnahme 7)
- Netzentgeltsystematik weiterentwickeln (Maßnahme 9)
- Verbreitung der Elektromobilität unterstützen (Maßnahme 11)

Abbildung 1: Für Speicher besonders relevante Maßnahmen im Weißbuch des BMWi

Außerdem werden in dem Weißbuch „Sowieso-Maßnahmen“ angekündigt, die in nächster Zeit ergriffen werden sollen und auch für Speicher Relevanz haben werden. Hierzu gehört insbesondere, dass die Teilnahme an den Regelenergiemärkten für neue Anbieter und insbesondere Speicher vereinfacht werden soll. Hierzu werden insbesondere kürzere Vorlaufzeiten und Produktzeitscheiben, aber beispielsweise auch ein Sekundärmarkt für Sekundärregelleistung diskutiert. Auch soll ein weiterentwickeltes Bilanz- und Ausgleichsenergiesystem die Anreize zur Bilanzkreistreue stärken. Angekündigt wird außerdem, dass die staatlich veranlassten Strompreisbestandteile mit dem Ziel eines flexiblen Reagierens auf die Preissignale überarbeitet werden sollen. Dabei soll auch die effiziente Kopplung der Sektoren ermöglicht werden.

Hinzuweisen ist allerdings auch darauf, dass die vermiedenen Netzentgelte für Neuanlagen mit Inbetriebnahme ab 2021 abgeschafft werden sollen. Vermiedene Netznutzungsentgelte können bei der Einspeisung von Strom aus dezentralen Speichern, für die keine Förderung nach dem Erneuerbare-Energien-Gesetz 2014 (EEG 2014 [2]) geltend gemacht wird, heute unter Umständen verlangt werden. Dies ist in der Rechtswissenschaft allerdings in Diskussion [3].

2 Netzausbau, EEG-Einspeisemanagement und Stromspeicher

Die Integration von einem höher werdenden Anteil erneuerbarer Energien, insbesondere von Windenergie und Solarenergie, macht unbestritten Anpassungen an der Energieinfrastruktur erforderlich. Dabei geht es – kurz gesagt – um ein Zusammenspiel von unterschiedlichen Flexibilitätsinstrumenten zur Integration der (erneuerbaren) Strommengen. Hier sollen im Folgenden Maßnahmen des Netzausbaus, des Einspeisemanagements sowie der Stromspeicherung angesprochen werden. In welchem Umfang diese Maßnahmen im Kontext der energiewirtschaftlichen Zieltrias aus Wirtschaftlichkeit, Umweltverträglichkeit und Versorgungssicherheit (siehe dazu die Zielbestimmung in § 1 des Energiewirtschaftsgesetzes, EnWG [4]) durchgeführt werden sollen, ist durchaus umstritten und wird in verschiedenen Studien diskutiert. Im Folgenden werden wichtige rechtliche Vorgaben skizziert, welche beim Einsatz von Speichern im Kontext von Netzausbau und Einspeisemanagement zu bedenken sind.

2.1 Netzausbau

Mit diversen regulatorischen Maßnahmen versucht der Gesetzgeber, dem erheblichen Bedarf an einem weiteren Ausbau der Energieversorgungsnetze nachzukommen. Hierzu sind Vorgaben insbesondere im EnWG selbst, aber auch mit dem Netzausbaubeschleunigungsgesetz Übertragungsnetz (NABEG [5]), dem Energieleitungsausbaugesetz (EnLAG [6]) und dem EEG getroffen worden.

Aus Sicht der Speicher stellt sich dabei die Frage, ob im Einzelfall anstelle einer Maßnahme des Netzausbaus möglicherweise eine Speichereinrichtung installiert werden kann. In welchen Fällen und in welchem Umfang dies angesichts grundsätzlicher Kostenvorteile des Netzausbaus sinnvoll erscheint, ist in der Diskussion und bedarf noch vertiefter Forschung. Aus Sicht des Regulierungsrechts ist eine Nutzung von Speichern anstelle einer Maßnahme des Netzausbaus nicht unproblematisch. Denn die Entflechtungsvorgaben des EnWG geben eine grundsätzliche Trennung der Bereiche von Netz und Vertrieb vor. Traditionell werden Stromspeicher dem Bereich des Stromvertriebs zugeordnet, so dass sich die Frage stellt, ob die Netzbetreiber überhaupt als Betreiber von Speichereinrichtungen in Betracht kommen. Viel spricht dafür, dass dies jedenfalls dann möglich ist, wenn der Speicher ausschließlich für Zwecke des Netzbetriebs eingesetzt wird [7]. Mit einer derartigen eingeschränkten Betriebsweise wird es aber jedenfalls bei den heutigen Kosten der Stromspeicherung aus Sicht eines Netzbetreibers schwierig sein, gegenüber anderen Optionen für das Netz die kostenwirtschaftlichste Möglichkeit darzustellen. In diesem Bereich ist daher wichtig, dass die Beteiligten (ggf. auch unter Weiterentwicklung regulatorischer Vorgaben) Wege finden, wie die unterschiedlichen Einsatzmöglichkeiten von Speichern im Regulierungsrahmen abgebildet werden können.

2.2 Einspeisemanagement

Eine ähnliche Beziehung stellt sich für Stromspeicher auch zu den Vorschriften des Einspeisemanagements im EEG dar. Nach der Vorschrift in § 14 EEG 2014 ist ausnahmsweise auch eine Regelung von EEG-Anlagen und KWK-Anlagen möglich, wenn anderenfalls in dem jeweils betroffenen Netzbereich die Gefahr von Netzengpässen besteht. Dabei ist der Vorrang für Strom aus erneuerbaren Energien so weit wie möglich zu wahren. Dennoch kommt es gerade in Norddeutschland angesichts des Ausbaus der regenerativen Stromerzeugungskapazitäten immer häufiger dazu, dass Windenergieanlagen aufgrund von Netzengpässen in ihrer Leistung reduziert werden müssen. Die Anlagenbetreiber haben in diesen Zeiträumen einen Anspruch auf im Grundsatz 95 Prozent der entgangenen Einnahmen aus der Härtefallregelung in § 15 EEG 2014.

Die Frage liegt nahe, ob in diesem Zusammenspiel Speicher zur Abnahme der „Stromüberschüsse" eingesetzt werden können, die infolge der Netzengpässe nicht eingespeist werden können, und damit die Leistungsreduzierung von EEG-Anlagen im Einspeisemanagement möglichst beschränkt werden kann. Einzelheiten hierzu bedürfen noch der rechtlichen Klärung [8]. Mit dem Weißbuch schlägt das BMWi nun eine Verringerung des Netzausbaubedarfs durch eine „Spitzenkappung" von Erneuerbare-Energien-Anlagen vor. Dies könne zu einem bedarfsgerechten und volkswirtschaftlichen Netzausbau beitragen [9]. Die Spitzenkappung steht eher in einem Kontrast zu dem Versuch, die überschüssigen Strommengen zu speichern (was angesichts der Kosten der Speicherung ohnehin nur unter bestimmten Umständen sinnvoll sein wird).

3 Kostenbelastungen für Strombezug der Speichereinrichtung

Ein wichtiges rechtliches Thema für Stromspeicher ist stets, welche gesetzlich vorgegebenen Kostenbelastungen für den Strombezug eines Stromspeichers anfallen. Hierbei geht es in erster Linie um die Netzentgelte, EEG-Umlage und die Stromsteuer.

Grundlegend könnte man bereits infrage stellen, ob diese Kostenbelastungen für den Strombezug eines Speichers überhaupt anfallen können, da der Strom doch „weiterhin vorhanden" ist und lediglich gespeichert wird. Der BGH hat jedoch in einer Entscheidung aus 2010 entschieden, dass die Entnahme von Strom für den Pumpvorgang in

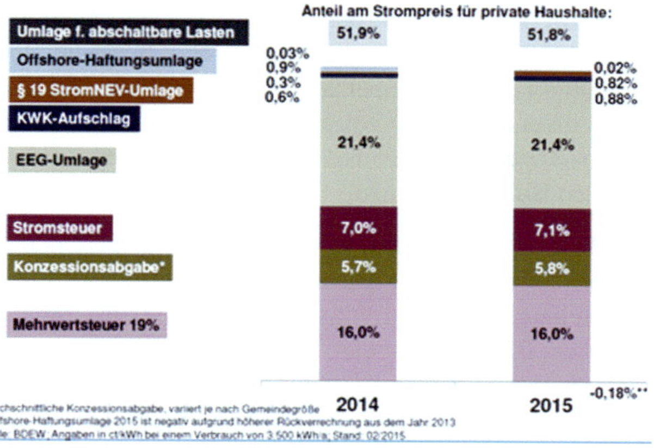

Abbildung 2: Übersicht Stromnebenkosten, Quelle: BDEW, Strompreisanalyse März 2015

einem Pumpspeicherkraftwerk als netzentgeltpflichtiger Letztverbrauch anzusehen ist [10]. Viel spricht dafür, dass diese Sichtweise grundsätzlich auch auf andere Speichertechnologien mit einer Umwandlung von Strom in einen anderen Energieträger übertragbar ist. Dies bedeutet als Ausgangsüberlegung, dass die Stromkostenbelastungen mit Netzentgelten, EEG-Umlage und Stromsteuer bei der Speicherung anfallen können.

Gleichwohl ist in einem ersten Schritt immer zu prüfen, ob der jeweilige Kostenbelastungstatbestand in der konkreten Betriebsweise und bei der jeweiligen Stromspeichertechnologie anfällt. Netzentgelte fallen beispielsweise nur dann an, wenn auch ein Energieversorgungsnetz genutzt wird. Soweit die Speicherung vor dem Energieversorgungsnetz stattfindet (dessen räumliche Abgrenzung im Einzelfall zu bestimmen ist), kann daher möglicherweise schon aus diesem Grund kein Netzentgelt anfallen.

Von hoher Bedeutung ist zu berücksichtigen, dass die EEG-Umlage und die Stromsteuer auch bei einer Lieferung von Strom an einen Speicher vor dem Netz für die allgemeine Versorgung anfallen können. Daher müssen Betreiber von Speichern diese Themen auch dann im Blick haben, wenn der Speicher z. B. unmittelbar über eine Direktleitung mit einer Stromerzeugungsanlage verbunden wird, aus welcher der Speicher versorgt wird.

In einem zweiten Schritt ist dann stets zu prüfen, ob ein Befreiungs- oder Reduzierungstatbestand für den Strombezug der Speichereinrichtung existiert. Hier sind für den Bereich der Netzentgelte und der EEG-Umlage seit 2010 gezielt Befreiungstatbestände für die Stromspeicherung geschaffen worden. Allerdings gelten diese nur unter jeweils im Einzelfall zu prüfenden Voraussetzungen.

So gewährt § 118 Abs. 6 Satz 1 EnWG eine Freistellung von den Netznutzungsentgelten für den Bezug der zu speichernden elektrischen Energie für nach dem 31.12.2008 neu errichtete Anlagen zur Speicherung elektrischer Energie mit Inbetriebnahme ab dem 04.08.2011 innerhalb von 15 Jahren. Diese Befreiung ist allerdings auf einen Zeitraum von (nach derzeitiger Rechtslage) 20 Jahren ab der Inbetriebnahme begrenzt. Außerdem wird in Satz 3 der Vorschrift weiter vorausgesetzt, dass die Freistellung nur gewährt wird, wenn die elektrische Energie zur Speicherung in einem elektrischen, chemischen, mechanischen oder physikalischen Stromspeicher aus einem Transport- oder Verteilernetz entnommen und die zur Ausspeisung zurückgewonnene elektrische Energie zeitlich verzögert wieder in dasselbe Netz eingespeist wird. Grundsätzlich bedarf es daher für die Befreiung von den Netzentgelten beim Strombezug eines Speichers einer Rückverstromung und Wiedereinspeisung in das Netz. Für Power-to-Gas-Einrichtungen können allerdings Ausnahmen greifen.

Für den Bereich der EEG-Umlage findet sich ein Befreiungstatbestand in § 60 Abs. 3 Satz 1 EEG 2014, dessen tendenziell engere Voraussetzungen ebenfalls im Einzelfall geprüft werden müssen. Für Strom, der zum Zweck der Zwischenspeicherung an einen elektrischen, chemischen, mechanischen oder physikalischen Stromspeicher geliefert oder geleitet wird, entfällt danach der Anspruch auf EEG-Umlage, „wenn dem Stromspeicher Energie ausschließlich zur Wiedereinspeisung von Strom in das Netz entnommen wird“. Hier wird daher sogar eine ausschließliche Entnahme von Strom aus dem Stromspeicher zur Wiedereinspeisung in das Netz für die allgemeine Versorgung vorausgesetzt, wenn eine Befreiung geltend gemacht werden soll.

Angesichts dessen, dass dieser Befreiungstatbestand nicht bei sämtlichen Betriebsweisen von Stromspeichern greifen wird, kommt auch dem so genannten Eigenversorgungsprivileg des EEG für den Strombezug eines Stromspeichers eine hohe Bedeutung zu. Dabei ist zu beachten, dass die Eigenversorgung ab dem 01.08.2014 grundsätzlich ebenfalls mit einer (ggf. reduzierten) EEG-Umlage belastet wird. Für Bestandsanlagen können allerdings Befreiungen eingreifen. Auch gibt es bei neuen Anlagen in bestimmten Fällen bei einer EEG-Eigenversorgung noch eine vollständige Befreiung von der EEG-Umlage, wenn die Voraussetzungen in § 61 Abs. 2 EEG 2014 eingehalten werden. Nach der Clearingstelle EEG kann die EEG-Umlage-Befreiung in § 61 Abs. 2 Nr. 4 EEG 2014 eingreifen, wenn eine Eigenversorgung aus einer Photovoltaik-Anlage bis 10 kW erfolgt und ein Speicher mit einer Leistung bis 10 kW eingebunden ist [11]. Allerdings sind die Empfehlungen der Clearingstelle EEG rechtlich nicht verbindlich. In jedem Fall müssen für die Befreiung/Reduzierung zudem die Anforderungen an eine Eigenversorgung gegeben sein.

4 Speicher und die Direktvermarktungsvorschriften des EEG 2014

Einsatzmöglichkeiten für Speicher können sich auch aus den Direktvermarktungsvorschriften des EEG ergeben. Mit dem EEG 2014 ist die Direktvermarktung für größere Neuanlagen zur Erzeugung von Strom aus erneuerbaren Energien grundsätzlich verpflichtend ausgestaltet worden. Für Speicher hat dies in zweierlei Hinsicht Relevanz [12]:

Zum einen können Speicher möglicherweise im Wege der Direktvermarktung von EEG-Stromerzeugungsanlagen beliefert werden oder ihrerseits zwischengespeicherten grünen Strom aus erneuerbaren Energien im Wege der EEG-Direktvermarktung an weitere Abnehmer liefern. Hierbei ist zu beachten, ob der Speicher vor dem Netz für die allgemeine Versorgung oder über das Netz betrieben werden soll. Wird der Speicher vor dem Netz betrieben, können bei der Ausspeisung unter Einhaltung der gesetzlichen Voraussetzungen Förderungen nach dem EEG für den zwischengespeicherten Strom geltend gemacht werden, beispielsweise im Wege der Direktvermarktung unter Inanspruchnahme der Marktprämie, wobei dann bei der Einspeicherung regelmäßig keine finanzielle Förderung nach dem EEG beansprucht worden sein darf.

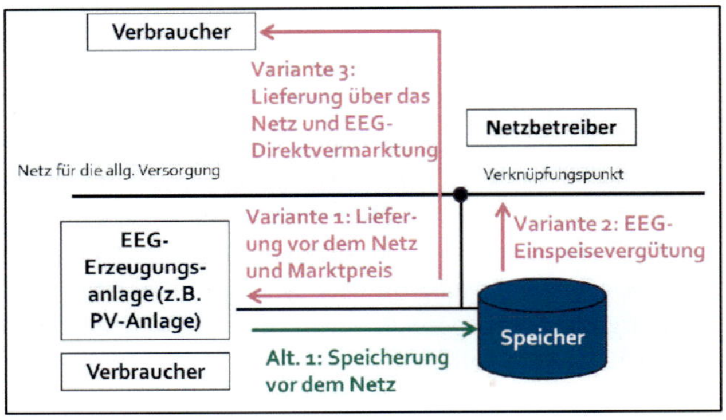

Abbildung 3: Speicherung *vor* dem Netz und Lieferoptionen nach EEG 2014

Zweitens haben aber die ausgeweiteten Vorgaben zur Direktvermarktung im EEG für Speicher vor allem auch dadurch Relevanz, dass sich EEG-Anlagenbetreiber immer mehr selbst Gedanken dazu machen müssen, wie sie den Strom aus erneuerbaren Energien möglichst sinnvoll vermarkten. Der Einsatz eines Speichers kann dabei jedenfalls perspektivisch eine Komponente darstellen. Dabei ist auch darauf hinzuweisen, dass sich zukünftig für große neue EEG-Anlagen die Förderung nach dem EEG auf null reduzieren kann, wenn an der Strombörse anhaltende negative Strompreise vorhalten (siehe § 24 EEG 2014). Dies allein wird den Einsatz eines Speichers kaum wirtschaftlich machen. Das Einsatzfeld für Speicher kann sich aber aus einer Zusammenschau der unterschiedlichen regulatorischen Vorgaben ergeben, die hier nur auszugsweise dargestellt werden konnten.

5 Fazit

Die regulatorischen Rahmenbedingungen ermöglichen den Einsatz von Stromspeichern mit unterschiedlichen Speicherkonzepten. Betreiber müssen sich aber genau überlegen, welche Einsatzmöglichkeiten eines Speichers genutzt werden sollen, rechtlich genutzt werden können und welche Auswirkungen dies insbesondere im Hinblick auf die Belastung mit den staatlich vorgegebenen Strompreisbestandteilen hat. Auch unter dem heutigen Rechtsrahmen lassen sich Speicherprojekte durchaus realisieren, wie diverse Projekte zeigen. Gerade die Themen Kostenbelastungen und Vermarktungsmöglichkeiten sind dabei in den Blick zu nehmen.

Das Weißbuch des BMWi macht deutlich, dass sich Speicher als eine der Flexibilitätsoptionen im Wettbewerb derselben behaupten müssen. Dabei ist insgesamt von großer Bedeutung, wie Netzausbau, Einspeisemanagement und Speicherung aufeinander abgestimmt werden und dass die regulatorischen Rahmenbedingungen hierbei die sinnvollste Lösung anreizen.

6 Literatur

[1] Bundesministerium für Wirtschaft und Energie, Ein Strommarkt für die Energiewende, Ergebnispapier des Bundesministeriums für Wirtschaft und Energie (Weißbuch), Juli 2015

[2] Erneuerbare-Energien-Gesetz vom 21.07.2014 (BGBl. I S. 1066), zuletzt geändert durch Artikel 1 des Gesetzes vom 29.06.2015 (BGBl. I S. 1010)

[3] Heller, H.: Optimierung der energierechtlichen Rahmenbedingungen durch den Einsatz moderner Stromspeichertechnologie. EWeRK 2013, 177 (184 f.)

[4] Energiewirtschaftsgesetz vom 07.07.2005 (BGBl. I S. 1970, 3621), zuletzt geändert durch Artikel 15 des Gesetzes vom 28.07.2015 (BGBl. I S. 1400)

[5] Netzausbaubeschleunigungsgesetz Übertragungsnetz vom 28.07.2011 (BGBl. I S. 1690), zuletzt geändert durch Artikel 4 des Gesetzes vom 20.12.2012 (BGBl. I S. 2730)

[6] Energieleitungsausbaugesetz vom 21.08.2009 (BGBl. I S. 2870), zuletzt geändert durch Artikel 3 des Gesetzes vom 23.07.2013 (BGBl. I S. 2543)

[7] Weyer, H., Lietz, F.: Entflechtungsvorgaben für den Betrieb von Stromspeichern – Teil 1. ZNER 2014, 241 (243)

[8] Thomas, H., Altrock, M.: Einsatzmöglichkeiten für Energiespeicher. ZUR 2013, 579 (585)

[9] BMWi, Weißbuch, S. 75 f. (siehe oben [1])

[10] Bundesgerichtshof (BGH), Urteil vom 17.11.2009 – EnVR 56/08. NVwZ-RR 2010, 431

[11] Clearingstelle EEG, Empfehlung vom 02.06.2015 – 2014/13

[12] Ausführlich Thomas, H., Altrock, M.: Einsatzmöglichkeiten für Energiespeicher. ZUR 2013, 579

Auswirkungen der Anreizsysteme für private PV-Anlagenbetreiber auf das lokale Verteilnetz

Verena Spielmann[1], Carola Bettinger[2], Katharina Skau[3],
Prof. Dr.-Ing. Hans-Peter Beck[1,4], Prof. Dr. Clemens Fuchs[3]
[1]Institut für Elektrische Energietechnik und Energiesysteme, TU Clausthal,
Leibnizstraße 28, 38678 Clausthal-Zellerfeld
[2]Institut für Bank-, Finanz- und Rechnungswesen, Leuphana Universität Lüneburg, Scharnhorststr.1, 21335 Lüneburg
[3]Fachbereiche Agrarwirtschaft und Lebensmittelwissenschaften, Hochschule Neubrandenburg,
Brodaer Str.2, 17033 Neubrandenburg
[4]Energie-Forschungszentrum Niedersachsen, Am Stollen 19, 38640 Goslar,
verena.schild@tu-clausthal.de

Kurzfassung

Erneuerbare Energien werden in Deutschland zu einem großen Teil in Verteilnetzen bereitgestellt. Hierdurch ändern sich die Anforderungen an die Verteilnetze, da der Leistungsfluss weniger gut prognostizierbar wird und dieser zunehmend durch Einspeisungen in den unteren Netzebenen und der damit einhergehenden Umkehrung des Lastflusses hin zur Rückspeisung geprägt ist. In der Vergangenheit wurden von verschiedenen Institutionen die Potenziale verschiedener Technologien zur Entlastung der Verteilnetze und des Gesamtsystems ermittelt, hierunter auch die Möglichkeiten des systemdienlichen Erzeugungsmanagements. Wie gut dieses Potential genutzt werden kann, hängt maßgeblich davon ab, ob es sich unter den aktuellen rechtliche-regulatorischen Rahmenbedingungen für den einzelnen Akteur betriebswirtschaftlich lohnt, seine Anlage netzdienlich zu betreiben. Dieses Paper untersucht innerhalb einer zweistufigen – wirtschaftlichen und technischen – Modellierung und Simulation, wie private PV-Anlagenbetreiber, welche ihre Einspeisung durch die zusätzliche Installation eines Speichers flexibilisieren, unter den aktuellen Anreizmechanismen des Rechts- und Förderrahmens handeln. Es ermittelt, welche Effekte dies auf die resultierende Austauschleistung eines lokalen Verteilnetzes hat. Die Arbeiten sind innerhalb des vom Bundesministerium für Bildung und Forschung geförderten Projekts „Effiziente Nutzung erneuerbarer Energien durch regionale ressourcenoptimierte ‚intelligente' Versorgungs- und Verbrauchsnetze (Smart Microgrids)" (FKZ: 03EK3524 B, C und D) entstanden.

1 Motivation

Anlagen zur Bereitstellung von elektrischer Energie aus regenerativen Quellen (EE-Anlagen) werden in Deutschland vornehmlich in Verteilnetzen installiert. Nach [1] sind Anlagen mit einer Gesamtleistung von etwa 61 GW in den verschiedenen Spannungsebenen des Verteilnetzes angeschlossen. In der Niederspannungsebene sind fast ausschließlich Photovoltaikanlagen installiert, die in Bezug auf die Anzahl den größten Anteil der EE-Anlagen in den Verteilnetzen ausmachen.

Durch den weiteren Ausbau dezentraler Anlagen ändern sich die Anforderungen an die Verteilnetze, da der Leistungsfluss durch die große Anzahl von Anlagen mit dargebotsabhängiger Einspeisung weniger gut prognostizierbar und in seinem Verlauf unbekannt wird. Somit kommt es zu häufigen Änderungen der Belastungssituation in den Verteilnetzen, wodurch sich deren Betrieb komplizierter darstellt als im klassischen, lastgeprägten Fall [2].

Im Rahmen verschiedener Studien (vgl. [3], [1], [2]) werden für die entstehenden Herausforderungen im Verteilnetz Lösungsansätze, wie etwa konventioneller Netzausbau, Einsatz von Speichern und neuen Technologien des Netzausbaus, aber auch Erzeugungs- und Lastmanagementansätze untersucht. Um das Potential eines systemdienlichen Erzeugungsmanagements zu heben wurden in der Vergangenheit immer wieder rechtliche Änderungen insbesondere am Erneuerbaren Energien Gesetz (EEG) vorgenommen (vgl. [4]). Es ist zu klären, wie sich dezentrale Einspeiser unter dem Einfluss der Anreizsysteme von Förderungen und gesetzliche Regelungen verhalten und ob hierdurch ein netzdienliches Verhalten angereizt wird.

Dies wird hier am Beispiel von privaten PV-Aufdach-Anlagen in Kombination mit Batteriespeichern untersucht. [5] zeigt, dass aus technischer Sicht solche Systeme zur Stützung des Gesamtsystems in der Lage sind, jedoch neben technologischen Lösungen auch rechtliche Rahmenbedingungen für eine ausreichende Honorierung netzdienlichen Verhaltens vorhanden sein müssen. Es werden die Anreizmechanismen für diese Anwendung ermittelt und das resultierende Verhalten im Hinblick auf die Stützung des lokalen Netzes eines Beispieldorfs in Mecklenburg-Vorpommern bzw. des Gesamtsystems untersucht.

2 Methodik

Um die Auswirkungen des Verhaltens einzelner Akteure auf das Gesamtsystem zu untersuchen, wird auf eine zweistufige Simulation zurückgegriffen. Dabei wird zunächst innerhalb einer wirtschaftlichen Simulation das betriebswirtschaftlich getriebene Verhalten der einzelnen Akteure ermittelt. Dieses schlägt sich darin nieder, dass der einzelne PV-Anlagenbetreiber eine für sich wirtschaftlich lohnenswerte Speichergröße installiert und diesen Speicher innerhalb eines Viertel-Stunden-Fahrplans zur Steigerung seiner jährlichen Überschüsse bewirtschaftet. Hierfür ist es notwendig, zunächst die Anreizmechanismen aus dem aktuell geltenden Förder- und Rechtsrahmen zu analysieren und daraus Betriebsstrategien für die Be- und Entladung des Speichers bzw. zur Lastverschiebung zu ermitteln.

In einer zweiten technischen Simulation werden alle Akteure im Beispieldorf betrachtet. Es wird ermittelt, wie im Hinblick auf Netzdienlichkeit wünschenswerte Viertel-Stunden-Fahrpläne gestaltet sein müssen und welche Speicherkapazität hierfür notwendig ist. Die Netzdienlichkeit wird hierbei mit Hilfe eines wünschenswerten Ziels für den sich ergebende Saldo aus Erzeugung, Verbrauch und Be-/Entladung des Speichers beschrieben.

Die Summe der Fahrpläne und installierten Speicherkapazitäten der einzelnen PV-Anlagenbetreiber im Dorf, welche innerhalb der wirtschaftlichen Simulation ermittelt wurden, werden mit dem innerhalb der technischen Simulation ermittelten Gesamt-Fahrplan verglichen. Es wird untersucht, ob die Summe der wirtschaftlich motivierten Fahrpläne der einzelnen Akteure im Vergleich zu besseren Ergebnissen im Hinblick auf Netzdienlichkeit führt als bei einer völlig ungeplanten Ein- und Ausspeisung.

3 Vorgehen für die Betrachtung privater PV-Anlagenbetreiber

3.1 Rahmenbedingungen

In dem betrachteten Ort in Mecklenburg-Vorpommern gibt es 151 Haushalte mit einem jährlichen Energiebezug von insgesamt 533 MWh. Es sind elf Photovoltaikanlagen mit einer installierten Leistung von insgesamt 54 kW_p mit Größen zwischen 2 kW_p und 10 kW_p je Anlage vorhanden [6]. Aufgrund der Größenordnung sowie durch eine weitere Satellitenbild-Betrachtung der Adressen kann darauf geschlossen werden, dass es sich bei allen Anlagen um Anlagen auf Einfamilienhäusern handelt. Für die weitere Betrachtung wird im Folgenden davon ausgegangen, dass sich diese elf Anlagen auf durchschnittliche private Betreiber verteilen, so dass der durchschnittliche Anlagenbetreiber eine PV-Anlage von 4,91 kW_p betreibt.

Die Anlagen wurden zwischen 1995 und 2011 installiert[1], die Mehrheit der Anlagen in 2010. Deshalb wird im Weiteren der Rechtsrahmen für Anlagen, welche in diesen Jahren in Betrieb gegangen sind, untersucht.

3.2 Anreizsysteme und Betriebsstrategien

Aus technischer Sicht wird eine planbare Austauschleistung (entspricht dem Saldo aus Erzeugung, Verbrauch und Be-/Entladung des Speichers) zwischen der betrachteten Gemeinde und dem überlagerten Netz angestrebt. Nach [5] wird durch die Vergleichmäßigung der Austauschleistung – Reduktion der Einspeise- und Bezugsspitzen – potentiell eine Netzentlastung erreicht.

Der beispielhafte PV-Anlagenbetreiber kann zur Vergleichmäßigung des Einspeise- bzw. Bezugsprofils durch die zusätzliche Installation eines Speichers bzw. durch die Verschiebung seiner Lasten beitragen. Die Motivation bzw. finanzielle Möglichkeit dazu erhält er aus Anreizsystemen, welche sich aus dem rechtlichen Rahmen bzw. aus Förderungen ergeben. Anreizmechanismen für den privaten PV-Anlagenbetreiber sind dabei insbesondere (a) die Eigenverbrauchsbegünstigung, sowie (b) die Förderung von Speichern durch das Programm KfW275.

(a) Die Eigenverbrauchsbegünstigung besteht aus der Befreiung von allen netzbezugsabhängigen Abgaben und Umlagen (Netznutzungsentgelte, Konzessionsabgabe, Offshore-Umlage, §19-Umlage, KWK-Umlage, abLa-Umlage) (siehe hierzu [7]), aus der Befreiung von der Stromsteuer (siehe hierzu [7]) und aus der Befreiung der EEG-Umlage vor dem EEG 2014 (siehe hierzu [7]) bzw. der Teilbefreiung der EEG-Umlage ab dem EEG 2014 (siehe hierzu [8]). Zusätzlich wurde innerhalb des EEG 2009 (für Anlagen, die zwischen dem 01.01.2009 und dem 31.03.2012 errichtet wurden) eine feste Vergütung für den eigenverbrauchten Strom gewährt (§33Abs. (2) EEG 2009 [9]). Mit der PV-Novelle 2010 wurde eine gestaffelte Vergütung des eigenverbrauchten PV-Stroms umgesetzt, welche höhere Anteile an Eigenverbrauch gesondert belohnte (§33 Abs. (2) Nr. 1 EEG 2009 und §33 Abs. (2) Nr. 2 EEG 2009 [10]). Seit der Novellierung des EEG in 2012 gibt es jedoch keine Vergütung des eigenverbrauchten Stroms mehr. Zweck der damals im EEG verankerten Eigenverbrauchbegünstigungen, waren dabei im Übrigen die Dezentralisierung sowie die Entlastung der Netze von der volatilen PV-Einspeisung [11].

(b) Die KfW-Bank fördert die Errichtung von Speichern, welche in Kombination mit PV-Anlagen bis zu einer Größe von 30 kWp errichtet werden, bzw. zu PV-Anlagen nachgerüstet werden, mit einem Förderkredit sowie einem Tilgungszuschuss von 30 % der Investitionskosten (mit einer Obergrenze von 660 €/kWp). Diese Förderung ist daran gekoppelt, dass der Speicher dazu genutzt wird, die Einspeiseleistung der PV-Anlage auf 60 % der installierten Peak-Leistung zu beschränken. Damit wird insbesondere „die bessere Integration von kleinen bis mittelgroßen Photovoltaikanlagen in das Stromnetz“ gefördert [12].

Die beiden Anreizmechanismen ((a) und (b)) führen zu unterschiedlichen Geschäftsmodellen mit unterschiedlichen Betriebsstrategien zur Steigerung der Erlöse und können ebenfalls kombiniert werden (c).

Für alle Geschäftsmodelle gilt, dass der Energiestrom, welcher durch den Betreiber selbst verbraucht wird, wirtschaftlich anders zu bewerten ist, als der Energiestrom, der in das Netz eingespeist wird. Dabei wird der eigenverbrauchten Energie ein fiktiver Strompreis gutgeschrie-

[1] Daten für die Beispiel-Gemeinde aus energymap.info (Bündelung und Aufbereitung der Daten des Anlagenregisters)

ben, welcher den Ersparnissen des anderenfalls aus dem Netz bezogenen Stroms entspricht (25,67 ct./kWh[2]). Da hier der durchschnittliche Anlagenbetreiber in der Beispielgemeinde mit einer in 2010 installierten Anlage betrachtet wird, wird von einer zusätzlichen Eigenverbrauchsvergütung in Höhe von 12,36 ct./kWh bei einem Eigenverbrauch unter 30 % und in Höhe von 16,74 ct./kWh bei einem Eigenverbrauch über 30 % gemäß EEG 2010 berücksichtigt. In der Summe wird dem Eigenverbrauch daher ein Erlös in Höhe von 38,03 ct./kWh (EV < 30 %) bzw. 42,41 ct./kWh (EV > 30 %) gutgeschrieben. Der in das öffentliche Netz eingespeisten Energie wird lediglich die EEG-Vergütung in Höhe von 27,33 ct./kWh gutgeschrieben. In allen drei Geschäftsmodellen erscheint es deshalb zunächst als sinnvoll, in der PV-Anlage produzierte Energie direkt selbst zu nutzen, sofern sie in diesem Moment benötigt wird. Für die Betriebsstrategien zu Speicherbewirtschaftung ergibt sich für die Geschäftsmodelle Folgendes:

(a) Die Eigenverbrauchsbegünstigung durch Befreiung von Umlagen und Abgaben führt zu einer Betriebsstrategie, in welcher die Jahressumme eigenverbrauchter Energie erhöht wird. Dazu kann der Speicher genutzt werden, um die Energie dann aufzunehmen, wenn sie nicht vollständig selbst genutzt werden kann. Wird mehr Energie benötigt als bereitgestellt, wird die Differenzmenge zunächst aus dem Speicher entnommen und erst nach dessen Entleerung aus dem Netz bezogen.

(b) Die notwendige Kappung der Spitzeneinspeisung, an welche die KfW-Förderung gebunden ist, führt dazu, dass der Speicher dann genutzt wird, wenn trotz Eigenverbrauch noch so viel Leistung durch die PV-Anlage bereitgestellt wird, dass mehr als 60 % der installierten Peak-Leistung eingespeist werden würde.

(c) Ein Geschäftsmodell aus Eigenverbrauchserhöhung und Spitzenlastkappung führt zu einer kombinierten Betriebsstrategie: Im Speicher wird immer so viel Kapazität frei gelassen, dass sichergestellt werden kann, dass täglich die Spitzenerzeugung, die sonst die 60 %-Grenze der erlaubten Einspeisung überschreitet, aufgenommen werden kann. Bis zum Erreichen der dadurch festgelegten Ladezustandsgrenze wird der Speicher für die Eigenverbrauchserhöhung genutzt.

3.3 Wirtschaftliche Modellierung

In der wirtschaftlichen Betrachtung wird auf den in Abschnitt 3.2 beschriebenen Geschäftsmodellen (a) bis (c) aufgebaut. Die aufgezeigten Entscheidungslogiken wirken sich auf die wirtschaftlichen Ergebnisse aus. Die jährlichen Nettoerlöse und Nettokosten basieren auf dem Betrachtungsjahr 2010 und spiegeln die zu diesem Zeitpunkt gängigen Werte wider. Die verwendete PV-Einspeisezeitreihe entstammt den im Projekt vorliegenden Daten der Stadtwerke Neustrelitz und wird auf die betrachteten Größen skaliert. Zur Abbildung der Haushaltslast werden generische Lastprofile[3] verwendet.

Im ersten Schritt wird die PV-Einspeisezeitreihe für ein Jahr abzüglich des direkten Verbrauches in jeder Viertelstunde betrachtet. Dies beruht auf der Entscheidungsregel, dass es das erste Ziel des Erzeugers ist, den eigenen Verbrauch auf direktem Weg zu decken. Gelingt dies, entsteht jeweils ein Überschuss, gelingt dies nicht, muss ein Bezug aus anderer Quelle stattfinden. Dieser Bezug kann aus dem Speicher entnommen werden, wenn dieser im betrachteten Zeitpunkt den Bedarf decken kann. Ist im Speicher nicht genügend Energie vorhanden, muss Energie aus dem öffentlichen Netz bezogen werden.

Ist ein Überschuss vorhanden, der aktuell nicht zur Deckung des Eigenbedarfs benötigt wird, kann dieser gespeichert oder ins vorgelagerte Netz gespeist werden. Die Entscheidung hängt vom Geschäftsmodell ab und orientiert sich an den oben beschriebenen Betriebsstrategien:

(a) Zur Steigerung des Eigenverbrauchs wird überschüssige Energie gespeichert. Dies gelingt nur, wenn im Speicher noch genügend freie Kapazität vorhanden ist. Ist dies nicht der Fall, wird die Energie verkauft.

(b) Zur Spitzenlastkappung wird bereitgestellte Energie eingespeichert, wenn anderenfalls die Einspeisung ins Netz 60 % der installierten Peak-Leistung der PV-Anlage überschreiten würde.

(c) In einem kombinierten Geschäftsmodell wird Energie entweder zur Spitzenlastkappung eingespeichert oder aber, wenn die Produktion den Bedarf übersteigt und gleichzeitig noch genügend freie Speicherkapazität vorhanden ist, um die Lastkappung zu ermöglichen. Hierfür wird aus der bekannten Einspeisezeitreihe für das gesamte Jahr ermittelt, wie hoch die maximalen täglichen Energiemengen sind, welche nach einer Spitzenkappung auf 60 % der installierten Peak-Leistung gespeichert werden müssten. Diese Menge wird stets im Speicher für die Spitzenlastkappung freigehalten.

Die aus diesen Entscheidungsregeln ermittelten Energieflüsse werden in der Wirtschaftsbetrachtung mit den zugehörigen spezifischen Preisen multipliziert. Je nach Geschäftsmodell ergeben sich verschiedene Erlöse für die Kenngrößen „Verkauf in das vorgelagerte Netz" und „Erlös aus Eigenverbrauch". Abgezogen werden die jährlichen Kosten bestehend aus Abschreibungen der Investitionskosten auf 20 Jahre, Reparaturkosten, Arbeitskosten und Sonstigen Kosten. Die Investitionskosten für die betrachtete PV-Anlage werden nach [13] kalkuliert. Die kapazitätsbezogenen Investitionskosten für den Speicher werden mit 175 €/kWh einbezogen. Es wird eine Blei Batterie betrachtet [16], für die eine Lebensdauer von zehn Jahren angenommen wird. Somit wird nach zehn Jahren von einer Ersatzinvestition ausgegangen.

---

[2] Strompreis privat 2012, BDEW-Strompreisanalyse 2014

[3] Die generischen Haushaltslastprofile wurden durch den Lehrstuhl LEEA der OvGU Magdeburg generiert und für die hier durchgeführten Betrachtungen zur Verfügung gestellt.

Die leistungsbezogenen Kosten werden auf 20 Jahre mit 175 €/kW angenommen [16]. Wie in Abschnitt 3.4 beschrieben, wird bei den Berechnungen der Wirkungsgrad der Batterie berücksichtigt.

In den Geschäftsmodellen (b) und (c) wird die KfW-Förderung eines Batteriespeichers berücksichtigt. Unter den in Abschnitt 3.2 beschriebenen Voraussetzungen wird die Installation eines Speichers finanziell unterstützt. Die Höhe dieser Unterstützung hängt davon ab, ob die Kosten des Speichers bezogen auf die installierte Leistung der PV-Anlage einen Wert von 2000 €/kW$_p$ über- oder unterschreiten. Liegt eine Unterschreitung vor, liegt der Tilgungszuschuss der KfW-Förderung bei 30 % der spezifischen Speicherkosten, bei einer Überschreitung der Grenze werden 30 % des Höchstsatzes gefördert [12].

3.4 Technische Modellierung

Die Einspeisung der PV-Anlagen wird mittels einer gemessenen Einspeisezeitreihe der Stadtwerke Neustrelitz, die im Rahmen des Projekts verfügbar ist, modelliert. Der Einspeisung gegenüber steht die Verbrauchergruppe private Haushalte (Standardlastprofile der Stadtwerke Neustrelitz und generische Haushaltslastprofile der OvGU Magdeburg). Vor dem Hintergrund der gewünschten Vergleichmäßigung der Austauschleistung wird in der technischen Betrachtung ein zentraler Batteriespeicher so dimensioniert und betrieben, dass die Austauschleistung zwischen dem betrachteten Energiesystem und dem überlagerten Netz jeweils über einen Tag konstant gehalten wird. Auf Basis der bekannten Einspeise- und Lastzeitreihe wird die jeweils täglich konstante Austauschleistung zum Ausgleich der Energiebilanz im Vorhinein bestimmt und als Zeitreihe im Modell berücksichtigt.

Während der in Zeitschritten von 15 Minuten durchgeführten Jahressimulation wird die Lade- bzw. Entladeleistung des Speichers gemäß der aktuellen Leistungsbilanz innerhalb des Energiesystems bestimmt.

Als zentraler Speicher wird ein Blei-Säure-Akkumulator mit einem Zyklenwirkungsgrad von 81 % (vgl. [14] und [15]), welcher sich auf die Gesamtenergiebilanz des Systems auswirkt, eingesetzt. Zunächst werden jedoch die Speicherverluste nicht in die Energiebilanz zur Bestimmung der täglich aus dem überlagerten Netz zu beziehenden Energie berücksichtigt.

Um die Speicherverluste auszugleichen, wird für jeden betrachteten Tag die Differenz zwischen dem festgesetzten Anfangsladezustand und dem Ladezustand am Ende des Tages und daraus die zum Ladezustandsausgleich benötigte konstante Leistung, die mit dem vorgelagerten Netz ausgetauscht werden muss, bestimmt. Gegenüber der ohne Berücksichtigung der Speicherverluste bestimmten täglich konstanten Austauschleistung wird die Austauschleistung somit erhöht, um die Speicherverluste auszugleichen.

4 Ergebnisse

4.1 Wirtschaftliche Simulation

Wie in Abschnitt 3.3 bereits erläutert, werden die jährlichen Überschüsse, die durch die Erzeugung, den Verbrauch und den Verkauf von elektrischer Energie einer PV-Anlage entstehen, als Vergleichsgröße zwischen verschiedenen Varianten der oben beschriebenen Geschäftsmodelle herangezogen. Auf diesem Weg wird die Installation verschiedener Speichergrößen wirtschaftlich beurteilt.

Dabei zeigt sich, dass es sich unter den getroffenen Annahmen aus wirtschaftlicher Sicht in keinem der Geschäftsmodelle lohnt, einen Speicher zu installieren.

Verfährt der Anlagenbetreiber so, dass er seine Anlage ohne Speicherung für den direkten Eigenverbrauch nutzt, ist der jährliche Überschuss in dem Geschäftsmodell (a) und (c) am höchsten, wie **Bild 1** zeigt.

Somit wird deutlich, dass die KfW-Förderung unter den hier getroffenen Annahmen, der Freihaltung der Speicherkapazität zur Aufnahme der maximalen Einspeisespitze, für die Installation eines Speichers nicht ausreichend finanziellen Vorteil bietet, um die zusätzlichen Investitionskosten für den Speicher zu tragen.

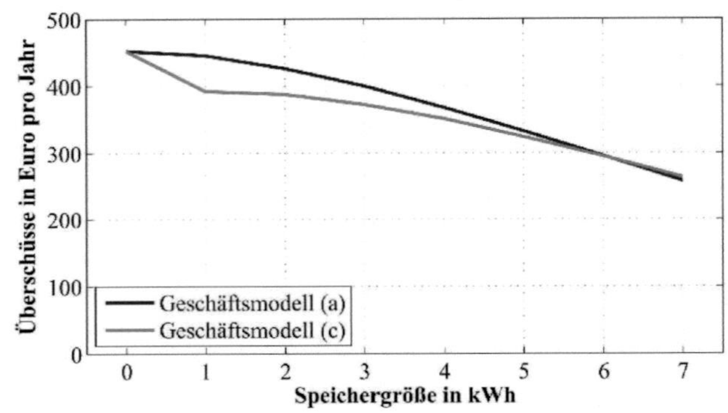

Bild 1: Jährliche Überschüsse Anwendung verschiedener Speicherkapazitäten nach den Geschäftsmodellen (a) und (c).

Auch wenn der Akteur nur einen Speicher errichtet, um damit die Lastspitzen zu kappen, also die Voraussetzungen für die KfW-Förderung zu erfüllen und keine Eigenverbrauchssteigerung durch den Speicher verfolgt wird, liegt der jährliche Überschuss in diesem Geschäftsmodell (b) mit Überschüssen von ca. 390 €/a unter dem des direkten Eigenverbrauchs ohne Speicher.

Wenn das Ziel der Investition in einen Speicher ist, den Eigenverbrauch zu erhöhen und somit die eigene Unabhängigkeit von steigenden Energiepreisen zu steigern, ist es in Geschäftsmodell (a) aus ökonomischer Sicht zu empfehlen, einen Speicher mit einer Kapazität von 3 kWh zu errichten. Der Anteil an eigenverbrauchter Energie lässt sich bis zu dieser Speichergröße maßgeblich auf ca. 47 % steigern. Bei größeren Speichern sind nur noch geringe Steigerungen des Eigenverbrauchanteils zu verzeichnen.

In Geschäftsmodell (c) wäre in diesem Fall ein Speicher mit einer Kapazität von 5 kWh zu installieren. Der Eigenverbrauchsanteil ließe sich bei einer Kombination aus Spitzenlastkappung und Steigerung des Eigenverbrauchs auf ebenfalls ca. 47 % erhöhen. Werden für diese Fälle die jährlichen Überschüsse miteinander verglichen, liegt dieser im Geschäftsmodell (a) und einem Speicher von 3 kWh um ca. 12 % über dem Wert in Geschäftsmodell (c) und einem Speicher von 5 kWh, wie in **Bild 1** dargestellt.

4.2 Technische Simulation

Durch die PV-Anlagen werden mit dem hinterlegten Einspeiseprofil im betrachteten Jahr 51,7 MWh zur Verfügung gestellt, bei einer installierten Leistung von 54 kW_p entspricht dies 958 Volllaststunden. Durch die regional vorhandenen Erzeugungsanlagen werden somit 9,7 % des jährlichen Energiebedarfs gedeckt. Die verbleibenden 482 MWh müssen durch den Bezug aus dem vorgelagerten Netz bereitgestellt werden, wobei die Speicherverluste dabei nicht berücksichtigt sind.

Wird die ohne Berücksichtigung der Speicherverluste berechnete Austauschleistung verwendet, entsteht durch die nicht gedeckten Verluste im betrachteten Jahr ein Defizit von etwa 17 MWh. Wird die Austauschleistung abhängig vom Speicherladezustand angepasst, reduziert sich dieses Defizit auf etwa 47 kWh. Die benötigte nutzbare Speicherkapazität beträgt für den letztgenannten Fall 347,3 kWh. Die Maximalwerte des täglichen Ladezustands bewegen sich zwischen 49 und 100 % der nutzbaren Kapazität, wobei der maximale Wert nur an einem Tag erreicht wird. Die maximale Ladeleistung des Speichers beträgt dabei 45,5 kW und die maximale Entladeleistung 53,7 kW. **Bild 2** zeigt die Dauerlinien der Austauschleistung mit dem vorgelagerten Netz.

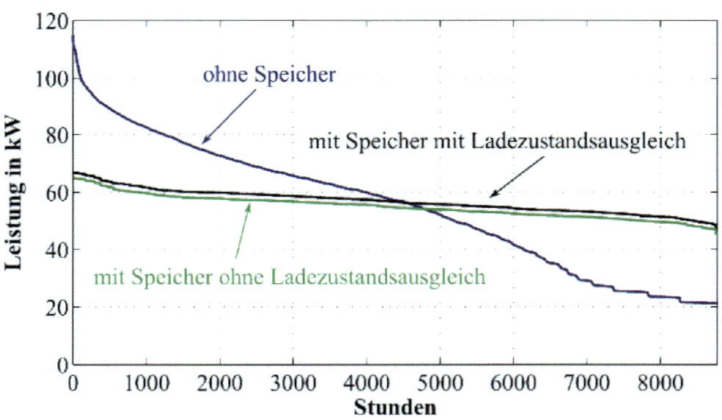

Bild 2: Dauerlinie der Austauschleistung zwischen dem betrachteten Dorf und dem vorgelagerten Netz mit und ohne Einsatz eines zentralen Speichers.

Durch den Einsatz des Speichers verringert sich die maximale Bezugsleistung von 115 kW auf 67 kW (vgl. **Bild 2**). Durch den zusätzlichen Energiebezug zum Ausgleich der Speicherverluste erhöht sich die täglich konstante Austauschleistung um Werte zwischen 1,5 kW und 2,7 kW. Während in den Fällen ohne Speichereinsatz bzw. ohne Ladezustandsausgleich 482 MWh aus dem vorgelagerten Netz bezogen werden, erhöht sich die bezogene Energie auf 499 MWh, wenn der Ladezustand des Speichers täglich ausgeglichen wird. Die Batterie durchläuft in dieser Betriebsart im betrachteten Jahr 230 äquivalente Vollzyklen.

4.3 Vergleich von netzdienlichem und erlösorientiertem Verhalten

Aus technischer Sicht wird eine täglich konstante Austauschleistung angestrebt, um Lastschwankungen zu vermeiden. Die dafür notwendige Speicherkapazität wurde in Abschnitt 4.2 angegeben. Die Gesamtspeicherkapazität, die die einzelnen Anlagenbetreiber aus wirtschaftlicher Sicht installieren würden beträgt 0 kWh und trägt damit nicht zur technisch notwendigen Kapazität bei.

Wie in Kapitel 4.1 dargestellt, wäre es denkbar, dass Akteure zur Steigerung ihrer Unabhängigkeit ihren Eigenverbrauchsanteil durch die Installation eines Speichers steigern wollen. In Geschäftsmodell (a) würde aus wirtschaftlicher Sicht in diesem Fall ein Speicher mit einer Kapazität von 3 kWh installiert werden, in Geschäftsmodell (c) eine Speicherkapazität von 5 kWh. Daraus ergibt sich in Geschäftsmodell (c) eine insgesamt installierte Speicherkapazität in den allen privaten Haushalten mit PV-Anlagen von 55 kWh, was 16 % der aus technischer Sicht benötigten Speicherkapazität entspricht.

Die Auswirkungen des Speichereinsatzes nach Geschäftsmodell (c), im Vergleich zur Austauschleistung ohne Speichereinsatz bzw. zur täglich konstanten Austauschleistung sind in **Bild 3** dargestellt.

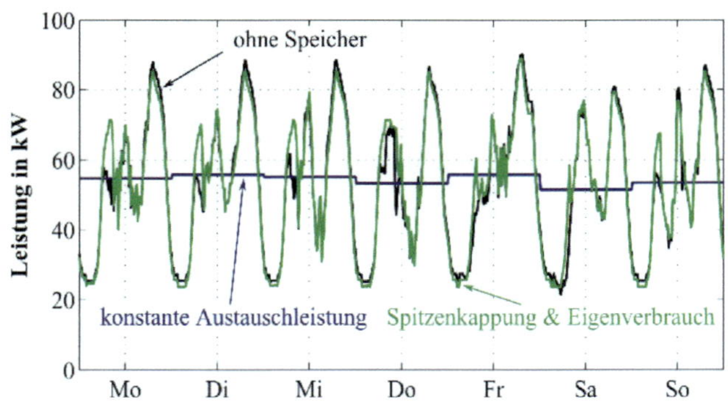

Bild 3: Vergleich der Austauschleistung mit dem vorgelagerten Netz für den aus technischer Sicht angestrebten Speichereinsatz und dem sich ergebenden Verlauf für das Geschäftsmodell (c) für die Beispielgemeinde in der ersten Woche im Juni im Betrachtungsjahr 2012.

Im Vergleich zur unbeeinflussten Austauschleistung („ohne Speicher“ in **Bild 3**) erhöht sich diese durch Anwendung des Geschäftsmodells (c) um die Mittagszeit eines jeden Tages, da die zur Erhöhung des persönlichen Eigenverbrauchs eingesetzten Speicher in diesen Stunden noch geladen werden. Ohne Speichereinsatz wird die verfügbare PV-Energie direkt zur Bedarfsdeckung genutzt, weshalb die Austauschleistung geringer ist.

Die durch die KfW-Förderung angestrebte Netzentlastung durch Reduzierung der Mittagsspitzen kommt am Netzübergabepunkt zum vorgelagerten Netz nicht zum Tragen, da der Stromverbrauch so viel höher ist die PV-Erzeugung ist, dass die Netzbelastung durch die Beispielgemeinde ausschließlich durch die Strom-Abnahme und nicht durch Rückspeisung der PV-Anlagen gekennzeichnet ist.

Über das Jahr betrachtet verringert sich die maximale Bezugsleistung aus dem vorgelagerten Netz für den hier betrachteten Fall (Geschäftsmodell (c)) gegenüber der unbeeinflussten Austauschleistung nicht. Sie beträgt für den Fall ohne Speichereinsatz 115 kW. Für den Speicherbetrieb nach Geschäftsmodell (c) wird die Austauschleistung auf 116 kW geringfügig erhöht. Dies liegt darin begründet, dass die Einspeisung der PV-Anlagen zur Steigerung des Eigenverbrauchsanteils der Betreiber zunächst zur Ladung der Speicher eingesetzt wird, folglich nicht gleichzeitig zur Bedarfsdeckung in der Region eingesetzt werden kann und somit der Bezug aus dem Netz erhöht wird.

5 Zusammenfassung und Ausblick

Am Beispiel von privaten PV-Anlagenbetreibern von kleiner als 10 kW_p wurde gezeigt, dass als Anreizmechanismen für die Milderung negativer Effekte der ungeplanten Einspeisung im Hinblick auf die Entlastung lokaler Netze bereits die EEG-Umlagen(teil-)befreiung bei Eigenverbrauch sowie die KfW-Speicherförderung mit der Kopplung an die Spitzenlast-Kappung etabliert wurden. Eine im EEG 2009 eingeführte gesonderte feste Vergütung für eigenverbrauchten Strom wurde ab der Novellierung in 2012 dagegen wieder abgeschafft.

Für den beispielhaften PV-Anlagenbetreiber mit einer in 2010 installierten Anlage wurde gezeigt, dass beide Mechanismen nicht ausreichend Anreiz bieten, um einen Speicher zur Erhöhung des Eigenverbrauchs bzw. zur Spitzenlastkappung zu installieren. Durch die Abschaffung der Eigenverbrauchsvergütung in 2012 ist davon auszugehen, dass bei späterer Anlageninstallation die Wirtschaftlichkeit eines Speichers noch schlechter ausfällt.

Es wurde hergeleitet, wie sich durch die Be- und Entladestrategie in den verschiedenen Geschäftsmodellen die Austauschleistung des gesamten Dorfs verändert. Hierbei ist in dem hier betrachteten Beispiel zu beobachten, dass sich die Netzbelastung durch Abnahmespitzen durch einen Speichereinsatz für Eigenverbrauchssteigerung noch erhöhen würden. Gleichzeitig könnte der Speichereinsatz für die Reduktion der Einspeisespitzen, welche durch das KfW275-Programm angestrebt wird, ebenfalls die Netzbelastung in der Beispielgemeinde nicht reduzieren, da die PV-Einspeisung in dem betrachteten Dorf im Vergleich zur Stromabnahme so gering ist, dass es auch ohne Speicher zu keinem Zeitpunkt zu einer Rückspeisung in das vorgelagerte Netz kommt. Für die Entlastung des Gesamtsystems durch die Abmilderung der gesamten in Deutschland entstehenden Mittagsspitze durch PV-Einspeisung findet jedoch allemal eine Verbesserung statt.

Die hier vorgestellte Modellierung gibt eine Annäherung zu den Auswirkungen des wirtschaftlich motivierten Verhaltens einzelner Akteure auf das lokale Netz wider. Es zeigt, wie individuell eine netzentlastende Wirkung von Anreizen zum Eigenverbrauch und zur Lastspitzenkappung von den Gegebenheiten im Netzgebiet abhängt. Zusätzliche Betrachtungen in weiteren Beispielregionen mit abweichenden Netzsituationen – hierbei besonders ein durch Rückspeisung durch eine hohe Anzahl der installierten PV-Anlagen geprägtes Netzgebiet – wären für eine weitere Abschätzung der Wirkung der genannten Anreizsysteme sinnvoll. Das hier vorgestellte zweistufige Modell betrachtet die Verhältnisse im Viertel-Stunden-Takt. Eine differenzierte Betrachtung wäre vor allen Dingen deshalb sinnvoll, da die Akteure aus wirtschaftlicher Sicht durch den rechtlichen Rahmen an der Viertel-Stunden-Betrachtung festhalten, aus technischer Sicht jedoch für den Netzzustand eine wesentlich höhere zeitliche Auflösung notwendig ist.

6 Literatur

[1] E-Bridge (Hrsg.): „Moderne Verteilernetze für Deutschland“ (Verteilernetzstudie) – Studie im Auftrag des Bundesministeriums für Wirtschaft und Energie (BMWi), Bonn, 2014

[2] P3-Energy (Hrsg.): Technologieoptionen für den Verteilungsnetzausbau Technologieoptionen für den Verteilungsnetzausbau in Deutschland - Marktanalyse und Bewertung, Aachen, 2013

[3] Deutsche Energie-Agentur GmbH (Hrsg.): „Ausbau- und Innovationsbedarf der Stromverteilnetze in Deutschland bis 2030“, Berlin, 2012

[4] Wissenschaftliche Dienste Deutscher Bundestag (Hrsg.): „Die Förderung der Stromerzeugung aus erneuerbaren Energien in Deutschland – Hintergründe und Entwicklungen“, Berlin, 2014

[5] Fraunhofer-Institut ISE (Hrsg.): Kurzgutachten zur Abschätzung und Einordnung energiewirtschaftlicher, ökonomischer und anderer Effekte bei Förderung von objektgebundenen elektrochemischen Speichern, Freiburg, 2013

[6] Trigenius GmbH (Hrsg.): Machbarkeitsstudie Bioenergiegemeinde Userin, Wismar, 2013

[7] Herms M.; Brahms F.: „Konzepte zur Umsetzung dezentraler Energieerzeugung aus Erneuerbaren Energien“, in UmweltWirtschaftsForum Vol.21, S.219 – 224, Berlin, 2013

[8] Loibl H.: „Die Eigenstromnutzung nach dem EEG2014“, in Zeitschrift für neues Energierecht Heft 5 2014, S. 437 – 440, Bochum 2014

[9] „Gesetz zur Neuregelung des Rechts der Erneuerbaren Energien im Strombereich und zur Änderung damit zusammenhängender Vorschriften", Bundesgesetzblatt Jahrgang 2008 Teil I Nr.49, Bonn 2008

[10] Erneuerbare-Energien-Gesetz vom 25. Oktober 2008 (BGBl. I S. 2074), das zuletzt durch Artikel 1 des Gesetzes vom 12. April 2011 (BGBl. I S. 619) geändert worden ist

[11] „Die Eigenverbrauchsregelung in §33Abs.2EEG nach der Photovoltaik-Novelle 2010", in Zeitschrift für neues Energierecht Heft 6 2010, S.558-563, Bochum 2010

[12] Kreditbank für Wiederaufbau, „Merkblatt Erneuerbare Energien – KfW-Programm Erneuerbare Energien „Speicher" 275", Frankfurt, 2015

[13] Bundesverband Solarwirtschaft e.V. (Hrsg.): „Statistische Zahlen der deutschen Solarstrombranche (Photovoltaik)", Berlin, 2014

[14] Energietechnische Gesellschaft im VDE (Hrsg.): VDE-Studie: Energiespeicher in Stromversorgungssystemen mit hohem Anteil erneuerbarer Energieträger, Frankfurt am Main, 2009

[15] Lunz, B. und Sauer, D.: Überblick über den aktuellen Stand der modernen Speichersysteme, Forum für Zukunfstenergien: Moderne Speichersysteme – Welche wirtschaftlichen und rechtlichen Voraussetzungen sind erforderlich?, Berlin, 2014

[16] Sauer, U.: „Marktanreizprogramm für dezentrale Speicher insbesondere für PV-Strom", Kurzgutachten, RWTH Aachen, Aachen, 2013

Wirtschaftlichkeitsbetrachtung von Haushaltsspeichern und realdatengestützte Untersuchung des elektrischen Eigenverbrauchs von PV-Strom

C. Stegner[1], J. Bogenrieder[1], S. Auer[2], P. Luchscheider[1], R. German[3], C. Brabec[1]
[1]ZAE Bayern e. V., c/o Stadtwerke Hof GmbH, Unterkotzauer Weg 25, 95028 Hof, Deutschland
[2]iisys Hof, Alfons-Goppel-Platz 1, 95028 Hof/Saale, Deutschland
[3]FAU Erlangen-Nürnberg, CS 7, Martensstrasse 3, 91058 Erlangen, Deutschland
christoph.stegner@zae-bayern.de

Kurzfassung

Im Jahr 2014 lieferten Erneuerbare Energien bereits mehr als ein Viertel der Bruttostromerzeugung, wovon wiederum rund 22 % auf Photovoltaik-Anlagen entfielen. Diese Technologie ist von einer hohen Dezentralität geprägt. Von den ca. 1,4 Millionen Anlagen in Deutschland sind ca. 97 % in der Niederspannung angeschlossen – das entspricht ca. 80 % der installierten Leistung. Mit den sinkenden Einspeisevergütungen für Photovoltaik-Strom rückt der elektrische Eigenverbrauch in den Fokus der Anwendung. Jedoch sind die Leistungsprofile der Erzeugung und vor allem des Verbrauchs hoch volatil. Dies erfordert kurze Zeitintervalle bei der Messung bzw. bei der Simulation der elektrischen Leistungen, wenn diese realitätsnah abgebildet werden sollen. Smart-Meter sind prinzipiell zu solchen Messungen fähig, jedoch ist die Datenerfassung und Übertragung nicht trivial. Zu berücksichtigen sind die regulatorischen Vorgaben, die vorhandene Infrastruktur für Datenkommunikation, ihre Sicherheit und Zuverlässigkeit, die Akzeptanz der beteiligten Haushalte oder sonstigen Endkunden sowie die Kosten. Im Projekt Smart Grid Solar werden seit November 2014 in einem ländlich geprägten Ortsteil der Stadt Hof mit hoher PV-Durchdringung über 30 Smart-Meter in sehr kurzen Messintervallen von 15 Sekunden ausgelesen. Die 12 erfassten PV-Anlagen weisen eine hohe Vielfalt an Ausrichtungen auf, welche es ermöglicht, deren Auswirkung auf den Eigenverbrauch zu untersuchen. Gleichzeitig werden Erkenntnisse darüber gewonnen, wie stark die Werte für Eigenverbrauch abweichen, wenn längere Zeitintervalle als 15 Sekunden als Grundlage der Berechnung dienen. Die bisherigen Ergebnisse zeigen, dass sich der Fehler bei der Betrachtung von 15 min-Intervallen bereits auf 5 – 10 % beläuft. Zusätzlich wurde in einer detaillierten techno-ökonomischen Betrachtung die Wirtschaftlichkeit verschiedener elektrischen Solarspeicher-Systeme untersucht. Alle notwendigen technischen Parameter und die aktuellen regulatorischen Rahmenbedingungen sowie die Preisgestaltung der Komponenten wurden berücksichtigt, inklusive einer Abschätzung der Genauigkeit anhand einer Monte-Carlo-Simulation. Das Model zeigt, dass unter aktuellen Bedingungen PV-Anlagen wirtschaftlich sind und die optimale Anlagengröße sich zu kleinen Leistungen hin verschiebt. Selbst die günstigsten Speicher reduzieren hingegen im Moment noch die Rendite.

1 Projektstandort Epplas

Im untersuchten Ortsteil Epplas haben sich nahezu alle Endkunden zur Teilnahme am Forschungsprojekt Smart Grid Solar entschlossen. Epplas ist geprägt von einer hohen Durchdringung mit PV-Anlagen, in Summe ca. 286 kWp. Die meisten Anlagen sind Volleinspeiser. Insgesamt stehen 12 erfasste PV-Anlagen 18 Bezugspunkten gegenüber. Die in Kooperation mit den lokalen Stadtwerken Hof installierten Smart-Meter wurden mit zusätzlicher Rechentechnik ausgestattet, um eine zuverlässige Verschlüsselung und Übertragung der Messdaten in der gewünschten zeitlichen Auflösung zu gewährleisten.

1.1 Eingesetzte Technik

Verfügbare Datenübertragungsmodule für die verwendeten EMH-Zähler leisten als kürzestes Messintervall 1 min. Außerdem verfügen sie nur über eine begrenzte Möglichkeit zur Pufferung der Messwerte. Ein Verbindungsabbruch führte somit zum Datenverlust für den betroffen Zeitraum. Aus diesem Grund wurde eine eigene Lösung in Form eines zusätzlichen Kleinrechners realisiert. Darin werden die Messdaten zwischengespeichert, verschlüsselt und übertragen. Die zeitliche Auflösung der Messung stellt dabei ein Novum dar. Bisherige Projekte mit Smart Metern bewegten sich im Bereich von 10 min-Intervallen [10]. Der Aufbau ist in **Bild 1** dargestellt.

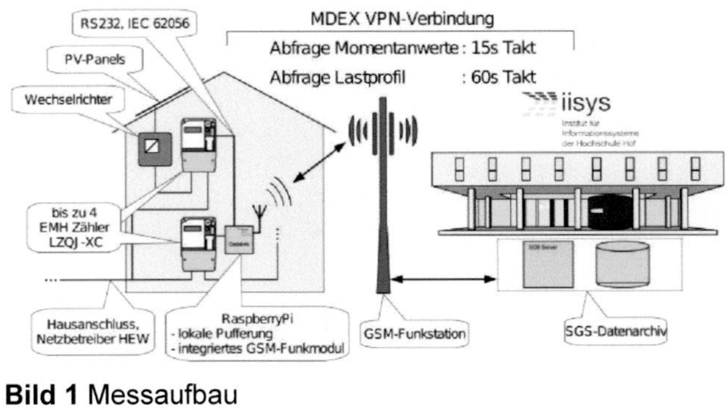

Bild 1 Messaufbau

Als Rechnertechnik wird ein Raspberry PI verwendet, der über die RS232 Schnittstelle des EMH-Zählers auf den internen Bus des Zählers zugreift. Die Kommunikation findet über das IEC 62056-Protokoll statt. Der PI liest die aktuellen Werte im 15 s-Takt und die Lastprofil-Werte im 60 s-Takt aus. Hierbei können bis zu vier EMH-Zähler an eine Recheneinheit angeschlossen werden. Festspeicher des PI ist eine 32 GB SD-Karte, die ein angepasstes Linux Betriebssystem und eine Datenbank zur Speicherung der ausgelesenen Werte enthält. Als Kommunikationsschnittstelle zum zentralen Datenarchiv dient ein integriertes GSM-Funkmodul, das die gewonnenen Daten auf Anfrage des Servers über das GSM-Netz versendet.

1.2 Ablauf der Datenakquise

Da der EMH-Zähler nur rund 1.800 Datensätze speichern kann und somit bei Aufzeichnung im Minutentakt nach 30 h die bereits gespeicherten Datensätze überschreiben müsste, wurde mit den PIs eine Möglichkeit gefunden, alle gemessenen Daten zu speichern. Hierbei werden die Momentanwerte alle 15 s vom EMH-Zähler abgefragt und direkt auf der SD-Karte gespeichert. Diese kurzen Zeitintervalle sind möglich, weil die lokale Rechentechnik unmittelbar auf den internen Bus des Zählers zugreift und die benötigten Messdaten durch Parametrierung zeitversetzt abfragt, statt darauf zu warten. Dieses Vorgehen ist außerdem notwendig, um bei einer Übertragungsunterbrechung zwischen EMH-Zähler und PI wieder mit dem Abholen der Datenpakete aufholen zu können um ein Überschreiben der noch nicht übertragenen Daten zu verhindern. Bei der Anfrage des Lastprofiles werden die Daten hingegen alle 60 s abgefragt. Dass die lokale Pufferung der abgerufenen Daten über den gesamten Zeitraum von Nöten ist, zeigt der hohe Anteil an nicht erfolgreichen Übertragungen von ca. 22 %. Dank der Pufferung wird jedoch seit Anfang November durchgehend eine Datenverfügbarkeit von über 95 % erreicht wird.

2 Untersuchung des möglichen Eigenverbrauchs

Die meisten PV-Anlagen in Epplas wurden als Volleinspeiser installiert. Da zur Ertragsmaximierung meist die gesamte verfügbare Dachfläche genutzt wurde haben die gemessenen Anlagen eine relativ hohe durchschnittliche Leistung von ca. 24 kWp. Um eine Aussage über die Eigenverbrauchsquote, englisch *self consumption rate* (SCR), bei unterschiedlicher Anlagenkonfiguration zu ermöglichen, wurden für jeden Tag die 18 zur Verfügung stehenden Lastprofile jeweils mit allen 12 PV-Anlagenprofilen überlagert und die SCR untersucht. Das heißt, die Werte für Eigenverbrauch sind bei den Volleinspeisern theoretischer Natur, basieren jedoch auf realen Profilen. Unvollständige Tagesprofile wurden nicht berücksichtigt. Da nur Profile des gleichen Tages kombiniert werden, stehen eventuelle klimatische Auswirkungen auf den Stromverbrauch somit im Einklang mit dem entsprechenden PV-Ertrag. Pro Tag ergeben sich im Schnitt über 100 Kombinationen, für die der Eigenverbrauch berechnet wird. Das Verhältnis von erzeugter PV-Energie zu Energie-Bedarf wird in Anlehnung an den englischen Begriff *prosumer* – ein Letztverbraucher der auch Energie erzeugt – als *prosume ratio* (PSR) bezeichnet. Ein PSR von 1 würde bedeuten, dass ein Haushalt bilanziell im betrachteten Zeitraum so viel Energie erzeugt wie er verbraucht. Ohne Angabe des entsprechenden PSR sind Aussagen über den absoluten Wert der SCR meist wenig hilfreich, da der Eigenverbrauch sehr stark vom PSR abhängt.

2.1 Ausrichtung der PV-Anlagen

Der Einfluss der Ausrichtung von PV-Anlagen auf den Eigenverbrauch wurde unter anderem in [4] simulativ untersucht, mit dem Ergebnis, dass bei optimaler Neigung und einer mittleren Speichergröße die Auswirkungen der Ausrichtung nach Ost oder West im Bereich weniger Prozent liegen. In Epplas ist eine Vielzahl von Photovoltaikanlagen mit unterschiedlichen Ausrichtungen, zum Teil sogar in Nordrichtung, installiert. Ein Beispiel für die zeitlich verschobenen Erträge der unterschiedlich ausgerichteten Anlagen ist in **Bild 2** zu sehen. Jedoch sind oft mehrere Ausrichtungen zusammen an einen Zähler angeschlossen, so dass eine anlagenscharfe Messung der Einspeiseleistung nicht möglich ist.

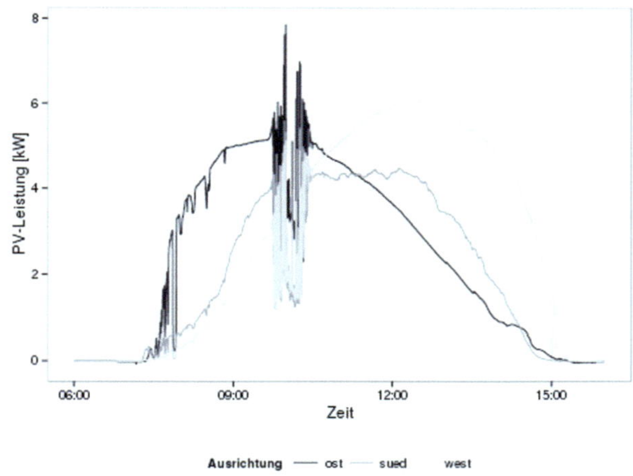

Bild 2 Erzeugungsprofile dreier unterschiedlich ausgerichteter PV-Anlagen an einem Beispieltag

In **Tabelle 1** ist die Verteilung der installierten Leistung über die Ausrichtung der einzelnen Dachflächen zu sehen. Die Anlagen wurden in fünf unterschiedliche Klassen eingeteilt, wobei 0° Süden entspricht:

Tabelle 1 Anlagenklassen in Epplas

Klasse	Ausrichtung	Inst. Leistung
1. Süd	-30° bis 30°	87 kWp
2. Ost-West	-90° bis -30° und 30° bis 90°	98 kWp
3. West	30° bis 90°	29 kWp
4. Ost	-90° bis -30°	43 kWp
5. Nord	<-90° und >90°	29 kWp

Anschließend wurde für die einzelnen Klassen die Abhängigkeit der SCR von PSR betrachtet. Die einzelnen Verteilungen des Eigenverbrauchs für die fünf Klassen sind in Bild 3 zu sehen. Es ist zwischen den einzelnen Plots kein grundsätzlich unterschiedliches Verhalten zu erkennen, was hauptsächlich durch die Größe der Anlagen begründet ist. Durch die vergleichsweise hohe installierte Leistung spielt die Ausrichtung eine weniger wichtige Rolle, da bereits bei geringer Einstrahlung eine hohe Leistung erzeugt wird, die den größten Teil des Bedarfs deckt. Zudem ist zu erkennen, dass bei einem Verhältnis von erzeugter zu verbrauchter Energie von ungefähr 1 eine starke Schwankung der SCR bei allen Konfigurationen zu erkennen ist: Wenn die Tagesproduktion an elektrischer Energie bilanziell dem Bedarf entspricht, bewegt sich die Eigenbedarfsquote zwischen 15 % und 50 %.

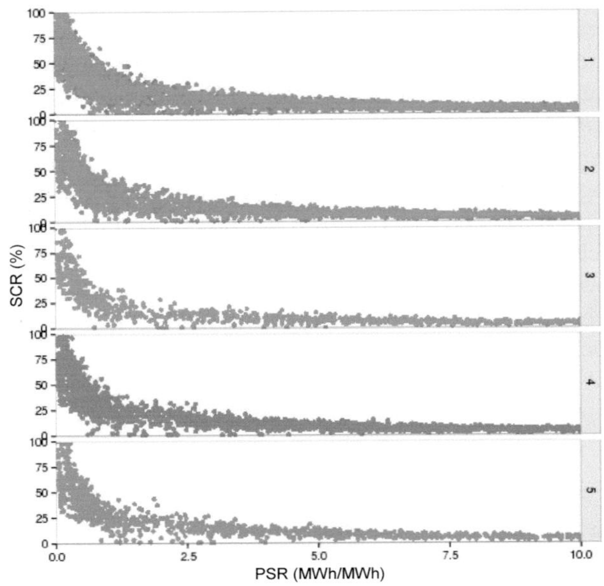

Bild 3 SCR als Funktion von PSR bei unterschiedlichen Klassen von PV-Ausrichtungen

2.2 Zeitliche Auflösung der Messdaten

In verschiedenen Studien wurde auf den Einfluss der zeitlichen Auflösung bei Leistungsprofilen eingegangen. Die wenigsten betrachten dabei Intervalle kleiner 1 min. Eine Ausnahme stellt [2] dar. Nicht immer sind reale Messdaten verfügbar, oder es wird auf eine Mischform von Messdaten und Simulation zurückgegriffen.

Dank der zeitlich hochaufgelösten Messung in Epplas stehen nun in statistisch relevantem Umfang reale Daten zur Verfügung. **Bild 4** steht beispielhaft für den Vergleich eines Lastprofils in verschiedener zeitlicher Auflösung.

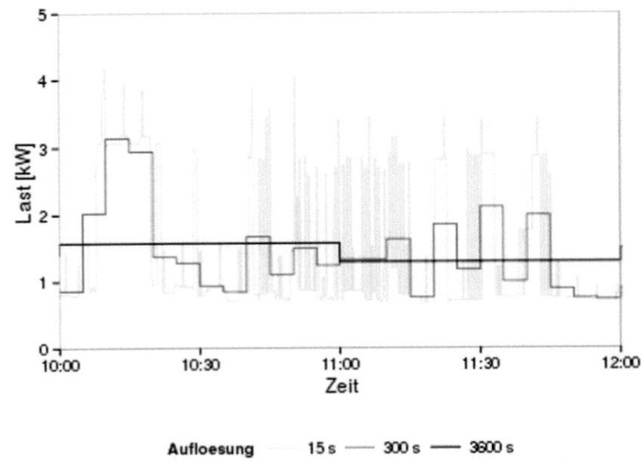

Bild 4 Darstellung des gleichen gemessenen Ausschnitt eines Lastprofils in drei zeitlichen Auflösungen

Es wird ersichtlich, wie häufig Lastsprünge von mehr als 50 % der Maximallast bei der 15 s-Auflösung auftreten, die bei gröberer Auflösung komplett in der Mittelwertbildung verschwinden.

In [3] wird ausführlich der Fehler bei der Berechnung von Eigenverbrauch bei Intervallen von 5, 10, 15, 30 und 60 Minuten gegenüber 1 Min beschrieben. In einem ähnlichen Ansatz werden in dieser Studie die SCR für die höchste zeitliche Auflösung von 15 s als Referenz bzw. Realwerte betrachtet. Die oben beschriebene Berechnung der täglichen SCR wird wiederholt mit dem Unterschied, dass vorher die Leistungswerte je über 1, 5, 15 und 60 min gemittelt werden. Die Abweichungen der SCR bei längeren Intervallen werden in ein Verhältnis zu den 15 s-Werten gesetzt und als relativer Fehler bezeichnet.

$$Fehler_{Intervall} = \frac{SCR_{Intervall}}{SCR_{15\,s}} - 1 \qquad (1)$$

In **Bild 5** sind die Fehler für verschiedene zeitliche Mittelungen zusammengefasst. Der Mittelwert wurde zusätzlich dargestellt und als x-Achse das Verhältnis von erzeugter Energie zu Verbrauch gewählt.

Bis auf sehr wenige Ausnahmen, führt eine zeitliche Auflösung mit längeren Intervallen stets zu hohen SCR, zu erkennen an den nahezu durchgängig positiven Werten für den relativen Fehler. Eine zeitlich gröbere Betrachtung überschätzt somit den Eigenverbrauch. Während sich der Fehler für 1 min noch im Bereich weniger Prozent bewegt, wächst er bei 15 min, was ein häufig verwendetes Simulationsintervall darstellt, auf 5 – 10 %. Eine stündliche Betrachtung überschätzt den Eigenverbrauch im Mittel sogar um 10 – 20 %.

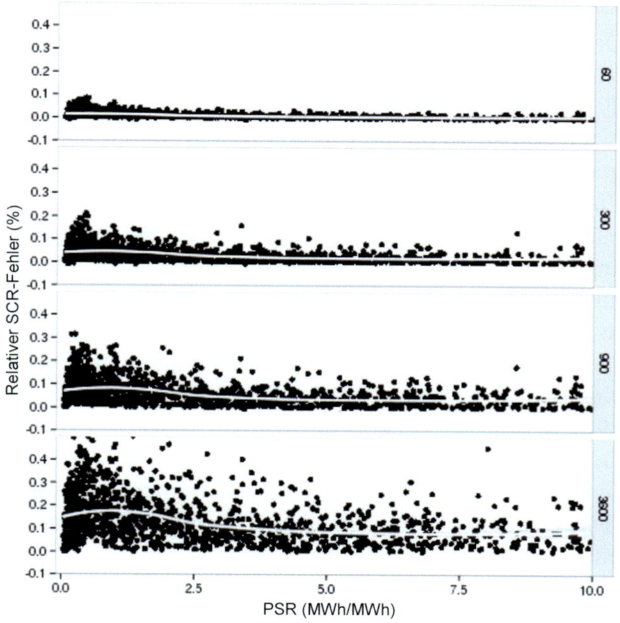

Bild 5 relativer Fehler des Eigenverbrauchs als Funktion von PSR bei Intervallen von 1 min, 5 min, 15 min und 1 h gegenüber 15 s für ein zufälliges Lastprofil

3 Wirtschaftlichkeitsbetrachtung von Kleinspeichern

Das techno-ökonomische Model zur Betrachtung der Wirtschaftlichkeit von elektronischen Speichersystemen (ESS) für Haushalte besteht aus drei Teilen: Einer Marktrecherche zur Ermittlung aktueller Kosten für PV- und Speicher-Systeme, einer Residuallastsimulation zur Ermittlung der SCR und einer Discounted Cash-Flow Berechnung.

3.1 Marktrecherche

Die ermittelte Kostenfunktionen für die PV-Anlage setzt sich aus den Fixkosten (900 €) und einem leistungsabhängigen Teil zusammen (1,4 €/Wp). Der Anteil der Wechselrichterkosten wird ebenfalls über eine leistungsabhängige Funktion abgebildet. Die Lebensdauer der PV-Anlage wird mit 20 Jahren angenommen während die Umrichter (PV und ESS) je nach 10 Jahren ersetzt werden müssen. Die Marktrecherche zu ESS umfasst über 75 verfügbare Systeme bzw. Setups. Für die verschiedenen Technologien Blei-Säure/Gel-Batterien (Pb), Lithium-Ionen (LI), Lithium-Titanat (LTO), thermische Natrium-Salz-Batterien (ZEBRA) und Redox-Flow-Batterien (RFB) wurden die Parameter Nennkapazität (E_{nom}), Entladetiefe (DoD), Zyklenzahl (N_{cyc}), kalendarische Lebensdauer (t_{life}), elektrischer Wirkungsgrad (η_{tor}) und Kosten ($CAPEX_{stor}$) erfasst und Mittelwerte gebildet.

Für jede Technologie wurde eine CAPEX-Kostenfunktion mit fixem sowie kapazitätsabhängigem Anteil ermittelt. Die Preise von LI-ESS variieren stark. Deswegen wurden die fünf günstigsten ermittelten ESS auf Lithium-Basis in der separaten Technologieklasse LI* zusammengefasst. Mit den oben genannten Parametern lassen sich die spezifischen Speicherkosten (C_{stor}) in €/kWh bestimmen:

$$C_{stor} = \frac{CAPEX_{stor}}{E_{nom} \cdot DoD \cdot N_{cyc} \cdot \eta_{stor}} \qquad (2)$$

Diese Formel ist nur gültig, wenn die Zyklenzahl die begrenzende Größe in der Betrachtung ist und nicht die kalendarische Lebensdauer vorher erreicht wird. LTO- und RFB-System erreichen bei dieser Betrachtung besonders günstige Werte, welche in ihrer hohen Zyklenzahl von >15.000 begründet liegt. Folglich sind solche Systeme vor allem für Anwendungen mit hoher zyklischer Auslastung geeignet, was bei der Eigenverbrauchsoptimierung nicht zwingend der Fall ist. Im vorgestellten Model werden beide Werte, kalendarische und zyklische Lebenserwartung, berücksichtigt. Die Kostenanteile für die Akkumulatoren sowie den Batterie-Umrichter wurden ebenfalls aus den Marktzahlen ermittelt. Für alle Preise gilt, dass sie einer starken Preisdynamik unterliegen und deswegen nur eine Momentaufnahme für den Zeitraum Q3 2015 darstellen. Außerdem sei darauf hingewiesen, dass für die Technologien LTO, ZEBRA und RFB jeweils nur ein Anbieter gefunden wurde, der über Speichersysteme in der betrachteten Dimension verfügt.

3.2 Residuallastsimulation

Da sich der Zeitraum, für den Messdaten von den installierten Smart-Metern zur Verfügung stehen, noch nicht über ein ganzes Jahr erstreckt, wurde zur Erzeugung von PV-Erzeugungs- und Lastprofilen für jeweils ein Jahr auf eine Simulation zurückgegriffen. Dem PV-Erzeugungsprofil liegen simulierte, standort- und zeitabhängige Einstrahlungswerte zu Grunde. Die erzeugte Leistung wird von der saisonalen Bewölkungssimulation sowie der Größe und Ausrichtung der Module beeinflusst [1]. Zur Simulation der Last wird eine Gamma-Verteilung, wie in [5] und [6] beschrieben, über das H0-Standardlastprofil [13] gelegt. Das Simulationsintervall beträgt 15 min. Die ermittelte SCR ergibt sich als Funktion der erzeugten PV-Energie und der verfügbaren Kapazität des ESS, jeweils im Verhältnis zum Jahresbedarf eines Haushalts, siehe **Bild 6**. In [4] wurde ein ähnliches Diagramm vorgestellt, mit anderen Eingangsparametern, v. a. bzgl. der Lastprofile. Dort werden im Fall ohne Speicher geringere Quoten erreicht, dafür wirkt sich der positive Einfluss eines Speichers stärker aus.

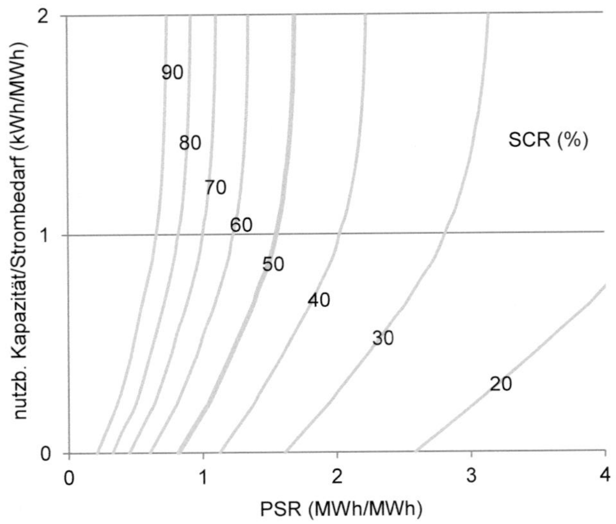

Bild 6 SCR in Abhängigkeit von PSR und nutzbaren Batteriekapazität

3.3 Discounted Cash-Flow Model

Für die ökonomische Betrachtung werden über die Projektlaufzeit von 20 Jahren die Nettobarwerte des Cash-Flow aufsummiert:

$$NB = CAP + \sum_{n=0}^{20} \frac{AO_n}{(1+DR)^n} - \frac{RV}{(1+DR)^{20}} \tag{3}$$

NB entspricht dem gesamten Nettobarwert des Projekts, CAP den Installationskosten im ersten Jahr, DR dem nominalen Kalkulationszinssatz (entspricht dem realen Kalkulationszins plus Inflationsrate), RV dem Restwert bei Projektende und AO_n den jährlichen Einnahmen und Ausgaben nach folgender Formel:

$$\begin{aligned} AO_n = EFI_n \cdot FIT - ESC_n(RPE_n - EEG_n) \\ + [OP + IRC_n(1+IPT)^n \\ + BRC_n(1+BPT)^n](1+IR)^n \end{aligned} \tag{4}$$

EFI_n ist der eingespeiste Anteil der PV-Energie, FIT die Einspeisevergütung, ESC_n der eigenverbrauchte Anteil der PV-Energie, RPE_n der Strompreis, EEG_n die reduzierte EEG-Umlage für Eigenverbrauch, OP sind die jährlichen Betriebskosten für PV und ESS, IRC_n die Umrichter-Austauschkosten für PV und ESS, IPT die entsprechenden Umrichter-Preistrends, BRC_n die Akkumulator-Austauschkosten, BPT ist der Akkumulator-Preistrend und IR die Inflationsrate. Austauschkosten für Umrichter fallen jeweils nur im zehnten Jahr an und Austauschkosten für Akkumulatoren immer, sobald entweder die kalendarische oder zyklische Lebenserwartung erreicht ist. Der Restwert der PV-Anlage wird mit Null betrachtet, der des ESS besteht in dem BOS-Anteil sowie dem Restwert der Akkumulatoren skaliert auf die verbleibende Lebenserwartung. Für die Berechnung der Rendite, auf Englisch internal rate of return (IRR), wird DR variiert bis NB Null ergibt. Liegt DR über dem erwarteten nominalen Kalkulationszins (3,2 %/a), ist das Projekt wirtschaftlich. Die Eingangsparameter werden im Folgenden diskutiert.

3.3.1 Ökonomische Parameter

Die Werte für den Strompreis und dessen Entwicklung wurden aus [12] abgeleitet, was einem Preis von 29,7 ct/kWh im Januar 2016 entspricht. Es ist optional möglich, einen Förderkredit aus dem KfW-Speicherprogramm zu berücksichtigen, welcher besonders günstige Zinsen und einen Tilgungszuschuss beinhaltet [14]. Im Fall eines Kredites wird eine Eigenfinanzierung von 20 % angenommen. Die Laufzeit beträgt 10 Jahre und die Annuitätszahlung wird in den jährlichen Ausgaben mit abgerechnet. Die Inflationsrate basiert auf Schätzungen des IWF für Deutschland [15]. Für den Kalkulationszinssatz wurde als Vergleichswert eine risikofreie Anlage in 20-jährige, deutsche Bundesanleihen zu Grunde gelegt [16]. Der Vergleichswert für den IRR ist mit 3,2 %/a Jahr relativ gering, stellt aber angesichts der Investorengruppe (Privathaushalte) und dem aktuell niedrigem Zinsniveau einen plausiblen Wert dar.

3.3.2 Regulatorische Parameter

Für den Einspeisetarif wurden die aktuellen Vergütungen und eine zukünftige, monatliche Degression von 0,5 % pro Monat angelegt, was zum Januar 2016 für Anlagen < 10 kWp zu einer Vergütung von 12,07 ct/kWh führt. Bei Anlagen > 10 kWp wird die reduzierte Einspeisevergütung in Rechnung gestellt, inklusive der Reduzierung auf 35 % statt 40 % für das Jahr 2016.

3.3.3 PV Parameter

Wie in [8] wurde der PV-Ertrag mit 1.000 kWh/kWp/a abgeschätzt. Die jährliche Leistungsdegression der PV-Module wurde im Einklang mit [8] und [7] auf 0,5 %/a gesetzt. Ebenso stellen die OPEX-Kosten von 1,25 % der PV-CAPEX einen Mittelwert der üblichen Literaturwerte dar [4] und [8].

3.3.4 ESS Parameter

Die Parameter, die nicht oder nur in geringer Zahl aus der Marktrecherche ermittelt werden konnten, wurden mit Werten aus [9] abgeglichen und ergänzt. Aus [9] stammen auch die Werte für die ESS-OPEX. In [17] wird von einer Kostenreduktion für Li-Akkumulatoren von 60 % bis 2030 ausgegangen. Dem entsprechend wurde der Preistrend für Li-Akkumulatoren auf -5 %/a gesetzt. Für die anderen Technologien wurden die Werte geschätzt. **Tabelle 2** fasst die Parameter zusammen.

Tabelle 2 ESS Parameter

Parameter	Pb	LI	LI*	LTO	ZEBRA	RFB
OPEX (€/kWh/a)	22	19	19	19	45	43
kalend. Lebenserwartung (a)	8.5	11.5	15	20	12	20
Akkumulator Preistrend[a] (%/a)	-1	-5	-5	-3	-2	-5

[a] nominal

Die ESS-Kapazität wird über die gesamte Lebenserwartung als konstant betrachtet, was eine optimistische Annahme darstellt. Auf der anderen Seite werden die Akkumulatoren bei Erreichen der zyklischen oder kalendarischen Lebenserwartung ausgetauscht. Zu diesem Zeitpunkt verfügen die meisten Akkumulatoren noch über signifikante Kapazität, so dass der Austausch eine konservative Maßnahme darstellt, die die vorher genannte optimistische Annahme wieder ausgleicht.

3.3.5 Monte Carlo Simulation

Die Parameter die im Model keine Stellgrößen sind, unterliegen einer unterschiedlich ausgeprägten Ungenauigkeit. Diese wurde für jeden Parameter abgeschätzt und zwar in Form des Bereiches, in dem der Parameter mit 99 %iger Wahrscheinlichkeit liegt. Als Mediane dienen jeweils die in den vorangegangenen Abschnitten diskutierten Werte, siehe auch **Tabelle 3**. Als Verteilungsfunktion wurde entweder ein Gamma oder eine Normalverteilung gewählt. Mit den definierten Verteilungen wurden für jedes Szenario 10.000 Simulationen berechnet. Die Verteilung der Ergebnisse wird für jedes Szenario als Box-Plot dargestellt, in dem der Median, der Bereich in dem die Hälfte der Ergebnisse liegen (Balken) sowie der Bereich in dem 99 % der Ergebnisse liegen (Whisker) abgebildet sind.

Tabelle 3 Verteilungen der stochastischen Parameter

Parameter	Median	99 %-Bereich
Jährlicher Energiebedarf (kWh/a)	5.000	Stellgröße
Nutzbare ESS-Kapazität (kWh)	5	Stellgröße
Strompreis 01/2016 (ct/kWh)	29,7	±5 % Median
Strompreistrend[a] (%/a)	+3	±10 % Median
Inflation (%/a)	1,7	±25 % Median
Kalkulationszinssatz[b] (%/a)	1,5	±50 % Median
PV Fixkosten (€)	900	±10 % Median
PV Relativkosten (€/Wp)	1,4	±10 % Median
PV OPEX (€/kWp/a)[c]	19,6	8.8 … 35
PV-WR-Preistrend[a] (%/a)	-2	±50 % Median
PV Ertrag (kWh/kWp/a)	1.000	±15 % Median
PV Degradation (%/a)[c]	0,5	0.1 … 1.0
Zeitpunkt der Installation	01/2016	±6 Monate
mittlere EEG-Umlage (ct/kWh)	6,2	±50 % Median
ESS CAPEX	t.-a.	±25 % Median
ESS OPEX[c]	t.-a.	t.-a.
ESS Installationskosten (€)	1.000	±50 % Median
Akkumulatorpreistrend[a]	t.-a.	±50 % Median
ESS-Umrichterpreistrend[a] (%/a)	-4	±50 % Median
ESS Zyklenzahl	t.-a.	±10 % Median
ESS kalend. Lebenserwartung	t.-a.	±25 % Median
ESS DoD	t.-a.	±10 % Median
ESS Wirkungsgrad	t.-a.	±10 % Median
KFW Zins[a] (%/a)	1,85	±25 % Median

[a]nominal, [b]real, [c] Gamma-Verteilung, t.-a. = technologie-abhängig

3.4 Ergebnisse

Zuerst wurde das Referenzszenario berechnet sowie der Fall einer PV-Anlage ohne Speicher mit variierender installierter Leistung betrachtet. Die Ergebnisse sind wie zuvor beschrieben als Verteilungen in **Bild 7** zusammengefasst. Im Referenzszenario gibt es keine investitionsbedingten Einnahmen oder Ausgaben und der Cash-Flow reduziert sich auf Null. Der Vergleichswert, an dem sich die Rendite einer Investition misst, ist der nominale Kalkulationszinssatz von 3,2 %/a, der in **Bild 7** und **8** als graue Striche dargestellt ist. Für den Fall der PV-Anlage ohne ESS wurden verschiedene installierte Leistungen betrachtet. Das 10,5 kWp Szenario verdeutlicht die regulatorischen Auswirkungen des Überschreitens von 10 kWp auf die Rendite, in Zahlen eine Verringerung um 0,9 %, wobei 0,4 % auf die geringere EEG-Vergütung und 0,5 % auf die nun fällige, reduzierte EEG-Umlage entfallen. Das Optimum wäre in diesem Modell mit einem IRR von 12,5 % bei einer 2 kWp-Anlage erreicht. Bei solch kleinen Werten ist es jedoch fraglich, ob die PV-Installations-Kostenfunktion die Realität noch präzise abbildet. Anlagen, bei denen der PV-Ertrag dem Strombedarf entspricht – in diesem Fall das 5 kWp-Szenario – erreichen ebenfalls eine hohe Rendite von über 10 %.

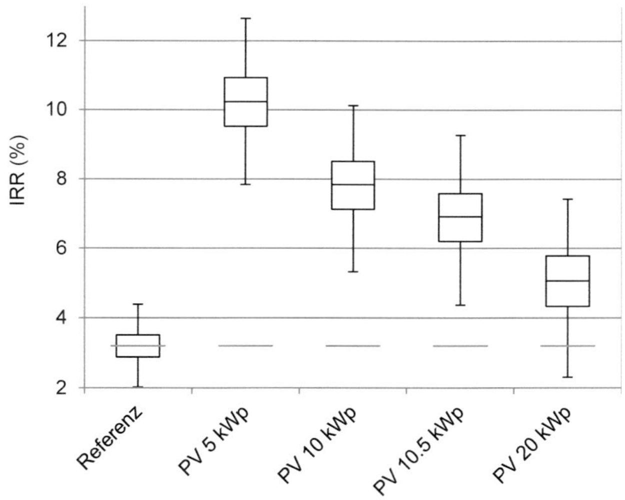

Bild 7 IRR des Referenzfalls und bei unterschiedlichen PV-Leistungen

Für den zweiten Fall wurde ein ESS mit nutzbarer Kapazität von 5 kWh, was einer der gängigsten Speichergrößen am Markt entspricht, sowie eine installierte PV-Leistung von 10 kWp gewählt, da sich diese Kombination bei den verschiedenen Betrachtungen mit am wirtschaftlichsten erwiesen hat. Anschließend wurde die Speichertechnologie variiert (Subszenario A). Außerdem wurde in den Subszenarien B der oben beschriebene KfW-Kredit in Anspruch genommen. Die Ergebnisse sind in **Bild 8** dargestellt. Nur die günstigsten Lithium-System LI* lassen sich wirtschaftlich darstellen. Bei allen anderen Technologien ist selbst bei Inanspruchnahme der KfW-Förderung unter aktuellen Bedingungen kein wirtschaftliches Ergebnis zu erzielen. Außerdem verschlechtert bei allen betrachteten Szenarien ein ESS immer die Rendite gegenüber der reinen PV-Anlage ohne ESS.

4 Diskussion und Ausblick

Bisher wurden Messwerte für ca. sechs Monate erfasst. Dieser Zeitraum ist noch nicht ausreichend, um Aussagen über ein ganzes Jahr zu treffen. In den kommenden Monaten werden die Aufzeichnungen fortgeführt und die Analysen vertieft. Saisonale Unterschiede sollen herausgearbeitet und in für Simulationen verwertbaren Zusammenhängen formuliert werden. Anhand der aufgezeigten relativen Fehler der SCR bei der Betrachtung längerer Zeitintervalle können vorhandene Werte korrigiert sowie der Nutzen einer höheren zeitlichen Auflösung abgeschätzt werden.

Obwohl andere Autoren zu dem Ergebnis kommen, dass ESS bereits wirtschaftlich sind [8], liefert das vorgestellte Model kein wirtschaftliches Szenario für ESS zur Eigenverbrauchserhöhung. Auch wenn sich wirtschaftliche Setups aus PV-Anlage inklusive ESS finden lassen, würde sich ein rein profit-orientierter Investor immer nur für die PV-Anlage entscheiden. Nach unseren Einschätzungen müssten sich ESS-Preise der am besten abschneidenden LI*-Systeme noch auf ca. die Hälfte reduzieren, damit sich wirtschaftlicher Vorteil und Mehrausgaben die Waage halten.

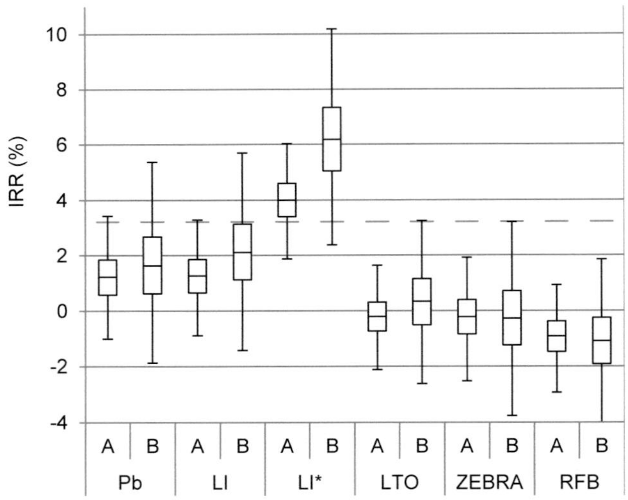

Bild 8 IRR für unterschiedliche Technologien, Subszenarien B nehmen den KfW-Förderkredit in Anspruch

5 Anerkennung

Das Forschungsvorhaben „Smart Grid Solar“ wird von der europäischen Union aus dem Europäischen Fonds für Regionale Entwicklung und vom Freistaat Bayern kofinanziert.

6 Literatur

[1] Luchscheider, P., Bazan, P., German, R.: Zeitlich hochaufgelöste Simulation von Solarstrahlung zur Bewertung von Smart Grids. 29. Symposium Photovoltaische Solarenergie (OTTI), Bad Staffelstein (2014)

[2] Hoevenaars, E.J., Crawford, C.A.: Implications of temporal resolution for modeling renewables-based power systems. Renew. Energy 41, S. 285–293. (2012). doi:10.1016/j.renene.2011.11.013

[3] Cao, S., Sirén, K.: Impact of simulation time-resolution on the matching of PV production and household electric demand. Appl. Energy 128, S. 192–208. (2014). doi:10.1016/j.apenergy.2014.04.075

[4] Weniger, J., Tjaden, T., Quaschning, V.: Sizing of Residential PV Battery Systems. Energy Procedia, 8th International Renewable Energy Stor-age Conference and Exhibition (IRES 2013), S. 78–87 Vol 46. (2014). doi:10.1016/j.egypro.2014.01.160

[5] Cagni, A., Carpaneto E., Chicco G., Napoli R.: Characterisation of the aggregated load patterns for extraurban residential customer groups. S. 951–954 Vol.3. IEEE, (2004). doi:10.1109/MELCON.2004.1348210

[6] Carpaneto, E., Chicco, G.: Probability distribu-tions of the aggregated residential load. S. 1–6. IEEE (2006). doi:10.1109/PMAPS.2006.360235.

[7] Darling, S.B., You, F., Veselka, T., Velosa, A.: Assumptions and the levelized cost of energy for photovoltaics. Energy Environ. Sci. 4, S. 3133–3139. (2011). doi:10.1039/C0EE00698J

[8] Hoppmann, J., Volland, J., Schmidt, T.S., Hoffmann, V.H.: The economic viability of battery storage for residential solar photovoltaic systems – A review and a simulation model. Renewable and Sustainable Energy Reviews 39, S. 1101–1118. (2014). doi:10.1016/j.rser.2014.07.068

[9] Battke, B., Schmidt, T.S., Grosspietsch, D., Hoffmann, V.H.: A review and probabilistic model of lifecycle costs of stationary batteries in multiple applications. Renewable and Sustainable Energy Reviews 25, S. 240–250. (2013). doi:10.1016/j.rser.2013.04.023

[10] Stetz, T., Wolf, H., Probst, A., Eilenberger, S., Saint Drenan, Y.-M., Kämpf, E., Braun, M., Schöllhorn, D., Schmidt, S.: Stochastische Analyse von Smart-Meter Messdaten, VDE-Kongress 2012. VDE VERLAG GmbH. (2012)

[11] Häberlin, H.: Photovoltaik: Strom aus Sonnenlicht für Verbundnetz und Inselanlagen, Auflage: 2., VDE VERLAG, Fehraltorf. (2010)

[12] BDEW: BDEW-Strompreisanalyse März 2015. https://www.bdew.de/internet.nsf/id/9D1CF269C1282487C1257E22002BC8DD/$file/150409%20BDEW%20zum%20Strompreis%20der%20Haushalte%20Anhang.pdf. Zugegriffen: 17. August 2015

[13] ED Netze GmbH, VDEW Lastprofil Haushalt http://www.ednetze.de/cms/medienboard/excel/lastprofile/VDEW-Lastprofile-Haushalt.xls. (1999) Zugegriffen: 12. Januar 2014

[14] KfW: Konditionen-Anzeiger. https://www.kfw-formularsammlung.de/KonditionenanzeigerINet/KonditionenAnzeiger. (2014) Zugegriffen: 13. Januar 2015

[15] IMF: World Economic Outlook Database October 2014. http://www.imf.org/external/pubs/ft/weo/2014/02/weodata/index.aspx. (2014). Zugegriffen: 9. Januar 2015

[16] Deutsche Bundesbank: Tägliche Zinsstruktur für Pfandbriefe http://www.bundesbank.de/Redaktion/DE/Downloads/Statistiken/Geld_Und_Kapitalmaerkte/Zinssaet-ze_Renditen/stat_zinsstruktur_Pfandbriefe.pdf. (2015). Zugegriffen: 9. Dezember 2014

[17] Solarpraxis: Preis einer gespeicherten Kilowattstunde Strom kann bis 2030 von über 20 auf 5 Eurocent sinken. http://www.solarpraxis.de/newsliste-von-startsei-te/newsdetail/article/Preis_einer_gespeicherten_Kilowattstunde_Strom_kan/. (2014) Zugegriffen: 17. November 2014

Modellierung und Bewertung einer autarken Notstromversorgung der Bevölkerung unterhalb der KRITIS-Schwelle unter Einsatz von Photovoltaik-Systemen

L. Lang, M. Dietmannsberger, A. Lücken, D. Schulz
Helmut-Schmidt-Universität, Fachgebiet Elektrische Energiesysteme, 22043 Hamburg

Kurzfassung

Im Rahmen dieser Publikation wurde analysiert, wie eine Notstromversorgung nicht-kritischer Infrastrukturen durch Photovoltaik-Systeme verwirklicht werden kann. Dazu wurden vier Szenarien entwickelt und modelliert, die eine Abschätzung des Leistungs- und Energiebedarfs ermöglichen. Dies beinhaltete die Untersuchung relevanter Verbraucher in den einzelnen Szenarien, die Größe des Solargenerators sowie eines Zwischenspeichers. Es wurde zwischen Sommer und Winter sowie der verfügbaren Solarstrahlung in drei Sektoren Deutschlands differenziert. Abschließend fand eine Analyse des Flächen- und des Kostenbedarfs sowie eine Diskussion der Realisierbarkeit in Anbetracht der räumlichen und finanziellen Möglichkeiten in den einzelnen Szenarien statt. Es wurde ermittelt, dass Unterschiede zwischen dem nordwestlichen Sektor und dem südlichen Sektor von bis zu 45 % beim Flächenbedarf existieren. Eine Bewertung der benötigten Flächen ergab, dass Notstromversorgungssysteme allein mit Photovoltaik-Systemen oft aus Platzgründen nicht realisierbar sind. Im Anschluss wurden verschiedene Kostenpunkte einer Photovoltaik-Anlage separat betrachtet. Die Kosten variieren bis zu 41 % zwischen den betrachteten Bereichen Deutschlands.

1 Einleitung

Aufgrund des Kurswechsels, den Deutschland in den letzten Jahren mit der Energiewende vollzogen hat, wird angenommen, dass die Wahrscheinlichkeit großflächiger Blackouts und kleinerer Stromausfälle stark zunimmt [1]. Für kritische Infrastrukturen (KRITIS), Organisationen und Einrichtungen mit wichtiger Bedeutung für das staatliche Gemeinwesen, existieren bereits Notfallpläne und Versorgungssysteme um im Falle eines Blackouts weiterhin ihre Funktion erfüllen zu können. Dazu zählen beispielsweise Krankenhäuser, Polizei und Feuerwehr. In der Stromversorgung normaler Haushalte existieren jedoch keine ausgearbeiteten Notfalllösungen. Zwar veröffentlicht das Bundesamt für Bevölkerungsschutz und Katastrophenhilfe (BBK) Broschüren zum richtigen Umgang mit Stromausfällen sowie Hinweisen für eine richtige Vorbereitung darauf, jedoch muss angenommen werden, dass ein Großteil der Bevölkerung im Ernstfall nicht in der Lage ist, sich selbst über mehrere Tage hinweg ohne ein funktionierendes Stromnetz zu versorgen.

Ausgangspunkt der Untersuchungen sind Photovoltaik-Anlagen, die bereits eine hohe Verbreitung in den KRITIS haben. Photovoltaik (PV)-Anlagen werden heutzutage vermehrt mit Batteriesystemen ausgestattet um beispielsweise den Eigenverbrauch zu optimieren. Damit werden vereinzelt bereits sehr hohe Autarkiegrade erreicht, jedoch fehlt bis dato eine Abschätzung des absoluten Minimalstandards für den Notfall. Im Rahmen dieser Publikation soll mit einer wissenschaftlichen Untersuchung geklärt werden, wie eine Notstromversorgung (NSV) für die normale Bevölkerung mittels Photovoltaik realisierbar ist. Die Versorgung der wichtigsten elektrischen Verbraucher soll mit netzunabhängigen Insel-Anlagen sichergestellt werden. Die Erkenntnisse dienen zur Planung von NSV-Systemen für Privatpersonen, jedoch auch für entsprechende öffentliche Einrichtungen, die Informationsmaterial für die Bevölkerung erstellen und verteilen müssen. Die gewählte Methodik in dieser Publikation erlaubt eine Bewertung, ob die Photovoltaik generell geeignet ist, für autarke NSV-Systeme unterhalb der KRITIS-Schwelle eingesetzt zu werden.

In Kapitel 2 wird ein Modell festgelegt, dass sich zur Untersuchung der im Folgenden betrachteten Aspekte eignet. Dazu werden zuerst die Verbraucher analysiert, die in den jeweiligen Szenarien notwendig sind und deren Energiebedarfe berechnet. Zudem werden allgemein die Erzeuger und Zwischenspeicher modelliert. Die Ergebnisse der Leistungsberechnung und Flächenabschätzung, ebenso wie eine Kostenbetrachtung der entwickelten Lösungen werden in Kapitel 3 dargestellt. Eine Zusammenfassung der Ergebnisse findet sich in Kapitel 4.

2 Modellbildung

Um einen repräsentativen Durchschnittsbedarf der benötigten Leistungs- und Energiemenge einer Notstromversorgung unterhalb der KRITIS-Schwelle zu erhalten, wurden die in **Tabelle 1** gezeigten Szenarien entwickelt.

Tabelle 1 Untersuchte Szenarien unterhalb der KRITIS-Schwelle.

Szenario	Definition
1	Alleinerziehende Person mit zwei Kindern
2	Heizungsanlage eines Einfamilienhauses
3	Mehrfamilienhaus
4	Kleine Siedlung

Zur weiteren Untersuchung müssen zunächst die Verbraucher in den verschiedenen Szenarien definiert, analysiert und ihr Energie- und Leistungsbedarf berechnet werden. Bei der Auswahl der benötigten Verbraucher werden nur die zum Leben notwendigen Geräte in Betracht gezogen. Anschließend werden die Randbedingungen für ein PV-Inselsystem festgelegt. Zudem wird zwischen Sommer (April bis September) und Winter (Oktober bis März) unterschieden.

2.1 Verbrauchermodellierung

2.1.1 Szenario 1: Alleinerziehende Person mit zwei Kindern

Szenario 1 berücksichtigt die Besonderheiten, die bei kleinen Haushalten mit Kindern auftreten. Da je nach Alter der Kinder unterschiedliche Energiemengen für deren Versorgung benötigt werden, wird davon ausgegangen, dass zumindest eines der Kinder unter zwei Jahren ist. Daraus resultiert ein erhöhter Energiebedarf, da eine Notwendigkeit der Fläschchen-/Babynahrungserwärmung und für Wärmelampen beim Wickeln besteht.

Ein Wechselrichter (WR), der die Gleichspannung der PV-Anlage auf 230/400 V transformiert, wird auf den größten gleichzeitig auftretenden Leistungsbedarf ausgelegt, der sich aus Kühlschrank sowie Wickelheizstrahler zusammensetzt und ca. 700 W beträgt. **Tabelle 2** zeigt eine Übersicht über die Verbraucher in diesem Szenario.

Tabelle 2 Verbraucher des Szenarios 1.

Kategorie	Leistung [W]	Täglicher Verbrauch [Wh]	
		Sommer	Winter
Lebensmittel-Zubereitung	150	75	75
Lagerung/ Kühlung	95	540	0
Heizen	600	0	375
Licht	5–12	120	360
Kommunikation	15	225	225
Wechselrichter	5	115	77
Summe	**max. 700**	**1.075**	**1.112**

Es ergibt sich ein Minimalverbrauch von 1.075 Wh pro Tag im Sommer und von 1.112 Wh pro Tag im Winter. Je nach Zusammenstellung der Verbraucher und Wechselrichter kann dieser Wert variieren.

2.1.2 Szenario 2: Heizungsanlage eines Einfamilienhauses

In der Modellierung werden nur Gas- und Ölheizungen betrachtet, da eine Raum- und Warmwassererwärmung mit Strom energetisch nicht abzudecken ist. Die Steuerung der Heizungssysteme sowie die Umwälzpumpe (UWP) müssen mit elektrischer Energie versorgt werden. Die Heizwärme wird dabei aus fossilen Brennträgern gewonnen.

Für Neubauten gilt nach derzeitiger Bauordnung ein Grenzwert für den jährlichen Heizwärmebedarf pro Quadratmeter zwischen 70 kWh/m^2 und 100 kWh/m^2, bei Niedrigenergiehäusern sogar zwischen 30 kWh/m^2 und 60 kWh/m^2 [2]. Die UWP muss die Heizwärme für die Wohnung in Form von erhitztem Wasser(-dampf) durch die Rohrleitungen zu den Heizkörpern befördern. Je nach benötigter Wärmeleistung erhöhen sich dementsprechend der Massenstrom und dadurch die benötigte Leistung der Pumpe. Die Spitzenleistung der Pumpe sollte auf die maximale Leistungsanforderung ausgelegt sein. Der Bedarf an Warmwasser bleibt über das Jahr fast konstant bei etwa 4 % in einem Einfamilienhaus (EFH) und 4,5 % in einem Mehrfamilienhaus (MFH) der maximalen Wärmeleistung [2]. Deshalb wird die Modellierung der UWP nach dem Raumwärmebedarf vorgenommen. Viele UWP sind mit Leistungen von 80 W oder mehr überdimensioniert. Bei einem ausbalancierten Heizungssystem genügt im Durchschnitt eine Leistung von 35 W [3], jedoch wird im Folgenden auf Grund der Verbreitung mit ersterem Wert kalkuliert.

Der Durchschnittswert für den jährlichen Heizwärmebedarf pro Quadratmeter E_{m^2} ermittelt sich mithilfe des jährlichen Heizwärmebedarfs Q und dem prozentualen Anteil des zugehörigen Baujahres p (relevant für Energieeffizienz) nach Gleichung (1). In Gleichung (2) ist beispielhaft die Berechnung für EFH gezeigt [2].

$$E_{\mathrm{m}^2} = \sum Q \times p \qquad (1)$$

$$\begin{aligned} &190\,\frac{\mathrm{kWh}}{\mathrm{m}^2}\cdot 25\,\% + 226\,\frac{\mathrm{kWh}}{\mathrm{m}^2}\cdot 22\,\% \\ &+160\,\frac{\mathrm{kWh}}{\mathrm{m}^2}\cdot 25\,\% + 99\,\frac{\mathrm{kWh}}{\mathrm{m}^2}\cdot 28\,\% \approx 165\,\frac{\mathrm{kWh}}{\mathrm{m}^2} \qquad (2)\end{aligned}$$

Für EFH ergibt sich demnach ein Wert von ca. 165 kWh/m^2. Die durchschnittliche Fläche eines EFH kann mit 127,5 m^2 angenommen werden. Damit ergibt sich der Heizwärmebedarf eines EFH während der Wintermonate pro Tag von 116 kWh.

Vereinfacht lässt sich die benötigte tägliche Energie der UWP mit 0,798 kWh berechnen [2]. Da dies ein Durchschnittswert ist, liegt die Anforderung an besonders kalten Tagen höher. Betrachtet man die täglich verbrauchte Energie im Laufe des Jahres, so zeigen sich im Januar Werte von bis zu 0,7 % pro Tag im Bezug auf den Jahresgesamtwärmebedarf. Rechnet man, inklusive eines zusätzlichen Puffers, mit 0,8 %, ergibt sich eine benötigte Energie von 1,164 kWh pro Tag.

Im Sommer wird die UWP nur für das Warmwasser benötigt. Der Verbrauch liegt dann bei 34,56 Wh pro Tag [2]. Zur Leistung der UWP muss die Leistung der Steuerung addiert werden. Im Standby-Betrieb kann der Energieverbrauch pro Tag für ein Heizungssystem des Baujahres 1998 mit einem Wert von 126 Wh angenähert werden. Es kann eine mittlere elektrische Leistungsaufnahme der Steuerung von 79 W angenommen werden. Die Zusammenfassung dieser Ergebnisse ist in **Tabelle 3** gezeigt.

Tabelle 3 Verbraucher des Szenarios 2.

Kategorie	Täglicher Verbrauch [Wh]	
	Sommer	Winter
Umwälzpumpe	35	1.164
Steuerung	126	1.149
Summe	**161**	**2.313**

Abschließend ergibt sich für den Winter ein täglicher Verbrauch von 2.313 Wh, während er im Sommer bei 161 Wh liegt. Die Heizung stellt damit eine antizyklische Last in Bezug auf die verfügbare Sonnenstrahlung dar.

2.1.3 Szenario 3: Mehrfamilienhaus

Ein durchschnittliches MFH in Deutschland besteht aus 4,5 Wohnungen. Für die Berechnung der Verbraucher wird angenommen, dass von diesen Haushalten einer eine alleinerziehende Person mit zwei Kindern (Szenario 1) darstellt. Der Rest besteht aus Haushalten mit älteren Kindern oder Paaren. Im Folgenden werden der Energieverbrauch der Haushalte und des Heizungssystems berechnet. Hierfür werden zum Teil Ergebnisse aus den beiden vorangegangenen Szenarien verwendet oder leicht abgewandelt.

Für den Haushalt mit alleinerziehendem Elternteil werden die Werte aus Szenario 1 übernommen. Haushalte ohne Kind lassen sich durch Modifizieren dieses Szenarios kalkulieren. Babykost-Erwärmer sowie Wickelheizstrahler werden nicht benötigt, daher reduziert sich der Bedarf um die Posten Zubereitung und Heizen. Dadurch ergeben sich tägliche Verbräuche von 1.000 Wh im Sommer sowie 662 Wh im Winter.

Der jährliche Heizwärmebedarf pro Quadratmeter von MFH ist niedriger als der von EFH. Nach Gleichung (1) kann ein Wert für MFH von ca. 105 kWh/m^2. Allerdings wird dieser Wert mit einem Aufschlag von 10 % versehen, da Studien zeigen, dass der tatsächliche Wert höher liegt. Damit ergibt sich ein Heizwärmebedarf von ca. 115 kWh/m^2. Unter Berücksichtigung der Personenanzahl und einer angenommenen Fläche von 41,3 m^2 pro Person [4], ergibt sich ein täglicher Heizwärmebedarf von ca. 229 kWh. Vereinfacht wird angenommen, dass sich der elektrische Energiebedarf der Heizung proportional zum Heizwärmebedarf verhält. Dadurch beläuft sich der tägliche elektrische Bedarf einer Heizung im Sommer auf ca. 318 Wh und im Winter auf ca. 4.576 Wh. Der Gesamtverbrauch des MFH ist in **Tabelle 4** gezeigt.

Tabelle 4 Verbraucher des Szenarios 3.

Kategorie	Anzahl	Täglicher Verbrauch [Wh]	
		Sommer	Winter
Haushalt mit Kleinkind	1	1.075	1.112
Haushalt ohne Kleinkind	3,5	3.500	2.317
Heizung	1	318	4.576
Summe	**4,5**	**4.893**	**8.005**

Der gesamte tägliche Verbrauch eines durchschnittlichen MFH berechnet sich zu ca. 4.893 Wh im Sommer bzw. ca. 8.005 Wh im Winter.

2.1.4 Szenario 4: Kleine Siedlung

Eine kleine Siedlung wird mit einer Einwohnerzahl von 100 definiert. Die Zusammensetzung der Siedlung aus Ein- und Mehrfamilienhäusern wird aus dem Durchschnitt im bundesweiten Vergleich erstellt. 65,1 % der Gebäude sind EFH, der Rest MFH. Gleichung (3) zeigt, wie sich die Anzahl (n_{EFH} bzw. n_{MFH}) durch Betrachtung der Einwohner je Wohnung n_{Ew}, der durchschnittlichen Anzahl an Wohnungen pro MFH n_{Wo} sowie der Einwohnerzahl der kleinen Siedlung E_{KS} berechnen lässt. In Deutschland leben in jeder Wohnung durchschnittlich 1,94 Einwohner.

$$n_{\mathrm{EFH}} \cdot n_{\mathrm{Ew}} + n_{\mathrm{MFH}} \cdot n_{\mathrm{Wo}} \cdot n_{\mathrm{Ew}} = E_{\mathrm{KS}} \tag{3}$$

Wird dieses Gleichungssystem gelöst, ergibt sich $n_{\mathrm{EFH}} \approx 15$ und $n_{\mathrm{MFH}} \approx 8$.

Mit den Annahmen der vorherigen Szenarien sind das in Summe 51 Haushalte. Ca. 2,5 % der Bevölkerung sind jünger als drei Jahre. Bei hundert Einwohnern ergeben sich aufgerundet drei Kinder und dadurch drei Haushalte die ungefähr den Verbrauch von Szenario 1 aufweisen. Der Rest der EFH-Haushalte wird entsprechend dem in Szenario 3 ermittelten modifizierten Energiebedarf behandelt. MFH werden äquivalent zu Szenario 3 berechnet. Entsprechend der Anzahl der Gebäude gibt es fünfzehn Heizungen für EFH und acht Heizungen für MFH. Die Ergebnisse für eine kleine Siedlung sind in **Tabelle 5** zusammengefasst.

Es ergeben sich Verbräuche von ca. 56,1 kWh im Sommer bzw. 111,7 kWh im Winter.

Tabelle 5 Verbraucher des Szenarios 4.

Kategorie	Anzahl	Täglicher Verbrauch [kWh]	
		Sommer	Winter
Haushalt mit Kleinkind	3	3,2	3,3
Haushalt ohne Kleinkind	48	48,0	37,1
Heizung (EFH)	15	2,4	34,7
Heizung (MFH)	8	2,5	36,6
Summe	**51**	**56,1**	**111,7**

2.2 Modellierung des Solargenerators

Die Fläche des Solargenerators ist ausschlaggebend für die erzeugte elektrische Leistung oder Energie. Dabei sind einige Faktoren zu beachten.

Da die Sonne im Sommer einen höheren Lauf zeigt als im Winter, unterscheidet sich der optimale Neigungswinkel β. Durch den niedrigen Sonnenstand wird in den folgenden Berechnungen für den Winter ein Neigungswinkel von 90° bzw. für den Sommer von 30° angenommen [5]. Bei der NSV steht eine hohe Versorgungssicherheit besonders im Fokus. Die Versorgungssicherheit wird durch den solaren Deckungsgrad f_{s} beschrieben und sollte bei 100 % oder höher liegen. Dies ist in Gleichung (4) veranschaulicht.

$$f_{\mathrm{s}} = \frac{E_{\mathrm{PV}}}{E_{\mathrm{Bedarf}}} \geq 1 \tag{4}$$

E_{PV} beschreibt die durch die PV-Anlage erzeugte Energie, während E_{Bedarf} für den mittleren Energiebedarf steht. Um einen Wert von 100 % oder höher zu erreichen, muss E_{PV} demnach genauso groß oder größer als E_{Bedarf} sein. Um diesen Zustand zu erreichen, müssen die Fläche des Generators und der Zwischenspeicher aufeinander abgestimmt sein [6]. Mit dem in Kapitel 2.1 berechneten Energiebedarf kann die benötigte Fläche an Solarmodulen kalkuliert werden. Für die Auslegung der Fläche A_M gilt Gleichung (5).

$$A_{\mathrm{M}} = \frac{k_L \cdot E_{\mathrm{Bedarf}}}{\eta_{\mathrm{M}} \cdot k_{\mathrm{G}} \cdot k_{\mathrm{T}} \cdot R(\beta) \cdot H} = \frac{k_L \cdot E_{\mathrm{Bedarf}}}{\eta_{\mathrm{M}} \cdot K_{\mathrm{H}}} \qquad (5)$$

Da zwischen Verbrauchern/Laderegler und Solarmodul Leitungsverluste auftreten, muss die Erzeugung durch k_L, einen Korrekturfaktor, der diese Leistungsverluste (ungefähr 6 %) berücksichtigt, um diesen Wert erhöht werden [7]. η_M beschreibt den Modulwirkungsgrad. k_G ist ein Generator-Korrekturfaktor, der dazu dient Abweichungen und Verluste des Solargenerators zu berücksichtigen. Unter anderem wären dies zum Beispiel eine ungeeignete Abstimmung der Module, Glasreflektion, Modulverschmutzung oder auch Minderleistung der Module im Gegensatz zu Herstellerangaben. Dieser Wert wird mit 0,9 angenommen [6]. k_T ist ein Korrekturfaktor bezüglich der Zelltemperatur. Er berücksichtigt, dass sie bei unterschiedlicher Einstrahlungsstärke von der Normtemperatur abweicht und somit der Wirkungsgrad verändert wird. Hier wird ein Wert zwischen 0,86 (Sommer) und 1,02 (Winter) angenommen [7]. $R(\beta)$ beschreibt den Korrekturfaktor für die Abweichung der horizontalen Neigung der PV-Zellen vom optimalen Winkel und ist sowohl Standort- als auch Jahreszeit-abhängig. H ist die solare Strahlungsmenge. Um die Formel für weitere Berechnungen zu vereinfachen, wurden einige der Faktoren in K_H zusammengefasst. K_H wurde für verschiedene Winkel, Regionen und Jahreszeiten anhand genannter Quellen errechnet und ist in **Tabelle 6** dargestellt.

Aufgrund unterschiedlicher solarer Bedingungen in Deutschland wurden drei Regionen mit unterschiedlich starken Einstrahlungen definiert, wie in **Abbildung 1** dargestellt.

Tabelle 6 Tägliche solare Strahlung inklusive Korrekturfaktoren (außer k_L), zusammengefasst in K_H.

K_H [Wh/m²]					
NW		NO		S	
Sommer	Winter	Sommer	Winter	Sommer	Winter
3.356	808	3.607	1.019	3.828	1.459

Je nach Wirkungsgrad des Solarmoduls ändern sich die benötigten Flächen erheblich, besonders in den Szenarien mit hohem Leistungsbedarf. Für die kommerziell erhältlichen Module lassen sich für den Wirkungsgrad Durchschnitts- und Höchstwerte feststellen [9]. Bei den folgenden Berechnungen wurde ein Wirkungsgrad von 15 % angenommen. Dieser Wert befindet sich am oberen Ende der Skala für heute kommerziell verfügbare Photovoltaik-Module.

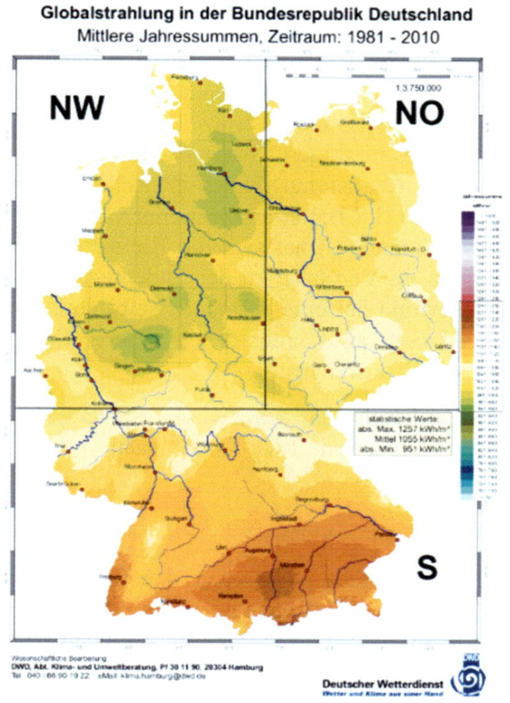

Abbildung 1 Einteilung Deutschlands in drei Sektoren (NW, NO, S). Modifiziert nach [8].

Zudem ist zu beachten, dass von einer optimalen Ausrichtung nach Süden der Solarzellen ausgegangen wird. Ist dies nicht der Fall, so ist mit einer Erhöhung der benötigten Fläche und damit auch der Kosten zu rechnen.

2.3 Modellierung des Zwischenspeichers

Der Energiespeicher stellt sicher, dass Verbraucher auch zu Zeiten genutzt werden können, in denen der PV-Generator keine oder nicht genügend Leistung abgeben kann. Die wesentlichen Auslegungskriterien des Energiespeichers sind dessen maximale Leistung sowie die Speicherkapazität.

Im ungünstigsten Fall ist anzunehmen, dass der gesamte Energieverbrauch während der Nachtstunden geschieht. Dies hätte zur Folge, dass während der Zeiten solarer Einstrahlung die Energie des PV-Generators vollständig in den Speicher geladen und bis zum nächsten Morgen wieder entladen werden müsste. In den meisten Fällen und Szenarien wird es jedoch zu Überschneidungen der Erzeugungs- sowie Verbrauchsphasen kommen. In diesen Fällen liegt die Speicherkapazität deutlich unterhalb der benötigten Gesamtenergie, da sie nur den Verbrauch abdecken muss, der momentan nicht vom PV-Generator gedeckt werden kann. Dieser Effekt tritt vor allem dann ein, wenn mehrere Haushalte (z. B. in den Szenarien 3 und 4) durch eine gemeinsame NSV versorgt werden. Durch die stochastische Verteilung der Verbrauchsgewohnheiten ergeben sich mehr Spielräume für die Dimensionierung des Energiespeichers.

Der Energiebedarf, der durch das PV-System gedeckt werden muss, ergibt sich durch die ermittelten Verbraucher. Dabei lässt sich die verbrauchte Energie E_V nach Gleichung (6) berechnen.

$$E_v = \sum_{i=1}^{n} \int_0^{24\,h} P_i \, dt \qquad (6)$$

P_i ist die benötigte Leistung eines Verbrauchers. Auf Grund der schnellen Startzeit und der langen Speichermöglichkeit sind Batterien, trotz ihrer geringen Lebensdauer und ihres niedrigen Nutzungsgrades, eine sinnvolle Auswahl zur Realisierung eines Zwischenspeichers für NSV-Systeme. Bei Batterien existieren verschiedene Typen. Für Solaranlagen hat sich die Bleibatterie durchgesetzt. Sie ist im Vergleich zu den anderen Batterien kostengünstiger und hat eine geringere Selbstentladung. Die Zyklenzahl liegt zwar unter denen anderer Energiespeicher, ist jedoch ausreichend.

Die benötigte Kapazität des Zwischenspeichers C in den jeweiligen Szenarien lässt sich nach Gleichung (7) bestimmen.

$$C = E_{\text{Bedarf,antizkl.,konst.}} = t \cdot P_{\text{Bedarf,antizykl.,konst.}} \qquad (7)$$

Hierbei beschreibt $E_{\text{Bedarf,antizkl.,konst.}}$ den Energiebedarf der antizyklischen und konstanten Lasten, die ihre Energie nur oder zum Teil von der Batterie beziehen. Die Zeit t gibt die Dauer an, in der die Energie nur aus der Batterie entnommen wird, also die Zeit zwischen Sonnenaufgang und -untergang. Da sich die Länge des Tages nur geringfügig (im Bereich von $\pm$ 15 min vom Mittelwert) unterscheidet, wird für alle drei Regionen mit der gleichen Tageslänge für die entscheidenden Monate August (14,4 h) und Januar (8,6 h) gerechnet. Die Zeit in der die benötigte Energie nur aus der Batterie bezogen wird beträgt im Winter folglich 15,4 h und im Sommer 9,6 h.

An antizyklischen Lasten existieren die Beleuchtung sowie der Wickelheizstrahler, da erfahrungsgemäß einmal auch während der Nacht oder am späten Abend gewickelt werden muss.

Für den benötigten Energiespeicher wird jeweils der Bedarf eines Wintertags veranschlagt (siehe Tabelle 2 bis Tabelle 5), da dies einen sicheren Puffer für Tage mit wenig Einstrahlung darstellt und die maximal empfohlene Entladung auf 20 % berücksichtigt.

3 Ergebnisse

3.1 Flächenbedarf der betrachteten Szenarien

Die Flächenbedarfe für die Szenarien, berechnet nach Gleichung (5), sind in **Tabelle 7** zusammengetragen.
Es ist klar zu erkennen, dass die geringere solare Einstrahlung im Winter zu einem größeren Flächenbedarf führt. Auffallend ist, dass der Abstand zwischen Süden und Norden im Winter überproportional zunimmt. Während im Sommer im NW nur ca. 10 % mehr Fläche benötigt werden als im S, sind es im Winter ca. 80 %.

Insgesamt erscheinen die notwendigen Flächen in den einzelnen Szenarien als nicht realisierbar. Eine Familie in einer Wohnung kann wahrscheinlich nicht bis zu 10 m² PV-Fläche akquirieren. MFH könnten zumindest ihren Heizungsbedarf decken (Szenario 2), eine Versorgung des kompletten Strombedarfs (Szenario 3) ist wohl nur im Sommer und eingeschränkt im Winter im Süden zu realisieren. Die Machbarkeit bei kleinen Siedlungen lässt sich pauschal nicht abschätzen, da es hier stark auf die Struktur des Gebiets und eventuelle Freiflächen ankommt.

Tabelle 7 Mindestflächenbedarfe für die verschiedenen Szenarien. Sz.: Szenario.

Sz.	Flächenbedarf [m²]					
	NW		NO		S	
	Sommer	Winter	Sommer	Winter	Sommer	Winter
1	2,3	9,7	2,1	7,7	2,0	5,4
2		20		16		11
3	10	70	9,6	56	9,0	39
4	118	976	110	775	104	541

3.2 Kosten eines PV-Notstromversorgungssystems

Der Preis pro Quadratmeter Solarmodul $K_{\text{M,m}^2}$ lässt sich unter Berücksichtigung des Preises pro Watt p, der erbrachten Leistung pro Quadratmeter unter Standard-Testbedingungen P_{STC} (1000 W/m²) und des Modulwirkungsgrades η_{M} nach Gleichung (8) berechnen.

$$K_{\text{M,m}^2} = p \cdot P_{\text{STC}} \cdot \eta_{\text{M}} \qquad (8)$$

Im Folgenden werden Werte von 0,9 €/W für p [9] und 15 % für η_{M} angenommen.

Die Kosten pro Kilowattstunde der Batterie K_{Sp} können mit ca. 150 € angenähert werden [10]. C ist die täglich benötigte Energie an einem Wintertag. Zusätzlich sind der WR mit seinen Kosten K_{WR} von ca. 0,75 €/W_{AC} und seiner Leistung P_{WR} sowie der Laderegler mit seinen Kosten K_{LR} von ca. 0,7 €/W_{M} (Euro pro Watt Leistung des PV-Moduls) und seiner Leistung P_{LR} miteinzubeziehen. A_{M} beschreibt die Fläche der Anlage. Die Gesamtkosten berechnen sich daher nach Gleichung (9).

$$K_{\text{ges}} = K_{\text{M,m}^2} \cdot A_{\text{M}} + K_{\text{Sp}} \cdot C + K_{\text{WR}} \cdot P_{\text{WR}} + K_{\text{LR}} \cdot P_{\text{LR}} \qquad (9)$$

Die folgend in **Tabelle 8** ermittelten Kosten gelten jeweils für Anlagen, die den Flächenbedarf des Winters decken.

Tabelle 8 Kostenverteilungen für die verschiedenen Szenarien. Angenommene Leistung der WR: Szenario 1: 700 W; Szenario 2: 160 W; Szenario 3: 3.150 W; Szenario 4: 7.785 W. Angaben in Euro. Sz.: Szenario.

Sz.	Posten	Region		
		NW	**NO**	**S**
1	PV-Anlage	1.584	1.257	882
	Batterie	167	167	167
	WR	525	525	525
	Laderegler	1.120	889	624
	Gesamt	**3.396**	**2.838**	**2.198**
2	PV-Anlage	3.300	2.614	1.829
	Batterie	347	347	347
	WR	120	120	120
	Laderegler	2.333	1.848	1.294
	Gesamt	**6.100**	**4.929**	**3.590**
3	PV-Anlage	11.435	9.066	6.338
	Batterie	1.201	1.201	1.201
	WR	2.363	2.363	2.363
	Laderegler	8.085	6.410	4.481
	Gesamt	**23.084**	**19.040**	**14.383**
4	PV-Anlage	159.430	126.597	88.373
	Batterie	16.755	16.755	16.755
	WR	5.839	5.839	5.839
	Laderegler	112.728	89.513	62.486
	Gesamt	**294.752**	**238.704**	**173.453**

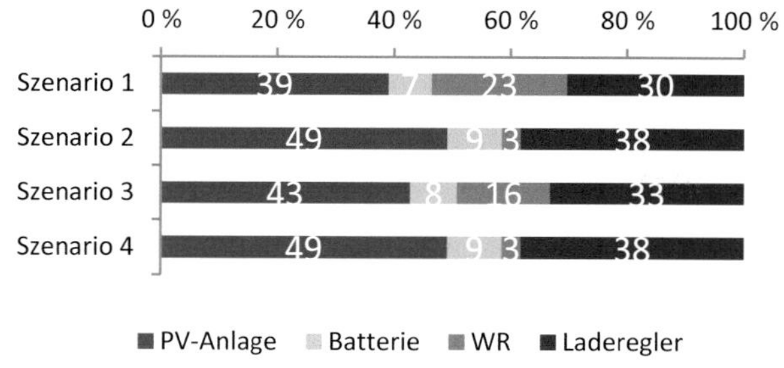

Abbildung 2 Anteil der veranschlagten Kosten bei der Planung einer PV-Anlage im südlichen Sektor. Angaben in Prozent.

Der prozentuale Anteil der Posten im Süden ist für die verschiedenen Szenarien in **Abbildung 2** dargestellt.

Da im Nordwesten die PV-Fläche für die gleiche Energiemenge größer sein muss, ist der Kostenanteil der PV-Anlage sowie des Ladereglers an den Gesamtkosten höher (maximal 6,4 % respektive 5,0 % in Szenario 1). Im Süden steigen die Anteile von Batterie und WR (3,3 % in den Szenarien 2 und 4 bzw. 7,1 % in Szenario 1).

4 Zusammenfassung und Ausblick

Die Analyse des Realisierungspotentials der unterschiedlichen Szenarien konnte aufzeigen, dass PV-Systeme unter Verwendung eines Energiespeichers theoretisch in der Lage sind, eine NSV sicherzustellen. Durch die Modellierung der Verbraucher sowie des Speichers kann eine Technologie-unabhängige NSV geplant und projektiert werden, welche die Bedürfnisse der Bevölkerung berücksichtigt. Die vorgestellte Methodik ist auf andere Szenarien sowie Verbraucher übertragbar. Das abstrahierte Speichermodell ermöglicht die flexible Anpassung der Modelle an andere Technologien.

Es zeigt sich, dass PV-Systeme unter bestimmten Rahmenbedingungen für NSV-Systeme geeignet sind. Insbesondere in Sommermonaten und im Süden kann eine durchgehende Versorgung mit vertretbarem Platzbedarf realisiert werden. Für kleine Siedlungen besteht je nach geographischen Gegebenheiten ein hohes Potential, auch zur Versorgung in Wintermonaten. Die Ergebnisse zeigen jedoch auch, dass PV-Systeme aufgrund mangelnder Solarstrahlung in Nordwesten Deutschlands sowie im Winter nicht für alle Szenarien geeignet sind. Zudem ist es in vielen Fällen aus Platzgründen nicht möglich eine entsprechende Anlage zu installieren. Dies gilt insbesondere in Szenario 1 (Alleinerziehende Person mit zwei Kindern).

Finanziell betrachtet ist die Anschaffung einer NSV für den privaten Haushalt mit einer hohen Investition verbunden. Hierbei gilt es abzuwägen, ob eine PV-Anlage mit den gezeigten Anschaffungskosten beispielsweise für eine alleinerziehende Person im Nordwesten Deutschlands überhaupt finanzierbar ist.

Als Alternativen zu PV-Technologien sind für Kleinst- und Kleinverbraucher beispielsweise handliche Kurbelgeneratoren denkbar. Mittelgroße Verbraucher können je nach geographischen Gegebenheiten neben PV-NSV-Systemen auch mit Kleinst-Windkraftanlagen oder geeigneten Batterietechnologien versorgt werden. Größere Lösungen könnten mit Brennstoffzellentechnologie realisiert werden. Eine Untersuchung dieser Alternativen ist Teil des Forschungsvorhabens.

5 Danksagung

Die Autoren danken dem Bundesministerium für Bevölkerungsschutz und Katastrophenhilfe für die gute Zusammenarbeit und Unterstützung.

6 Literatur

[1] Schlandt, J.: „Schneller Kollaps". BIZZ energy today, Ausgabe Februar (2014)

[2] Brauner, G., Pöppl, G., Tiefgraber, D.: Verbraucher als virtuelles Kraftwerk: Potentiale für Demand Side Management in Österreich im Hinblick auf die Integration von Windenergie. Wien (2006)

[3] Stamminger, R.: Synergy Potential of Smart Appliances: D2.3 of WP 2 from the Smart-A project. Universität Bonn (2008)

[4] Statistisches Bundesamt: Bauen und Wohnen: Mikrozensus – Zusatzerhebung 2010. Fachserie 5, Heft 1. Wiesbaden (2012)

[5] Wesselak, V., Voswinckel, S.: Photovoltaik: Wie Sonne zu Strom wird. Springer, Berlin (2012)

[6] Häberlein, H.: Photovoltaik: Strom aus Sonnenlicht für Verbundnetz und Inselanlagen. Westermann Druck Zwickau GmbH, Zwickau (2007)

[7] Morsch, R.: PV-Inselanlagen – Komponenten und Auslegung. Berlin (2001)

[8] Deutscher Wetterdienst: Globalstrahlung in der Bundesrepublik Deutschland: Mittlere Jahressummen, Zeitraum: 1981 - 2010.
http://www.dwd.de/bvbw/generator/DWDWWW/Content/Oeffentlichkeit/KU/KU1/KU12/Klimagutachten/Solarenergie/Straka__Mittel__8110,templateId=raw,property=publicationFile.pdf/Straka_Mittel_8110.pdf (2012). Zugegriffen: 3. November 2014

[9] Wirkungsgrad von Modulen auf dem Markt. Photon (2012)

[10] Fuchs, G., Lunz, B., Leuthold, M., Sauer, D. U.: Technologischer Überblick zur Speicherung von Elektrizität: Überblick zum Potenzial und zu Perspektiven des Einsatzes elektrischer Speichertechnologien. textbild digitaldruck und Verlag, Berlin (2012)

Netzintegration

Effiziente Bewertung des Anschlusspotenzials von Niederspannungsnetzen für dezentrale Erzeugungsanlagen: Klassifikation mit Methoden der Computational Intelligence

Dipl.-Ing. Dipl.-Wirtsch.-Ing. M.Sc. Sebastian Breker,
EnergieNetz Mitte GmbH, 34131 Kassel, Deutschland, Sebastian.Breker@EnergieNetz-Mitte.de
Prof. Dr. rer. nat. Bernhard Sick,
FG Intelligent Embedded Systems, 34131 Kassel, Deutschland, bsick@uni-kassel.de

Kurzfassung

Die Entwicklung optimierter Verteilnetze erfordert einen hohen Bedarf an Investitionsmitteln. Insbesondere auf der Niederspannungsebene gestaltet sich die Steuerung der im regulierten Netzgeschäft verfügbaren Investitionsmittel zur Verstärkung von Netzen auf Grund ihrer sehr hohen Zahl und historisch in Abhängigkeit von lokalen Gegebenheiten gewachsenen Netzstrukturen schwierig. Durch eine Klassifikation der Netze hinsichtlich ihres Aufnahmevermögens für dezentrale Erzeugungsanlagen, wie in diesem Artikel vorgestellt, kann die Menge der zu beobachtenden Netze automatisiert auf eine geringere Zahl „schwacher" Netze reduziert werden. Weiterhin wird eine strukturierte Möglichkeit zur Auswahl von relevanten NS-Netzen für detaillierte Untersuchungen geschaffen (z. B. Smart Grids oder Flexibilität im Verteilnetz: gelbe Ampelphase). Es wird anhand von 300 realen, ländlichen und vorstädtischen Netzen gezeigt, dass durch Einsatz von Support Vector Machines (SVM) die Klassifikation effizient gestaltet werden kann, da nur eine Teilmenge der zu klassifizierenden Netze bewertet werden muss. Der restliche (große) Teil der Netze kann dann anhand von Merkmalen der Netze automatisch bewertet werden. Der Anteil falsch klassifizierter Netze ist mit 18 % gering.

1 Einleitung

1.1 Klassifikation von Niederspannungsnetzen im Kontext der Herausforderungen in der Verteilnetzplanung

Zur Erreichung energiepolitischer Zielsetzungen werden optimierte Infrastrukturen zur Stromversorgung erforderlich, deren Aufbau von einem hohen Bedarf an Investitionsmitteln begleitet sein wird. Auf der Niederspannungs-(NS-)ebene gestaltet sich die Steuerung der im regulierten Netzgeschäft verfügbaren Investitionsmittel zur Verstärkung von Netzen auf Grund ihrer sehr hohen Zahl und der historisch in Abhängigkeit von lokalen Gegebenheiten gewachsenen Netzstrukturen schwierig. Eine Klassifikation der Netze hinsichtlich ihres Aufnahmevermögens für dezentrale Erzeugungsanlagen (DEA) in geordnete Klassen (z. B. von 1 „sehr schwach" bis 5 „sehr stark") kann diese Aufgabenstellung erleichtern. Die Beobachtung von Betriebsmittelbelastungen und Spannungsniveaus bei fortschreitendem DEA-Zubau reduziert sich auf eine ausgewählte Menge an (schwachen) NS-Netzen. Weiterhin wird eine strukturierte Möglichkeit zur Auswahl von relevanten NS-Netzen für detaillierte Untersuchungen geschaffen (z. B. Smart Grids oder Flexibilität im Verteilnetz).

Allerdings ist auch die Klassifikation von NS-Netzen aufwendig (z. B. hohe Kosten für Expertenbefragungen). Der Aufwand kann durch den Einsatz von Methoden der Computational Intelligence (CI), wie z. B. hier von Support Vector Machines (SVM), stark reduziert werden. Zum Training von SVM wird nur eine Teilmenge der Netze benötigt. Der restliche (große) Teil der Netze kann anhand ihrer Merkmale (z. B. Anz. Hausanschlüsse, Anz. Trafostationen) mit der SVM automatisch (ohne z. B. eine Expertenbefragung für diese Netze) bewertet werden, da diese das Bewertungsverhalten der Beispielbewertungen in der zum Training genutzten Teilmenge erlernt hat.

1.2 Stand der Technik & Innovationsgehalt

In den letzten Jahren wurden einige Arbeiten zum Einsatz von CI in Planung und Betrieb von Verteilnetzen untersucht. Die Anwendung konzentrierte sich auf die Bereiche Lastvorhersage [1, 2, 3], Netzbetrieb und -bewertung [4, 5, 6], Beurteilung der Systemzuverlässigkeit und -sicherheit sowie Klassifikation von Betriebszuständen und Systemfehlern im Netz [7, 8, 9]. Weiterhin wurden einige initale Arbeiten zur Einbindung von CI bei der Entwicklung von Smart Grids veröffentlicht [10, 11, 12].

Die Ergebnisse dieses Artikels lassen sich in den Bereich des Einsatzes von CI zur automatischen Bewertung von Netzstrukturen einordnen. Es werden erstmalig effiziente Methoden zur Bewertung gesamter Netztopologien und nicht nur einzelner Betriebsmittel wie in [4, 6] vorgestellt. Um den Anwendungsbezug zu untermauern, wurde im Gegensatz zu anderen Studien eine Evaluation anhand von 300 realen, ländlichen und vorstädtischen NS-Netzen vorgenommen.

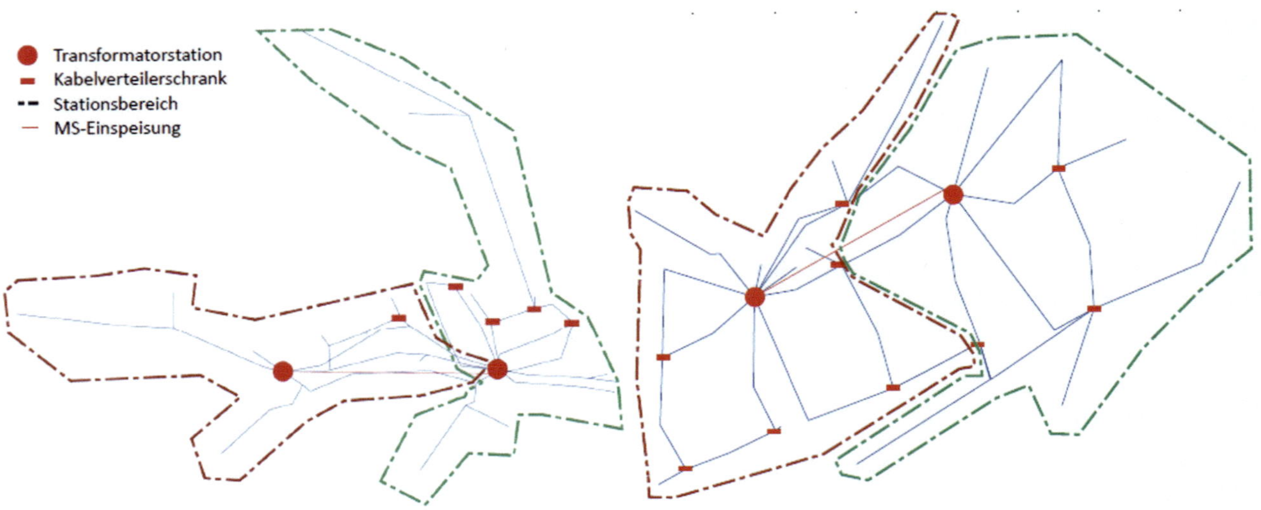

Bild 1 Struktur zweier realer, ländlicher Niederspannungsnetze, Netz I (links) und Netz II (rechts) [13].

2 Klassifikation von Niederspannungsnetzen

2.1 Merkmale zur Charakterisierung der Netze

Die Erhebung von netzspezifischen Merkmalen, die elektrotechnische Größen von NS-Netzen berücksichtigen, benötigt Anwendungswissen und kann aufwendig sein. Vielfach liegen Informationen zur NS-Ebene noch nicht vollständig in digitaler Form vor. Netzspezifische Merkmale bringen jedoch Vorteile mit sich, wenn NS-Netze z. B. anhand von menschlichen Experten bewertet werden sollen. In Zusammenarbeit mit fünf Experten aus der Verteilnetzplanung wurden folgende netzspezifische Merkmale zur Charakterisierung von NS-Netzen identifiziert: 1) Anz. Hausanschlüsse, 2) Anz. Trafostationen, 3) Summe Trafonennleistungen, 4) Anz. Kabelverteilerschränke (KVS), 5) Summe Leitungslänge, 6) Summe vermaschte Leitungslänge[1], 7) Vermaschungsgrad[2], 8) Anteil NAYY 4x150mm²[3], 9) max. Stationsradius und 10) durchschn. Stationsradius[4].

Tabelle 1 Beispiele zu den netzspezifischen Merkmalen für zwei reale, ländliche NS-Netze [13].

Netzspezifisches Merkmal	Netz I	Netz II
1 Anz. Hausanschlüsse [1]	92	100
2 Anz. Trafostationen [1]	2	2
7 Vermaschungsgrad [%]	24	58
9 Max. Stationsradius [m]	1082	354
10 Durchschn. Stationsradius [m]	840	344

Um die Relevanz dieser Merkmale hinsichtlich der Charakterisierung von NS-Netzen zu veranschaulichen, erfolgt nun eine Interpretation dieser Merkmale für zwei reale NS-Netze. Die Struktur dieser Musternetze ist in **Bild 1** gezeigt. Darüber hinaus sind in Tabelle 1 einige Daten dieser Musternetze ausgewiesen. Es ist ersichtlich, dass sich beide Netze nicht nur stark in ihren Strukturen, sondern auch in den Werten ihrer Merkmale unterscheiden. Wie Merkmal 2 zeigt, bestehen beide Netze aus jeweils zwei Stationsbereichen und versorgen mit jeweils ungefähr 100 Hausanschlüssen eine ähnliche Anzahl an Endkunden (Merkmal 1), aber ein Vergleich der Merkmale 9 und 10 beider Netze lässt erkennen, dass die durch Netz I versorgte Fläche eine ungefähr um den Faktor 2,5-mal größere räumliche Ausdehnung aufweist als die von Netz II. Die galvanische Auftrennung der beiden Netze in jeweils zwei Stationsbereiche erfolgt in der Praxis in den in **Bild 1** gezeigten KVS, die sich in der Nähe der parallel verlaufenden Kanten der Polygone um die Stationsbereiche befinden. Hinsichtlich der unterschiedlichen räumlichen Dimensionierung der versorgten Flächen tritt – bei gleicher Anzahl von Stationsbereichen – im Netz II ein höherer Vermaschungsgrad als in Netz I auf. Dies wird direkt durch die beiden Ausprägungen des Merkmals 7 deutlich. Die Begründung für die unterschiedliche Vermaschung liegt in der ökonomischen Effizienz bei der Versorgung der Endkunden. Im Gegensatz zu Netz II besteht das Netz I nur aus einem vermaschten Bereich in dessen Zentrum, wohingegen die in den Außenbereichen liegenden Endkunden durch strahlenförmige Leitungszüge versorgt werden. Die Interpretation der netzspezifischen Merkmale der beiden Musternetze zeigt, dass Netz II eine wesentlich robustere Struktur zur Integration von DEA aufweist als Netz I [13].

Eine effizientere Erhebung einer Datengrundlage für eine große Anzahl von Netzen, die kein Anwendungswissen erfordert, kann durch Interpretation der Strukturen eines NS-Netzes als Graph durchgeführt werden. Auf Basis der Knoten und Kanten der NS-Netze wurden weitere 22 Merkmale identifiziert, die aus Basismerkmalen, pfadbasierten Merkmalen und spektralen Eigenschaften der Graphen bestehen, z. B. durchschn. Knotengrad, proz. Anteil zentraler Knoten, spektraler Radius usw. (siehe z. B. [14]).

2.2 Simulations- und expertenbasierte Klassifikation

Zur Generierung von Beispielbewertungen zum späteren Training von SVM werden hier zwei Ansätze verwendet. Der erste Ansatz basiert auf stochastischen Simulationen. Er ist an den normativen Stand der Technik zur Bewertung der Aufnahmekapazität von NS-Netzen für DEA angelehnt, der zur Integration von DEA in der NS-Ebene angewendet wird. Da eine starke Abhängigkeit zwischen der Aufnahmefähigkeit eines NS-Netzes für DEA und der DEA-Konfiguration im Netz (Position Netzanschluss, Nennleistung) existiert, erlaubt der Ansatz die Berücksichtigung vieler zufallsgenerierter DEA-Konfigurationen. Dies führt zu einer differenzierten Bewertung der DEA-Kapazität eines Netzes mit Wahrscheinlichkeiten für die Klassenzuordnung. Das Vorgehen bei der simulationsbasierten Klassifikation der Netze in fünf Klassen (von 1 „sehr schwach" bis 5 „sehr stark") kann in folgende Schritte zusammengefasst werden: 1) stochastische Simulation bezüglich der Aufnahmekapazität für DEA, 2) parametrische stochastische Modellierung der Aufnahmekapazität mit Weibullverteilungen und 3) Klassifikation auf Basis der ermittelten Weibullmodelle [13].

Da die „wahre" Klasse eines NS-Netzes nicht bekannt ist (i. S. einer „ground truth", ein realer DEA-Ausbau zu Testzwecken kann nicht erfolgen) wurde zur umfassenderen Bewertung eines NS-Netzes zusätzlich ein expertenbasierter Ansatz erarbeitet. Um die Zuverlässigkeit zu erhöhen, wurde die Befragung mit fünf Experten aus der Verteilnetzplanung (VNB-Mitarbeiter) durchgeführt. Das Design des Fragebogens orientierte sich an den Messqualitätskriterien „Validität", „Reliabilität" und „Objektivität" [15]. Den Experten wurden zur Klassifikation für jedes Netz die Ausprägungen der netzspezifischen Merkmale aus Abschnitt 2.1, ein Netzplan sowie Zusatzinformationen bereitgestellt (z. B. Netzprototyp für jede Klasse, persönliche Klassifikationsindikatoren) [13].

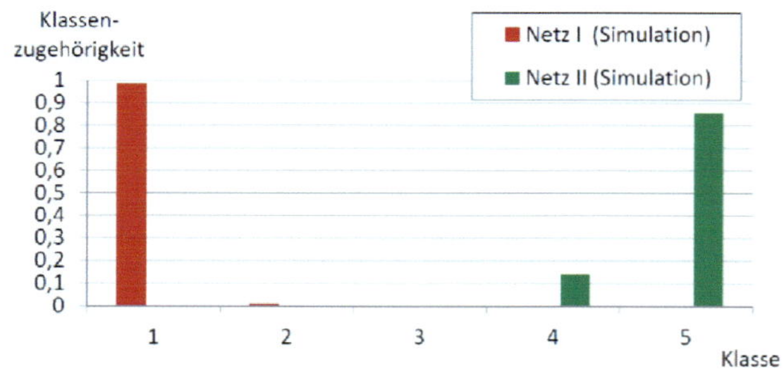

Bild 2 Verteilung der Klassenzugehörigkeiten für zwei reale NS-Netze [13].

Die **Bilder 2** und **3** zeigen, dass beide Ansätze für die beiden Musternetze aus **Bild 1** zu ähnlichen Ergebnissen führen und sich nicht widersprechen. Differenzen resultieren aus unterschiedlichen Schwerpunkten der Ansätze. Anhand eines statistischen Tests kann gezeigt werden, dass die Ergebnisse aller 300 Netze konkordant sind [13].

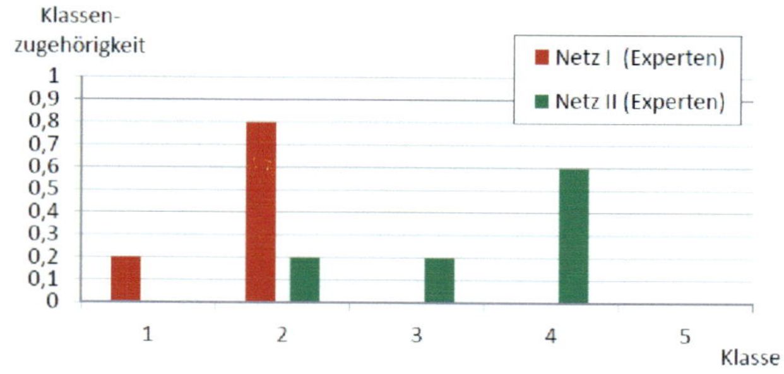

Bild 3 Verteilung der Expertenaussagen für zwei reale NS-Netze [13].

Aus den absoluten Häufigkeiten der Klassifikationsergebnisse für alle 300 NS-Netze lässt sich ableiten, dass ein Großteil der bewerteten Netze eine grundsätzlich robuste Struktur zur Integration von DEA aufweist. Es tritt allerdings jeweils eine geringe absolute Häufigkeit von Netzen der Klasse 5 auf, die auch zeigt, dass reale NS-Netze innerhalb der letzten Jahrzehnte zur ökomischen Versorgung von Kunden optimiert worden sind und zur Integration einer hohen DEA-Leistung verstärkt werden müssten. **Bild 4** zeigt beispielhaft die Verteilung der Klassifikationsergebnisse für die Experten 1, 4 und 5.

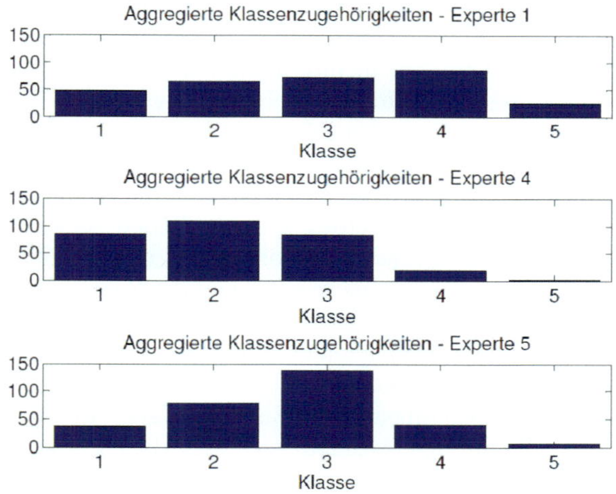

Bild 4 Aggregierte Verteilung der Klassenzugehörigkeiten für zwei reale NS-Netze bei Klassifikation durch verschiedene Experten.

Da in beiden Ansätzen unterschiedliche Schwerpunkte gelegt wurden und die Entscheidung eines Experten für eine ordinale Klasse auf Grund subjektiver Einflüsse (z. B. unterschiedlich strenge Bewertung) mehr oder weniger unsicherheitsbehaftet ist, sollte eine finale Klassifikationsentscheidung in einer Anwendung nicht anhand eines einzelnen Urteils getroffen werden. Die genutzten verschiedenen Urteile können dabei, wie in **Bild 2** und **3** gezeigt, nicht übereinstimmen. Um eine umfassendere Gesamtaussage zu erhalten, wurden die einzelnen Klassifikationsergebnisse der Ansätze mit einer neuen Kombinationsregel fusioniert, die für die untersuchte Anwendung drei bekannten Kombinationsregeln (z. B. Kombinationsregel nach Dempster-Shafer [16]) überlegen ist, wenn nur eine geringe Anzahl von Aussagen verfügbar ist (<<10).

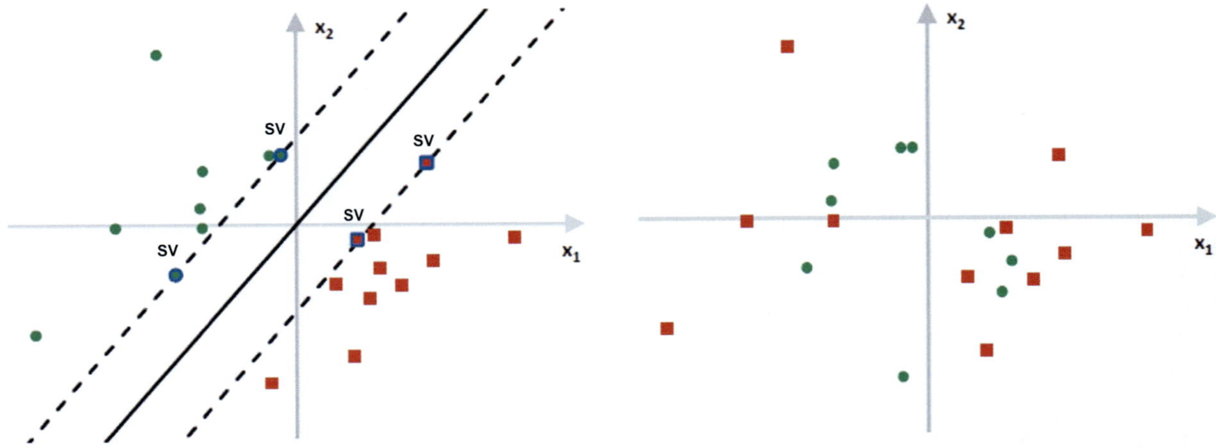

Bild 5 Schematische Darstellung eines linear trennbaren (links) und eines nicht linear trennbaren (rechts) Klassifikationsproblems mit Beispieldatenpunken von zwei Klassen (Kreis, Quadrat) in einem zweidimensionalen Eingaberaum.

3 Effizienzsteigerung durch Support Vector Machines

3.1 Grundlagen Support Vector Machines

Grundlegend können SVM zur Klassifikation mit Hilfe eine Hyperebene zur Separierung von zwei Klassen in einem (möglicherweise mehrdimensionalen) Eingaberaum genutzt werden. In **Bild 5** sind ein im Eingaberaum linear trennbares und ein im Eingaberaum nicht linear trennbares Klassifikationsproblem schematisch gegenübergestellt. Für Klassifikationsprobleme, die nicht im Eingaberaum linear getrennt werden können, finden SVM eine Lösung durch nichtlineares Abbilden aus dem originalen Eingaberaum in einen hochdimensionalen Raum, der als Featureraum bezeichnet wird. Die optimale Hyperebene zur Trennung der Datenpunkte wird dann im Featureraum gesucht. Dabei wird diejenige Hyperebene als optimal erachtet, die einen maximalen Rand aufweist, wobei der Begriff „Rand" den Abstand zwischen Hyperebene und dem nächstliegenden Datenpunkt im Featureraum beschreibt, dem Support Vector (SV). Die Abbildung wird i. d. R. mit einer nichtlinearen Kernelfunktion realisiert, die im Prinzip die Ähnlichkeit zwischen einer Eingabe an der SVM und einem SV misst. Häufig werden zur Abbildung Gaußkernel oder polynomiale Kernel genutzt. Die üblicherweise eingesetzten Varianten, *C*-SVM und ν-SVM, erlauben und minimieren Fehlklassifikationen. Über die Parameter *C* bzw. wird die Bestrafung von Fehlklassifikationen beim Training der SVM gesteuert. Eine Verallgemeinerung auf Mehrklassenklassifikationsprobleme erfolgt in der Literatur anhand der Ansätze „One-against-all SVM", „Pairwise SVM", "Error-correcting Output Code SVM" und „All-at-once SVM". Hier wurde der Ansatz „Pairwise SVM" genutzt. Darin wird das Mehrklassenklassifikationsproblem jeweils in mehrere Zweiklassenklassifikationsprobleme (für C Klassen in C(C-1)/2 Zweiklassenklassifikationsprobleme) zerlegt. Diese Zweiklassenklassifikationsprobleme werden zunächst wie oben beschrieben gelöst. Im Anschluss wird auf Basis der Ausgaben der paarweisen Klassifikatoren die Ausgabe der Mehrklassen-SVM bestimmt [17, 18].

Zur Modellierung der SVM wurde hier die LIBSVM-Standardbibliothek mit einem Gaußkernel verwendet [19].

3.2 Anwendung Support Vector Machines

Um für weitere NS-Netze auf eine aufwendige Bewertung mit den Klassifikationsansätzen aus Abschnitt 2.2 sowie die anschließende Kombination der Ergebnisse verzichten zu können und so eine Effizienzsteigerung zu generieren, wurden einige SVM-Klassifikationskonzepte empirisch untersucht. Die auf die in Abschnitt 2 beschriebene Weise erhobenen Daten und Klassifikationsergebnisse wurden dabei als Beispieldaten zum Training und zur Evaluation der SVM-Klassifikatoren genutzt. Das Ziel bestand darin, eine bestmögliche Klassifikationsprognose hinsichtlich der kombinierten Klassifikationsergebnisse (kombinierte Label) zu erreichen. In diesem Zuge wurde ebenfalls untersucht, ob unter Verwendung der graphspezifischen Merkmale aus Abschnitt 2.1 ähnlich gute Prognoseergebnisse erreicht werden können. Unter dieser Voraussetzung könnte eine weitere Effizienzsteigerung bei der Klassifikation bewirkt werden, da diese Merkmale leicht automatisiert und ohne Anwendungswissen für eine große Anzahl von NS-Netzen erhoben werden können.

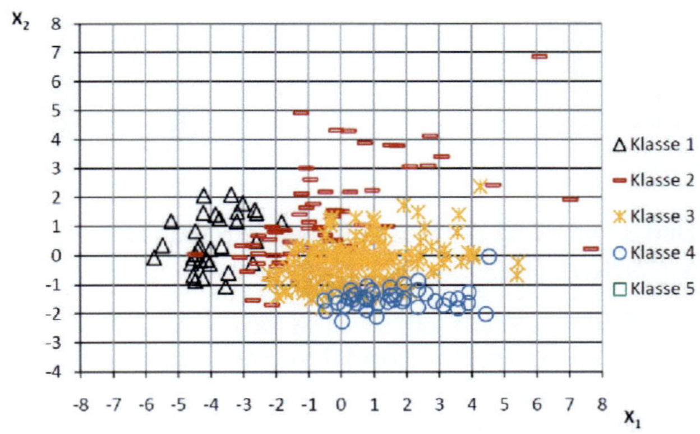

Bild 6 Darstellung der kombinierten Label auf den beiden Hauptachsen aller *z*-transformierten netzspezifischen Merkmale.

Vor der Anwendung der SVM wurden die Merkmale jeweils mit einer z-Tranformation skaliert [20]. **Bild 6** zeigt die kombinierten Label projiziert auf die ersten beiden Hauptachsen, die mit einer Hauptachsentransformation aller netzspezifischen Merkmale ermittelt wurden. Auf Basis der durch die Hauptachsentransformation vorgenommenen Dimensionsreduktion lässt sich ein (wenn auch auf Grund des mit der Dimensionsreduktion einhergehenden Informationsverlustes nicht vollständiger) visueller Eindruck über die Verteilung der Daten gewinnen. Auffällig ist, dass die Kombination der einzelnen Klassifikationsergebnisse keine Datenpunkte der Klasse 5 zuordnet. Die Ursache liegt in der bereits bei den einzelnen Labeln auftretenden geringen Anzahl von NS-Netzen der Klasse 5 (vgl. Abschnitt 2.2). Weiterhin ist ersichtlich, dass die Anwendung der neuen Kombinationsregel bereits in zwei Dimensionen zu einer guten Separierbarkeit der kombinierten Label führt (hier z. B. besser als die der Kombinationsregel nach Dempster-Shafer). Eine gute Separierbarkeit der Label ist hinsichtlich einer späteren hohen Generalisierungsfähigkeit eines SVM-Klassifikators wichtig. Aufgrund der visuell guten Separierbarkeit der kombinierten Label in **Bild 6** lässt sich bereits eine gute Prognostizierbarkeit dieser Label mit einer SVM vermuten.

Zur Abschätzung guter Parameter für die SVM, und um zusätzlich eine Überanpassung auf bestimmte Datenpunkte bzw. Klassen zu vermeiden, wurde in den Experimenten zunächst jeweils eine stratifizierte Kreuzvalidierung durchgeführt. Für die Gittersuche wurde eine zusätzliche unterlagerte Kreuzvalidierung durchgeführt, um eine robuste Parametersuche zu erhalten, die lediglich die jeweiligen Trainingsdaten der übergeordneten Kreuzvalidierung nutzt. Eine Besonderheit bei der Beurteilung der Klassifikationsgüte liegt darin, dass die Klassen eine Ordnung aufweisen, in einer SVM jedoch als nominale Klassen interpretiert und verarbeitet werden. Daher wurde die Optimierung der Parameter in der Gittersuche nicht nur anhand des Klassifikationsfehlers, sondern anhand der gewichteten Übereinstimmung (Cohen's Kappa-Statistik [21]) zwischen der SVM-Ausgabe und den korrekten Klassen durchgeführt. Die auf diese Weise bestimmten, optimalen Parameterkombinationen weisen insgesamt nur marginale Unterschiede auf. Daher konnte für die Experimente die Parameterkombination $C = 10$ und $\gamma = 0{,}03$ ausgewählt und für alle Experimente genutzt werden. Damit die Ergebnisse statistisch aussagekräftig sind, wurde mit diesen Parametern in allen Experimenten jeweils zehnmal eine 3-fache Kreuzvalidierung durchgeführt. Die zehn Wiederholungen unterscheiden sich in der Zusammensetzung der jeweils per Zufall gezogenen, stratifizierten Mengen, die zur Kreuzvalidierung genutzt wurden. Ausgewertet wurde auf dieser Basis der mittlere Klassifikationsfehler $\mu(e)$ (mittlerer Anteil falsch klassifizierter Netze).

In den Experimenten wurden drei verschiedene SVM-Klassifikationskonzepte untersucht. Die besten Prognoseergebnisse stellten sich bei direktem Training der SVM mit den fusionierten Labeln ein. Eine schematische Darstellung dieses Klassifikationskonzeptes zeigt **Bild 7**.

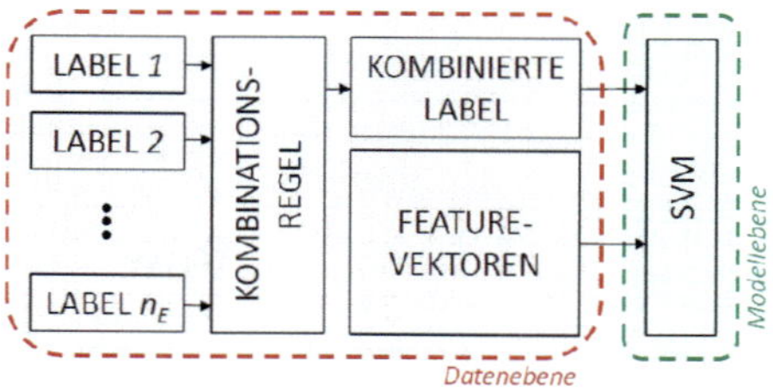

Bild 7 Angewandtes SVM-Klassifikationskonzept.

Die n_E vielen einzelnen Label für ein Netz, die mit den Klassifikationsansätzen aus Abschnitt 2.2 bestimmt worden sind, werden bereits auf der Datenebene mit einer Kombinationsregel vor dem Training der SVM kombiniert. Anschließend werden die kombinierten Label zusammen mit den Merkmalen aus Abschnitt 2.1 genutzt, um die SVM zu trainieren und zu evaluieren. Zur Analyse des Einflusses der verwendeten Merkmale auf das Prognoseergebnis wurde die SVM jeweils mit den Merkmalen aus Abschnitt 2.1 trainiert. Zusätzlich wurde die SVM auch mit im Rahmen der simulationsbasierten Klassifikation erhobenen weiteren Merkmalen (simulationsspezifische Merkmale, z. B. Parameter der dreiparametrigen Weibullverteilung) trainiert.

Bild 8 zeigt die jeweils unter Verwendung der Merkmale resultierenden mittleren Klassifikationsfehler. Das beste Prognoseergebnis wird beim Training der SVM mit den netzspezifischen Merkmalen erreicht (18 % falsch klassifizierte Netze). Bei Verwendung der graphspezifischen Merkmale lassen sich ähnlich gute Ergebnisse erzielen (20 % falsch klassifizierte Netze). Das Training der SVM mit den simulationsspezifischen Merkmalen führt zu signifikant schlechteren Prognoseergebnissen.

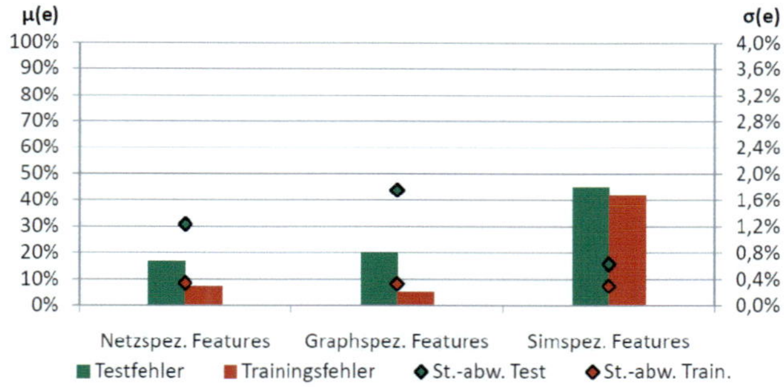

Bild 8 Klassifikationsfehler beim Training der SVM mit verschiedenen Merkmalsmengen.

3.3 Kombination verschiedener Merkmale

In einem zusätzlichen Experiment wurde analysiert, ob sich das Prognoseergebnis durch kombinierte Verwendung von Merkmalen weiter optimieren lässt. Dazu wurde zunächst mit zwei Methoden der Merkmalsselektion (Relief-F-Algorithmus, χ^2-Methode [22]) eine Rangfolge aller Merkmale hinsichtlich ihrer Relevanz bei der Klassifikation erstellt. Auf dieser Grundlage wurde die Anzahl der zum Training der SVM verwendeten Merkmale sukzessive in Schritten zu je fünf Merkmalen bis auf zehn Merkmale reduziert.

Bild 9 zeigt, dass sich die Prognoseergebnisse nur gering unterscheiden. Gegenüber der Nutzung der graphspezifischen Merkmale wird bei kombinierter Verwendung von

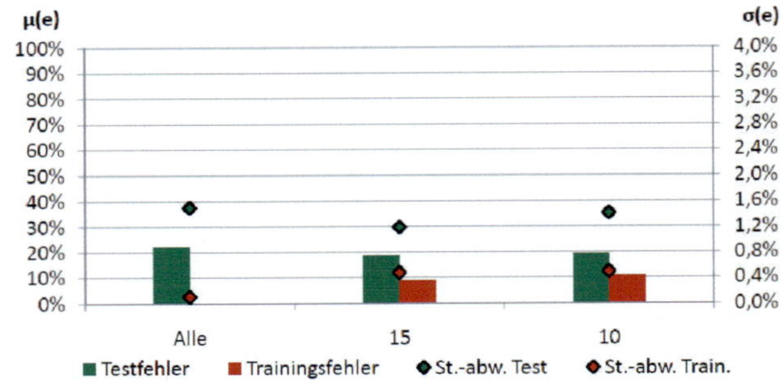

Bild 9 Klassifikationsfehler beim Training der SVM mit verschiedenen Merkmalsanzahlen.

Merkmalen ein leicht verbessertes Ergebnis erzielt (19 % falsch klassifizierte Netze). Insgesamt variiert das Prognoseergebnis je nach verwendeten Merkmalen nur geringfügig zwischen 18 % und 20 % falsch klassifizierter Netze. Das beste Ergebnis wird mit den netzspezifischen Merkmalen erreicht.

4 Fazit und Ausblick

Es wurde an einer hohen Zahl (300) realer NS-Netze gezeigt, dass eine Klassifikation hinsichtlich ihres Aufnahmevermögens für DEA effizient mit SVM durchgeführt werden kann. Die Zahl der Fehlklassifikationen stellt sich in den Experimenten als gering heraus. Der alternative Einsatz von graphspezifischen Merkmalen der Netze beim Training der SVM hat nur einen geringen Einfluss auf das Prognoseergebnis. Vor diesem Hintergrund kann die Klassifikation noch effizienter durchgeführt werden, da graphspezifische Merkmale leicht, automatisiert und ohne Anwendungswissen erhoben werden können.

Die Ergebnisse zeigen, dass CI ein hohes Potenzial zur Verbesserung der Netzplanung und des Netzbetriebs haben kann. Unsere Forschungsarbeit konzentriert sich auf weitere Pilotkonzepte zum Einsatz von CI im Verteilnetz, z. B. zur Verbesserung der Spannungsregelung.

5 Literatur

[1] Li, B.; Gandahar, S.; Cheng, S.; Verma P. K.: Predicting User Comfort Level Using Machine Learning for Smart Grids. *IEEE PES Inn. Smart Grid Techn. (2011)*, pp. 1-6.

[2] Yuanzhe, C.; Quing, X.; Chengquiang, W.; Fangcheng, L.: Short-Term Load Forecasting for City Holidays Based on Genetic Support Vector Machines. *Int. Conf. on Electrical and Control Engineering (2011)*, pp. 3144-3147.

[3] Chandler, S. A.; Hughes, J. G.: Smart Grid Distribution Prediction and Control Using Computational Intelligence. *IEEE Conf. on Techn. for Sustainability (2013)*, pp. 86-89.

[4] Rudin, C.; Waltz, D.; Anderson, R. N.; Boulanger, A.: Machine Learning for the New York City Power Grid. *IEEE Trans. on Pattern Analysis and Machine Intelligence 34 (2012)*, No. 2, pp. 328–345.

[5] Stark, M.; Krost, G.: Design Tool for Isolated Micro-Grids Based on Methods of Computational Intelligence. *Int. Conf. on Intelligent System Application to Power Systems (2011)*, pp. 1–6.

[6] Gross, P.; Salleb-Aouissi, A.; Dutta, H.; Boulanger, A.: Ranking Electrical Feeders of the New York Power Grid. *Int. Conf. on Machine Learning and Applications (2009)*, pp. 359–365.

[7] Pisica, I.; Eremia, M.: Making Smart Grids Smarter by Using Machine Learning. *Proc. of Int. Universities' Power Engineering Conf. (2011)*, pp. 1–5.

[8] Shahid, N.; Aleem, S. A.; Naqvi, I. H.; Zaffar, N.: Support Vector Machine Based Fault Detection & Classification in Smart Grids. *IEEE Globecom Workshops (2012)*, pp. 1526–1531.

[9] Haruna, Y.; Bakare, G. A.; Aliyu, U. O.: Adaptive Static Load Shedding Schemes for Nigerian Grid System Based on Computational Intelligence Techniques. *Power Systems Conf. (2014)*, S. 1–6.

[10] Haibo, H.: Toward a Smart Grid: Integration of Computational Intelligence Into Power Grid. *Int. Joint Conf. on Neural Networks (2010)*, pp. 1–6.

[11] Cartes, D.; Chow, J. H.; McCaugherty, D.; Widergren, S.: The IEEE Computer Society Smart Grid Vision Project Opens Opportunites for Computational Intelligence. *IEEE Conf. on Evolving and Adaptive Intelligent Systems (2013)*, pp. 144–150.

[12] Byung-Gook, K.; Yu, Z.; Schaar, M. van d.; Lee, L. J.: Dynamic Pricing for Smart Grid with Reinforcement Learning. *IEEE Conf. on Computer Communications Workshops (2014)*, pp. 640–645.

[13] Breker, S.; Claudi, A.; Sick, B.: Capacity of Low-Voltage Grids for Distributed Generation: Classification by Means of Stochastic Simulations. *IEEE Trans. on Power Systems 30 (2015)*, No. 2, pp. 689–700.

[14] Li, G.; Semerci, M.; Yener, B.; Zaki, M. J.: Graph Classification via Topological and Label Attributes. *MLG, San Diego (2011)*, pp. 1–9.

[15] Bortz, J.; Lienert, G. A.; Boehnke, K.: Verteilungsfreie Methoden in der Biostatistik. *Springer Heidelberg (2008).*

[16] Shafer, G.: A Mathematical Theory of Evidence. *Princeton University Press (1976).*

[17] Abe, S.: Support Vector Machines for Pattern Classification. *Adv. in Pattern Recognition, Springer London (2010).*

[18] Steinwart, I.; Christmann, A.: Support Vector Machines. *Information Science and Statistics, Springer London (2008).*

[19] Chang, C.-C.; Lin, C.-J.: LIBSVM: A Library for Support Vector Machines. *ACM Trans. on Intelligent Systems and Technology 2 (2011)*, pp. 1-27. *Software available at http://www.csie.ntu.edu.tw/~cjlin/libsvm (last access: 02/13/2015)*

[20] Chang, C.-C.; Lin, C.-J.: A Practical Guide to Support Vector Classification. *Technical Report, National Taiwan University. (2010), pp. 1–16.*

[21] Bortz, J.; Lienert, G. A.; Boehnke, K.: Verteilungsfreie Methoden in der Biostatistik. *Springer Heidelberg (2008).*

[22] Liu, H.; Motoda, H.: Computational Methods of Feature Selection. *Chapman & Hall (2008).*

Bewertung verschiedener Spannungsregelungskonzepte in einem einspeisegeprägten Mittelspannungsnetz und Ausblick auf neue Konzepte basierend auf Methoden der Computational Intelligence

M.Eng. Johannes Rudolph, EnergieNetz Mitte GmbH, Kassel, Deutschland,
Johannes.Rudolph2@EnergieNetz-Mitte.de
Dr.-Ing. Sebastian Breker, EnergieNetz Mitte GmbH, Kassel, Deutschland,
Sebastian.Breker@EnergieNetz-Mitte.de
Prof. Dr. rer. nat. Bernhard Sick, FG Intelligent Embedded Systems, Universität Kassel, Kassel, Deutschland,
bsick@uni-kassel.de

Kurzfassung

In Mittelspannungsnetzen (MS-Netzen) wird die Netzspannung häufig direkt über den im Netztransformator untergebrachten Stufenschalter einschließlich Regelungseinheit geregelt. Bis vor wenigen Jahren erfolgte die Spannungssollwertvorgabe als statischer Wert an der Mittelspannungssammelschiene des Umspannwerkes (UW), unabhängig von der vorhandenen Netzsituation. Mit der Entwicklung „intelligenter" Netze werden derzeit neue Konzepte im Bereich der Spannungsregelung entwickelt, die eine Erhöhung der Netzanschlusskapazität gewährleisten können und zu einer höheren Netzeffizienz beitragen sollen. Besonders Ansätzen für eine dynamische Sollwertanpassung an die gegenwärtige Netzsituation durch Erfassung verschiedener elektrischer Netzparameter werden hohe Entwicklungschancen zugesprochen. Dieser Artikel befasst sich mit der simulativen Untersuchung neu entwickelter Regelungsarten für die Spannungsregelung in MS-Netzen mittels Stufenschalter. Ziel ist es, das Verhalten neuer Spannungsregelungskonzepte auf Basis einer realen MS-Netzstruktur nachzubilden und die Auswirkungen bzgl. Netzaufnahmekapazität und Spannungsqualität zu bewerten. Neben der Bewertung der erzielten Ergebnisse wird ein Ausblick auf aktuelle Forschungstätigkeiten aus dem Gebiet innovativer Spannungsregelungskonzepte gegeben, die u. a. Methoden aus dem Bereich der *Computational Intelligence (CI)* berücksichtigen.

1 Einleitung

In Deutschland bilden den Großteil der dezentralen Erzeugungsanlagen (DEA) die Photovoltaik (PV) und die Windenergieanlagen (WEA), die überwiegend in der Verteilnetzebene angeschlossen sind und deren Wirkleistungseinspeisungen – vorrangig in ländlichen Netzregionen – zu Rückspeisungen und lokalen Spannungsanhebungen führen. Besonders die Einhaltung des normierten Spannungsbandes nach DIN EN 50160 [1] zeigt sich oftmals als begrenzendes Kriterium für die Netzaufnahmekapazität von DEA. Um ein normgerechtes Spannungsniveau auch bei sich ändernden Netzverhältnissen zu gewährleisten, müssen vom Verteilnetzbetreiber (VNB) adäquate Gegenmaßnahmen geplant und umgesetzt werden. Hierzu zählen neben konventionellen Netzausbaumaßnahmen auch neue Methoden aus den Bereichen der Netzberechnung und der Spannungsregelung. Im Rahmen dieses Beitrages wird untersucht, welchen Beitrag die neu entwickelten Spannungsregelungskonzepte zur Erhöhung der Netzaufnahmekapazität liefern können. Die Simulationen zeigen ein hohes Potential bei der Überwachung der Netzspannung in den selektiven Knoten des Netzes.

Der nachfolgende Abschnitt 2 gibt einen Überblick über den Stand der Technik der heute in MS-Netzen eingesetzten Spannungsregelungsverfahren. Neben der Beschreibung der für die Simulation zugrunde gelegten Daten wird in Abschnitt 3 die Wirksamkeit der verschiedenen Spannungsregelungsverfahren anhand einer realen MS-Netzstruktur für unterschiedliche Betrachtungszeiträume beschrieben. Zum Abschluss wird in Abschnitt 4 eine Zusammenfassung der Simulationsergebnisse und ein Ausblick auf die Möglichkeit der Einbindung der *CI* in neue Spannungsregelungsverfahren gegeben.

2 Stand der Technik

Maßnahmen zur Spannungsregelung in der Mittelspannung werden in direkte und indirekte Verfahren unterteilt. *Indirekte Spannungsregelungsverfahren* basieren auf der dezentralen Bereitstellung von Blindleistung. Diese sind nicht Teil der Untersuchungen, weshalb hier z. B. auf [2] verwiesen wird.

Bei den *direkten Spannungsregelungsverfahren* wird für gewöhnlich das Spannungsniveau durch den im HS/MS-Transformator untergebrachten Spannungsregler zentral im UW geregelt. Als Stellglied dieser Regelung fungiert ein im Transformator installierter Stufenschalter, der unter Last das Transformatorübersetzungsverhältnis stufenweise verändern kann. Die Sollwertvorgabe der Regelung erfolgt dabei statisch oder zunehmend auch dynamisch.

2.1 Statische Sollwertregelung

Bei der statischen Sollwertregelung (nachfolgend als Variante 1 bezeichnet) dient als Regelgröße die Sammelschienenspannung. Die einmalige, feste Auslegung des Sollwertes hat zur Folge, dass keine Optimierung des

Sollwertes entsprechend der vorliegenden Netzsituation erfolgt. Hierdurch wird das verfügbare Spannungsregelungspotential mit Stufenschalter nur teilweise ausgenutzt [3].

2.2 Dynamische Sollwertanpassung

Heutzutage ist, bedingt durch den Wandel der Erzeugungsstruktur, ein Trend zur dynamischen Sollwertvorgabe erkennbar. Hierbei werden in der Praxis derzeit die nachfolgend beschriebenen zwei Ansätze verfolgt.

Bei der *wirkleistungsabhängigen Sollwertvorgabe* (nachfolgend als Variante 2 bezeichnet) wird der Sollwert als Funktion der über den Transformator übertragenen Wirkleistung vorgegeben. Von der Richtung und der Höhe des Wirkleistungsflusses lassen sich Rückschlüsse auf die gegenwärtige Netzsituation ziehen.

Ein Wirkleistungsfluss in Richtung des vorgelagerten 110 kV-Netzes deutet auf eine hohe Erzeugungsleistung und gewöhnlich hohe Spannungspegel im MS-Netz hin. Hier führt ein Absenken der Sammelschienenspannung durch Verringerung des Sollwertes zu einer Verringerung der Spannungspegel im MS-Netz und damit zu einer Erhöhung der Aufnahmekapazität.

Umgekehrt lässt ein Transformatorwirkfluss in Richtung des MS-Netzes auf ein dominierendes Lastverhalten im Netz schließen. Um hier eine ausreichende Spannungshöhe am Verbraucher zu gewährleisten, wird der Sollwert der Sammelschienenspannung angehoben.

Voraussetzung für die wirkleistungsabhängige Sollwertvorgabe ist eine homogen verteilte Last- und Einspeisestruktur im MS-Netz [4], die anhand archivierter Spannungsmesswerte des MS-Netzes überprüft werden kann. Das Konzept der *Spannungsweitbereichsregelung* (nachfolgend als Variante 3 bezeichnet) basiert auf der zyklischen Erfassung von Spannungsmesswerten im MS-Netz, die an die zentrale Regelungseinheit des HS/MS-Transformators übertragen werden (**Bild 1**). Abhängig von der Anzahl der verteilten Messpunkte können Rückschlüsse auf die vorhandene Netzsituation gezogen werden.

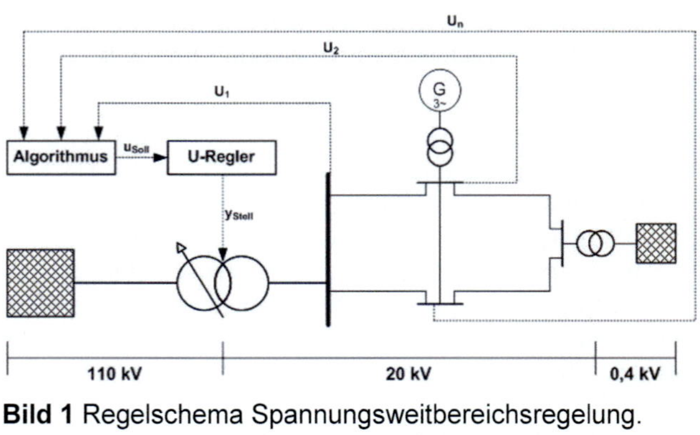

Bild 1 Regelschema Spannungsweitbereichsregelung.

Für jeden gemessenen Punkt (vorwiegend selektive Knoten) können vom VNB individuelle Spannungsgrenzen und Prioritäten festgelegt werden, wodurch auch eine inhomogene Last- und Erzeugungsstruktur durch Ausnutzen des kompletten Spannungsbandes berücksichtigt werden kann [3]. Wird das obere Spannungsregelungsziel (U_o) an einem der Messpunkte überschritten, wird aus einer beliebigen Anzahl von im Regler hinterlegten Sollwerten (in diesem Fall 5 Sollwerte: 20,00 kV, 20,20 kV, 20,40 kV, 20,60 kV und 20,80 kV) der nächstniedrigere Sollwert als der aktuell eingestellte Sollwert ausgewählt. Entsprechend erfolgt bei einer Verletzung des unteren Spannungsregelungsziels (U_u) die Auswahl des nächsthöheren Sollwertes. Netzsituationen, in denen das Ausregeln einer Grenzwertverletzung zu einer weiteren Grenzwertverletzung führen würde, werden durch Einhaltung der Ungleichung

$$U_{max} - U_{min} + \text{Trafostufenspannung} < U_o - U_u \qquad (1)$$

ausgeschlossen, wobei U_{max} und U_{min} den maximalsten und minimalsten Wert aller aktuell an den Messpunkten erfassten Spannungswerte repräsentieren. Für die Netzzustände, in denen die Ungleichung nicht eingehalten wird, erfolgt ein Umschalten auf die Variante 2.

Die Ungleichung gilt nur für den Fall, das an allen beobachteten Messpunkten die gleichen Regelungsziele vorgegeben werden, andernfalls muss eine Anpassung der Ungleichung erfolgen.

3 Simulation verschiedener Sollwertregelungsarten

3.1 Grundlagen der Simulation

Ausgangspunkt der simulativen Untersuchungen bilden iterative Leistungsflussberechnungen auf Basis des *Newton-Verfahrens*, die für die drei Spannungsregelungsvarianten in einem realen MS-Netzmodell durchgeführt werden.

3.1.1 Netzstruktur des Mittelspannungsnetzes

Um eine praxisnahe Simulation zu gewährleisten, werden die verschiedenen Spannungsregelungsverfahren auf Basis einer MS-Netzstruktur (**Bild 2**) durchgeführt.

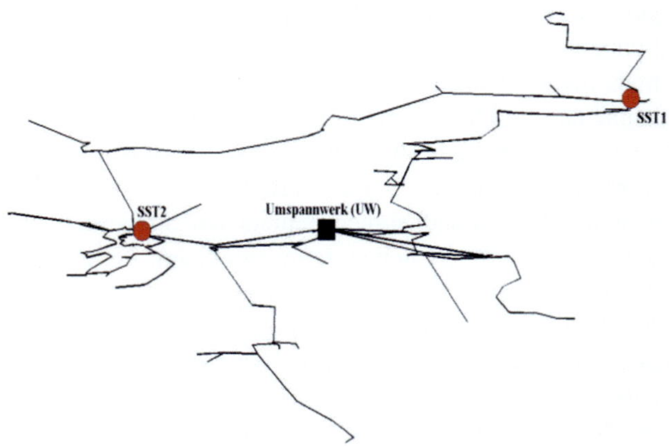

Bild 2 Netzstruktur des 20-kV-Mittelspannungsnetzes.

In den Schaltstationen (SST) erfolgt neben der Realisierung der Schutzfunktionen auch die Erfassung der Spannungswerte für die Spannungsweitbereichsregelung.

Das MS-Netz mit einer gesamten Leitungslänge von etwa 126 km dient der Versorgung von 9 Ortschaften und 2

Kleinstädten. Die vergleichsweise hohe Konzentration an Leitungen im Bereich der SST2 deutet dabei auf eine der beiden Kleinstädte hin. Die zweite Kleinstadt befindet sich in unmittelbarer Nähe des UW. Der überwiegende Teil des MS-Netzes besitzt jedoch eine ländlich geprägte Netzstruktur.

Gekennzeichnet ist das MS-Netz durch eine hohe dezentrale Erzeugungsleistung (**Tabelle 1**) sowie geringe Lasten. Die max. Rückspeisung im Jahr 2014 betrug 17 MW. Demgegenüber steht eine max. Bezugslast von 10 MW.

Tabelle 1 Installierte Erzeugungsleistung im Netzgebiet.

	PV	Wind	KWK[2]	Wasser
NS[1] [MW]	8,7	-	0,3	0,2
MS [MW]	4,4	14,4	3,3	-

[1] Niederspannung [2] Kraft-Wärme-Kopplungsanlagen

3.1.2 Eingangsparameter der Simulation

Für die Modellierung der verschiedenen Spannungsregelungsverfahren werden in den Lastflussberechnungen reale Erzeugungs- und Bezugsleistungswerte leistungsgemessener DEA und Verbraucher angesetzt. Leistungen nicht gemessener DEA (vorwiegend PV-Anlagen in den NS-Netzen) werden anhand ihrer installierten Leistungen und eines ortsnetzbezogenen Skalierungsfaktors in der Simulation berücksichtigt. Tariflasten, die i. d. R. nicht durch eine Leistungsmessung erfasst werden, sind durch Differenzbildung zwischen den gemessenen Leistungswerten am HS/MS-Transformator abzüglich der Summe aus gemessenen Erzeugungs-/Verbrauchsleistungen und nicht gemessenen DEA ermittelt worden.

Die in die Simulation eingehenden Verbrauchs- und Erzeugungsdaten sowie die Simulationsergebnisse besitzen eine zeitliche Auflösung von 0,25 h.

3.2 Bewertung der Simulationsergebnisse über ein Wochenintervall

Die Untersuchungen der verschiedenen Spannungsregelungskonzepte werden zunächst auf Basis der Erzeugungs- und Lastdaten für den Zeitraum eines Wochenintervalls durchgeführt. **Bild 3** veranschaulicht für den betrachteten Zeitraum die Gleichzeitigkeit zwischen der PV- und Windenergieeinspeisung anhand von zwei Referenzanlagen. Der dargestellte normierte Lastgang des HS/MS-Transformators gibt Aufschluss über die vorhandenen Last- und Einspeiseverhältnisse im gesamten Netz.

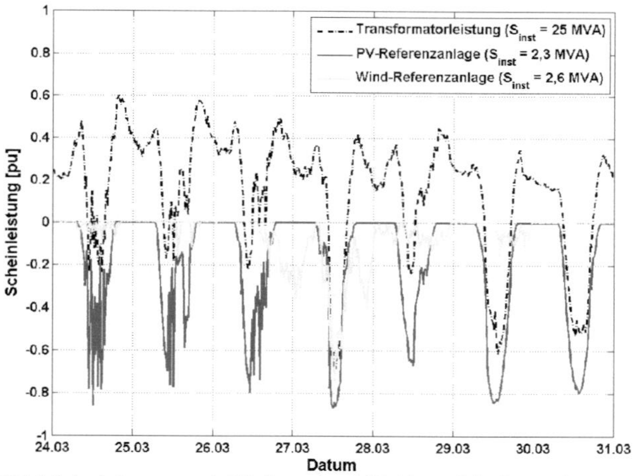

Bild 3 Leistungsverhältnisse im MS-Netz (Verbraucherzählpfeilsystem).

Vorwiegend zur Mittagszeit sind hohe Rückspeisungen in die vorgelagerte Netzebene zu beobachten, die auf eine hohe PV-Durchdringung in der MS- und NS-Ebene schließen lassen. Verschärft wird diese Situation bei gleichzeitig hoher Einspeiseleistung aus WEA, wie sie etwa in der Mittagszeit der Wochenmitte des betrachteten Zeitintervalls auftritt. Insbesondere in den Zeiträumen hoher Rückspeisungen in das 110 kV-Netz ist von hohen Spannungspegeln im MS-Netz auszugehen, die die Einhaltung vorgegebener Spannungsgrenzwerte gefährden können.

Sowohl der maximale als auch der minimale Spannungsgrenzwert (SG_{max} und SG_{min}), die vorzugsweise an allen Punkten im Netz eingehalten werden sollen, können abhängig vom Netzbetreiber und dessen Betriebsphilosophie variieren. Unter Berücksichtigung der allgemein anerkannten *Regeln der Technik* sollten SG_{max} und SG_{min} einen gewissen Abstand zu den normierten Grenzen des Spannungsbandes [1] aufweisen, um eine ausreichende Spannungsbandreserve für die NS-Ebene zu gewährleisten. Hier wird die Spannungshaltung in einem Bereich zwischen den Spannungsgrenzen 19,80 kV (SG_{min}) und 21,20 kV (SG_{max}) im gesamten MS-Netz angestrebt.

Bild 4 zeigt den Verlauf der sich in der Simulation einstellenden Spannungsverhältnisse an den SST und dem UW im betrachteten MS-Netz bei fester Sollwertvorgabe von 20,30 kV.

In Variante 1 wird ohne Berücksichtigung der Last- oder Spannungsverhältnisse die Sammelschienenspannung auf ca. 20,30 kV geregelt. In der betrachteten Woche wird SG_{min} auf Grund der moderaten Last nicht verletzt. Für Zeiträume mit geringer Einspeiseleistung und hohem Bezug (vorrangig in den Wintermonaten) liegt eine Reserve von ungefähr 0,20 kV bis zur Verletzung von SG_{min} vor. Demgegenüber steht die Nichteinhaltung von SG_{max} im Einspeisefall bei hoher Gleichzeitigkeit zwischen den Primärenergieträgern Sonne und Wind. Diese Situation wird verschärft durch eine mögliche Erhöhung der Anlagenauslastungen – insbesondere bei den WEA – bei annähernd gleichbleibender Gleichzeitigkeit.

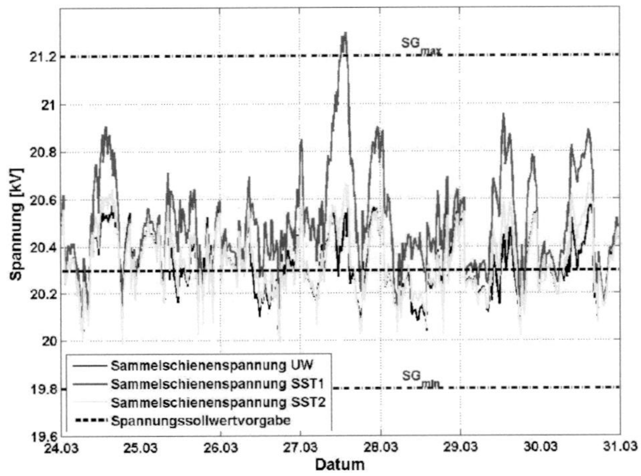

Bild 4 Spannungsverhältnisse bei fester Sollwertvorgabe (Variante 1).

Die Überschreitung von SG_{max} beträgt rd. 0,10 kV. Ein Absenken des dargestellten Sollwertes auf 20,20 kV führt näherungsweise zu einem Herabsetzen der dargestellten Spannungsverläufe um ca. 0,10 kV und damit zur Einhaltung von SG_{max} für den betrachteten Zeitraum, ohne jedoch neue Kapazitäten für neue DEA zu generieren. Auch reduziert diese Maßnahme die Reserve für die Einhaltung von SG_{min} in Zeiten von schwacher Einspeisung und starkem Bezug. Dieses Regelungskonzept ist daher für den praktischen Einsatz in dem zugrundegelegten MS-Netz nicht mehr langfristig geeignet.

Die Spannungsverläufe für Variante 2 sind in **Bild 5** dargestellt. Ab einer Bezugsleistung von 5 MW wird ein Spannungssollwert von 20,60 kV vorgegeben, ab einer Rückspeisung von -5 MW hingegen ein Spannungssollwert von 20,10 kV. Zwischen den genannten Wirkleistungsgrenzen erfolgt die Sollwertvorgabe auf Basis eines linearen Zusammenhanges.

Im Unterschied zur Variante 1 werden in Variante 2 unter Berücksichtigung der genannten Parametrierung die festgelegten Spannungsgrenzwerte eingehalten. Erwartungsgemäß erfolgt besonders in den rückspeisestarken Mittagszeiten eine Absenkung des Sollwertes. Die vorher aufgetretene Spannungsspitze an der SST1 in der Wochenmitte wird auf ca. 21,10 kV gesenkt, wodurch im Unterschied zur Variante 1 ein geringer Mehrwert bzgl. der Netzanschlusskapazität von Erzeugungsanlagen bei annähernd gleichbleibender Starklast-Reserve (rd. 0,20 kV) geschaffen wird.

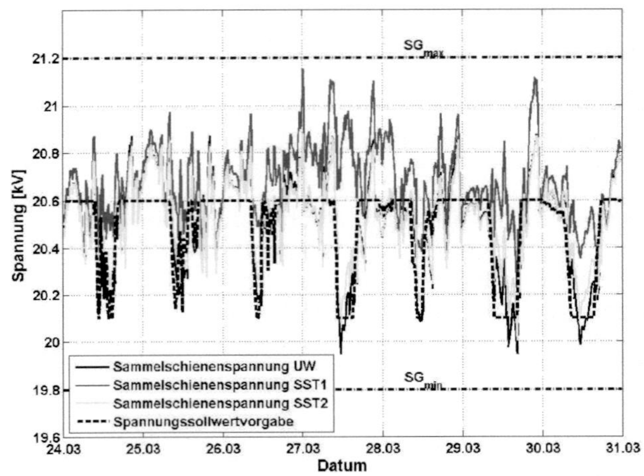

Bild 5 Spannungsverhältnisse bei wirkleistungsabhängiger Sollwertvorgabe (Variante 2).

Die Simulationsergebnisse für die Variante 3 (für beide SST gilt: U_u = 20,00 kV und U_o = 21,00 kV) zeigt **Bild 6**. Die Einhaltung des festgelegten Regelungsziels in Variante 3 bewirkt gleichzeitig auch die Einhaltung der definierten Spannungsgrenzwerte.

Die hohen Spannungswerte in der SST1, die in der Variante 1 noch zu einem überschreiten von SG_{max} führten, werden in Variante 3 im zulässigen Spannungsband gehalten. Die Überschreitung von U_o in der Wochenmitte bewirkt eine Änderung des Sollwertes auf 20,00 kV. Bedingt durch die Sollwertabsenkung erfolgt eine Stufung des Stufenschalters von Stufe 7 in 8, wodurch die Netzspannung entsprechend um die Stufenspannung absinkt, sodass die Regelungsziele eingehalten werden. Für ca. 2,5 h wird die Ungleichung (1) verletzt (Bandbreite der zu beobachteten Spannungen im Netz). In diesem Zeitraum schaltet der Regelalgorithmus auf Variante 2 um. Im

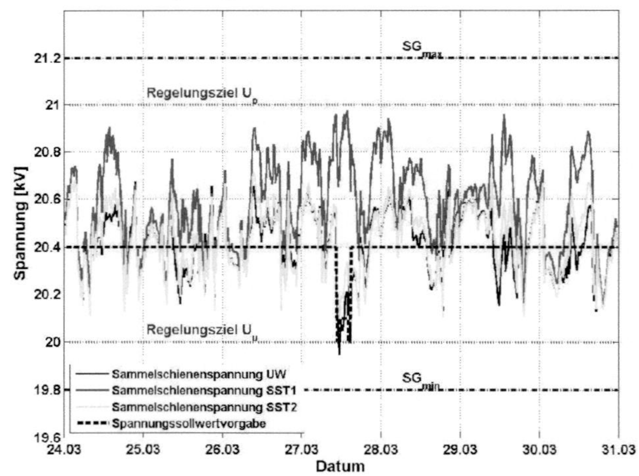

Bild 6 Spannungsverhältnisse bei der Spannungsweitbereichsregelung (Variante 3).

weiteren Verlauf führen die Spannungsverhältnisse an den SST (U_{max} und U_{min}) wieder zu einer Einhaltung der Ungleichung (1), jedoch wird an der SST2 U_u erreicht, sodass nun eine Anhebung des Sollwertes auf 20,40 kV erfolgt. Die Sollwertanhebung führt zu einer Schaltung von Stufe 8 nach 7. Die Folge ist eine Anhebung der Netzspannung um die Stufenspannung, wodurch die Regelungsziele nicht verletzt werden.

Die Einhaltung des oberen Regelungsziels bewirkt, dass die Spannung an keinem Zeitpunkt des betrachteten Intervalls einen Wert von 21,00 kV überschreitet. Somit wird auch für die Zeiten hoher Erzeugung eine Reserve von mindestens 0,20 kV sichergestellt, wodurch die Netzanschlusskapazität im Vergleich zu Variante 2 erhöht wird. Die Regelungsziele werden bei der Variante 3 nur für die SST vorgegeben, nicht aber für die UW-Sammelschienenspannung. Dies ist daran erkennbar, dass in der Wochenmitte bedingt durch den niedrigen Sollwert die UW-Spannung unterhalb von U_u absinkt. Dennoch wird im MS-Netz weiterhin eine Starklast-Reserve von ca. 0,20 kV vorgehalten.

Erkennbar ist auch, das trotz der wechselnden Last- und Einspeisesituation der Sollwert wenig variiert, was auf die Auswahl von U_o und U_u und die hinterlegten Sollwerte zurückzuführen ist. Der Verlauf des Sollwertes deutet ferner daraufhin, dass nur in der Wochenmitte eine Umschaltung auf die Variante 2 in dem betrachteten Zeitraum erfolgt.

3.3 Simulationsergebnisse aus der Langzeituntersuchung

Um die Wirkung auch unter Berücksichtigung der von den Jahreszeiten abhängigen Last- und Einspeiseverhältnisse präzisieren zu können, wurde für alle drei Spannungsregelungsarten der Simulationszeitraum auf ein Jahr erhöht. Die Regelungsparameter und -ziele wurden indes beibehalten.

Zur übersichtlichen Darstellung der Ergebnisse wurden für alle drei Varianten die simulierten Spannungswerte an der jeweiligen SST in *Klassen* mit einer Breite von je $\Delta U = 50$ V unterteilt. **Bild** 7 zeigt die Verteilung der Spannungswerte an der SST1 für die drei Varianten.

Bezogen auf den Zeitraum von einem Jahr befindet sich für die Variante 1 die höchste Klassenhäufigkeit in einem Bereich um die Klassenmitte von 20,40 kV. Das bedeutet, dass für ca. 10 % des Jahreszeitraumes an der SST1 ein Spannungswert im Bereich von 20,40 kV vorliegt. Ferner wird die bereits festgestellte Verletzung von SG_{max} bei Anwendung der Variante 1 bestätigt, wobei jedoch die Häufigkeiten der Überschreitungen von SG_{max} sehr gering sind. Die niedrigsten Spannungspegel an der SST1 liegen in einem Bereich um 20,00 kV, wodurch eine Reserve von rd. 0,20 kV bis zum Erreichen von SG_{min} vorliegt.

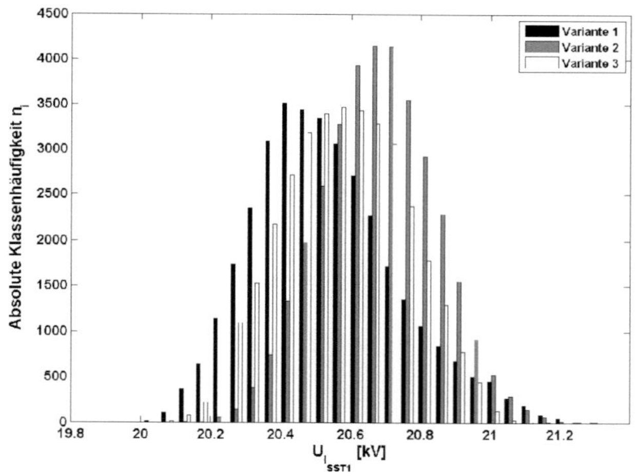

Bild 7 Absolute Häufigkeit der Spannungspegel an SST1 je Spannungsregelungsart.

In Bezug zu Variante 1 sind bei Variante 2 die Häufigkeiten auf der Abszisse deutlich nach rechts verschoben, wodurch sich die größten Klassenhäufigkeiten in einem Bereich zwischen 20,65 kV und 20,70 kV befinden. Die Überschreitung von SG_{max} in der Variante 2 sind denen der Variante 1 in Häufigkeit und Spannungswert sehr ähnlich. Der niedrigste Spannungswert an der SST1 liegt bei ca. 20,10 kV, wodurch die Reserve für den Lastfall an diesem Knoten rd. 0,30 kV beträgt.

Die größte Häufung der Spannungspegel an der SST1 befindet sich bei Variante 3 um die Klassenmitte von 20,55 kV. Verglichen mit den Varianten 1 und 2 werden bei Variante 3 sowohl SG_{min} als auch SG_{max} eingehalten. Der höchste Spannungswert liegt im Bereich der Klassenmitte von 21,15 kV, der jedoch nur für eine sehr kurze Zeit im Jahr erreicht wird. Daraus folgt, dass an diesem Knoten noch eine Reserve von 0,05 kV für den Einspeisefall zur Verfügung steht. Weiterhin liegt eine Starklast-Reserve von 0,30 kV vor. Ferner konnte bei der Jahresbetrachtung beobachtet werden, dass nicht bei allen aufgetretenen Last- und Einspeiseverhältnissen das obere Regelungsziel U_o eingehalten wurde (bedingt durch Umschaltung auf Variante 2, siehe Abschnitt 2.2.2).

An der SST1 ist bei allen drei Varianten eine annähernd symmetrische Verteilung der Spannungswerte erkennbar. Analog zur SST1 ist in **Bild 8** die Verteilung der Spannungswerte an der SST2 dargestellt.

Wie bereits in den Wochenverläufen erkennbar, lässt die Häufigkeitsverteilung der Spannungswerte der SST2 auf einen lastgeprägten Teilnetzbereich schließen. Dies ist anhand der Verschiebung der Häufigkeitsverteilungen der Spannungswerte der SST2 erkennbar, die im Vergleich

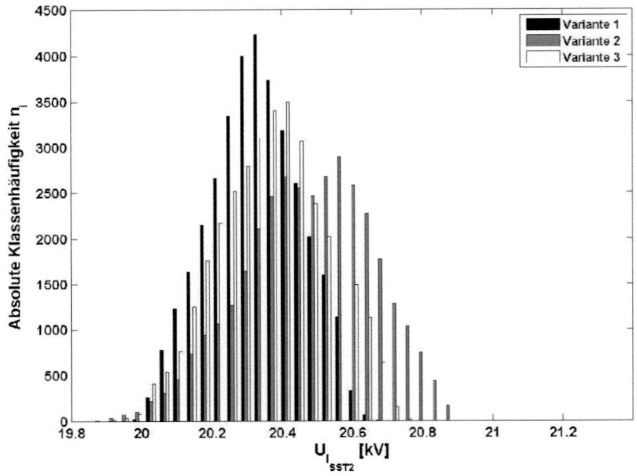

Bild 8 Absolute Häufigkeit der Spannungspegel an SST2 je Spannungsregelungsart.

mit den Verteilungen an der SST1 für alle Varianten etwas nach links auf der Abszisse verschoben sind.

Die Maxima der Häufung der Spannungswerte der SST2 befinden sich für Variante 1 bei 20,30 kV, für Variante 2 bei 20,55 kV und für Variante 3 bei 20,40 kV.

An der SST2 werden bei allen drei Varianten SG_{min} und SG_{max} eingehalten. Die niedrigsten Spannungswerte liegen bei ca. 19,90 kV und werden bei den Varianten 2 und 3 erreicht. Dadurch beträgt die Starklast-Reserve an diesem Knoten für diese beiden Varianten nur rd. 0,10 kV, für Variante 1 indes ca. 0,20 kV. Auch wird für Variante 3 U_u von 20,00 kV nicht jederzeit eingehalten.

Die Häufigkeitsverteilungen der verschiedenen Regelungsvarianten deuten bei beiden SST auf eine Normalverteilung der Spannungspegel hin. Beispielhaft für die SST1 ist für Variante 3 die Normalverteilung der Spannungswerte in **Bild 9** abgebildet.

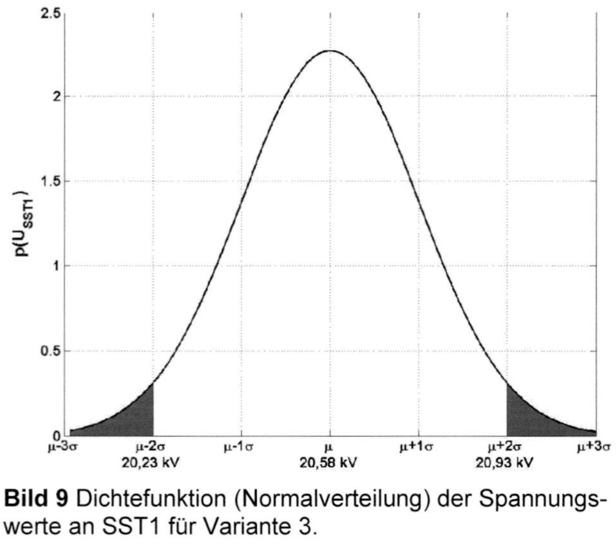

Bild 9 Dichtefunktion (Normalverteilung) der Spannungswerte an SST1 für Variante 3.

Unter Berücksichtigung von [5] gelten für eine normalverteilte Messgröße verschiedene σ-Bereiche. In einem Intervall von [μ - 2σ; μ + 2σ] befinden sich bspw. rd. 95,5 % aller Messwerte. Dieses Intervall ist für die Dichtefunktion der Spannungswerte an der SST1 für die Variante 3 in **Bild 9** abgebildet. Demnach folgt bei Einsatz der Variante 3, dass rd. 95,5 % aller Spannungswerte eines Jahres in einem Bereich zwischen 20,23 kV und 20,93 kV liegen (weißer Bereich unterhalb des Funktionsverlaufes zwischen den beiden blauen Flächen).

	Variante 1	Variante 2	Variante 3
Stufungen	1039	1829	1197
Max. Bandbreite [kV]	1,348	1,475	1,327
μ ± 2 [kV]	0,826 (1) 0,512 (2)	0,646 (1) 0,732 (2)	0,703 (1) 0,613 (2)
μ ± 3 [kV]	1,238 (1) 0,767 (2)	0,969 (1) 1,099 (2)	1,055 (1) 0,920 (2)

Tabelle 2 Kennwerte aus der Jahresbetrachtung ((1) SST1, (2) SST2)

In **Tabelle 2** sind verschiedene Ergebnisse aus den Simulationen und der Wahrscheinlichkeitsrechnung abgebildet. Erkennbar ist, dass bei Variante 2 deutlich mehr Stufenschaltungen erfolgten als bei den Varianten 1 und 3. Die max. Regelbandbreite, die sich aus der Differenz des höchsten und des niedrigsten Spannungswertes aller Spannungswerte der beiden SST ergibt, ist bei der Variante 2 ebenfalls größer als bei den Varianten 1 und 3.

Die Betrachtung der σ-Bereiche zeigt, dass das Regelverhalten maßgeblich vom kritischsten Knoten bestimmt wird. In dem hier untersuchten MS-Netz ist der kritischste Knoten die SST1, die durch eine hohe Erzeugungsleistung geprägt ist. Die Regelbandbreite wird bei Variante 3 im Vergleich zu Variante 1 an der SST1 schmaler, gleichzeitig wird die Regelbandbreite aber am weniger kritischen Knoten SST2 breiter (siehe σ-Bereiche).

4 Zusammenfassung und Ausblick

Die Ergebnisse über die verschiedenen Zeiträume zeigen, dass in dem betrachteten MS-Netz nur die Variante 3 zur Einhaltung des definierten Regelungszieles führt bei vergleichbar geringer Zahl von Stufenschaltungen.

Eine breite Bandbreite wirkt sich nachteilig auf die Netzanschlusskapazität aus, da sie die Abstände zu SG_{min} und SG_{max} verringert. Durch die schmale max. Bandbreite konnte bei Variante 3, wenn auch hier nur geringfügig, die Netzanschlusskapazität für neue DEA erhöht werden. Die -Bereiche für Varianten 1 und 3 zeigen, dass das Regelverhalten vorwiegend durch den kritischsten Knoten bestimmt wird. Dies kann zu einer Erhöhung der Bandbreite an weniger kritischen Knoten führen.

Um eine repräsentative Aussage zu den hier ermittelten Ergebnissen zu tätigen, bedarf es einer höheren Stichprobe an MS-Netzen. Dies wird aktuell vorbereitet.

Die Spannungsweitbereichsregelung wird im praktischen Pilotbetrieb bereits bei der *EnergieNetz Mitte GmbH* eingesetzt. Erste Ergebnisse aus dem Pilotbetrieb werden aktuell ausgewertet und sollen bei der Verifizierung der Simulationsergebnisse mit berücksichtigt werden.

Infolge der vorgestellten Simulationsergebnisse der Spannungsweitbereichsregelung und der Problematik von ungleichmäßig verteilten Lasten und DEA, muss die Möglichkeit einer dezentralen Blindleistungsbereitstellung geprüft und der Mehrwert für die Spannungsweitbereichsregelung untersucht werden. Hierdurch kann evtl. auch ein Umschalten auf die Variante 2 reduziert bzw. verhindert werden. Erste Ansätze sind bereits in [2] beschrieben.

Aktuell werden die Möglichkeiten untersucht, Techniken der *CI*, die ihren Schwerpunkt auf maschinelles Lernen mit Hilfe von Beispieldaten legt [6], in das Konzept der Spannungsweitbereichsregelung zu integrieren. Im Vordergrund steht die Realisierung und Analyse einer selbstlernenden Spannungsregelung, die automatisch durch verschiedene Methoden der Zeitreihenprognose eine Selbstadaption an sich ändernde Netzzustände (z. B. weiterer Zubau von DEA, Umschaltungen) vornehmen soll. Besonders dem Lernen anhand von historischen Spannungsmesswerten aus dem MS-Netz wird eine hohe Relevanz bei der Zeitreihenprognose zugesprochen. Durch die Forderung an die Spannungsregelung, sich auch proaktiv an zukünftige Netzsituationen anzupassen, werden Methoden des überwachten Lernens (bspw. Support Vector Machines) zur Realisierung von Regressionsmodellen für Zeitreihenprognosen untersucht [7].

Auf der Basis von prognostizierten Netzzuständen und vorangegangener Messwerte für einen definierten Zeitraum soll ein Sollwert vorhergesagt werden, durch den u. a. die Zahl der Stufenschaltungen des Transformatorstufenschalters minimiert werden soll. Dies soll auch dann geschehen, wenn bedingt durch Fehlfunktionen keine bzw. nur eingeschränkt aktuelle Messwerte aus dem Netz bereitgestellt werden können. Neben der Berücksichtigung elektrischer Parameter sollen in die Sollwertfindung z. B. auch Zeitreihen von Windgeschwindigkeiten und die solare Bestrahlungsstärke mit einfließen.

5 Literatur

[1] Norm DIN EN 50160: *Merkmale der Spannung in öffentlichen Elektrizitätsversorgungsnetzen.* Berlin: Beuth, 2011

[2] Österreichische Forschungsförderungsgesellschaft mbH (FFG): *Energie der Zukunft - Beitrag zum aktiven Verteilernetzbetrieb durch Innovative Spannungsregelung - BAVIS.* Endbericht, 2013

[3] Körner, C.; Oechsle, F.; Braun, M; Probst, A: *Leistungsabhängige Spannungsregelung im Mittelspannungsnetz.* Mainz: ETG/FNN-Tutorial Schutz- und Leittechnik, 2012

[4] EP Photovoltaik des AK Verteilernetze: *Abschlussbericht des EP Photovoltaik.* Version 2015-1.1

[5] Papula, L: *Mathematik für Ingenieure und Naturwissenschaftler Band 3.* 4., verbesserte Auflage, Vieweg Verlag, 2001

[6] Kroll, A: *Computational Intelligence - Eine Einführung in Probleme, Methoden und technische Anwendungen.* München: Oldenbourg Verlag, 2013

[7] Pree, H.; Herwig, B.; Gruber, T.; Sick, B.; David, K.; Lukowicz, P: *On general purpose times series similarity measures and their use as kernel functions in support vector machines,* in: Information Sciences Volume 281 (2014), 478-495.

Einsatzmöglichkeiten regelbarer Ortsnetztransformatoren zur Spannungshaltung und Netzverlustreduzierung in Niederspannungsverteilnetzen

Dipl.-Ing. Raimund Schnieder [1,2], Technische Universität Clausthal, Clausthal-Zellerfeld, Deutschland, raimund.schnieder@tu-clausthal.de
Dr.-Ing. Ernst-August Wehrmann [1], Prof. Dr.-Ing. Hans-Peter Beck[1,2]
[1] Institut für Elektrische Energietechnik und Energiesysteme, Leibnizstraße 28, 38678 Clausthal-Zellerfeld
[2] Energie-Forschungszentrum Niedersachsen, Am Stollen 19A, 38640 Goslar

Kurzfassung

Regelbare Ortsnetztransformatoren (rONT) wurden bisher als Netzbetriebsmittel zur Beseitigung der Spannungsbandproblematik bei dezentraler Einspeisung im Niederspannungsverteilnetz und folglich zur Reduzierung konventioneller Netzausbaumaßnahmen eingesetzt. Bei entsprechendem Regelbereich entkoppeln diese die Niederspannungsnetzebene vollständig von der Mittelspannungsebene, sodass sich weitere netzplanerische Freiheiten bezüglich der zulässigen Spannungstoleranzen ergeben. Im Monosensorbetrieb werden die Spannungsistwerte ausschließlich an der Niederspannungsverteilung erfasst, beim Multisensorbetrieb werden dem Spannungsregler weitere Netzknotenspannungen zur Verfügung gestellt. Beide Betriebsarten zielen darauf ab, den durch dezentrale Erzeugungseinheiten verursachten Spannungshub bzw. durch zusätzliche Lasten, wie beispielsweise Elektrofahrzeuge, generierten Spannungsfall, auszuregeln. In diesem Aufsatz geht es darum, dem Spannungsregler noch zusätzlich die an der Ortsnetzstation gemessene Solarstrahlung als Eingangsgröße zur Verfügung zu stellen, sodass der Spannungssollwert an der Ortsnetzstation bedarfsgerecht eingestellt werden kann und bei Lastfällen die Spannung zur Netzverlustreduzierung entsprechend hochgesetzt werden kann.

1 Übersicht über die Betriebsarten regelbarer Ortsnetztransformatoren

Zunächst wird eine kurze Übersicht über die verschiedenen Betriebsarten regelbarer Ortsnetztransformatoren (rONT) zur Spannungsregelung in der Niederspannungsnetzebene gegeben. Bei allen Regelarten handelt es sich beim Ausgang jeweils um die entsprechende Stufenposition des rONT, um unterspannungsseitig die Spannung einzustellen. Die Wirkung regelbarer Ortsnetztransformatoren auf das vorgelagerte Mittelspannungsnetz bei flächendeckendem Einsatz entlang eines Mittelspannungsrings [5] wird hier nicht betrachtet, sondern ausschließlich die Wirkung auf das Niederspannungsverteilnetz.

Bei allen Betriebsarten ist zu berücksichtigen, dass das Reglertotband (s. u. a. **Bild 1**) so gewählt wird, dass nach einem Stufvorgang des rONT die Spannung wieder in dieses geregelt wird, um ein Aufschwingen durch entgegengesetzte Stufrichtungen zu vermeiden.

1.1 Regelbare Ortsnetztransformatoren im Monosensorbetrieb

Beim Monosensorbetrieb erfolgt die Spannungsistwerterfassung an der Niederspannungsverteilung, also direkt an der Transformatorenunterspannungsseite. Das Ziel dieser Betriebsart ist die Entkopplung des Niederspannungsnetzes von der Mittelspannungsebene, sodass das nach DIN EN 50160 [1] zulässige Spannungstoleranzband bezüglich des Reglertotbandes komplett für das Niederspannungsnetz genutzt werden kann. In dieser Betriebsart ist der Spannungssollwert fest, meist auf Nennspannung U_N (1,0 p. u.), eingestellt.

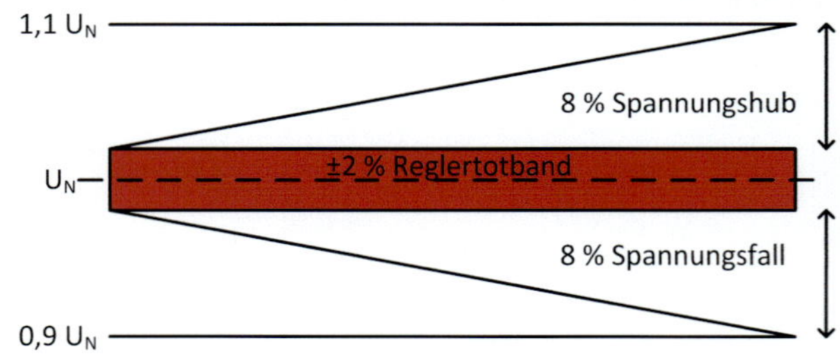

Bild 1 Mögliche Spannungstoleranzen beim Monosensorbetrieb [3],[4],[6],[9]

1.2 Regelbare Ortsnetztransformatoren im Multisensorbetrieb

Beim Multisensorbetrieb werden der Reglereinheit mehrere Spannungsistwerte, also zusätzlich zum Spannungswert der Niederspannungsverteilung auch von ausgewählten Niederspannungsnetzknoten, zur Verfügung gestellt. Hierbei ergibt sich die Möglichkeit, die Spannung unter oder über die Nennspannung an der Ortsnetzstation zu regeln, sodass die gemessenen Knoten auch in den zulässigen Toleranzbereich geregelt werden. Folglich ergeben sich in dieser Betriebsart aufgrund der zusätzlichen Informationen größere netzplanerische Spannungsfreiheiten. In Untersuchungen zum Multisensorbetrieb an Netzstrukturen mit überwiegenden Leitungslängen bis etwa 700 m stellte sich diese Betriebsart in den betrachteten Szenarien als Einzelfallentscheidung heraus, da nach einem rONT-Einsatz im Monosensorbetrieb die thermischen Belastungsgrenzen der Leitungen erreicht bzw. überschritten wurde und daher ein konventioneller Netzausbau durchgeführt werden muss. [3],[4]

1.3 Regelbare Ortsnetztransformatoren im Monosensorbetrieb bei variabler Sollwertvorgabe

Bei der nachfolgend beschriebenen Betriebsart wird der Spannungssollwert bedarfsgerecht angepasst. Die variable Sollwertvorgabe kann entweder bei Kenntnis des Lastflusses über dem Transformator [2] oder bei Berücksichtigung der Strahlungsmessung erfolgen. Bei beiden Möglichkeiten wird eine entsprechende Kennlinie verwendet, um Sollwertsprünge, und ein damit verbundenes Nachstufen des rONT, zu vermeiden.

Die sich bei variabler Sollwertvorgabe nach **Bild 2** ergebenden Spannungsfreiheiten sind in **Tabelle 1** dargestellt.

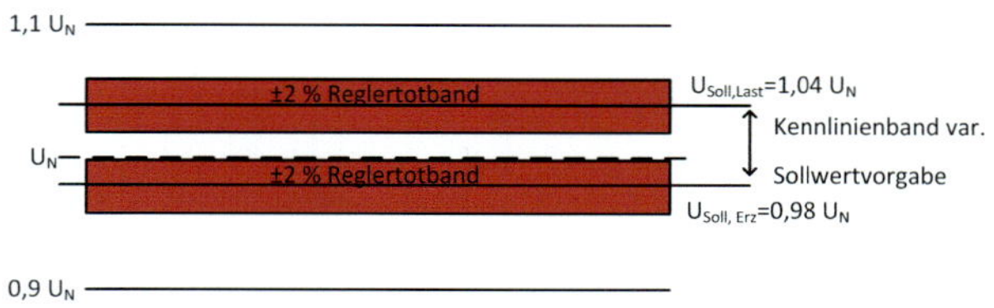

Bild 2 Toleranz- und Reglertotbänder bei variabler Sollwertvorgabe

Tabelle 1: Spannungsfreiheiten (bezogen auf Bild 2)

Szenario:	U_N in p.u.	Spannungshub:	Spannungsfall:
Erzeugung	0,98	10 % U_N	6 % U_N
Last	1,04	4 % U_N	12 % U_N

1.3.1 Variable Sollwertvorgabe in Abhängigkeit des Lastflusses

Bei dieser Betriebsart wird zusätzlich zur Sammelschienenspannung die Stromstärke und -richtung (Lastfluss) an der Ortsnetzstation mit gemessen. **Bild 3** zeigt eine mögliche Kennlinie zur variablen Sollwertvorgabe in Abhängigkeit des Lastflusses über der Ortsnetzstation. Bei einer Rückspeisung ins Mittelspannungsnetz (Erzeugungsfall im Niederspannungsnetz) wird die Spannung an der Ortsnetzstation runtergesetzt, im Lastfall entsprechend hoch.

Der Parameter p_{grenz} ist in diesem Fall symmetrisch ausgelegt und setzt die Kennliniengrenzen fest. Dieser ist normiert angegeben und bezieht sich auf die Transformatorennennscheinleistung. [2],[3]

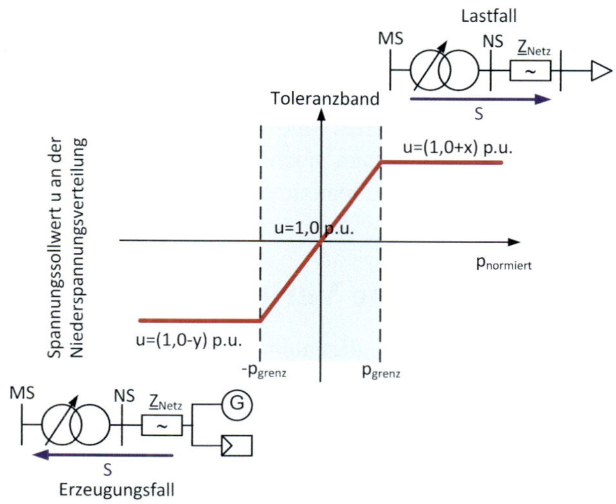

Bild 3 Kennlinie für die variable Spannungssollwertvorgabe an der Ortsnetzstation bei Berücksichtigung des Lastflusses über die Ortsnetzstation (vgl. [2],[3])

1.3.2 Variable Sollwertvorgabe in Abhängigkeit der Solarstrahlung

Bei dieser Variante wird zusätzlich zur Sammelschienenspannung die gemessene Solarstrahlung mit berücksichtigt. Da in Niederspannungsverteilnetzen hauptsächlich Photovoltaikanlagen als dezentrale Erzeugungseinheiten vorhanden sind, deren ins Netz eingespeiste elektrische Leistung mit der Solarstrahlung korreliert, kann durch eine zusätzliche Messung der Strahlungsdichte ein Rückschluss auf die Knotenspannungen im Netz getroffen und der Spannungssollwert an der Ortsnetzstation bedarfsgerecht angepasst werden. In der Nacht (hauptsächlich Lasten im Niederspannungsnetz) kann durch eine Erhöhung des Spannungssollwerts die Spannung im Netz hochgesetzt werden, sodass folglich kleinere Ströme durch die Betriebsmittel fließen und sich die Verluste reduzieren.

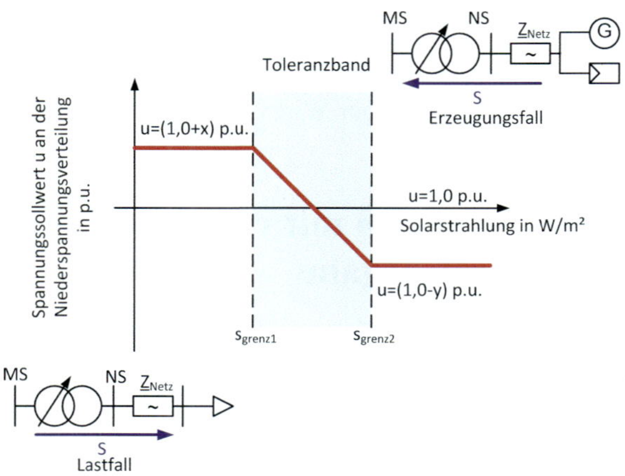

Bild 4 Kennlinie für die variable Spannungssollwertvorgabe bei Messung der Solarstrahlung an der Ortsnetzstation

Bild 4 zeigt eine mögliche Kennlinie für die Vorgabe eines variablen Spannungssollwerts an der Niederspannungsverteilung der Ortsnetzstation in Abhängigkeit der an der Station gemessenen Solarstrahlung. Die Grenzen des Spannungssollwertes sind durch die Parameter s_{grenz1} und s_{grenz2} festgelegt, um noch eine Toleranz zum nach Norm [1] zulässigen Spannungstoleranzband zu berücksichtigen.

2 Allgemeine Verlustbetrachtung

Zunächst erfolgt eine allgemeine Betrachtung der sich ergebenden Verluste im Niederspannungsverteilnetz bei unterschiedlichen Spannungen, wie sie durch einen rONT an der Ortsnetzstation eingestellt werden können.

Es gilt die allgemeine Definition der Wirkleistung für Drehstromnetze:

$$P = \sqrt{3} \cdot U_1 \cdot I_1 \cdot cos\varphi_1 = \sqrt{3} \cdot U_2 \cdot I_2 \cdot cos\varphi_2 \qquad (2.1)$$

Mit der Annahme konstanter Wirkleistung und eines konstanten Verschiebungsfaktors cosφ, sowie der Erhöhung der Spannung auf 1,04 p. u. (Indizes 2) anstatt 1,0 p. u. (Indizes 1) an der Ortsnetzstation durch den rONT ergibt sich der nachfolgende Zusammenhang:

$$P = \sqrt{3} \cdot U_1 \cdot I_1 = \sqrt{3} \cdot 1{,}04 \cdot U_1 \cdot I_2 \qquad (2.2)$$

Hieraus ergibt sich das Verhältnis der Leiterströme in diesem Fall zueinander:

$$\frac{I_1}{I_2} = 1{,}04 \qquad (2.3)$$

Für die Verlustleistung ergibt sich das folgende Verhältnis:

$$\frac{P_{V1}}{P_{V2}} = \frac{I_1^2 \cdot R}{\left(\frac{1}{1{,}04} \cdot I_1\right)^2 \cdot R} = 1{,}0816 \qquad (2.4)$$

Bei einer Erhöhung der Spannung um 4 % an der Ortsnetzstation ergeben sich mit den getroffenen Annahmen theoretisch ca. 8 % weniger Wirkverluste im Niederspannungsnetz.

3 Simulationen mit variabler Sollwertvorgabe

In diesem Abschnitt werden ausgewählte Ergebnisse der durchgeführten Szenarien vorgestellt. Hierbei wird in erster Linie der Unterschied zwischen den Szenarien im Winter- und im Sommer aufgezeigt.

Die nachfolgende Abbildung zeigt die Solarstrahlung über einen Tag jeweils für einen exemplarischen Sommertag (15.06.14) und Wintertag (15.01.14). Die Daten stammen aus dem Projektportal des e-home Energieprojektes [8].

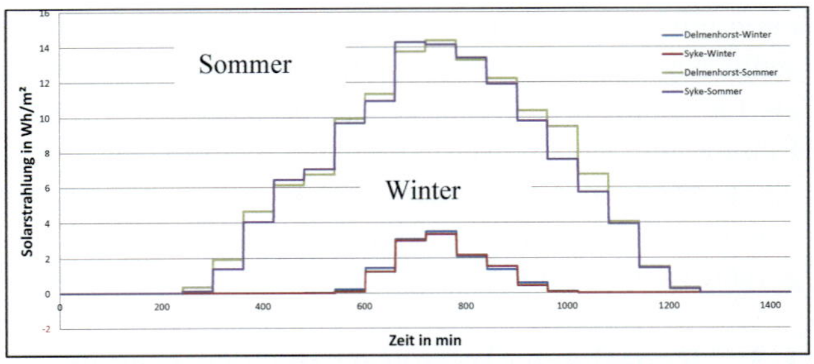

Bild 5 Solarstrahlungsmessdaten für einen Sommer- und Wintertag aus dem e-home Datenportal [8]

Für die untersuchten Szenarien ist die unten dargestellte Niederspannungsnetzstruktur verwendet worden. Diese wurde bewusst durch lange Leitungslängen oder geringe Leiterquerschnitte entsprechend schwach ausgelegt um eine Rückwirkung auf das System für den Funktionstest des Reglers mit variabler Sollwertvorgabe zu erhalten. Hierbei handelt es sich um kein reales Netz, sondern um ein für die Untersuchungen generiertes Netzmodell.

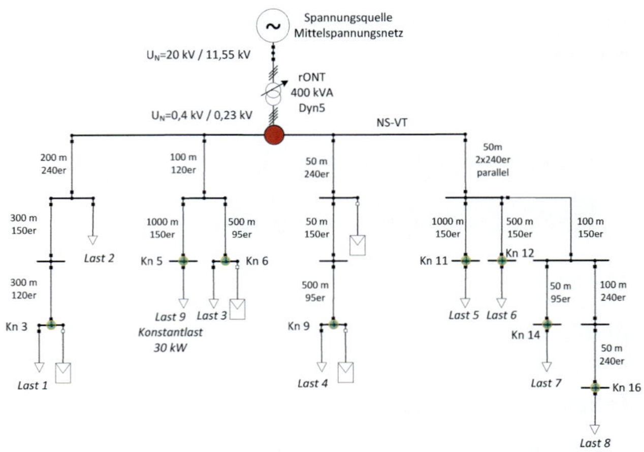

Bild 6 Niederspannungsnetzstruktur für die Untersuchungen (dicker Punkt (rot): Spannungsistwerterfassung rONT; transparente Punkte (gelb-grün): Knotenpunkte, die in Simulationsergebnissen dargestellt werden)

Die eingezeichneten Photovoltaik-Anlagen sind vom Netz getrennt (Schalter geöffnet), da die aus dem Datenportal stammenden überlagerten Gesamtprofile (Last und Erzeugung) über die Lastmodelle in das System eingeprägt werden. Die Lasten sind nichtlinear modelliert, sodass die aufgenommenen Ströme bei Erhöhung der Spannung geringer ausfallen (P(U,I)=const). Dies ist notwendig, um eine Bilanzierung des Niederspannungsnetzes hinsichtlich der Netzverluste durchführen zu können. Ein Vergleich der maximalen auftretenden Wirkleistungen der Winterszenarien (lastgeprägtes Niederspannungsnetz) ist in **Bild 7** dargestellt. **Tabelle 2** beinhaltet die zugehörigen Parameter der variablen Sollwertvorgabe (Grenzen der variablen Spannungssollwertvorgabe). Die Parameter des rONT-Reglers (Reglertotband ±2 % der Nennspannung, Verzugszeit bis zur Stufung T_i=15 s) waren in allen Szenarien gleich gewählt. Die Kennliniengrenzen wurden normiert auf s_{grenz1}=0,1 p. u. und s_{grenz2}=0,2 p. u. eingestellt, um Sollwertänderungen zu erhalten.

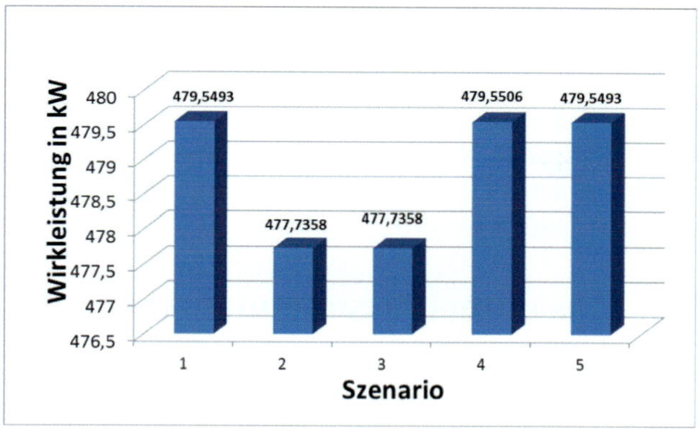

Bild 7 Maximale Wirkleistungen über den rONT bei den unterschiedlichen Szenarien

Tabelle 2 Grenzen der Kennlinie (vgl. Bild 4) für die in Bild 7 dargestellten Szenarien

Szenario:	$U_{Soll,max}$ in p.u.	$U_{Soll,min}$ in p.u.
1	1,02	0,98
2	1,04	0,98
3	1,04	1,0
4	1,0	1,0
5	rONT aus	

Aus **Bild 7** geht hervor, dass bei einer Anhebung der Spannung auf 1,04 p. u. durch den rONT in der „Lastzeit" (Szenario 2+3) die gesamte maximale Wirkleistung über die Unterspannungsseite des rONT geringer ausfällt. Dies ist mit den kleineren Strömen bei konstanter Leistung (P(U,I)=const) und den sich dadurch ergebenden geringeren Netzverlusten verbunden. In den vorgestellten Szenarien (Winter) fallen die Netzverluste um etwa 2 kW geringer aus (Szenarien 2+3 im Vergleich zu den Szenarien 1, 4 und 5).

Die Verläufe der Knotenspannungen, des Spannungssollwertes für den rONT, sowie die Stufenpositionen sind in den nachfolgenden Abbildungen für einige ausgewählte Szenarien dargestellt. Der Simulationszeitraum bezieht sich jeweils auf einen Tag (24 h).

Bild 8 und **Bild 9** beziehen sich dabei auf Szenarien mit Daten aus dem Winter (15.01.14), **Bild 10**, **Bild 11** und **Bild 12** auf Sommerszenarien (15.06.14).

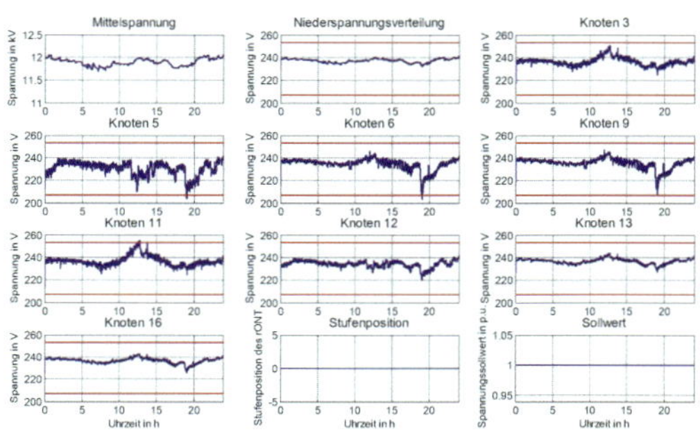

Bild 8 Kotenspannungen Winterszenario bei deaktiviertem rONT

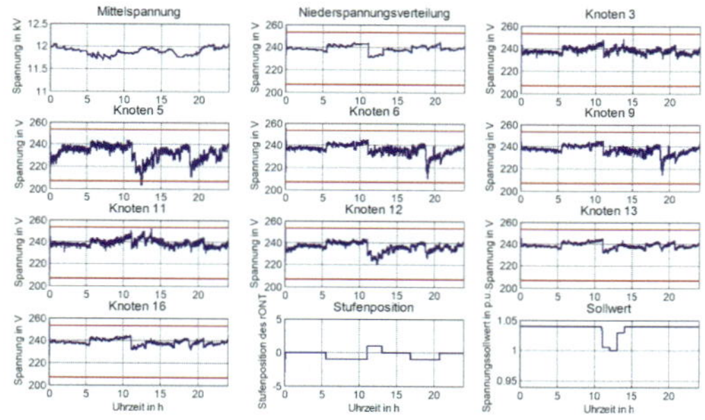

Bild 9 Kotenspannungen Winterszenario bei aktiviertem rONT mit variabler Sollwertvorgabe in den Grenzen 1,0 und 1,04 p. u.

In **Bild 9** ist ersichtlich, dass das Zeitfenster, in welchem der Spannungssollwert aufgrund der entsprechenden Strahlungsdichte heruntergesetzt wird, sich auf wenige Stunden am Tag beschränkt. Die meiste Zeit wird die Spannung durch die Vorsteuerung (variable Sollwertvorgabe) hochgesetzt, um die Netzverluste zu reduzieren. Dies ist möglich, da im Winter die Lasten in Photovoltaik geprägten Niederspannungsnetzen dominieren.

Die folgenden drei Abbildungen beziehen sich auf die Szenarien, welche mit Messdaten aus dem Sommer durchgeführt worden sind. In **Bild 10** sind die Ergebnisse bei deaktiviertem rONT-Regler (entspricht starrer Kopplung von Mittel- und Niederspannung wie bei einem konventionellem Transformator) dargestellt. Hier kommt es in diesem Szenario an den Knoten 3, 5, 6, 9 und 11 zu Verletzungen des zulässigen Spannungstoleranzbandes nach [1].

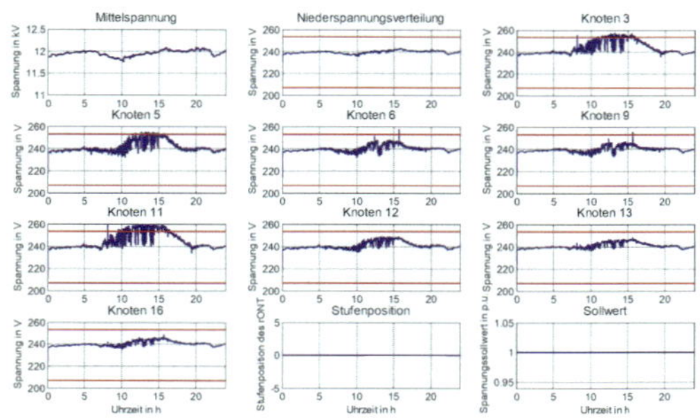

Bild 10 Kotenspannungen Sommerszenario bei deaktiviertem rONT

In **Bild 11** sind die Berechnungsergebnisse bei rONT-Einsatz im Monosensorbetrieb (Spannungsistwerterfassung an der Niederspannungsverteilung) bei einem eingestellten Sollwert von 1,0 p. u. dargestellt. Hierbei ist zu sehen, dass es am Knoten 11 weiterhin zu einer Verletzung des nach Norm [1] zulässigen Spannungstoleranzbandes kommt.

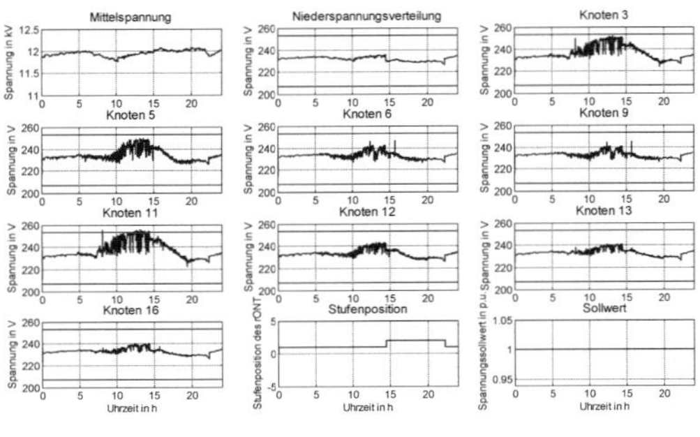

Bild 11 Kotenspannungen Sommerszenario bei rONT im Monosensorbetrieb bei einem Spannungssollwert von 1,0 p. u.

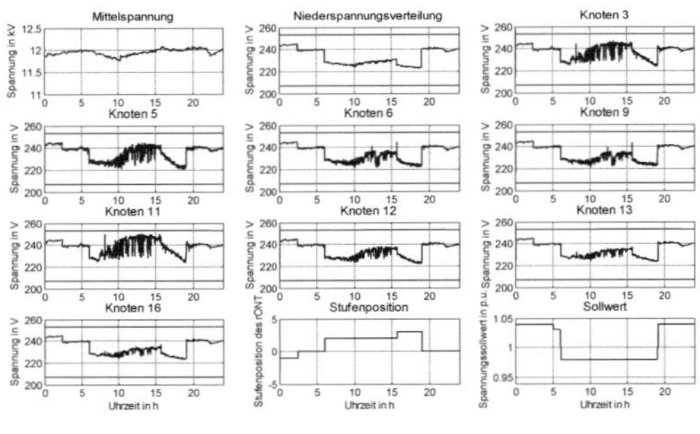

Bild 12 Kotenspannungen Sommerszenario bei rONT-Einsatz im Monosensorbetrieb mit variabler Sollwertvorgabe in den Grenzen 0,98 und 1,04 p. u.

Bild 12 beinhaltet die Berechnungsergebnisse bei rONT-Einsatz im Monosensorbetrieb unter Berücksichtigung der zuvor vorgestellten variablen Sollwertvorgabe in Abhängigkeit der gemessenen Solarstrahlung in den Grenzen 0,98 p. u. (Erzeugungsfall) und 1,04 p. u. (Lastfall). Die zuvor bestehende Spannungsbandverletzung aus **Bild 11** wird hier beseitigt. Des Weiteren ist beim Vergleich der Abbildungen **Bild 9** und **Bild 11** ersichtlich, dass im Sommer das Zeitfenster, in welchem der Spannungssollwert anhand der Kennlinie (**Bild 4**) reduziert wird, deutlich größer ausfällt. Lediglich über Nacht wird der Spannungssollwert in diesem Szenario erhöht.

4 Verluste des regelbaren Ortsnetztransformators

Der regelbare Ortsnetztransformator besitzt im Vergleich zu einem konventionellen Ortsnetztransformator zusätzlich eine Reglereinheit in der Ortsnetzstation, sowie den Laststufenschalter. Diese Zusatzkomponenten führen zu höheren Verlusten. Des Weiteren sind die Hersteller verpflichtet, durch geeignete Materialien die Verteilnetztransformatoren verlustärmer zu gestalten. [7]

Durch die Schaltungen kommt es zu geringfügigen zusätzlichen Verlusten. Diese Zusatzverluste im Vergleich zu konventionellen Verteilnetztransformatoren müssen den durch den rONT bei entsprechend intelligenter Regelung eingesparten Netzverlusten durch eine Spannungserhöhung an der Niederspannungsverteilung zukünftig noch gegenübergestellt werden.

5 Zusammenfassung und Ausblick

In den vorgestellten Szenarien der vorangegangenen Abschnitte ist ersichtlich, dass eine Berücksichtigung der Solarstrahlung zur variablen Sollwertvorgabe sinnvoll erscheint. Durch einen minimalen zusätzlichen Aufwand an Sensoren (Solarstrahlungsmessung an der Ortsnetzstation) ergibt sich für den rONT im Monosensorbetrieb somit zusätzlich die Möglichkeit, die Spannung mit dem Ziel der Netzverlustminimierung zu Lastzeiten hochzusetzen. Dies ist hauptsächlich im Winter der Fall und die Voraussetzung ist, dass im Netz überwiegend nichtlineare Lasten vorhanden sind, die bei Erhöhung der Spannung keine größeren Ströme ziehen.

In den Sommermonaten fällt die Solarstrahlungsintensität und damit der Spannungshub über der Netzimpedanz durch dezentrale Einspeisung aus Photovoltaikanlagen größer aus, sodass in diesen Zeiten der rONT ausschließlich zur Spannungsregelung und der Einhaltung des Spannungsbandes nach DIN EN 50160 [1] genutzt wird. Um eine abschließende Aussage über den Zusatznutzen, mit dem rONT Netzverluste reduzieren zu können, zu erhalten, werden derzeit noch weitere Untersuchungen im Rahmen eines Forschungsprojektes [8] durchgeführt, um die Zusatzverluste des rONT den eingesparten Netzverlusten gegenüberstellen zu können.

Die Ergebnisse stammen zum Großteil aus dem Verbundforschungsprojekt „e-home Energieprojekt 2020“ [8], das von der Avacon AG und dem Energie-Forschungszentrum Niedersachsen (EFZN) gefördert wird. Im Rahmen dieses Projektes werden die aktuellen Anforderungen an Niederspannungsverteilnetze unter Berücksichtigung neuer Technologien (rONT, Speichersysteme, e-KFZ etc.) untersucht. Hierzu wurden 30 Haushalte in zwei Niederspannungsnetzen bei der Anschaffung der zuvor genannten Technologien unterstützt und die Netze wurden durch die Avacon AG mit entsprechender Messinfrastruktur ausgestattet, auf deren Messwerte die Projektbearbeiter für die Untersuchung zugreifen können.

6 Literatur

[1] DIN EN 50160: Merkmale der Spannung in öffentlichen Elektrizitätsversorgungsnetzen; CENELEC 2010

[2] Schnieder, R.; Wehrmann, E.-A.; Beck, H.-P.; Untersuchung verschiedener Regelungskonzepte für regelbare Ortsnetztransformatoren zur Spannungshaltung in Niederspannungsverteilnetzen; NEIS-Konferenz 12.-13. September 2013, Hamburg, Konferenzband

[3] Becker, A.; Schnieder, R.; Beck, H.-P. et.al.; Netzausbauvarianten in Niederspannungsverteilnetzen – Regelbare Ortsnetztransformatoren in Konkurrenz zu konventionellen Netzausbaumaßnahmen; Cuvillier-Verlag; ISBN:978-3-95404-757-4, 2014

[4] Schnieder, R.; Wehrmann, E.-A.; Beck, H.-P.; Spannungsregelung in Niederspannungsverteilnetzen durch regelbare Ortsnetztransformatoren im Mono- und Multisensorbetrieb; NEIS-Konferenz 18.-19. September 2014, Hamburg, Konferenzband

[5] Schmiesing, J.; Beck, H.-P.; Modellhafte Dezentralisierung der Spannungsregelung eines Umspannwerksbereichs, ew Heft 06/2013; Jg. 112; S.68 – 72; März 2013

[6] Maschinenfabrik Reinhausen GmbH; Gridcon Transformer – Der regelbare Ortsnetztransformator; Produktbeschreibung 2012

[7] Amtsblatt der Europäischen Union; Verordnung (EU) Nr. 548/2014 der Kommission vom 21. Mai 2014 zur Umsetzung der Richtlinie 2009/125/EG des Europäischen Parlaments und des Rates hinsichtlich Kleinleistung-, Mittelleistungs- und Großleistungstransformatoren

[8] e-home Energieprojekt der Avacon AG in Kooperation mit dem Energie-Forschungszentrum Niedersachsen (www.ehomeprojekt.de)

[9] Ratsch, P.; Nachhaltige Netzentwicklung – Regelbare Ortsnetztransformatoren bei der E.ON Avacon AG; Energietechnische Gesellschaft im VDE; ETG-Mitgliederinformation; Seiten 7 – 10; Nr.2 Juni 2013

Netzdienlicher Einsatz von Speichersystemen in Verteilungsnetzen

Sören Schrader, M.Sc, P3 energy, Aachen, Germany, Soeren.Schrader@p3-group.com
Dipl.-Ing. Philipp Goergens, RWTH Aachen University, Aachen, Germany, Goergens@ifht.rwth-aachen.de
Dr.-Ing. Christian Hille, P3 energy, Aachen, Germany, Christian.Hille@p3-group.com
Univ.-Prof. Dr.-Ing. Armin Schnettler, RWTH Aachen University, Aachen, Germany, Schnettler@ifht.rwth-aachen.de

Kurzfassung

Im Rahmen dieses Beitrags wird der netzdienliche Einsatz von dezentralen Speichersystemen in Verteilungsnetzen analysiert. Dabei stellt das gesamte Niederspannungsnetz den Betrachtungsrahmen dar und es erfolgt eine Optimierung von systemischen Kenngrößen durch eine Flexibilisierung der Last bzw. Einspeisung mittels Speichersystemen. Anhand von zwei Niederspannungstypnetzen und zwei zukünftigen Last- und Erzeugungsszenarien werden unterschiedliche Dimensionierungen und Platzierungen von Speichersystemen untersucht. Dabei stehen insbesondere die Unterschiede zwischen einem Einsatz vieler Heimspeichersysteme versus einzelner größer dimensionierter Netzspeicher im Fokus der Untersuchungen. Zudem werden das Optimierungsmodell und die Vorgehensweise dargestellt und exemplarische Untersuchungen diskutiert.

1 Hintergrund und Zielsetzung

Verteilungsnetzbetreiber werden durch die stetig wachsende Integration dezentraler Energiewandlungseinheiten (DEA) vor neue Herausforderungen gestellt. Bereits heute kommt es in ländlichen Verteilungsnetzen zur Verletzung von zulässigen Spannungsbändern und zu hoch ausgelasteten oder zeitweise überlasteten Betriebsmittel bei zeitgleicher Einspeisung vieler dezentraler Erzeuger [1].

Klassischer Netzausbau durch Netzverstärkung stellt nicht mehr die einzige Lösungsoption zur Begegnung dieser Herausforderung dar. Viele neue z. T. intelligente und steuerbare Netzbetriebsmittel bieten ebenso das Potenzial, maximale (minimale) Knotenspannungen zu reduzieren (anzuheben) und Betriebsmittelüberlastungen zu vermeiden. Hierzu zählen u. a. Batteriespeichersysteme [1]. Durch die Fähigkeit des bidirektionalen Wirk- und Blindleistungsaustauschs mit dem Netz bieten diese vielfältige Möglichkeiten zum netzdienlichen Einsatz.

Ziel dieses Beitrags ist es daher, die Auswirkungen einer netzdienlich optimierten Speicherfahrweise im Verteilungsnetz zu untersuchen.

Im Fokus der Untersuchungen steht der Einfluss auf systemische Kenngrößen wie z. B. die Knotenspannung oder Verluste im Netzgebiet. Ebenso werden die Auswirkungen der Platzierung und Dimensionierung der Speichersysteme im Verteilungsnetz näher betrachtet und die Unterschiede zwischen vielen dezentralen Heimspeichersystemen versus einiger größer dimensionierter zentraler Speichersysteme (Netzspeicher) aufgezeigt.

2 Methodik

Zur technischen Analyse wurde das Verfahren der Probabilistischen Leistungsflussrechnung ([1], [2]) um einen Ansatz zur Optimierung des Gesamtsystems Niederspannungsnetz und eine zeitkoppelnde Betrachtung von Batteriespeichersystemen erweitert. Durch Modifikation der Zielfunktion können verschiedene bewertungsrelevante systemische Kenngrößen optimiert werden. Im Weiteren werden das entwickelte Verfahren und die Implementierung von zwei Zielfunktionen zur Bereitstellung von Netzdienstleistungen vorgestellt.

2.1 Probabilistisches Optimal-Power-Flow-Verfahren

In Anlehnung an das in [1] und [2] verwendete Verfahren der Probabilistischen Leistungsflussrechnung besteht das *Probabilistische Optimal-Power-Flow-Verfahren* aus drei wesentlichen Bestandteilen:

- Profilzuordnung: Allen Netzknoten werden entsprechend der Verteilung von Verbrauchern und Erzeugern Last- bzw. Erzeugungsprofile zugeordnet. Zusätzlich wird eine feste Platzierung und Konfiguration von Batteriespeichern im Netzgebiet definiert.
- Deterministische optimale Leistungsflussrechnung: Die Netzdaten inkl. der Verteilung von DEA, sowie Last- und Erzeugungsdaten inkl. Wetterdaten dienen als Input für die gemäß einer vorgegebenen Zielfunktion durchgeführten Optimierung mittels Optimal-Power-Flow-Algorithmus mit integriertem zeitgekoppeltem Batteriemodell.
- Ergebnisauswertung: Die Ergebnisse von insgesamt 100 durchgeführten Wiederholungen mit unterschiedlicher Profilzuordnung von Lasten und Erzeugern sowie mit wechselnden Standorten von zukünftigen DEA werden mittels Quantilen hinsichtlich ihrer Maximal- (95 %) und Minimalwerte (5 %) der Kenngrößen bewertet. Als Bewertungskriterien dienen die folgenden zeitpunktbezogenen Netzzustandskenngrößen:
    - max. / min. Knotenspannung
    - max. Betriebsmittelauslastung
    - max. Verlustleistung

und die daraus abgeleiteten zeitraumbezogenen Kenngrößen für Wirk-, Blind- und gesamte Verlustenergie des Netzes.

Es wird ein Netzanschluss der Batteriespeicher über einen Wechselrichter an das Niederspannungsnetz angenommen, wodurch ein vollständiger vier Quadranten Betrieb ermöglicht wird. Sowohl für die maximalen Ein- und Ausgangsleistungen P_e^{max} bzw. P_a^{max} als auch für die Ein- und Ausgangswirkungsgrade η_e bzw. η_a werden fixe Werte vorgegeben. Im Rahmen dieses Beitrags wird vereinfachend $\eta_e=\eta_a=1$ angenommen.

Weiterhin wurden zwei verschiedene Zielfunktionen zur Bereitstellung von Netzdienstleistungen implementiert, die im Weiteren vorgestellt werden.

Zielfunktion: Statische Spannungshaltung
Mithilfe dieser Zielfunktion kann das Potential von Batteriespeichersystemen zur statischen Spannungshaltung in Niederspannungsnetzen untersucht werden. Dazu wird eine Zielfunktion definiert, welche die Summe der Abweichungen von Knotenspannungen zur Nennspannung über einen gegebenen Betrachtungszeitraum an jedem Ort im Netz minimiert, um flache Spannungsverläufe im Niederspannungsnetz zu erhalten.

Implementiert wird die Zielfunktion als Linearkombination des Kostenkoeffizientenvektors C und dem Gewichtungsvektor w:

$$f(x) = C^T w = \sum_{t=1}^{n_t} \sum_{i=1}^{n_b} C_i * (V_{m,i}^t - 1)^2$$

Der Vektor w gewichtet dabei alle von der Netznennspannung V_m abweichenden Spannungsamplituden in Form einer Parabelverteilung. Die Kostenfunktion für einen Betrachtungszeitraum bestehend aus n_t diskreten Zeitschritten und setzt sich somit als lineare Funktion aus $n_b \times n_t$ (n_b ≙ Anzahl Netzknoten) quadratischen Strafkostentermen zusammen.

Zielfunktion: Verlustminimierung
Eine weitere betrachtete Zielfunktion stellt die Verlustminimierung dar. Mithilfe dieser Zielfunktion werden sowohl Wirk- als auch Blindleistungsverluste im gesamten Netzgebiet minimiert. Aufgrund der quadratischen Beziehung von Stromstärke zu Verlusten nach $P_V(I) = R * I^2$ kommt es dabei zur Reduzierung der Ströme, wodurch eine Verringerung maximaler Betriebsmittelauslastungen erreicht werden kann.

3 Szenariorahmen und Annahmen

Der Szenariorahmen beinhaltet eine Auswahl aus den in [1] vorgestellten Typnetzen und Durchschnitts-Lokalszenarien sowie Angaben zur Dimensionierung und Platzierung der Batteriespeichersysteme (BSS). Die Auswahl der untersuchten Typnetz-/Szenariokombination basiert auf den Ergebnissen von [1], welche durch ungültige Netzzustände (Knotenspannungen und Betriebsmittelüberlastungen) charakterisiert sind.

3.1 Last- und Erzeugungsszenario

Szenarien der zukünftigen Versorgungsaufgabe eines Verteilungsnetzbetreibers werden durch eine Disaggregation von deutschlandweiten Ausbauzielen für dezentrale Energiewandlungseinheiten auf Basis des Szenarios B des Netzentwicklungsplans für 2034 definiert. Das Ergebnis der Disaggregation stellt eine Anzahl an DEA unterschiedlichen Typs sowie deren durchschnittliche Leistung dar. Zu den betrachteten Anlagen zählen Photovoltaikanlagen (PV), Kleinwindenergieanlagen (KWEA), Wärmepumpen (WP), Elektrofahrzeuge (EV) und Kraft-Wärme-Kopplungsanlagen (KWK). Die den Untersuchungen zugrundeliegende Versorgungsaufgabe wird durch die in **Tabelle 1** dargestellten Durchdringungsraten und durchschnittlichen Anlagenleistungen charakterisiert.

Tabelle 1: Betrachtete Versorgungsaufgaben in Anlehnung an [1]

	Ø Durchdringungsraten in % (PV/KWEA/KWK/WP/EV)	**Ø Anlagenleistung** in kW (PV/KWEA/KWK/WP/EV)
Szenario 1	38,6 / 6,6 / 10 / - / -	15,4 / 1 / 4,1 / - / -
Szenario 2	43,0 / 6,6 / 10 / - / -	17,4 / 5 / 4,1 / - / -

3.2 Betrachtete Typnetze

Betrachtet werden sowohl ein ländliches als auch ein vorstädtisches Typnetz, welche als kritische Netze im Hinblick auf Spannungsbandverletzungen und Betriebsmittelauslastungen bei Anwendung der in Kapitel 3.1 vorgestellten Szenarien im Rahmen von [1] identifiziert wurden. Da Heimspeichersysteme vornehmlich in mäßig besiedelten Gebieten mit höheren PV-Durchdringungsraten installiert werden, wird der Einsatz dieser anhand des ländlichen Typnetzes 1 bewertet. In dichter besiedelten Gebieten sind die PV-Durchdringungsraten eher geringer, sodass folglich weniger Heimspeichersysteme installiert werden. Daher wird für die Betrachtung von Netzspeichern das Typnetz 2 herangezogen, welches eine vorstädtische Struktur aufweist.

3.2.1 Typnetz 1 – ländliche Struktur

Das untersuchte Typnetz 1 repräsentiert ein ländliches Netz bestehend aus 72 Netzknoten, welche sich auf drei Netzstränge mit insgesamt 55 Gebäuden aufteilen. Es befinden sich insgesamt drei landwirtschaftliche Betriebe jeweils am Ende eines Strangs. An dem nach unten abzweigenden Strang liegt ein landwirtschaftlicher Betrieb am Ende eines 800 m langen Ausläufers. Zusammen entfallen ca. 25 % des Jahresenergieumsatzes (JEU) auf die drei Betriebe. Die restlichen 75 % des JEU entfallen auf private Haushalte und einen im Zentrum des Netzgebietes liegenden Gewerbebetrieb. Die Netztopologie sowie die Platzierung der Batteriespeicher ist in **Bild** 1 dargestellt.

Dimensionierung und Platzierung von Batteriespeichersystemen
Zur Bewertung des Einflusses der Dimensionierung von BSS, werden drei verschiedengroße Batterietypen mit Kapazitätswerten von 8, 16 und 24 kWh bei einem festen Verhältnis von Energie zu Leistung von 1,6 modelliert.

Mit Blick auf das Jahr 2034 werden Durchdringungsraten bezogen auf die Anzahl der Gebäude im Netzgebiet von 5 % bzw. 15 % angenommen, woraus sich eine Anzahl, der installierten BSS von 3 bzw. 8 Batterien ergibt. Die Entscheidung zur Platzierung der Batterien geschieht aufgrund der Ergebnisse (Spannungsbandverletzungen und Betriebsmittelüberlastungen) aus [1] und den damit bekannten Betriebszuständen ohne zusätzliche Technologieoptionen.

Heimspeichersysteme (Typnetz 1)	
Durchdringungsraten	5 % (grüne BSS) 15 % (grüne + rote BSS)
Gesamtkapazität pro BSS (kWh)	8 / 16 / 24
Leistung pro BSS (kW)	5 / 10 / 15
Energie zu Leistung	1,6
Kapazität verfügbar	10 – 100 %

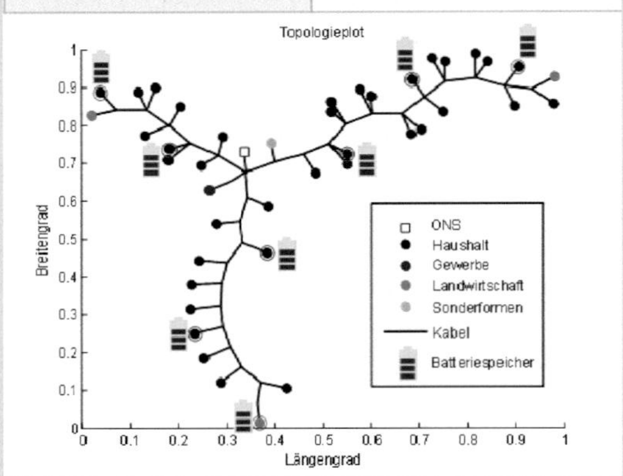

Bild 1: Verteilung und Dimensionierung der Heimspeicher

3.2.2 Typnetz 2 – vorstädtische Struktur

Typnetz 2 bildet einen städtischen Raum mit 282 Netzknoten ab, welche sich auf 4 Stränge mit insgesamt 155 Gebäuden aufteilen. Der Jahresenergiebedarf entfällt zu 97 % auf private Haushalte und zu 3 % auf zwei gewerbliche Lasten in der Nähe der Ortsnetzstation (ONS). Das Netz ist zu 100 % verkabelt und die Leitungen sind als 150 mm² Aluminiumkabeln ausgeführt. Die Netztopologie sowie die Platzierung der Batteriespeicher sind in **Bild 2** dargestellt.

Dimensionierung und Platzierung von Batteriespeichersystemen

Unter gleichbleibenden Annahmen zu Durchdringungsraten von BSS von 5 % bzw. 15 % bezogen auf die Gebäudeanzahl im Netzgebiet ergeben sich für die Batterietypen 2 (16 kWh) und 3 (24 kWh) eine aggregierte Kapazität von 128 kWh bzw. 192 kWh bei einer 5 %igen Durchdringung und 368 kWh bzw. 552 kWh bei einer 15 %igen Durchdringung für das gesamte Netz. Diese berechneten Gesamtkapazitäten bilden die Basis der vier möglichen BSS-Konfigurationen. Für jede Konfiguration wird die verfügbare Gesamtspeicherkapazität auf die Netzstränge entsprechend der relativen Gebäudeanzahl pro Netzstrang verteilt (Strang1 (oben): 35,7 %; Strang 2(rechts): 14,9 %; Strang 3/4 (unten/links): jeweils 24,7 %).

Übersteigt die Leistung eines Speichers die Leitungskapazität am Netzanschlusspunkt, wird die zugewiesene Gesamtkapazität des Strangs auf zwei Netzspeicher aufgeteilt, sodass es durch alleinige Batterieaktivität nicht zu Betriebsmittelüberlastungen kommen kann.

Netzspeichersysteme (Typnetz 2)	
Anzahl pro Netzstrang	1 / 2
Kapazität gesamt im Netz (kWh)	128 / 192 / 368 / 552
relative Kapazitätsverteilung pro Strang in %	35,7 / 14,9 / 24,7 / 24,7
Kapazität verfügbar	10 – 100 %
Energie zu Leistung	1,6
Platzierung der BSS	
Konfiguration 1	1, 4, 6, 8
Konfiguration 2	1, 4, 6, 8
Konfiguration 3	1, 3, 4, 6, 8
Konfiguration 4	2, 3, 4, 5, 6, 7, 8

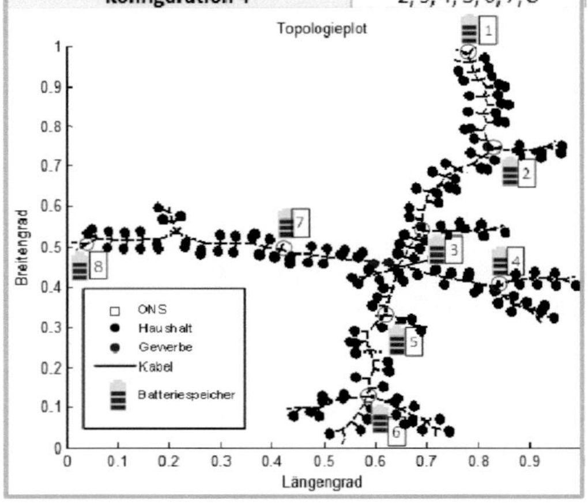

Bild 2: Verteilung und Dimensionierung der Netzspeicher

4 Exemplarische Ergebnisse

Im Folgenden werden exemplarische Ergebnisse anhand beider Optimierungszielfunktionen in beiden Typnetzen für eine exemplarische Sommer- und Winterwoche aufgezeigt und diskutiert. Zur Bewertung der Knotenspannungswerte dient der relative Längsspannungsfall (-hub) im betrachteten Niederspannungsnetz im Vergleich zum Referenzfall ohne BSS-Einsatz. Die Slack-Knotenspannung wird in allen Simulationen zu 1 p. u. angenommen. Zur Bewertung der Auslastung von Betriebsmitteln wird der Grenzwert auf 100 % der Nennscheinleistung festgelegt.

Als Referenz werden Simulationen ohne Speichereinsatz betrachtet.

Statische Spannungshaltung

Zur Betrachtung der statischen Spannungshaltung werden beide Typnetze im Szenario 2 ausgewertet. Die maximalen Spannungshübe in der Sommerwoche ohne Speichereinsatz sind in **Bild 3** für Typnetz 1 und in **Bild 4** für Typnetz 2 dargestellt.

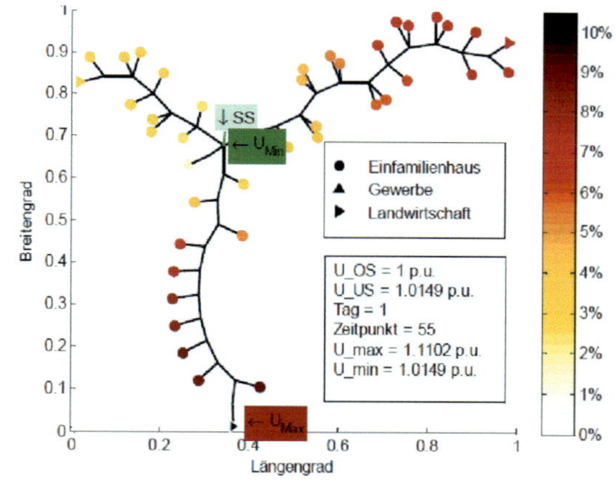

Bild 3: Maximale Spannungshübe im Referenzszenario in der Sommerwoche in Typnetz 1

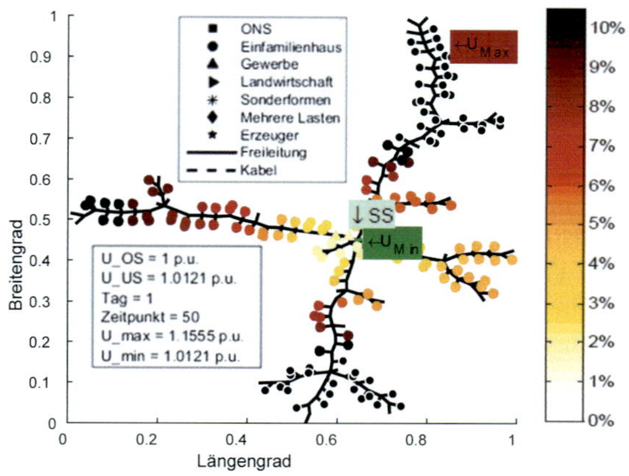

Bild 4: Maximale Spannungshübe im Referenzszenario in der Sommerwoche in Typnetz 2

Dabei werden maximale Werte von 1,11 p. u, im Typnetz 1 und 1,16 p. u. im Typnetz 2 erreicht. Bei alleiniger statischer Spannungshaltung durch BSS können die maximalen Knotenspannungen mit steigender Batteriekapazität und Leistung, sowie mit steigender Anzahl von Batterien in beiden Typnetzen gesenkt werden (vgl. **Bild 5**, **Bild 6**), während die minimalen Knotenspannungen angehoben werden.

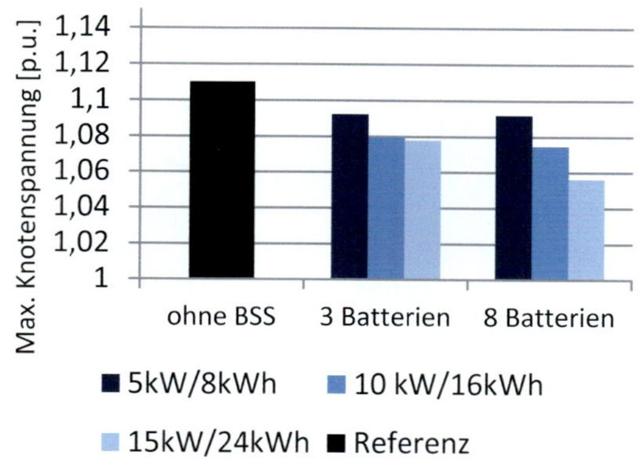

Bild 5: Potential zur Senkung der max. Knotenspannung durch BSS im Typnetz 1 ohne cosφ(P)-Steuerung der DEA

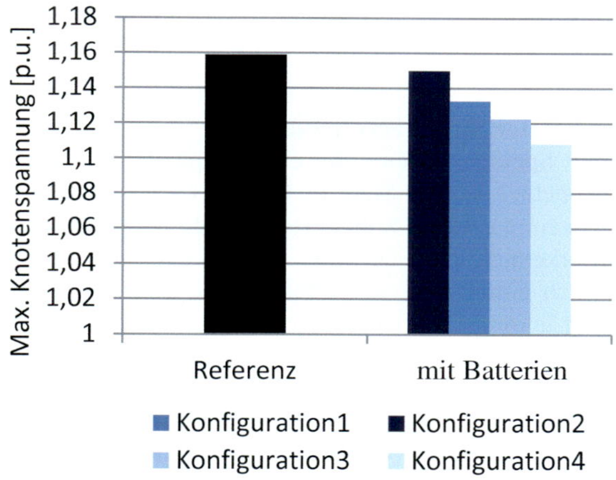

Bild 6: Potential zur Senkung der max. Knotenspannung durch BSS im Typnetz 2 ohne cosφ(P)-Steuerung der DEA

In beiden Typnetzen ist es somit durch Batteriespeichersysteme möglich, einen Beitrag zur statischen Spannungshaltung zu leisten. Dabei kann im Typnetz 1 der maximale Längsspannungshub um bis zu 5,4 %-Punkte reduziert werden, in Typnetz 2 um bis zu 5,1 %-Punkte. Werden zusätzlich zum Batterieeinsatz alle DEA mit einer wirkleistungsabhängigen Blindleistungsregelung nach einer Standardkennlinie implementiert, so lässt sich die maximale Knotenspannung im Typnetz 1 um bis zu 8,2 %-Punkte und im Typnetz 2 um bis zu 11,3 %-Punkte (vgl. **Bild 7**) reduzieren.

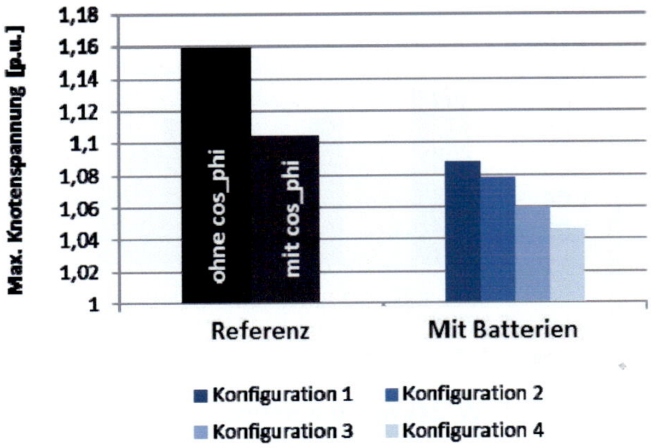

Bild 7: Potential zur Senkung der max. Knotenspannung durch BSS im Typnetz 2 mit cosφ(P)-Steuerung der DEA

Die minimale Knotenspannung in der Winterwoche hingegen kann in beiden Typnetzen von 0,95 p. u. auf Nennspannung und somit um ca. 5 %-Punkte angehoben werden.

Fahrplan der Speichersysteme

Für drei Speichersysteme mit einer Kapazität von 8 kWh im Typnetz 1 ist der State of Charge (SoC) Verlauf für einen exemplarischen Sommertag in **Bild 8** dargestellt. Es ist zu erkennen, dass der SoC Verlauf aller Speichersysteme sich ähnelt und alle Speicher ca. zwei Vollzyklen pro Tag fahren. Dabei wird der komplette zur Verfügung stehende SoC-Bereich ausgenutzt. Außerdem ist die durch die Probabilistik bedingte Ungewissheit des Speicherstands zu vielen Zeitpunkten sehr gering und erreicht einen Maximalwert von ca. 60 % zwischen Mitternacht und ca. 7:00 morgens.

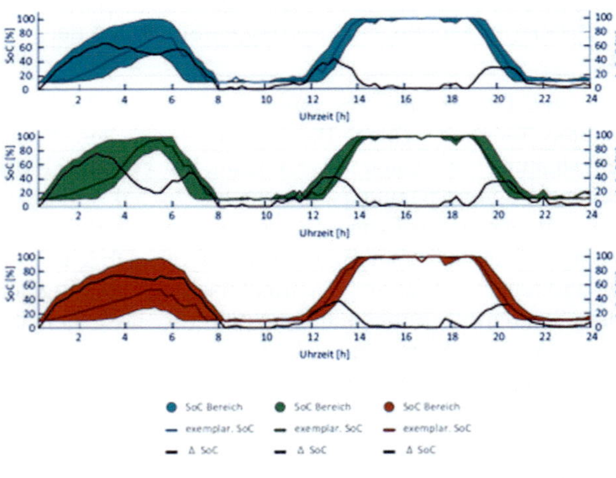

Bild 8: SoC-Verlauf dreier Speichersysteme an einem exemplarischen Sommertag im Typnetz 1 bei der Zielfunktion zur statischen Spannungshaltung

Verlustminimierung

Bei der Betrachtung der Optimierungszielfunktion zur Verlustminimierung und somit indirekt zur Reduzierung der Betriebsmittelauslastung bildet Szenario 1 ohne den Einsatz von Batteriespeichersystemen die Referenz. Im weiteren Verlauf der Ergebnisdiskussion wird detailliert auf das Typnetz 1 eingegangen. In der Ausgangssituation treten in der Sommerwoche aggregierte Verluste i.H.v. 1.468 kVAh auf (22 % der Wirk- und 69 % der Blindverluste treten am Transformator selbst auf) und es kommt es zu einer maximalen Betriebsmittelauslastung von 1,14 p. u. Die Auslastung aller Betriebsmittel ist für den Tag der höchsten Auslastung im Tagesverlauf in **Bild 9** dargestellt.

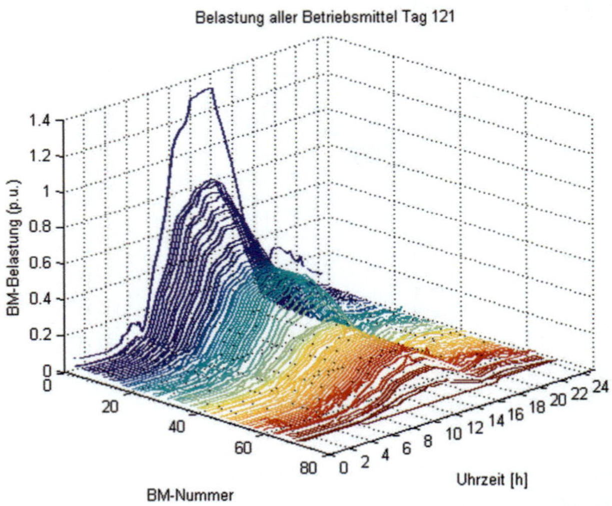

Bild 9: Maximale Auslastung aller Betriebsmittel in absteigender Reihenfolge

Das am höchsten ausgelastete bzw. überlastete Betriebsmittel stellt in diesem Fall der Ortsnetztransformator dar. Zur Verortung der hoch ausgelasteten Betriebsmittel im Netzgebiet ist für den Zeitpunkt der höchsten Auslastung an diesem Tag die Leitungsauslastung in der Netztopologie in **Bild 10** dargestellt. Zu sehen ist, dass für die drei von der Ortsnetzstation abgehenden Stränge gilt, dass je näher ein Leitungssegment an der Ortsnetzstation liegt, die Auslastung umso höher ist.

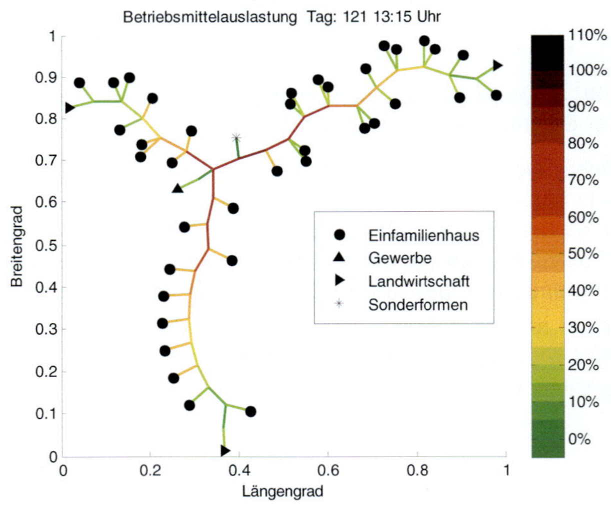

Bild 10: Leitungsauslastung zum Zeitpunkt der maximalen Auslastung in der Sommerwoche als Topologieplot

Zum dargestellten Zeitpunkt um 13:15 Uhr des ersten Tages der Sommerwoche erfolgt eine starke PV-Einspeisung, die zur Rückspeisung und in diesem Falle zur Überlastung des Ortsnetztransformators führt.

Durch den Einsatz von verteilten Speichern im Netz (vgl. **Bild 11**) lässt sich die maximale Betriebsmittelauslastung reduzieren, sodass keine Betriebsmittel überlastet werden.

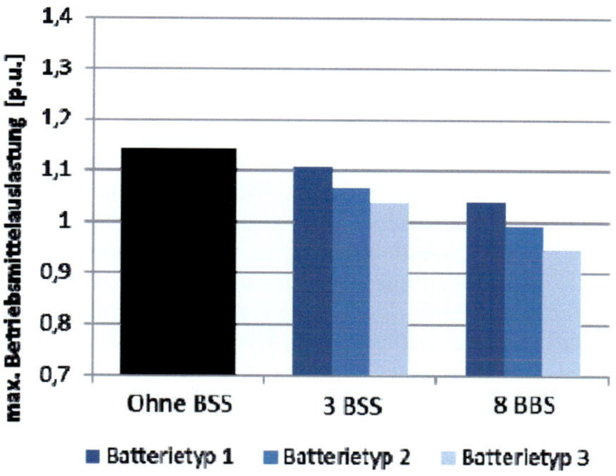

Bild 11: Übersicht der maximalen Betriebsmittelauslastungen aller BSS-Konfigurationen ohne cosφ(P)-Steuerung der DEA

Insgesamt ist ein abnehmender Trend der Belastung mit zunehmender Batteriedimensionierung und Anzahl zu erkennen.

Werden 8 Batteriespeicher vom Batterietyp 3 (15 kW, 24 kWh) eingesetzt, so kann die maximale Betriebsmittelauslastung auf 95 % und somit um 19,5 %-Punkte reduziert werden. Des Weiteren sind durch die verlustminimierende Fahrweise Einsparungen von ca. 254 kVAh (-17,3 %) in der betrachteten Sommerwoche möglich.
Bei zusätzlicher cos (P)-Steuerung der dezentralen Anlagen kommt es im Ausgangsszenario ohne Batteriespeicher zunächst zu höheren Verlusten i. H. v. 1851 kVAh sowie zu deutlich erhöhter Betriebsmittelbelastung.

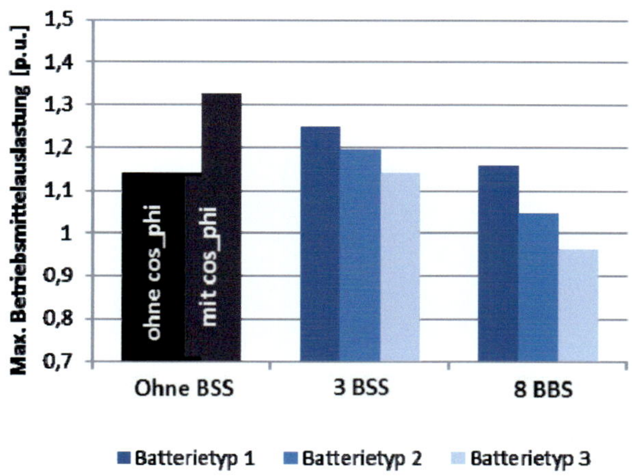

Bild 12: Übersicht der maximalen Betriebsmittelauslastungen aller BSS-Konfigurationen mit cosφ(P)-Steuerung der DEA

Diese können allerdings durch den Einsatz von 8 Batteriespeichern des Typs 3 auf ein ähnliches Niveau (1.258 kVAh) wie im Falle ohne cosφ(P)-Steuerung der dezentralen Anlagen reduziert werden. Ebenfalls lässt sich die Betriebsmittelauslastung auf ein Niveau wie im Falle ohne cosφ(P)-Steuerung der dezentralen Anlagen auf 96 % reduzieren.

Bei der Betrachtung der Fahrweise der Speichersysteme im Netz ist ein ähnlicher SoC-Verlauf wie bei der Optimierung nach der Zielfunktion zur „statischen Spannungshaltung" erkennbar (vgl. **Bild 13**).

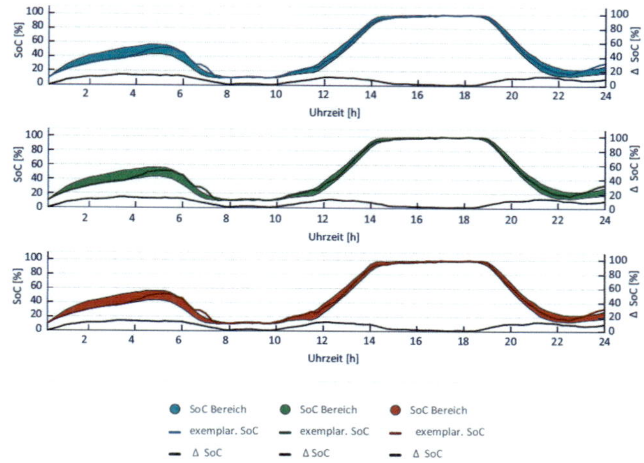

Bild 13: SoC Verlauf dreier Speichersysteme an einem exemplarischen Sommertag im Typnetz 1 bei der Zielfunktion Verlustminimierung

Es kommt ebenfalls zu einer Einspeicherung von Energie in den sehr frühen Morgenstunden (00:00 bis 06:00 Uhr) sowie in den Mittagsstunden (12:00 bis 14:00 Uhr) in denen die PV-Einspeisung ihren Maximalwert erreicht. Eine Ausspeicherung von Energie findet zu den Zeiten der Lastpeaks am Morgen zwischen 07:00 und 08:00 sowie in den Abendstunden zwischen 19:00 und 22:00 Uhr statt. Die durch die Probabilistik bedingte Unsicherheit des Speicherstandes, welche durch den SoC-Bereich ausgedrückt wird, ist im Gegensatz zur Zielfunktion zur statischen Spannungshaltung deutlich geringer und erreicht einen Maximalwert von ca. 20 %.

5 Fazit

Die durchgeführten Untersuchungen in zwei Niederspannungs-Typnetzen und zwei zukünftigen Versorgungsszenarien zur Optimierung des betrachteten Netzgebiets anhand von zwei Zielfunktionen zur Verlustminimierung und zur statischen Spannungshaltung haben gezeigt, dass der Einsatz von dezentral platzierten Speichern in Niederspannungsnetzen eine Option darstellt, um den neuen Herausforderungen im Rahmen der Energiewende zu begegnen.

Es wurde gezeigt, dass Speicher sowohl Betriebsmittelüberlastungen vermeiden als auch Längsspannungshübe- bzw. Längsspannungsfälle deutlich reduzieren können und somit einen Beitrag zur Reduzierung der konventionellen Netzausbaukosten leisten.

Während das Potenzial zur Senkung der maximalen bzw. Anhebung der minimalen Knotenspannungen in Niederspannungsnetzen durch die zunehmende Durchdringung von dezentralen Energiewandlungseinheiten (wie z. B. PV-Anlagen) sehr hoch ist, ist das Potenzial zur Verlustminimierung aufgrund der sehr geringen Verluste in Niederspannungsnetzen sehr gering. Indirekt ergibt sich allerdings durch eine verlustminimierende Fahrweise ein

Nutzen durch die Reduzierung von Betriebsmittelauslastungen. Bei einem netzdienlichen Einsatz der Speichersysteme werden pro Tag ca. 2 Vollzyklen genutzt und dabei der komplette zur Verfügung stehende SoC-Bereich ausgenutzt, was zu einer schnelleren Alterung der Batterie beiträgt. Netzspeicher führen im Vergleich zu Heimspeichersystemen zu ähnlichen netzstützenden Effekten, allerdings gilt es aufgrund der größer dimensionierten Speicher, die Anschlussleistung vor Ort zu beachten, sodass eine Speicheraktivität selbst nicht direkt zu Betriebsmittelüberlastungen führt und ein flexibler Speichereinsatz gewährleistet werden kann.

6 Literatur

[1] Hille, Christian; Schrader, Sören et al: Studie: „Technologieoptionen für den Verteilungsnetzausbau in Deutschland", 2013

[2] Gödde, Markus; Potratz, Fabian; Matrose, Claas; Schnettler, Armin; Brandl, Stephan; Draxler, Reinhard; Ortner, Manfred: „Statistische Analyse der Implikationen dezentraler Energiewandlungseinheiten auf Verteilungsnetze mit Hilfe probabilistischer Lastflussrechnung", VDE Kongress 2012 Smart Grid : Intelligente Energieversorgung der Zukunft, Kongressbeitäge 5. - 6. November 2012, Internationales Congresscenter Stuttgart (ICS), Berlin, 2012, ISBN: 978-3-8007-3446-7

[3] Hille, Christian; Schrader, Sören et al: Studie: „Technische und wirtschaftliche Potenziale von Speichersystemen in Verteilungsnetzen", 2013

Wirkungsgrad und Energieeffizienz-Kennzeichnung für LED-Lampen

Dr.-Ing. Reinhard Jaschke, Helmut-Schmidt-Universität/Leistungselektronik, 22043Hamburg, Germany,
reinhard.jaschke@hsu-hh.de

Kurzfassung

Heute ersetzen LED-Lampen die quecksilberhaltigen Leuchtstofflampen (FL) und Kompaktleuchtstofflampen (CFL). Wichtig sind dabei der Wirkungsgrad (Effizienz) der Stromversorgung und die Lichtausbeute (Efficacy) der LED-Beleuchtung. Herkömmliche „Leuchtstoffröhren“ (FL) können durch LED-Röhren ersetzt werden um den Leistungsumsatz zu halbieren. Umgangssprachliche Energiesparlampen (CFL) können durch Faden-LED ausgetauscht werden um die eingesetzte Wirkleistung zu dritteln. Zur Beurteilung dieser nachhaltigen Energieversorgung ist die Kenntnis der Energieeffizienz-Label A^{++}, A^{+}, A bis E erforderlich und es wird deren Berechnung erläutert und Missstände aufgezeigt.

1 Einleitung

Bei der LED-Röhre werden die Vorteile der erhöhten Lebensdauer (50.000 h), halbem Energieverbrauch, sofortiger voller Leuchtkraft und Quecksilberfreiheit genutzt. Eine LED-Röhre 20 Watt (Retrofit-Variante) wird mit der zu ersetzenden konventionellen „Leuchtstofflampe“ (FL) 36 Watt mit 50-Hz-Drossel als Ballast zur Strombegrenzung und Zündung mit Starter verglichen. Weiterhin wird der Wirkungsgrad und die Lichtausbeute von dimmbaren und nicht dimmbaren Faden-LED gemessen und berechnet.

2 LED ersetzen Leuchtstofflampen

In **Bild 1** ist die konventionelle Leuchtstofflampen-Schaltung (KVG) mit 50-Hz-Drossel (L_{Dr}=1 H; R_{Dr}=66 Ω) als Ballast zur Strombegrenzung und Zündung der Leuchtstoffröhre mit dem Starter am realen Wechselspannungsnetz (L_N=0,8 mH; R_N=0,4 Ω) dargestellt. Ein Kompensationskondensator C_1=4,5 µF kann wahlweise für die Messung zugeschaltet werden.

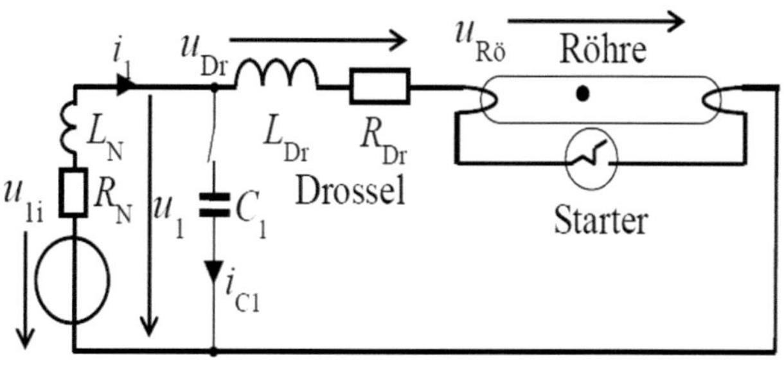

Bild 1 Konventionelle Leuchtstofflampe mit Drossel

In **Bild 2** sind die Messung der Zeitverläufe und die Wirkungsgrad-Messung gezeigt. Der Netzstrom i_1 eilt der Netzspannung u_1 um $\varphi_1 = 60°$ nach. Die nicht sinusförmige Röhrenspannung $U_{RöRMS} = 110.9$ V ist in Phase mit dem Röhrenstrom i_1. Die Drosselverluste betragen $P_{DR} = R_{DR}\, I^2_{1RMS} = 66\ \Omega\, (0{,}3715\ \text{A})^2 = 9{,}24$ W. Dies sind 21,8 % der aufgenommen Leistung $P_1 = 42{,}44$ W. Die nominal 36 W-Röhre nimmt $P_{Rö} = 33{,}16$ W bei der Netzspannung $U_{1RMS} = 230$ V auf.

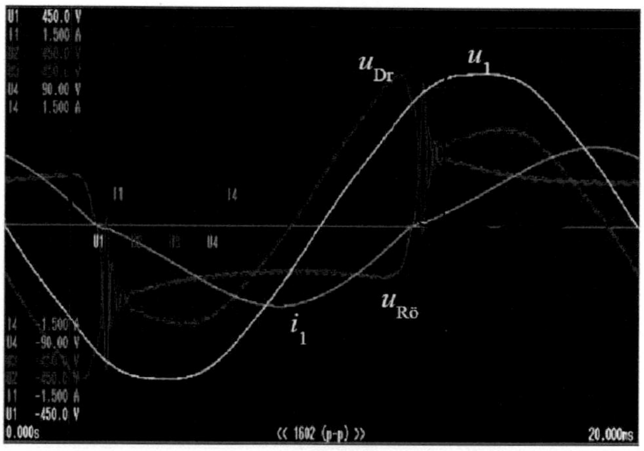

	Netz 1	Röhre Rö	Drossel Dr
U_{RMS}/V	230.10	110.85	182.26
U_{RMN}/V	207.02	106.51	163.78
I_{RMS}/A	371.47 m	371.22 m	371.45 m
I_{RMN}/A	326.10 m	325.88 m	326.08 m
P/W	42.44	33.16	9.24
Q/var	74.19	24.37	67.06
λ	0.4965	0.8057	0.1366
Φ/°	G60.23	G36.32	G82.15
η/%	P_i/P_{Netz}	78.2	21.8

Bild 2 Messung: konventionelle Leuchtstofflampe (KVG)

In **Bild 3** ist die Leuchtstoff-Röhre durch eine LED-Röhre ersetzt und der Starter aus **Bild 1** durch einen Blindstarter (Drahtbrücke oder Feinsicherung) ausgetauscht. Es sind vier parallele Strings von je 20 Reihen-LED verschaltet. Bei einem elektronischem Vorschaltgerät (EVG) darf diese Schaltung nicht ausgeführt werden.

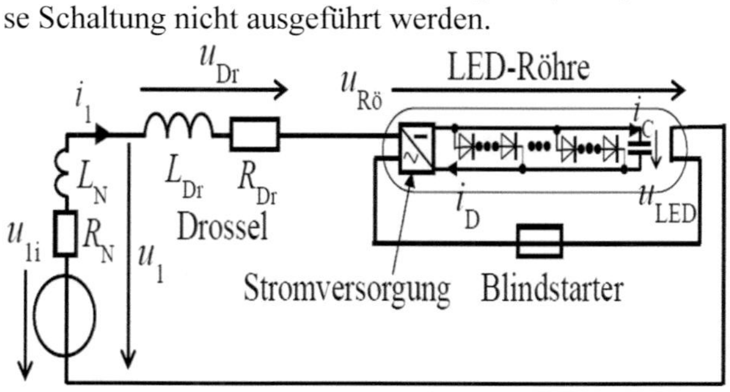

Bild 3 LED-Röhre mit Blindstarter und Drossel (KVG)

In **Bild 4** entstehen in der Drossel Stromwärmeverluste $P_{DR} = R_{DR} I^2_{1RMS} = 66$ □$(0{,}092\ A)^2 = 0{,}56$ W (0,66 W gemessen). Gegenüber **Bild 2** treten 1/4 Strom und damit 1/16 Verluste auf. Dies sind 2,7 % der aufgenommen Leistung $P_1 = 20{,}8$ W. Die nominal 20-W-LED-Röhre hat $P_{Rö} = 20{,}2$ W dem Netz bei $U_{1RMS} = 230$ V entnommen. Über den 80 LED wird die Leistung $P_{LED} = 18{,}75$ W gemessen. In einer einzelnen LED treten die Werte $U_{LEDi} = 61{,}8V/20 = 3{,}09$ V; $I_{LEDi} = 0{,}299A/4 = 74{,}8mA$ und $P_{LEDi} = 0{,}23$ W auf.

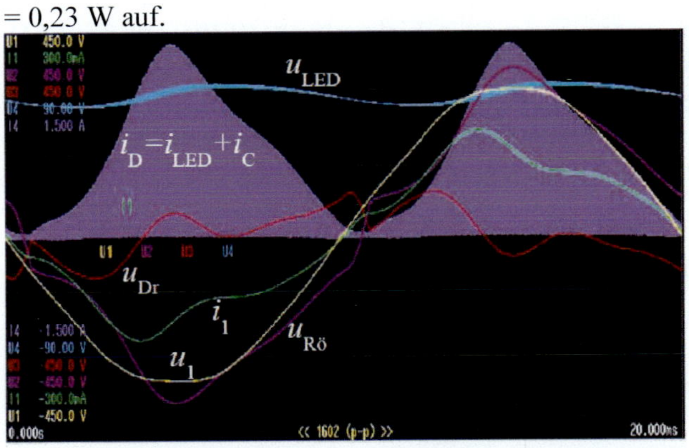

	Netz 1	Röhre Rö	Drossel Dr	LED
U_{RMS}/V	230.03	229.93	52.19	61.876
U_{RMN}/V	207.01	205.77	43.77	61.794
I_{RMS}/A	91.97 m	91.91 m	91.97 m	0.4340
I_{RMN}/A	81.81 m	81.76 m	81.81 m	0.2985
P/W	20.844	20.17	0.659	18.745
Q/var	3.619	-6.31	4.755	-19.229
λ	0.9853	0.9544	0.1372	0.6980
Φ/°	G9.85	D17.37	G82.11	D45.73
η/%	P_i/P_{Netz}	96.8	3.2	89.9

Bild 4 Messung: LED-Röhre mit verbliebener Drossel

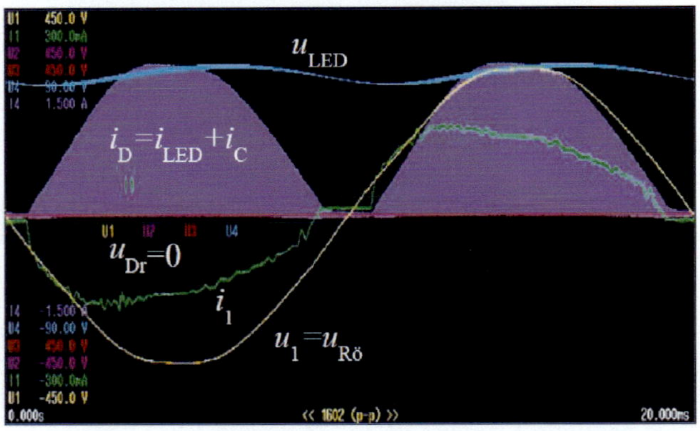

	Netz 1	Röhre Rö	Drossel Dr	LED
U_{RMS}/V	230.06	229.96	0.00	61.731
U_{RMN}/V	206.94	206.86	0.00	61.664
I_{RMS}/A	90.49 m	90.43 m	90.49 m	0.4202
I_{RMN}/A	80.23 m	80.18 m	80.23 m	0.2980
P/W	20.043	20.02	0.007	18.594
Q/var	-5.633	5.63	0.000	-18.089
λ	0.9627	0.9627	Error	0.7168
Φ/°	D15.70	G15.71	Error	D44.21
η/%	P_i/P_{Netz}	99.9	0.0	92.8

Bild 5 Messung: LED-Röhre ohne Drossel $U_{Dr} = 0V$

In **Bild 4** liegt an der LED-Röhre fast der Effektivwert der Netzspannung $U_{RöRMS} = U_{1RMS} = 230$ V, so dass die bei der industriellen Umrüstung verbliebene Drossel entfernt werden kann. In **Bild 5** ist die Spalte mit den Drosselwerten Null und es kann ein Wirkungsgrad $\eta = 92{,}8$ % der in der LED-Röhre integrierten Stromversorgung von $P_{LED} = 18{,}59$ W zu $P_1 = 20{,}04$ W gemessen werden.

In den **Bildern 4** und **5** ist der mit 40 kHz getaktete Strom i_D der Sperrdiode des Flyback-Converters gemessen. Der 100-Hz-Wechselanteil stellt den Kondensatorstrom i_C aus **Bild 3** dar, so dass durch die LED ein konstanter Strom I_{LED} fließt, der auch einen konstanten Lichtstrom erzeugt.

Aus **Bild 5** kann die Genauigkeit des 0,1 % Leistungsanalysators [1] überprüft werden. Der Betragsmittelwert RMN (Rectified Mean Norm) in der 4. Zeile beträgt $I_{RMN} = (80{,}23+80{,}18+80{,}23)mA/3 = 80{,}213$ mA. Mit der größten Abweichung $\Delta I_{RMN}\ /I_{RMN} = 80{,}18/80{,}213-1 = -0{,}04$ % wird die angegebene Messtoleranz eingehalten.

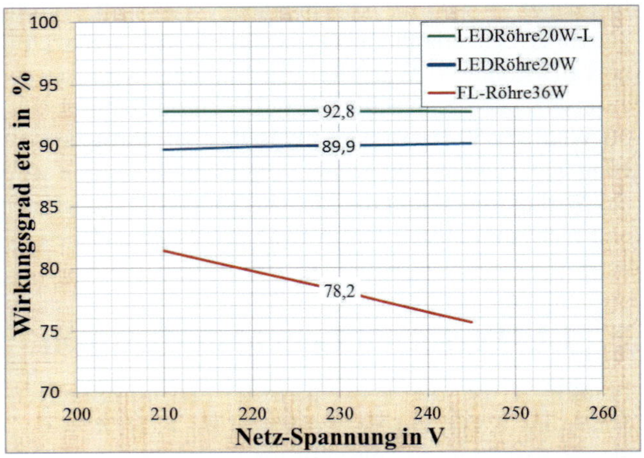

Bild 6 Messung: Wirkungsgrad über Spannungsänderung

In **Bild 6** sind die Wirkungsgrade über der 10 % schwankenden Netzspannung gemessen. Bei Nennspannung $U_{1RMS} = 230$ V hat die alte 36 Watt Fluoreszenz-Röhre einen Wirkungsgrad $\eta = 78{,}2$ % mit fallender Tendenz.

Mit dem Reihenersatz-Schaltbild 1 errechnet man unter linearen Bedingungen die Drosselverluste $P_{Dr} = R_{Dr} I_{RMS}^2$, die gesamte Wirkleistung $P_1 = (R_{Dr} + R_{Rö}) I_{RMS}^2$ und den Wirkungsgrad $\eta = 1 - R_{Dr}/(R_{Dr} + R_{Rö}) = 1\ /[1 + (R_{Dr}/R_{Rö})]$ der unabhängig von der Spannung wäre. Mit steigender Spannung verringert sich der nichtlineare Röhrenwiderstand $R_{Rö}$, sodass der Wirkungsgrad η fällt.

Bei der 20 Watt LED-Röhre mit 50-Hz-Drossel ist der Wirkungsgrad bei Nennspannung $\eta = 89{,}9$ %. Ohne 50-Hz-Drossel (–L) kann der Wirkungsgrad um ca. 3 % auf $\eta = 92{,}8$ % gesteigert werden. Bei der industriellen Umrüstung z. B. von Logistikhallen auf LED-Röhren-Beleuchtung verbleibt die 50-Hz-Drossel in der Leuchtstofflampenverdrahtung, weil dann ungeschultes Personal nur Röhre und Starter tauschen muss. Ein Entfernen der Drossel und ein Umverdrahten durch Fachhandwerker würde die Amortisationszeit zu stark verlängern.

3 Energieeffizienz-Kennzeichnung

Ab 09/2013 werden Beleuchtungsmittel nach der neuen EU Verordnung 874/2012 mit Lampen Energie-Verbrauchs-Kennzeichnungen: A^{++}; A^{+}; A; B; C; D und E gelabelt. Die bisherigen Label: A; B; C; D; E und F entfallen.

Entsprechend der Größe des Energie-Effizienz-Index *EEI* in **Tabelle 1** werden die Buchstaben der Klassen A^{++} bis E der Lampe zugeteilt [2].

Der Energie-Effizienz-Index $EEI = P_{cor} / P_{ref}$ ist das Wirkleistungsverhältnis der eingespeisten Leistung P_{cor} bezogen auf eine variable Referenzleistung P_{ref}, die von der Glühlampe abgeleitet ist. Für Lampen mit einem Lichtstrom kleiner $\Phi_{use} < 1300$ lm (Lumen) gilt die Wurzel- Berechnungsformel: $P_{ref} = 0{,}049\Phi_{use} + 0{,}88\sqrt{\Phi_{use}}$. Für Lampen mit einem Lichtstrom größer $\Phi_{use} \geq 1300$ lm folgt der lineare Zusammenhang: $P_{ref} = 0{,}07341\ \Phi_{use}$.

Die LED-Lampendaten in **Bild 7** ergeben mit 810 lm eine Referenzleistung: $P_{ref} = 0{,}049 \cdot 810 + 0{,}88\sqrt{810} = 64{,}74$ W. Mit der Einspeiseleistung $P_{cor} = 10$ W berechnet man einen Energie-Effizienz-Index *EEI* =10/64,74=15,45 %, der in **Tabelle 1** für ungebündeltes Licht den Label: A^{+} ergibt.

Eine Berechnung der Lichtausbeute oder Efficacy ergibt den übersichtlichen Wert $\varepsilon = \Phi_{use} / P_{cor} = 81$ lm/W.

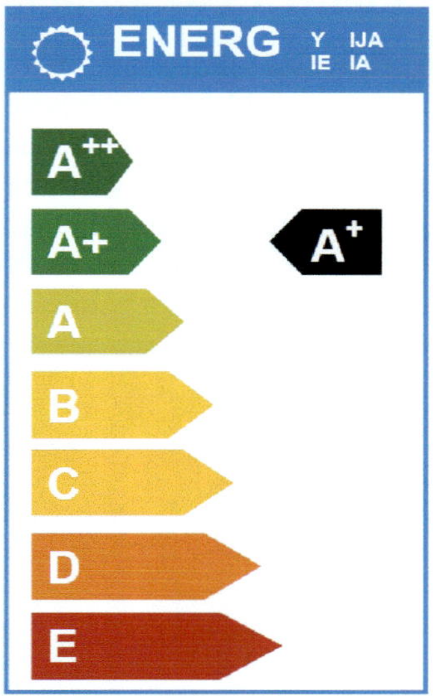

LED-Bulb
Volt: 220-240V
Watt: 10W
Lumen: 810lm
Winkel: 360°
Farbe: Ra > 80
Warm White
T/Kelvin: 2700K
t/h: 15.000h
On/Off: 40.000
Hg: 0,0mg
Temp:−20°bis+45°C
Lamp Base: E27
Label: A^{+}

Bild 7 Energie-Effizienz-Index *EEI* einer LED-Lampe

Tabelle 1: Energie-Effizienz-Klassen A^{++} bis E (Lampen)

Klasse	*EEI* ungebündeltes Licht	*EEI* gebündeltes Licht (90°;120°)
A^{++}	$EEI \leq 11\%$	$EEI \leq 13\%$
A^{+}	$11\% < EEI \leq 17\%$	$13\% < EEI \leq 18\%$
A	$17\% < EEI \leq 24\%$	$18\% < EEI \leq 40\%$
B	$24\% < EEI \leq 60\%$	$40\% < EEI \leq 95\%$
C	$60\% < EEI \leq 80\%$	$95\% < EEI \leq 120\%$
D	$80\% < EEI \leq 95\%$	$120\% < EEI \leq 175\%$
E	$EEI > 95\%$	$EEI > 175\%$

Wird die Wurzel-Berechnungsformel nach dem Lichtstrom Φ_{use} umgestellt, dann errechnet man für die verbotene $P_{ref} = 100$ W Glühlampe einen Lichtstrom:

$\Phi_{use} = \{[\sqrt{(0{,}049/0{,}44^2 P_{ref}+1)}-1]0{,}44/0{,}049\}^2$;
$\Phi_{use} = \{[\sqrt{(0{,}2531 P_{ref}+1)}-1]8{,}98\}^2 = 1375$ lm. Hier muss für > 1300 lm die zweite Berechnungsformel angewendet werden $\Phi_{use} = P_{ref}/0{,}07341 = 100\text{W}/0{,}07341 = 1362$ lm mit der Efficacy $\varepsilon = \Phi_{use}/P_{cor} = 1362\text{lm}/100\text{W} = 13{,}62$ lm/W.

Mit dem variablen Bezugswert soll berücksichtigt werden, dass eine leistungsschwächere Glühlampe eine zunehmend kleinere Lichtausbeute hatte. Es gilt für die15 W Glühlampe $\Phi_{use} = \{[\sqrt{(0{,}2531 \cdot 15+1)}-1]8{,}98\}^2 = 114{,}2$ lm mit der Efficacy $\varepsilon = \Phi_{use}/P_{cor} = 114{,}2\text{lm}/15\text{W} = 7{,}61$ lm/W. Warum diese Glühlampen-Bezugswerte Grundlage der EU-Verordnung sind, ist nicht zu verstehen, weil zukünftig die ineffizienten Glühlampen vom Markt verschwunden sind.

Die in Kapitel 2 vermessene LED-Röhre mit den Daten in **Bild 8** ergibt eine Glühlampen-Referenzleistung

$P_{ref} = 0{,}07341 \cdot 1500\text{lm} = 110{,}1$ W. Mit einem Energie-Effizienz-Index *EEI* = 20W/110,1W = 18,2 % wird mit der Klasse A laut **Tabelle 1** zertifiziert. Die Lichtausbeute oder Efficacy ist $\varepsilon = \Phi_{use} / P_{cor} = 75$ lm/W.

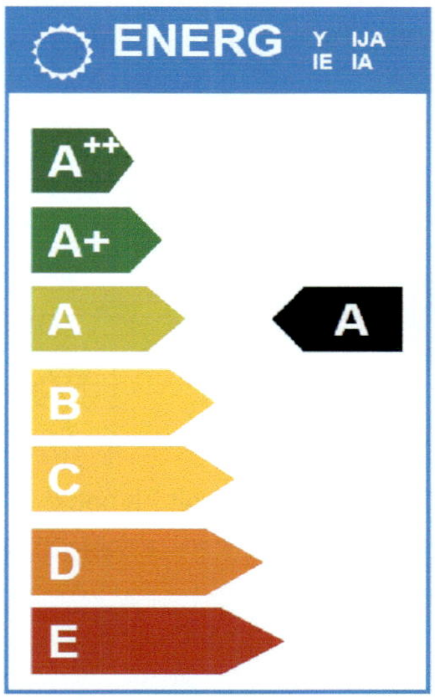

LED-TUBE
120cm
Volt: 220-240V
Watt: 20W
Lumen: 1500lm
Winkel: 150°
Farbe: Ra > 80
Warm White
T/Kelvin: 3000K
t/h: 30.000h
On/Off: 200.000
Hg: 0,0mg
Temp:−20°bis+45°C
Lamp Base: G13
Label: A

Bild 8 Energie-Effizienz-Index *EEI* der 20W-Röhre

Die ursprüngliche 36-W-Leuchtstofflampe hatte bei der Farbtemperatur 3000 K einen Lichtstrom von 3150 lm, die hier mit der LED-Röhre 1500 lm/150° also Lichtstromverhältnis 2,1 ersetzt wird. Im Prüflabor beleuchtet die LED-Röhre den Arbeitsplatz sogar heller. Ursache dafür ist der kleinere Raumwinkel 150° gegenüber der rundum 360° abstrahlenden 36-W-Röhre. Durch die EU-Verordnung 1194/2012 soll bei Lampen mit gerichtetem Licht der Nutzlichtstrom verglichen werden [3].

Tritt bei einem Kugelstrahler das Licht in einem Kegelwinkel φ aus, dann erhöht sich die Lichtstärke im Verhältnis des Vollkugel-Raumwinkels $\Theta_0 = 4\pi$ zum Kegel-Raumwinkels $\Theta = 4\pi \sin^2(\varphi/4)$.

Bei dem Winkel $\varphi = 150°$ aus **Bild 8** erhöht sich die Lichtstärke um $\Theta_0/\Theta = 1/\sin^2(\varphi/4) = 1/\sin^2(150°/4) = 2{,}7$-fach.

Da bei der LED-Röhre eine Abstrahlung aus einem Zylindersegment mit $\varphi = 150°$ auftritt, wird die Lichterhöhung mit einem Zylinder-Winkel-Verhältnis $\Theta_0/\Theta = 2\pi/\varphi = 360°/150° = 2{,}4$-fach wahrgenommen.

4 Elektronische Stromversorgung

Die Stromversorgung in der LED-Röhre in **Bild 3** besteht prinzipiell aus dem Eingangsfilter L_X, C_X, dem netzgeführten Gleichrichter D1-D4, dem Spannungszwischenkreis-Kondensator C_1 und einem Flyback-Wandler bestehend aus Schalttransistor T, Trafo L_1, L_2 und Sperrdiode D wie in **Bild 8** dargestellt [4]. Mit diesem LED-Treiber werden gute Wirkungsgrade $\eta = 92{,}8$ % in **Bild 6** erzielt. Eine einfachere Schaltung in **Bild 10** wird bei Faden-LED mit einem Linearregeltransistor für stabile Netzspannungen $U_1 = 230$ V benutzt. Hierbei werden über 100 LED in Reihe (**Bild 9**) geschaltet, um mit u_{LED} etwas unterhalb dem Minimum der Zwischenkreisspannung u_{C1} zu liegen. Die Spannungsdifferenz u_T wird im Regeltransistor T genutzt, um den LED-Strom i_{LED} konstant zu halten. In **Bild 11** ist zu erkennen, dass der LED-Strom i_{LED} linear einbricht, wenn die Zwischenkreisspannung u_{C1} die LED-Spannung u_{LED} unterschreitet. Im Lichtstrom ist diese 100-Hz-Schwankung nicht zu erkennen. Für die 5,5-W-Lampe werden Wirkungsgrade bis 95 % erreicht. Der Energie-Effizienz-Index $EEI = 5{,}5\text{W}/58{,}9\text{W} = 9{,}3$ % in **Tab. 1** ergibt den Label: A^{++}, wenn die Referenzleistung 58,9 W ist.

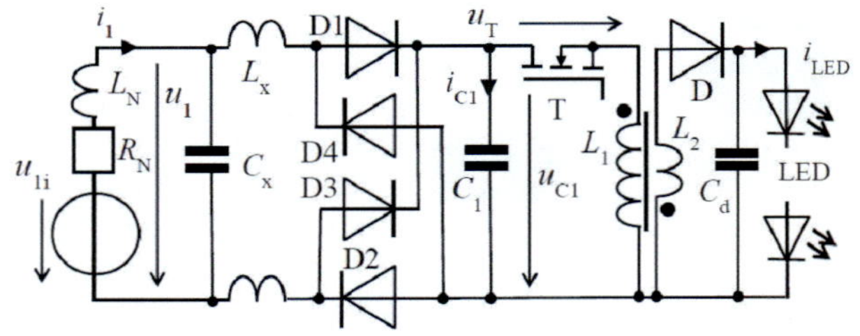

Bild 8 Stromversorgung einer LED-Lampe mit Flyback

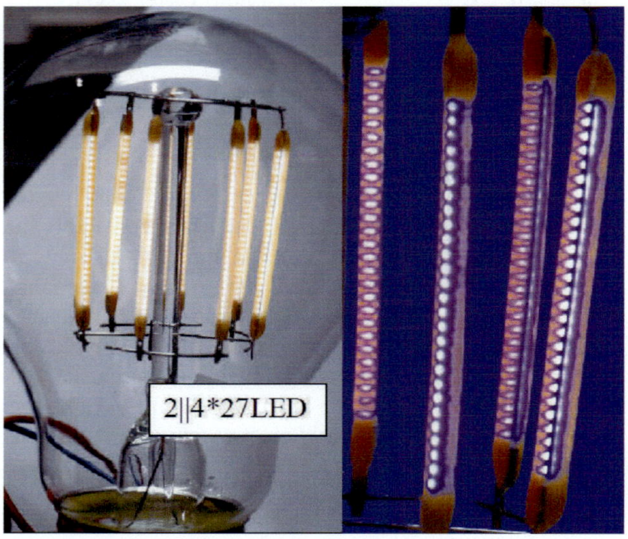

Bild 9 Faden-LED „Glühlampe" 720lm; 5,5W; 131lm/W

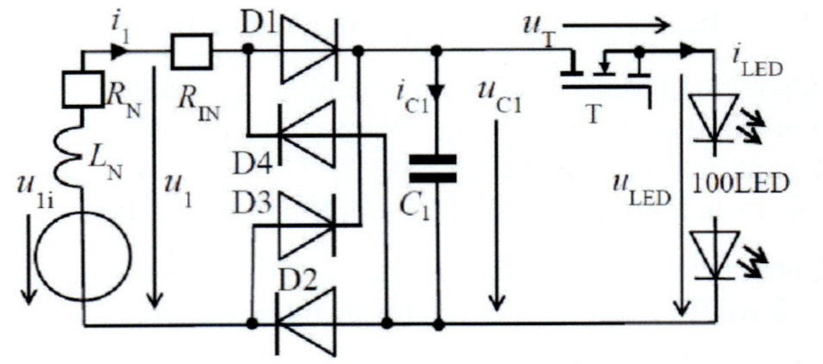

Bild 10 Stromversorgung der Faden-LED (Regeltransistor)

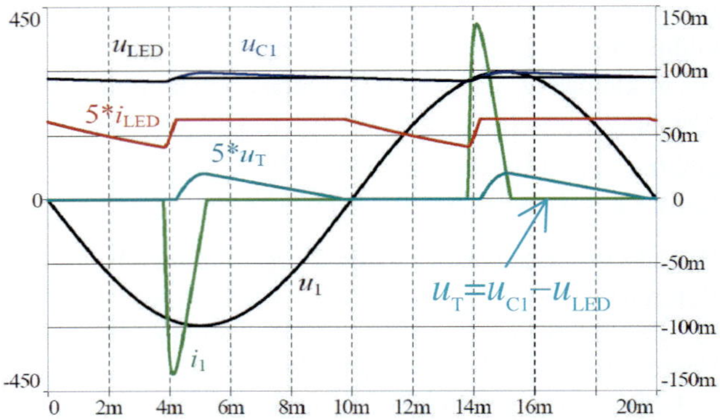

Bild 11 Simulation der Faden-LED mit Regeltransistor

5 Dimmbare Faden-LED-Lampen

In **Bild 12** sind die Faden-LEDs einer dimmbaren 6-W-Lampe abgebildet. Die Schaltung in **Bild 13** enthält keine Regeleinrichtung. Der große Reihenwiderstand $R_{IN} = 1{,}8$ kΩ und der Parallelwiderstand $R_P = 100$ kΩ stabilisieren den LED-Strom i_{LED}, wenn durch Temperaturerhöhung die LED-Spannung u_{LED} abnimmt und der Lichtstrom proportional zum LED-Strom schwanken würde.

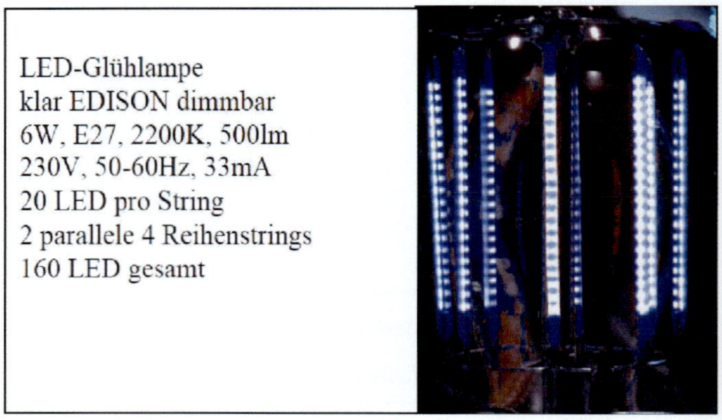

Bild 12 dimmbare Faden-LED: 500lm; 6W; 83lm/W

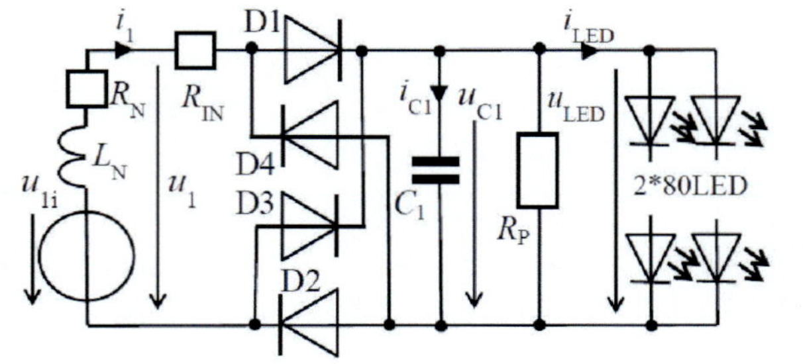

Bild 13 Schaltung einer dimmbaren Faden-LED

Mit dem Zwischenkreiskondensator C_1=4,7 µF ergibt sich die Zeitkonstante $\tau_{IN} = R_{IN}\, C_1$ = 1,8k 4,7us ≈ 8,5 ms. Der Widerstand R_{IN} bestimmt ein ohmsches Betriebsverhalten. Wird der Betrag der Netzspannung $|u_1|$ größer als die LED-Spannung u_{LED}, dann tritt die Sinuskuppe des Netzstromes i_1 in **Bild 14** auf. Die Entlade-Zeitkonstante $\tau_{LED} = R_{LED}\, C_1$= 0,65k 4,7us1,2 ≈ 3,6 ms mit dem LED-Widerstand R_{LED} = 650 Ω bestimmt den abklingenden LED-Strom i_{LED}, wenn der Netzstrom unterbrochen ist. Der LED-Strom besitzt eine starke 100-Hz-Stromschwankung. Die Simulation in **Bild 14** bestätigt die Messung in **Bild 15**. Die LED-Lampe soll am üblichen Haushalts-Dimmer (40W–400W) zur Helligkeitseinstellung benutzt werden können. Hierbei ist zu beachten, dass mindestens 7 LED-Lampen für die Grundlast notwendig sind. Anderenfalls kommt es zu Fehlfunktionen, wie **Bild 16** bei einem Steuerwinkel von $\alpha = 45°$ zeigt.

Die dimmbare 6-W-Faden-LED hat infolge des großen Widerstandes R_{IN} einen miserablen Wirkungsgrad η=65 %.

Bei dem Lichtstrom Φ= 500 lm mit der Referenzleistung P_{ref} = 44,2 W ist der EEI = 6W/44,2W = 13,6 %; Label: A+.

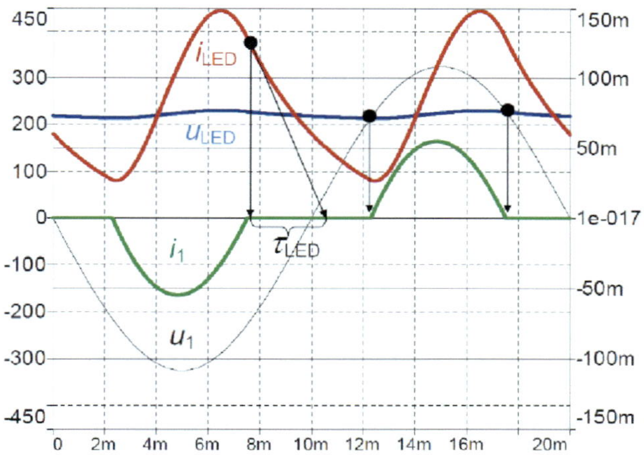

Bild 14 Simulation der dimmbaren Faden-LED 6W

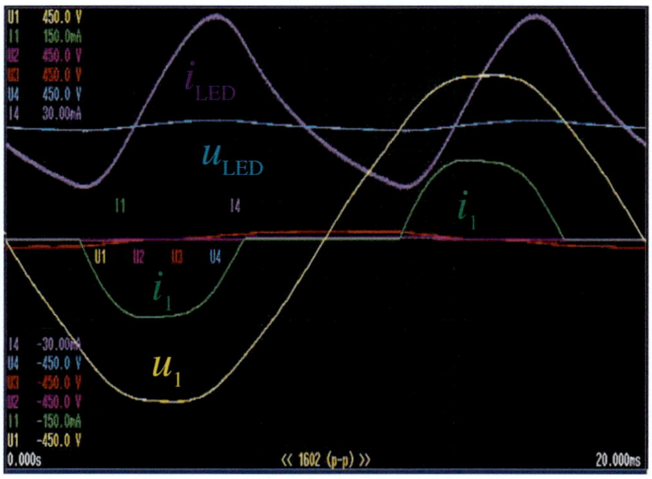

Bild 15 Messung der dimmbaren Faden-LED 6W

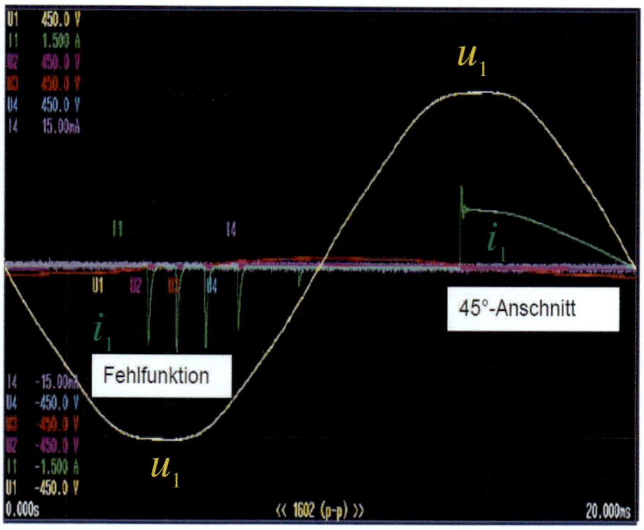

Bild 16 Fehler der dimmbaren Faden-LED 6W bei α=45°

6 Zusammenfassung

Die Schonung der Ressourcen kann bei der Beleuchtungstechnik in drei Schritten erfolgen:

1. gute Wirkungsgrade in LED-Stromversorgungen,
2. helligkeitsabhängige Beleuchtungssteuerung,
3. Lichtausschalten durch Bewegungsmelder.

Es wurde gezeigt, dass durch Substitution konventioneller „Leuchtstoffröhren" mittels „LED-Röhren" der Energieeinsatz halbiert werden kann. Weitere 3 % könnten durch Entfernen der 50-Hz-Drosseln eingespart werden.

Die Wirkungsgrade moderner LED-Treiber liegen laut Messung und Simulation bei 90 %. Dieser Wert ist auch in der Norm für externe Stromversorgungen zu berücksichtigen. Der Energie-Effizienz-Index berücksichtigt im Wesentlichen die erzeugte Lichtenergie und vergibt Effizienz-Label von A^{++}, A^{+}, A bis E. Da durch die Glühlampen-Referenzwerte kleine Leistungen bevorteilt werden, kann es zu Fehlern kommen. Zum Beispiel werden vier 5-W-A^{+}-Lampen besser wie eine 20-W-A-Lampe ausgewiesen, obwohl die 20-W-LED mehr Licht erzeugt.

Bei dimmbaren Lampen herrscht noch erheblicher Handlungsbedarf, weil die Funktion und ein guter Wirkungsgrad nicht erreicht werden.

7 Literatur

[1] YOKOGAWA Meters & Instruments Corporation: WT1800 Präzisionsleistungsanalysator, 1. Ausgabe

[2] Verordnung (EU) Nr. 874/2012 zur Energieverbrauchskennzeichnung von elektrischen Lampen und Leuchten. 26.9.2012

[3] Verordnung (EU) Nr. 1194/2012 zur umweltgerechten Gestaltung von Lampen mit gebündeltem Licht, LED-Lampen und dazugehörigen Geräten. 14.12.2012

[4] Jaschke, R.: Reduzierung der Netz-Oberschwingungsströme durch modifizierte Stromversorgungen für LED-Lampen kleiner 25 Watt. NEIS Konferenz 2014, Hamburg

Regelleistung

Auswirkungen des Network Code on Electricity Balancing auf den deutschen Regelleistungsmarkt

V. D. Kocevar[1], Prof. Dr.-Ing. J. Wendiggensen[2]
[1]Universität Flensburg, Auf dem Campus 1, 24943 Flensburg, kocevar.dario@gmail.com
[2]ZNES Zentrum für Nachhaltige Energiesysteme, FH Flensburg, Kanzleistraße 91-93, 24943 Flensburg
Mit Unterstützung von E.ON Global Commodities SE

Kurzfassung

Im Rahmen der Harmonisierung des europäischen Regelenergiemarktes veröffentlichte der Verband Europäischer Übertragungsnetzbetreiber den Network Code on Electricity Balancing. Dieser Network Code formuliert Rahmenbedingungen, aus denen sich ein neues europäisches Marktdesign für die Regelleistung ableitet. Das neue Markdesign umfasst unter anderem die Trennung der Angebote für Leistungsvorhaltung und Regelenergie, Marktöffnung für freiwillige Regelenergieangebote und ein Uniform-Price Auktionsmodell. Während die theoretischen Unterschiede von Auktionsmodellen bereits hinreichend dargestellt wurden, ist die Bedeutung dieser für den deutschen Regelleistungsmarkt nicht formuliert worden. Zudem kann der Regelleistungsmarkt nicht losgelöst betrachtet werden, da zwischen diesem, dem Bilanzkreismanagement und dem Intraday Handel dynamische Wechselwirkungen existieren. Die Rahmenbedingungen des neuen Marktdesigns beeinflussen die Preisbildung der Arbeitspreise nachhaltig und ändern somit die Struktur der Merit-Order der Arbeitspreise. Es zeigt sich, dass das neue Marktdesign die Teilnahme Erneuerbare Energien am Regelleistungsmarkt vereinfacht, die wirtschaftlichen Chancen für diese jedoch begrenzt sind.

1 Einleitung

Der Verband der Europäischen Übertragungsnetzbetreiber (ENTSO-E) formuliert mit der Unterstützung der Agentur für die Zusammenarbeit der Energieregulierungsbehörden (ACER) unterschiedliche Network Codes, welche die Vernetzung und die Effizienz der europäischen Elektrizitätsmärkte unterstützen sollen. Die Network Codes dienen als ein einheitliches Regelwerk, um die nationalen Märkte langfristig zu harmonisieren. Als essentieller Bestandteil einer sicheren und verlässlichen Elektrizitätsversorgung werden auch die Regelleistungsmärkte von der ENTSO-E aufgegriffen. Ziel ist es, einen grenzüberschreitenden Regelleistungsmarkt mit einer internationalen Merit-Order der Arbeitspreise zu entwickeln. Während der Network Code on Load-Frequency Control and Reserves [1] die technischen Eigenschaften festlegt, beschreibt der Network Code on Electricity Balancing [2] Rahmenbedingungen für ein europäisches Marktdesign. Die deutschen Regelleistungsprodukte entsprechen den erwarteten technischen Eigenschaften oder übererfüllen diese bereits heute. Das Marktdesign des deutschen Regelleistungsmarktes wird jedoch nach dem Network Code von fundamentalen Änderungen betroffen sein. Die parallelen Angebote von Leistungsvorhaltung und Regelenergie werden getrennt, sodass Übertragungsnetzbetreiber (ÜNB) Regelleistung längerfristig beschaffen können, der Handelsschluss für Arbeitspreisgebote jedoch erst kurz vor Erfüllung ist. Somit können Anbieter ihre Arbeitspreise der aktuellen Marktsituation anpassen. Der Regelleistungsmarkt wird zudem für zusätzliche, freiwillige Anbieter geöffnet. Während in der Regelleistungsauktion bezuschlagte Anbieter verpflichtet sind Arbeitspreisgebote bereitzustellen, können präqualifizierte Marktteilnehmer weitere Gebote offerieren. Mit der Umsetzung dieser Maßnahmen verfolgt die ENTSO-E zum einen das Ziel eines größeren Wettbewerbs auf dem Regelenergiemarkt, zum anderen soll die Teilnahme Erneuerbarer Energien am Regelleistungsmarkt gefördert werden. Das Auktionsverfahren wird zudem von Pay-As-Bid auf ein Uniform-Price Modell geändert. Alle aktivierten Anbieter erhalten den Preis des teuersten aktivierten Angebots innerhalb eines definierten Zeitraums.

Im Folgenden werden, nach einer kurzen Darstellung des aktuellen Regelleistungsmarktes, die Änderungen des Marktverhaltens speziell für diesen verdeutlicht und daraus die Bedeutung für die Marktteilnahme Erneuerbarer Energien abgeleitet.

2 Regelleistungsmarkt

Regelenergie wird benötigt, um Frequenzschwankungen abweichend von der Sollfrequenz auszugleichen und ist damit ein essentieller Bestandteil einer sicheren und verlässlichen Energieversorgung. Frequenzschwankungen können unterschiedliche Ursachen haben. Grundsätzlich besteht immer ein Ungleichgewicht zwischen erzeugter und verbrauchter elektrischer Energie. Das **Bild 1** zeigt anhand der Differenz von Sollleistung (rot) und Istleistung (schwarz) unterschiedliche Gründe für Abweichungen vom Sollwert der Netzfrequenz. Eine Übererfüllung der Istleistung führt zu einem Frequenzanstieg, eine Untererfüllung zu einem Frequenzabfall.

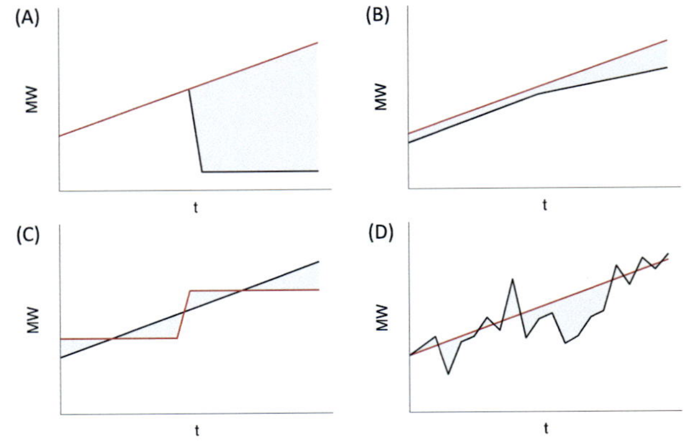

Bild 1 Gründe für Last- und Erzeugungsabweichungen [3]

Ausfälle von Kraftwerkseinheiten, HVDC Verbindungsleitungen oder großen Lasten resultieren in spontanen Änderungen der Frequenz (A). Statische stochastische Abweichungen der Frequenz werden durch langfristige Prognoseabweichungen von Lasten und Erzeugung verursacht (B). Da Kraftwerke dem treppenförmigen Fahrplan des Handels nicht folgen können, treten unvermeidbare deterministische Abweichungen auf (C). Volatilitäten von Lasten und Erzeugung verursachen kontinuierliche stochastische Abweichungen (D).

Um die Frequenz innerhalb eines festgelegten Toleranzbereiches zu stabilisieren, werden in Deutschland drei aufeinander abgestimmte Regelleistungsprodukte vorgehalten. Die Produkte haben jeweils unterschiedliche Funktionen und weisen verschiedene technische Eigenschaften auf [4]. Die Primärregelung (PRL) dient der Stabilisierung kurzfristiger Frequenzabweichungen. Die Sekundärregelung (SRL) löst die PRL ab und stellt das Gleichgewicht zwischen vereinbarten und tatsächlichen Austauschleistungen wieder her. Die Minutenreserve (MRL) löst bei großen Abweichungen die Sekundärregelung ab und hält die Frequenz auf dem Sollwert. Alle drei Regelleistungsarten werden für negative als auch positive Regelrichtung vorgehalten. Während die PRL gemeinsam von allen beteiligten Teilnehmern erbracht wird, werden SRL und MRL auf nationaler Ebene aktiviert.

2.1 Regelenergiebedarf

Ex post zeigt sich, dass der Regelenergiebedarf durch SRL und MRL trotz des Zubaus Erneuerbarer Energien deutlich abgenommen hat. **Bild 2** zeigt die Entwicklung der vergangenen drei Jahre.

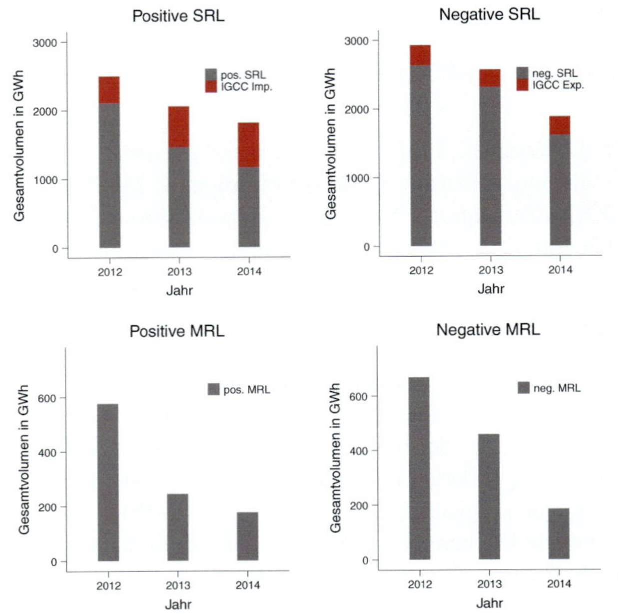

Bild 2 Entwicklung des Sekundär- und Minutenregelenergiebedarfes (Datenquelle: regelleistung.net, 2015)

Diese Entwicklung hat unterschiedliche Gründe. Der deutsche und der internationale Netzregelverbund (IGCC) ermöglichen das Imbalancenetting zur Vermeidung von SRL Aktivierung, d. h. Abweichungen einzelner Regelzonen werden untereinander ausgeglichen und Regelenergie muss lediglich für die verbleibende Abweichung abgerufen werden. IGCC Importe ersetzen positive SRL Aktivierung, IGCC Exporte entsprechend negative SRL Aktivierung. IGCC Exporte fallen dabei im Vergleich geringer aus, da die IGCC Flüsse durch freie Grenzkapazitäten begrenzt sind und die Exportauslastung höher ist. Die Bilanzkreisverantwortung wurde gesetzlich stärker gestützt, sodass Bilanzkreise besser ausgeglichen werden müssen und folglich weniger Imbalancen verursachen. Die Bilanzkreisverantwortung gilt auch für Portfolios aus Erneuerbaren Energien. Ein großer Anteil, z.B. 88,9 % (Stand August 2014) der installierten onshore Windleistung [5], wird inzwischen direkt vermarktet. Zudem besteht für Übertragungsnetzbetreiber, welche EEG Leistungen vermarkten, die Möglichkeit am Intraday Handel teilzunehmen. Direktvermarktete Erneuerbare Energien werden effizient in den Handel integriert, da durch Prognosefehler und resultierende Bilanzkreisabweichungen wirtschaftliche Verluste drohen. Ende 2012 wurde zusätzlich das Abrechnungssystem des Ausgleichsenergiepreises angepasst, sodass eine Optimierung gegenüber dem Intraday seitdem nicht mehr möglich sein soll. Zuvor war es zu bestimmten Zeitpunkten wirtschaftlicher den Ausgleichenergiepreis zu zahlen, als den eigenen Bilanzkreis am Intraday auszugleichen. Das verstärkte Interesse des Bilanzkreisausgleichs wird durch das gesteigerte Intraday-Handelsvolumen bestätigt. Die Einführung von Viertelstundenprodukten am Day-Ahead Spotmarkt verbessert die Genauigkeit der Bilanzkreisbewirtschaftung zusätzlich.

2.2 Abrufwahrscheinlichkeiten

Die Position in der Merit-Order der Arbeitspreise bestimmt maßgeblich die Abrufwahrscheinlichkeit eines Angebotes. Anhand der veröffentlichen Viertelstunden Mittelwerte der abgerufenen Sekundärregelleistung lässt sich die Verteilung der Sekundärregeleistung über einen gewünschten Zeitraum darstellen. **Bild 3** zeigt die Häufigkeitsverteilungen für die Jahre 2013 und 2014.

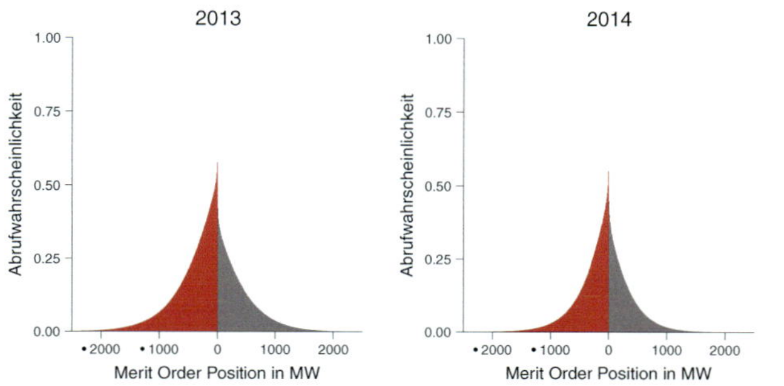

Bild 3 Häufigkeitsverteilung des SRL Bedarfs (Datenquelle: regelleistung.net, 2015)

Im direkten Vergleich der beiden Histogramme wird deutlich, dass die Verteilung schmaler geworden ist, der Gesamtbedarf der SRL hat abgenommen. Ein Anbieter, der mit seinem Portfolio hohe Abrufstunden generieren möchte, muss seine Arbeitspreise so wählen, dass er sich im vorderen Bereich der Meritorder positioniert.

3 Änderungen des Marktverhaltens

Die beschriebenen Neuerungen des Marktdesign verändern das Verhalten der Marktteilnehmer fundamental. Wie bereits von Tierney et al. [6] und Müsgens et al. [7] dargelegt, bewegt das Uniform Price Auktionsmodell die Teilnehmer dazu, Arbeitspreisgebote nahe ihrer marginalen Kosten zu offerieren. Fundamental gibt es keine Gründe, warum ein Anbieter seinen Arbeitspreis künstlich erhöhen sollte. Ein höherer Arbeitspreis reduziert die Abrufwahrscheinlichkeit, während die Einnahmen nicht erhöht werden, da das teuerste aktivierte Angebot den Preis für alle aktivierten Angebote setzt. Lediglich im Falle, dass das überhöhte Angebot den Preis setzt, entsteht für den Anbieter ein Vorteil. Es zeigt sich, dass die Arbeitspreise für die SRL derzeit oberhalb des Day-Ahead Clearing Preises liegen. Der Day-Ahead Clearing Preis stellt vereinfacht die Grenzkosten des gesamtdeutschen Kraftwerksportfolios für die entsprechende Stunde dar [8]. SRL wird zu einem großen Teil durch konventionelle Kraftwerksportfolios geliefert, die bereits auf dem Day-Ahead Markt vermarktet wurden, jedoch freie Kapazitäten zur Verfügung haben. Folglich liegen diese Kraftwerksportfolios innerhalb der Grenzkosten. Es ist also davon auszugehen, dass die derzeitigen Arbeitspreise oberhalb der marginalen Kosten der Anbieter liegen. Durch die Änderung des Auktionsmodells werden die Arbeitspreise für Sekundärregelleistung der beschriebenen Logik nach fallen. Der kurzfristige Handelsschluss der Regelenergieauktion ermöglicht den Teilnehmern die Arbeitspreise aus der Marktsituation und nicht aus einer Preisvorhersagekurve abzuleiten. Folglich reduziert sich das Risiko der Teilnehmer und Risikoaufschläge fallen geringer aus. Der kurzfristige Handelsschluss, kleinere Mindestleistungen und die Öffnung des Marktes für freiwillige Arbeitspreisgebote ermöglicht neuen Anbietern die Teilnahme am Regelleistungsmarkt. Diese Änderungen fördern den Preiswettbewerb.

4 Bedeutung für Erneuerbare Energien

Hirth und Ziegenhagen [9] beschreiben die prohibitiven Barrieren des aktuellen Marktdesigns des Sekundärregelleistungsmarktes für die Teilnahme von Erneuerbaren Energien. Die wöchentliche Ausschreibung in Zeitscheiben von mindestens acht Stunden überschreiten die Prognosemöglichkeiten, um Regelleistung aus Erneuerbaren Energien sicher anbieten zu können. Das neue Marktdesign greift diesen Kritikpunkt auf und forciert die Möglichkeit einer kurzfristigen Auktion von Regelenergie.

Das Fraunhofer IWES veröffentlichte einen Bericht [10] der unter anderem verschiedene Verfahren beschreibt, wie Regelenergie aus Windkraftportfolios erbracht werden kann. Um eine hohe Zuverlässigkeit zu gewährleisten, werden Windkraftportfolios mit einem Fahrplan betrieben. Der Fahrplan ist im Sinne des Bilanzkreisausgleiches durch entsprechende Abregelung der Anlagen und kurzfristige Stromgeschäfte gedeckt. Negative Regelenergie kann so vorgehalten werden, während zur Vorhaltung positiver Regelenergie das Windkraftportfolio weiter abgeregelt werden muss. Der gebotene Arbeitspreis muss die Mindereinnahmen durch die Abregelung kompensieren. Folgt man der von Speckmann [11] vorgeschlagenen Vorgehensweise zur Berechnung der Arbeitspreise wird deutlich, dass hohe Arbeitspreise geboten werden müssen. Hohe Arbeitspreise bedeuten jedoch in einem Markt, in welchem sowohl die Preise als auch die Nachfrage fallen, zudem das Angebot möglicherweise noch steigt, dass der Anbieter sich nicht lukrativ in der Merit-Order positionieren kann. Die wirtschaftlichen Chancen für Windkraftanlagen sind in Anbetracht der derzeitigen Zusammensetzung des gesamtdeutschen Kraftwerkparks als gering einzuschätzen. Mit dem Rückbau konventioneller Erzeugungseinheiten könnte sich die Struktur zukünftig so ändern, dass den Erneuerbaren Energien ein Marktanteil am Regelleistungsmarkt zukommt und sich zu einem wirtschaftlichen Sekundärabsatzmarkt entwickelt.

5 Literatur

[1] ENTSO-E: Network Code on Load Frequency Control & Reserves http://networkcodes.entsoe.eu/wp-content/uploads/2013/08/130628-NC_LFCR-Issue1.pdf (2014) Zugegriffen: 25. Februar 2015

[2] ENTSO-E: Network Code on Electricity Balancing – 09/2014 http://networkcodes.entsoe.eu/wp-content/uploads/2013/08/140806_NCEB_Resubmission_to_ACER_v.03.pdf (2014) Zugegriffen: 25. Februar 2015

[3] ENTSO-E: Supporting Document fort he Network Code on Load Frequency Control & Reserves https://www.entsoe.eu/fileadmin/user_upload/_library/resources/LCFR/130628-NC_LFCR-Supporting_Document-Issue1.pdf (2014) Zugegriffen: 25. Februar 2015

[4] UCTE (Hrsg.): Operation Handbook – Policy 1: Load-Frequency Control. https://www.entsoe.eu/fileadmin/user_upload/_library/publications/entsoe/Operation_Handbook/Policy_1_final.pdf (2009). Zugegriffen: 25. Februar 2015

[5] BMWi (Hrsg.): Laufende Evaluierung der Direktvermarktung von Strom aus Erneuerbaren Energien – 10. Quartalsbericht. http://www.erneuerbare-energien.de/EE/Redaktion/DE/Downloads/Berichte/direktvermarktung-quartalsbericht-10.pdf?__blob=publicationFile&v=4 (2014). Zugegriffen: 25. Februar 2015

[6] Tierney, S.; Schatzki, T.; Mukerji, R.: Uniform-pricing versus Pay-as-Bid in Wholesale Electricity Markets. Analysis Group. http://www.nyiso.com/public/webdocs/media_room/current_issues/uniformpricing_v_payasbid_tierneyschatzkimukerji_2008.pdf (2008). Zugegriffen: 12. Dezember 2014

[7] Müsgens, F.; Ockenfels, A.; Peek, M.: Economics and design of balancing power markets in Germany. International Journal of Electrical Power & Energy Systems, Volume 55, 392-401 (2014) doi: 10.1016/j.ijepes.2013.09.020

[8] Weber, C.; Woll, O.: Merit-Order-Effekte von Erneuerbaren Energien - zu schön um wahr zu sein?. EWL Working Paper No. 01/07. https://www.wiwi.uni-due.de/fileadmin/fileupload/BWL-ENERGIE/Arbeitspapiere/RePEc/pdf/wp0701_Weber_Woll_Merit-Order-Effekt_WP.pdf (2007). Zugegriffen: 25. Februar 2015

[9] Hirth, L.; Ziegenhagen, I.: Wind, Sonne und Regelleistung. Energiewirtschaftliche Tagesfragen, 63, Heft 10 (2013)

[10] Jansen, M.; Speckmann, M.; Siefert, M.; Baier, A.; Schwinn, R.; Jost, D.: Regelenergie durch Windkraftanlagen. Fraunhofer IWES, Kassel. (2014)

[11] Speckmann, M.: Standardprodukte für einen europäischen Regelenergiemarkt. In: 4. Jahreskonferenz Zukunftsperspektiven für den Regelenergiemarkt Strom 2014. IQPC. 24.-26.11.2014

Regelleistungspotenzial durch wetterabhängige erneuerbare Energieanlagen unter Einfluss regionaler Vergleichmäßigungseffekte

Annika Magdowski, Martin Kaltschmitt
Technische Universität Hamburg-Harburg (TUHH), Institut für Umwelttechnik und Energiewirtschaft (IUE),
Eißendorfer Straße 40, D-21073 Hamburg, Deutschland,
magdowski@tuhh.de, kaltschmitt@tuhh.de

Kurzfassung

Die Zielsetzung der Bundesregierung sieht einen Ausbau der Energieerzeugungsoptionen auf Basis erneuerbarer Energien bei einer Reduzierung des konventionellen Kraftwerkparks vor. Um das bisher gewohnte Maß an Versorgungssicherheit jederzeit zu gewährleisten, muss durch die bisher nur fehlerhaft zu prognostizierende Einspeisung aus wetterabhängigen erneuerbaren Energien (d. h. Windenergie, Solarstrahlung) ein steigender Anteil an Regelleistung vorgehalten werden. Damit ist es zwingend notwendig, dass zukünftig auch Energieanlagen auf der Basis wetterabhängiger, fluktuierender erneuerbarer Energien zur Ausbilanzierung des Stromversorgungssystems beitragen. Vor diesem Hintergrund ist das Ziel der folgenden Ausführung, das Potenzial von Windanlagen zur Erbringung von Regelleistung zu untersuchen. Dabei wird der Einfluss von Prognosegenauigkeiten der erzeugten Windleistung auf das Regelleistungspotenzial abgeschätzt. Zudem werden Vergleichmäßigungseffekte durch (über-)regionale Zusammenschlüsse von Windkraftanlagen hinsichtlich einer Potenzialerhöhung betrachtet. Abschließend werden die Effekte der Bereitstellung von Regelleistung durch wetterabhängige erneuerbare Energien auf das Gesamtstromversorgungssystem beurteilt.

Schlüsselwörter: Windenergie, Regelleistung, regionale Vergleichmäßigungseffekte, rekursive Faltung, Prognosefehler

1 Einleitung

Um einen zuverlässigen Betrieb des elektrischen Energiesystems in Deutschland zu gewährleisten, muss die Bilanz zwischen Erzeugung und Nachfrage zu jedem Zeitpunkt des Jahres ausgeglichen sein. Dafür wird Regelleistung in Abhängigkeit der jeweiligen Aktivierungsgeschwindigkeit in Form von Primär-, Sekundär- und Minutenreserveleistung vorgehalten. Bislang wird diese Ausgleichsleistung vor allem durch thermische Kraftwerke, Gasturbinen und Pumpspeicher bereitgestellt [1]. Wetterabhängige erneuerbare Energien erhöhen u. a. durch die bisher unzureichende Prognosegüte ihrer Fluktuationen den Bedarf an Regelleistung, leisten derzeit aber noch keinen Beitrag zur Regelleistungsbereitstellung. Dies führt dazu, dass auch in Stunden sehr hoher Einspeisung durch Wind- und PV-Anlagen konventionelle Kraftwerke am Netz sein müssen, um die notwendige Regelleistung sicher liefern zu können. Infolgedessen müssen bei weiter steigenden Anteilen einer Stromerzeugung aus regenerativen angebotsorientierten Energien, wie es nach den Vorgaben der Bundesregierung angestrebt wird, zukünftig auch Wind- und PV-Anlagen in Kombination mit z. B. Speichern zur Ausbilanzierung des Stromversorgungssystems beitragen, um auch weiterhin ein konstant hohes Versorgungsniveau zu gewährleisten und gleichzeitig die Kosten der Stromversorgung zu begrenzen.

Vor diesem Hintergrund werden nachfolgend diese Forschungsfragen beantwortet:

- Welche technischen und regulatorischen Voraussetzungen sind notwendig, um eine Regelleistungserbringung durch wetterabhängige erneuerbare Energien zu ermöglichen?
- Inwieweit kann ein Verbund von wetterabhängigen erneuerbaren Energien zur Bereitstellung von Regelleistung beitragen?
- Welche Effekte hat die Bereitstellung von Regelleistung durch wetterabhängige erneuerbare Energien auf das Stromversorgungssystem?

Der Fokus in dieser Arbeit liegt auf der Betrachtung der Windenergie als eine der wichtigsten wetterabhängigen erneuerbaren Energien in Deutschland.

Die Bereitstellung von Regelenergie durch Windkraftanlagen wird aktuell weltweit in vielfältigen Forschungsarbeiten untersucht (u. a. [2, 3]). Der Stand der Forschung wurde bei dieser Untersuchung berücksichtigt.

Für die Bereitstellung positiver Regelleistung müssen Windkraftanlagen zuvor auf einem gedrosselten Leistungsniveau gefahren werden. Diese Fahrweise verursacht durch Prognoseungenauigkeiten und dem geforderten Zuverlässigkeitsniveau der Regelleistungserbringung energetische und damit auch finanzielle Verluste durch Anlagenabgeregelung. Daher wird im Folgenden ausschließlich auf die Potenziale der Erbringung negativer Regelleistung durch Windkraftanlagen eingegangen.

2 Voraussetzungen

2.1 Regulatorische Voraussetzungen

Durch die aktuell gültigen Regularien des deutschen Regelenergiemarkts (z. B. wöchentliche Ausschreibungen in Zeitscheiben von mindestens 8 h unter geforderter Zuverlässigkeit von mindestens 99,99 %) ist der Marktzutritt von Anlagen basierend auf angebotsabhängigen erneuerbaren Energien nur schwer realisierbar [2, 4, 5]. Daher wird folgend das Regelleistungspotenzial unabhängig von den aktuell gültigen Ausschreibungszeiträumen und Produktlängen der Regelenergieangebote bestimmt.

2.2 Technische Voraussetzungen

Moderne Windkraftanlagen können aus technischer Sicht schon heute die Anforderungen zur Regelleistungsbereitstellung erfüllen. Vorrausetzung dafür ist ein drehzahlvariabler Antrieb bei gleichzeitiger Nutzung des Konzepts der Pitchregelung [2, 3]. Durch die Änderung des Pitchwinkels und der Schnelllaufzahl kann der Leistungsbeiwert und damit der Leistungsoutput im aktiven Betrieb angepasst werden. Diese Abhängigkeit kann durch eine für derartige moderne Windenergieanlagen gültige empirische Gleichung beschrieben werden (**Abb. 1**) [6]. Für die Änderung der Schnelllaufzahl muss die Rotordrehzahl über das Generatormoment geregelt werden. Grundsätzlich gilt, dass der Leistungsbeiwert mit steigendem Pitchwinkel abnimmt. So könnte negative Regelleistung durch eine Erhöhung des Pitchwinkels bereitgestellt werden; bereits eine Winkeländerung von 1 ° führt zu einer Leistungsreduktion von bis zu 10 %.

Eine modelltechnische Abbildung der Regelungskonzepte einer modernen Windkraftanlage nach [3] zeigt, dass bei gedrosseltem Betrieb innerhalb von maximal 30 sec eine signifikante Leistungssteigerung bzw. -minderung erreicht werden kann und damit die technischen Voraussetzung für die Teilnahme am Regelenergiemarkt grundsätzlich gegeben sind. Die Veränderung des Pitchwinkels dient dabei hauptsächlich als Stellgröße der Leistungsanpassung. Die Momentenregelung wird nur zur Steuerung genutzt, um optimale Rotordrehzahlen einzustellen. Die zeitliche Verzögerung der Leistungsänderung resultiert zum einen aus der Trägheit der Berechnung, Übertragung/Steuerung und Umsetzung der Pitchwinkeländerung (d. h. Anfahren der Blattwinkelverstellmotoren, maximalen Blattwinkelverstellgeschwindigkeit von 3 bis 10 °/sec [6]) und zum anderen aus der Anpassung der Drehzahl durch das Generatorgegenmoment. Die Rege-

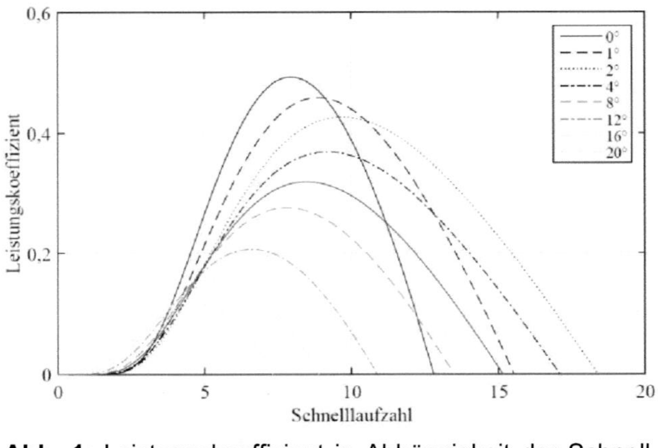

Abb. 1: Leistungskoeffizient in Abhängigkeit der Schnelllaufzahl bei verschiedenen Anstellwinkeln.

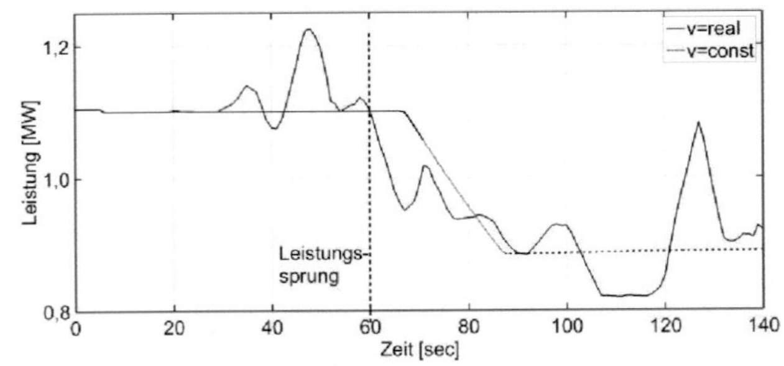

Abb. 2: Regelung von Windkraftanlagen; zeitliche Verzögerung bei geforderter Leistungsminderung bei konstanter und variabler Windgeschwindigkeit (v).

lung des Generatorgegenmoments erfolgt dabei leicht zeitverzögert, um einen konstanteren Leistungsoutput zu erreichen und Netzbelastungen zu vermeiden. Bei laufender Anpassung des Generatormoments bei jeder Windgeschwindigkeitsfluktuation würde sich ein stark schwankender Leistungsausgang an der Windkraftanlage ergeben, den es aus Systemgründen zu vermeiden gilt.

Exemplarisch ist diese zeitliche Verzögerung für eine Leistungsminderung von 200 MW bei konstantem und variablem Windeingang in **Abb. 2** dargestellt.

Unabhängig von den dargestellten technischen Möglichkeiten, kann eine windabhängige Regelleistungserbringung nur dann erfolgen, wenn die dafür benötigten meteorologischen Voraussetzungen erfüllt sind.

3 Vorgehensweise

Das Potenzial an durch moderne Windkraftanlagen bereitstellbarer Regelleistung ist hauptsächlich von folgenden Faktoren abhängig:

- meteorologische Randbedingungen,
- installierte Leistung des Erzeugungsportfolios,
- räumliche Verteilung der Anlagen und
- Prognosefehler der Einspeisung.

Für die folgenden Analysen wird eine räumlich und zeitlich differenzierte Einspeisung aus Windkraftanlagen unterstellt. Dafür werden meteorologische Daten der freiverfügbaren Wettermessstationen des Deutschen Wetterdienstes (DWD) [7] und die installierten Leistungen an Windkraftanlagen aus dem öffentlich zugänglichen Anlagenregister [8] jeweils für das Jahr 2013 genutzt.

Die Bestimmung des Regelleistungspotenzials kann auf der genannten Datenbasis abhängig von der ausgeschriebenen Produktlänge des Regelleistungsangebots (z. B. Angebotsdauern von 1 bis 24 h) für ein zugrunde gelegtes Versorgungsgebiet und ein gefordertes Zuverlässigkeitsniveau der Regelleistungserbringung erfolgen. Der Zeitpunkt der Abgabe des Regelleistungsangebots (z. B. wöchentliche, tägliche oder untertägige Ausschreibung) beeinflusst zudem das Regelleistungspotenzial durch die Prognoseungenauigkeit der Windleistungseinspeisung.

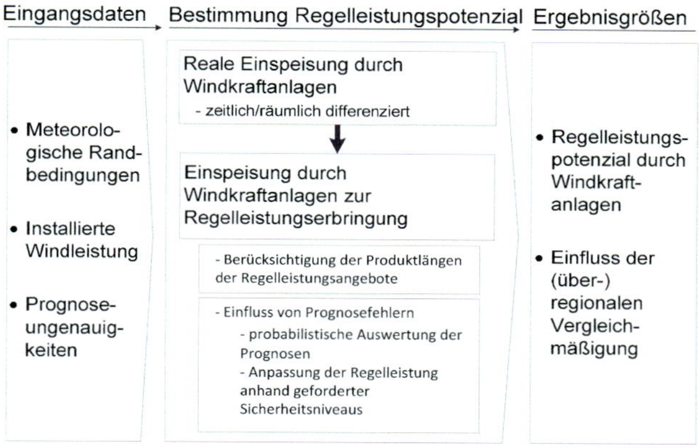

Abb. 3: Methodik zur Bestimmung des Regelleistungspotenzials

Die angewendete Methodik zur Potenzialabschätzung kann durch **Abb. 3** beschrieben werden:

1) Bestimmung der Einspeisung durch Windkraftanlagen auf Basis der meteorologischen Daten.
2) Anpassung der in 1) bestimmten Einspeisung für verschiedene Produktlängen der Regelleistungsangebote.
3) Anpassung der in 2) bestimmten Einspeisung unter Berücksichtigung des geforderten Zuverlässigkeitsniveaus der Regelleistungserbringung.

Räumliche Vergleichmäßigungseffekte durch regionale Zusammenschlüsse von Windkraftanlagen können zusätzlich das Regelleistungspotenzial positiv beeinflussen.

Ein Vergleich des Regeleistungspotenzials mit der deutschlandweit benötigten Regelleistung zeigt in welchem Maß Windkraftanlagen trotz fluktuierender Einspeisung einen Beitrag zur Systemstabilität leisten können.

4 Potenzialabschätzung

Aufbauend auf der in Abschnitt 3 beschriebenen Methodik erfolgt eine Abschätzung des Regelleistungspotenzials für verschiedene Produktlängen der Regelleistungsangebote [2]. **Abb. 4** (linke Säule je kategorisierter Produktlänge) zeigt das Regelleistungspotenzial normiert auf die installierte Windleistung für einen Windparkpool mit einer installierten Leistung von 1 GW unter durchschnittlichen meteorologischen Randbedingungen (Windgeschwindigkeit in 10 m Höhe = 3,6 m/s) und ohne Berücksichtigung des Fehlers der Windleistungsprognose. Die weiteren Säulen geben das Regelleistungspotenzial unter Berücksichtigung des Prognosefehlers an und werden folgend erläutert.

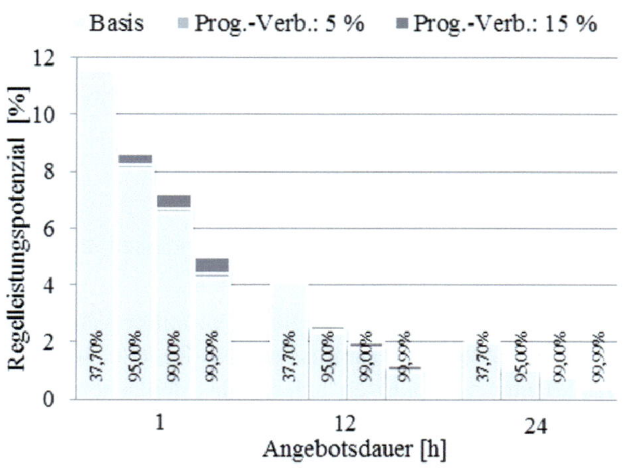

Abb. 4: Regelleistungspotenzial (normiert auf die installierte Windleistung) eines Windparks von 1 GW; für verschiedene Angebotsdauern, unter Berücksichtigung des Zuverlässigkeitsniveaus der Regelleistungserbringung (basierend auf dem Day-Ahead Prognosefehler) und Verbesserung der Prognosegüte.

4.1 Prognosefehler

Für die Teilnahme am Regelleistungsmarkt müssen Angebote mit einer definierten Vorlaufzeit erstellt werden. Demnach muss je nach Ausschreibungszeitraum für die Bestimmung des Regelleistungspotenzials der jeweilige Prognosefehler der Windleistungseinspeisung berücksichtigt werden. Üblicherweise wird der Prognosefehler als Differenz zwischen prognostizierter und Ist-Einspeisung und in Abhängigkeit vom Vorhersagehorizont durch den normierten mittleren quadratischen Fehler (eng. NRMSE: Normalized Root Mean Square Error) als Funktion der installierten Windanlagenleistung angegeben [9]. Der mittlere Day-Ahead Prognosefehler der Einspeisung aus Windkraftanlagen für die Übertragungsnetzbetreiber als auch für eine deutschlandweite Betrachtung für die Jahre 2012 bis 2014 zeigt **Tabelle 1**. Day-Ahead beschreibt hierbei einen Prognosezeitraum von 16 bis 40 h.

Tabelle 1: NRMSE der Prognosefehler der Einspeisung aus Windkraftanlagen [10–13].

	2012 [%]	2013 [%]	2014 [%]
50Hertz	5,1	4,6	4,0
TenneT	3,0	3,8	4,1
Amprion	6,2	6,0	5,4
Transnet BW	3,9	6,4	7,7
Deutschland	3,0	3,1	3,0

Demnach nimmt der Prognosefehler bei einer deutschlandweiten Betrachtung im Vergleich zu einer Analyse der Netzgebiete der Übertragungsnetzbetreiber ab, da sich fehlerhafte Einspeiseprognosen hervorgerufen durch kurzzeitige Fluktuationen innerhalb einer größeren Be-

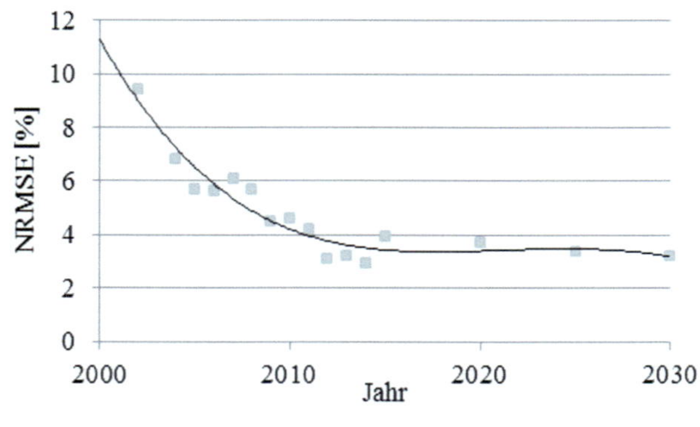

Abb. 5: Entwicklung der Prognosegüte (NRMSE) (normiert auf die installierte Windleistung) [1–5].

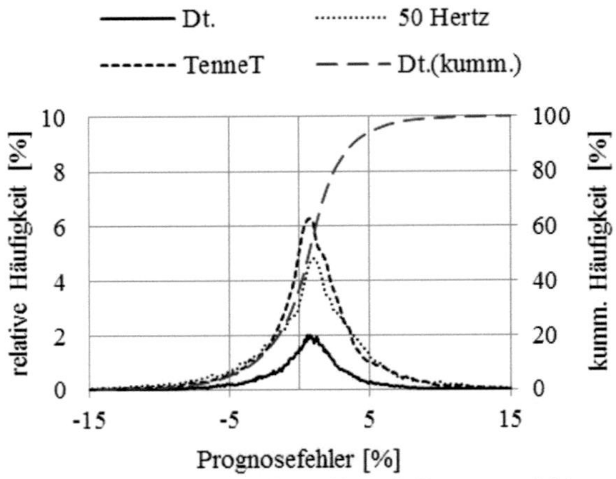

Abb. 6: Verteilung der Day-Ahead Prognosefehler der Windleistungseinspeisung (normiert auf die installierte Windleistung) (2013) [10–13].

trachtungsfläche ausgleichen. Zudem hat sich die Prognosegüte im Verlauf der dargestellten Jahre kontinuierlich verbessert; diese Tendenz gilt auch für einen längeren Betrachtungszeitraum (**Abb. 5**). Demnach hat sich in den letzten 10 Jahren die Prognosegüte des Day-Ahead Fehlers durch kontinuierlich verbesserte Prognosewerkzeuge um fast 50 % gesteigert [4].

Abb. 6 zeigt exemplarisch für das Jahr 2013 für die Übertragungsnetzbetreiber mit den größten Anteilen an installierten Windleistung und für Deutschland die Häufigkeitsverteilung des Day-Ahead Prognosefehlers der Windeinspeisung normiert auf die installierte Windleistung [10–13]. Zusätzlich wird hier die Verbesserung der Prognosegüte in weiträumigeren Gebieten ersichtlich. Demnach liegen maximale und minimale Fehler in den Gebieten der Übertragungsnetzbetreiber in einem Bereich von ±30 % der installierten Windleistung und deutschlandweit nur bei ±20 %.

Im Jahresdurchschnitt wird das Windangebot überschätzt, d. h. es wird weniger Windleistung eingespeist als prognostiziert. Exemplarisch für das Jahr 2013 zeigt **Abb. 6**, dass in ca. 62 % des Jahres gleichviel oder weniger Windleistung eingespeist wurde wie vortägig prognostiziert. Für die Erbringung von negativer Regelleistung folgt demnach, dass das zuvor bestimmte Regelleistungspotenzial nach unten korrigiert werden muss, um eine Überschätzung der angebotenen Windleistung mit geforderter Sicherheit ausschließen zu können. Mit steigendem Sicherheitsniveau der Regelleistungserbringung, können zunehmend geringere Anteile der einzuspeisenden Windleistung sicher vorhergesagt werden und das Regelleistungspotenzial sinkt signifikant [2]. Diese Abhängigkeit zwischen geforderter Zuverlässigkeit der Regelleistungserbringung und dem Anteil, um welchen die Windeinspeisung reduziert werden muss, um eine Überschätzung der angebotenen Windleistung ausschließen zu können, zeigt **Abb. 6** anhand der kumulierten Häufigkeitsverteilung des Prognosefehlers.

Demnach kann das zuvor bestimmte Regelleistungspotenzial mit einer Sicherheit von ca. 38 % erbracht werden. Regelleistungspotenziale für höhere Zuverlässigkeitsniveaus der Regelleistungserbringung zeigt **Abb. 4**.

Das Regelleistungspotenzial steigt zusätzlich durch eine Verbesserung der Prognosegüte. **Abb. 4** zeigt die Potenzialerhöhung für eine 5 % bzw. 15 % bessere Prognosegüte.

Demnach kann durch kürzere Produktlängen der Regelleistungsangebote, steigende Prognosegüte der Windleistungseinspeisung und geringere Zuverlässigkeitsniveaus der Regelleistungserbringung das Regelleistungspotenzial durch Windkraftanlagen vergrößert werden.

Im Jahresdurchschnitt (2013) können unter Einhaltung aktuell gefordertem Sicherheitsniveau der Regelleistungserbringung (99,99 %) bei einer Angebotsdauer von 1 h ca. 40 MW an negativer Regelleistung durch einen Windparkpool von 1 GW zur Verfügung gestellt werden.

4.2 Vergleichmäßigungseffekte

Das Regelleistungspotenzial verändert sich durch regionale Vergleichmäßigungseffekte, d. h. durch eine Kombination räumlicher verteilter Windparks [14]. Anhand einer Korrelation der Windgeschwindigkeiten regional verteilter Wettermessstationen kann nachgewiesen werden, dass eine räumliche Verteilung von Windkraftanlagen die Bereitstellung von Windleistung positiv beeinflusst. **Abb. 7** zeigt den Korrelationskoeffizient in Abhängigkeit der Entfernung der Wettermessstationen. Mit zunehmender Entfernung zwischen den Wettermessstationen sind immer geringere Korrelationen der Windgeschwindigkeiten zu verzeichnen. Dieser Effekt kann auf die Überlagerung kurzzeitiger Fluktuationen in der Einspeisung weit entfernter Windparks übertragen werden.

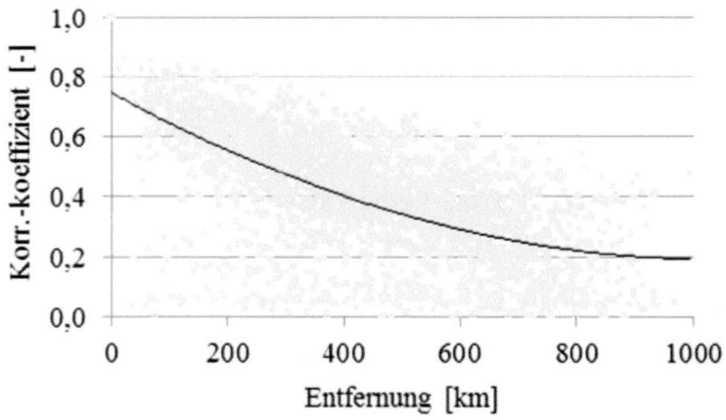

Abb. 7: Korrelation von Windgeschwindigkeit zwischen verschiedenen Wettermessstationen [4].

Der durch diese Kopplung regional verteilter Windparks zu erwartende positive Einfluss auf das Regelleistungspotenzial kann exemplarisch anhand des deutschen Windkraftanlagen-Portfolios abgeschätzt werden. Dafür wurde die in Kapitel 3 vorgestellte Methode unter denselben Randbedingungen genutzt. **Abb. 8** zeigt die resultierende Erhöhung des Regelleistungspotenzials im Vergleich zu dem bisher betrachteten 1 GW Windpark. Demnach wird bei Nutzung eines räumlich verteilten Windparkpools von

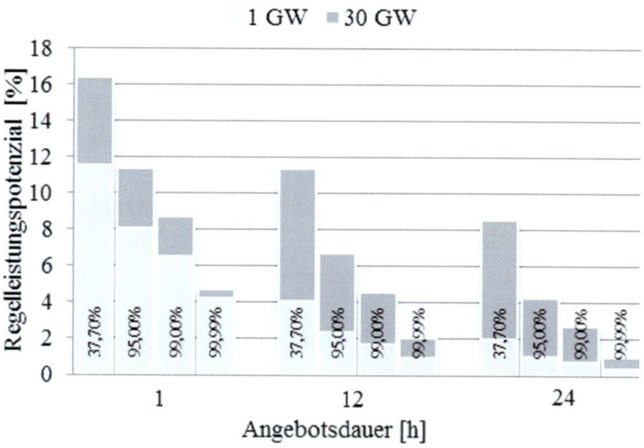

Abb. 8: Regelleistungspotenzial (normiert auf die installierte Windleistung) eines Windparks von 1 GW bzw. 30 GW; für verschiedene Angebotsdauern und unter Berücksichtigung des Zuverlässigkeitsniveaus der Regelleistungserbringung.

30 GW im Jahresdurchschnitt ein im Mittel etwa 3 % höheres Regelleistungspotenzial erreicht.

Abb. 9 zeigt die Regelenergiepotenziale in TWh (als Produkt der Leistung und der Produktlänge des Regelenergieangebots) für das untersuchte Jahr.

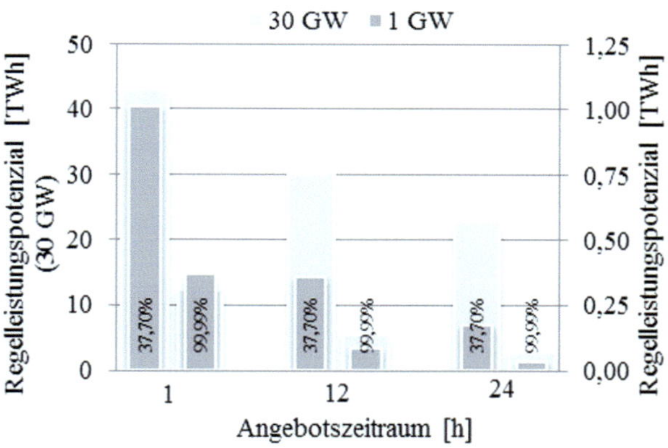

Abb. 9: Regelleistungspotenzial eines Windparks von 1 GW bzw. 30 GW für verschiedene Angebotsdauern und unter Berücksichtigung des Zuverlässigkeitsniveaus der Regelleistungserbringung.

5 Zusammenfassung und Ausblick

Mit Hilfe einer modelltechnischen Abbildung der Regelungskonzepte von Windkraftanlagen konnte gezeigt werden, dass diese technisch in der Lage sind Regelleistung anzubieten, sofern es die meteorologischen Gegebenheiten erlauben.

Dabei steigt üblicherweise das Regelleistungspotenzial mit kürzeren Längen der Regelenergieangebote und mit steigender Prognosegüte, d. h. kurzfristigeren Ausschreibungszeiträumen. Des Weiteren kann durch die Bündelung regional verteilter Windparks ein größeres Potenzial an Regelleistung angeboten werden. Die Kombination aller Windparks in Deutschland könnte unter dem aktuell geforderten Sicherheitsniveau im Jahresdurchschnitt eine maximale Regelleistung von 1,5 GW bereitstellen.

Um den Einfluss dieses Ergebnis auf das deutsche Stromversorgungssystem abzuschätzen, zeigt **Abb. 10** eine mögliche Entwicklung des Regelleistungsbedarfs an Sekunden- und Minutenreserve für eine zunehmende installierte Leistung von Windkraftanlagen [9]. Demnach dürfte sich der Bedarf an Regelenergie durch die zunehmende Integration von angebotsorientierten erneuerbare Energien vergrößern. Dieser Zunahme entgegen wirkt zum einen die sich verbessernde Prognosegüte, zum anderen die Verstetigung der eingespeisten Windleistung durch eine vermehrte Installation von Offshore-Windkraftanlagen [15]. Bei einer derartigen Entwicklung des Regelleistungsbedarfs könnten Windkraftanlagen schon heute im Jahresdurchschnitt die zu erwartende Zunahme der Regelleistung unter Einhaltung des geforderten Sicherheitsniveaus abdecken und damit den Must-Run-Sockel von konventionellen Kraftwerken im System reduzieren.

Für eine ökonomisch und ökologisch effiziente Nutzung von Windkraftanlagen zur Wirk- und Regelleistungserbringung, müssen jedoch die regulatorischen und marktwirtschaftlichen Voraussetzungen angepasst werden, und eine regelungstechnische Umsetzung geboten werden.

Für die Erbringung von Regelleistung durch einen regional verteilten Verbund an Anlagen, müssen die Kapazitäten des elektrischen Netzes gesondert untersucht werden, um so zusätzliche Netzbelastungen zu reduzieren.

Das Regelleistungspotenzial durch Windkraftanlagen kann durch einen Verbund mit PV-Anlagen und/oder Speichern zudem erhöht werden. Eine zusätzliche Untersuchung der zeitlichen Differenzierung der Regelleistungspotenziale kann Schlussfolgerungen liefern, welchen Anteil an Wirk- und Regelleistung wetterabhängige erneuerbare Energien in zeitlicher und räumlicher Auflösung gesichert bereitstellen können. Dem folgend kann der Einfluss auf das zugrunde liegende Stromversorgungssystem erneut betrachtet werden.

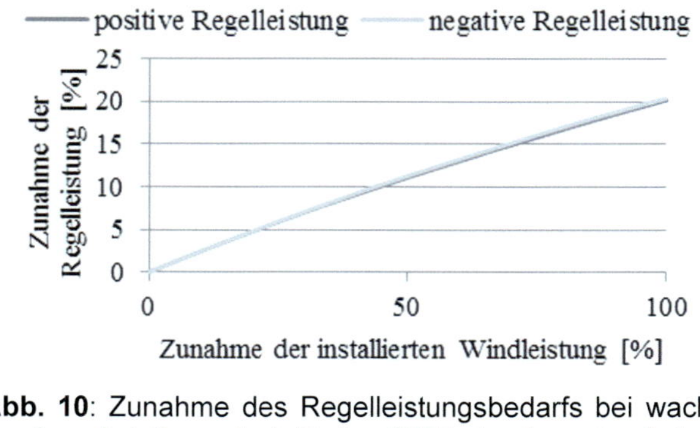

Abb. 10: Zunahme des Regelleistungsbedarfs bei wachsendem Anteil von installierten WEA; basierend auf dem wahrscheinlichkeitstheoretischen Ansatz der rekursiven Faltung (Graf/Haubrich-Verfahren) [1–3].

6 Literatur

[1] Spliethoff, H.; Hentschel, J.; et al.: Steigerung der Lastdynamik thermischer Kraftwerke; In: BWK-Das Energie-Fachmagazin 6 (2015).

[2] Fraunhofer IWES: Regelenergie durch Windkraftanlagen; Abschlussbericht (2014).

[3] Al-Awaad, A.-R. K.: Beitrag von Windenergieanlagen zu den Systemdienstleistungen in Hoch- und Höchstspannungsnetzen (2009).

[4] dena-Netzstudie II.; Integration erneuerbarer Energien in die deutsche Stromversorgung im Zeitraum 2015 - 2020 mit Ausblick 2025 (2010).

[5] Consentec GmbH; R2B Energy Consulting GmbH: Voraussetzungen einer optimalen Integration erneuerbarer Energien in das Stromversorgungssystem; Studie im Auftrag des Bundesministeriums für Wirtschaft und Technologie (2010).

[6] Heier, S.: Windkraftanlagen; Systemauslegung, Netzintegration und Regelung; 5., Aufl (2008).

[7] Deutschen Wetterdienstes: Climate Data Center; www.dwd.de.

[8] Bundesnetzagentur: Anlagenregister der installierten erneuerbaren Energieanlagen; www.bundesnetzagentur.de.

[9] Brückl, O.: Wahrscheinlichkeitstheoretische Bestimmung des Regel- und Reserveleistungsbedarfs in der Elektrizitätswirtschaft (2006).

[10] 50Hertz Transmission GmbH: Netzkennzahlen; www. 50hertz.com.

[11] Amprion GmbH: Netzkennzahlen; www.amprion.de.

[12] Tennet TSO GmbH: Netzkennzahlen; www.tennettso.de.

[13] TransnetBW GmbH: Netzkennzahlen; www. transnetbw.de.

[14] Mono, R.; Horn, F.; et al.: Ungleichzeitigkeit und Effekte räumlicher Verteilung von Wind- und Solarenergie in Deutschland; Eine Untersuchung der 100 prozent erneuerbar stiftung (2014).

[15] Rohrig, K.; Richts, C.; et al.: Energiewirtschaftliche Bedeutung der Offshore-Windenergie für die Energiewende.

Einsatzsimulation zur Erbringung von Primärregelleistung mit Batteriespeichern

P. Krüger, M. Nimtz, U. Fischer, H.-J. Krautz
Brandenburgische Technische Universität Cottbus–Senftenberg
Universitätsstraße 22 03046 Cottbus, perco.krueger@b-tu.de

Kurzfassung

Die Verbundpartner Energiequelle GmbH, 50Hertz Transmission GmbH und die BTU Cottbus – Senftenberg / Lehrstuhl Kraftwerkstechnik befassen sich im FuE-Vorhaben „Systemdienstleistungen und Energiespeicherung mittels Großbatterien zur Stabilisierung von Netzen mit hohen EE-Anteilen – Konzeption und Demonstration“ (kurz: SDL-Batt) mit der Weiterentwicklung eines hocheffizienten Batteriespeichers zum Regelkraftwerk. Die Erbringung von Regelleistung und weiteren Systemdienstleistungen sowie die hierfür notwendigen Entwicklungen und Erprobungen werden auf Basis eines innovativen 10-MW-Li-Ion-Batteriesystems, welches durch den Projektpartner Energiequelle GmbH in Feldheim errichtet und betrieben wird, betrachtet. Betriebsdaten, energetische Bilanzierung, rechtlich-regulatorische Randbedingungen und Simulation verschiedener Betriebsstrategien stehen im Vordergrund der Untersuchungen.
Im Rahmen des Forschungsprojektes wurde am Lehrstuhl Kraftwerkstechnik der BTU Cottbus–Senftenberg ein MATLAB/Simulink®-Modell zur Simulation der Primärregelleistungserbringung eines Batteriespeichers entwickelt. Ziel der Simulation ist es, Betriebsstrategien in einem breiten Spektrum von Parametervorgaben bereits vor der Inbetriebnahme zu testen, um ideale Einsatzfälle des Batteriespeichers ableiten zu können.
Die Simulation der ersten Betriebsstrategien zeigen, dass der Batteriespeicher als Primärregelleistungserbringer gut geeignet ist. Im Wesentlichen hängt die Verfügbarkeit des Batteriespeichers von der für eine Fahrplanverschiebung reservierten Leistung sowie von der Voraussetzung einer zeitnahen Fahrplanverschiebung ab.

1 Einleitung

Regelleistung, insbesondere die Primärregelleistung (PRL), wird bisher hauptsächlich durch konventionelle thermische Kraftwerke erbracht. In den meisten Fällen muss dafür eine Mindestleistung vorhanden sein, damit die Erbringung der Regelleistung sichergestellt werden kann. Mit der weiteren Integration von fluktuierenden Erneuerbaren Energien und der damit einhergehenden Flexibilisierung der Stromerzeugung sind die Reduzierung und das adäquate Ersetzen von diesen so genannten „must-run“-Kapazitäten von hoher Bedeutung. Einen wichtigen Baustein bilden hierbei Großbatteriespeicher, welche theoretisch ihr volles Leistungsband als Regelleistung anbieten können. Der wesentliche Nachteil von Batterien ist deren begrenzte Kapazität. Mit Hilfe eines im Rahmen des Forschungsprojektes entwickelten MATLAB/Simulink®-Modells werden erste Betriebsstrategien und mögliche Einsatzfälle für das sichere Betreiben einer Großbatterie in der Primärregelung charakterisiert.

2 Simulationsmodell

Zur Simulation der PRL-Erbringung einer Großbatterie oder eines anderen, kapazitiv begrenzten Erzeugers wurde ein MATLAB/Simulink®-Modell entwickelt, welches folgende Funktionsgruppen abbildet:

- Netzfrequenzverlauf und PRL-Randbedingungen (Totband, Übererfüllung)
- Batteriespeicher (Leistung, Kapazität etc.)
- Strommarkt (angebotene Leistung, Handelsperioden, Preise)
- Speicherregelung (Schaltpunkte/Grenzen etc.)

Das Modell ist frei parametrierbar und so beispielsweise an andere Speichergrößen oder Batterietypen anpassbar. Ziel der Untersuchungen ist es, für den Batteriespeicher in der Primärregelung den Last- und Speicherstandverlauf und den Einfluss verschiedener Nachladestrategien zu erfassen und zu bewerten.

2.1 Zeitliche Auflösung

Das MATLAB/Simulink®-Modell basiert auf historischen sekündlichen Messdaten der Netzfrequenz [1] sowie Marktergebnissen der wöchentlichen PRL-Ausschreibung [2] der Netzbetreiber.

In einer Zeitreihenanalyse wird der aktuelle Speicherstand (E_{S,t_1}) bilanziell nach Gleichung 1 über das Aufsummieren des vorherigen Speicherstands (E_{S,t_0}) und der Regel- ($E_{S,R}$) und Fahrplanarbeit ($E_{S,Fp}$) im Zeitraum t_0 bis t_1 ermittelt.

$$E_{S,t_1} = E_{S,t_0} + E_{S,R} + E_{S,Fp} \qquad \text{Gl. 1}$$

$$\Delta t = t_1 - t_0 = 1\ s \qquad \text{Gl. 2}$$

Die PRL ergibt sich Anhand der Netzfrequenz und wird gemäß dem Transmission Code 2003 Anhang D1 [3] berechnet. Die Leistungsabgabe, bzw. -aufnahme erfolgt außerhalb des Totbandes linear zur Netzfrequenz bis zu einer Abweichung von 0,2 Hz. Der Speicherstand wird über die Integration der erbrachten Leistung und unter Berücksichtigung der Systemwirkungsgrade berechnet.

2.2 Management des Speicherstands

Die Skalierung des Batteriespeichers erfolgt anhand der Parameter Leistungsgröße, Speichervermögen und Wirkungsgrad. In die Simulation wurden Prozeduren zum Management des Speichers implementiert. Diese halten den Speicher im Rahmen der erlaubten Speicherstände (SOC) und stellen so einen möglichst ununterbrochenen Betrieb sicher. Das feste Verhältnis von Kapazität zu Leistung führt früher oder später zu dem Fall, dass ein Speicher (ohne Eingriff) voll- oder leerläuft und dann PRL nur noch in eine Richtung erbringen kann. Der Grund hierfür sind die Wirkungsgrade beim Be- und Entladen sowie der Frequenzverlauf, welcher oft über längere Zeiträume eine Abweichung in dieselbe Richtung aufweist. Durch geeignete Eingriffe über ein Speicherstandmanagement können diese Fälle reduziert werden.

2.2.1 Fahrplanmanagement

Durch die Vorhaltung einer Leistungsreserve (bspw.: Leistungsreserve von 1 MW bei 10 MWh Speicherkapazität und PRL-Angebot von 9 MW) wird während der PRL-Erbringung bei unter- oder überschreiten eines SOC-Grenzwertes ein Handelsgeschäft ausgelöst. Die Fahrplanverschiebung (Fvs) erfolgt in Höhe der positiven und negativen Leistungsreserve.

Bild 1 zeigt die Verfahrensweise zum *Fahrplanmanagement*. Mit dem Erreichen des SOC-Grenzwerts wird nach einer veranschlagten Reaktionszeit (t_R, hier 5 Minuten) die Anforderung für ein Handelsgeschäft gegeben. Die Reaktionszeit stellt hier die Zeit dar, die vom Detektieren der Überschreitung des Grenzwertes bis zum Auslösen des Handelsgeschäftes benötigt wird. Da hier bereits das 15-Minuten-Handelsintervall überschritten war, muss bis zum nächsten 15-Minuten-Intervall gewartet werden. Dies stellt die Wartezeit (t_W) dar. Je nach vorgegebener Zeit ($t_{gate\ close}$) erfolgt dann die Ausführung des Handelsgeschäftes (hier 15 Minuten). Die Zeit bis zum Ausführen des Handelsgeschäftes wird im minimalen Fall mit 15 Minuten angenommen (direktes Handelsgeschäft zwischen zwei Partnern). Dies ist die minimale Vorlaufzeit nach §5 StromNZV [4], die zum Anmelden eines neuen Fahrplans beim Netzbetreiber eingehalten werden muss. Ebenso ist eine Vorlaufzeit von 45 Minuten denkbar, welche den Handelsschluss vor Lieferung von innertägigen Handelsgeschäften über die Strombörse EPEX Spot darstellt.

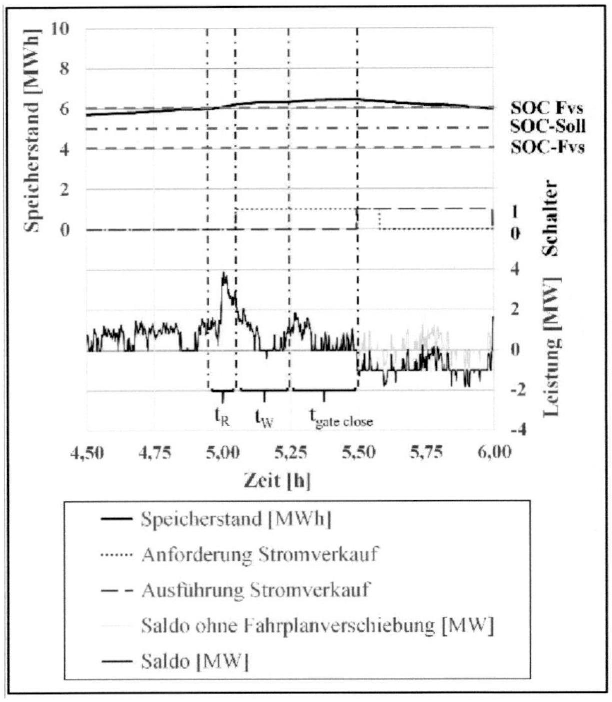

Bild 1 Ausschnitt zur Verfahrensweise beim Fahrplanmanagement (Okt. 2013)

2.2.2 Aussetzen des Speichers aus der PRL

Das Aussetzen des Speichers aus der PRL-Erbringung erfordert einen Partner (bspw. eine weitere Anlage im Pool), der in dem betreffenden Zeitraum das Aussetzen besichert. Bei Erreichen eines vorgegebenen SOC-Grenzwertes wird der Speicher ähnlich wie beim Verfahren *Fahrplanmanagement* unter Beachtung der gewählten Vorlaufzeiten aus der PRL genommen. Durch ein Handelsgeschäft wird anschließend der Speicherstand mit einer vorgegebenen Leistung des Speichers über ein Handelsintervall zurückgeführt. Nach der Beendigung der Speicherrückführung erfolgt das Umschalten in den PRL-Betrieb.

2.2.3 Asynchroner Betrieb im Pool

Wird der Speicher in einem Pool betrieben, besteht die Möglichkeit bei Erreichen eines vorgegebenen SOC-Grenzwertes die PRL nur in eine Richtung zu erbringen. Die gegensätzliche Richtung wird durch eine andere Anlage im Pool erbracht. Dies wird solange aufrechterhalten, bis ein Abschalt-SOC erreicht wird.

Aufgrund der Möglichkeit, im Pool jederzeit die Anlagen zur Erbringung der PRL wechseln zu können, ist für den Umschaltvorgang nur die Reaktionszeit im Pool von Bedeutung (hier 1 Minute).

2.2.4 Rückführung im Totband

Ist die Frequenzmessung des Speichers mit einer Genauigkeit deutlich kleiner des Toleranzbereiches des Totbandes ausgelegt, ist eine Nutzung des Totbandes seitens des ÜNB gestattet [5]. Das Totband darf allerdings nur systemkonform (Leistung immer entsprechend der Richtung der Frequenzabweichung) genutzt werden.

Bei Aktivierung dieses Freiheitsgrades in der Simulation erfolgt, ab einem vorgegeben SOC-Grenzwert, das Erbringen von PRL im Bereich des Totbandes von 49,99 bis 50,01 Hz.

2.2.5 Rückführung durch Übererfüllung

Einen weiteren Freiheitsgrad zum Speichermanagement ist die Möglichkeit der Übererfüllung. Diese ermöglicht es, dass bis zu 20 % über der geforderten PRL erbracht werden dürfen. In der Simulation wird der Freiheitsgrad bei einem definierten SOC-Grenzwert aktiviert. Bei Unterschreitung des unteren SOC-Grenzwertes wird bis zu 20 % mehr negative PRL erbracht, bei Überschreitung des oberen Grenzwertes wird bis zu 20 % mehr positive PRL erbracht.

3 Untersuchung verschiedener Szenarien

3.1 Grundlegende Parameter

Tabelle 1 zeigt die grundlegenden Parameter des Speichers. Diese lehnen sich an die projektierten Parameter des RRKW (Regionalen Regelkraftwerkes) Feldheim an. Das Verhältnis von Speicherkapazität zu Speicherleistung am Netzanschlusspunkt (Erbringungsort der PRL) wird mit 1:1 festgelegt. Der Wirkungsgrad AC zu AC wird mit 85 % angenommen. Der Start-SOC wird für jeden Monat auf 0,5 gesetzt.

Tabelle 1 Parameter des Speichers

Größe	Einheit	Wert
Kapazität	[MWh]	10
Soll-SOC	[-]	0,5
Ladeleistung	[MW]	10*
Entladeleistung	[MW]	10*
Wirkungsgrad AC/AC	[-]	0,85

*am Netzanschlusspunkt

Zur Vermarktung des Speichers wurde eine Vergütung der PRL mit dem gewichteten Mittelwert aller bezuschlagten Gebote (LP Ø) der jeweiligen Wochen angenommen.

Die für den Handel und den Besicherungsfall angesetzten Werte können der **Tabelle 2** entnommen werden. Die Kosten für den Besicherungsfall sowie der asynchronen Betriebsweise sind vorläufige Schätzungen. Es wird davon ausgegangen, dass lediglich für die Absicherung der Netzeinspeisung eine Zahlung notwendig ist. Für den Bezug aus dem Netz wird keine Zahlung angenommen, da es hier, bei den einspringenden konventionellen Anlagen, zu Brennstoffkosteneinsparungen kommt. Vor allem im Poolbetrieb ist es möglich, dass die Besicherung durch einen prozentualen Abschlag auf den PRL-Erlös abgerechnet wird. Im Falle des kompletten Aussetzens des Speichers aus der PRL werden für diesen Zeitraum keine Vergütung und die Zahlungen für das Aussetzen berechnet.

Tabelle 2 Wirtschaftliche Randbedingungen

Größe	Einheit	Wert
Zahlung Aussetzen aus der PRL	[€/(h·MW)]	100
Zahlung asynchron POS (PRL NEG Absicherung)	[€/(h·MW)]	0
Zahlung asynchron NEG (nur Laden)	[€/(h·MW)]	150
Stromeinkauf	[€/MWh]	90
Stromverkauf	[€/MWh]	-30*

*Einnahmen

3.2 Szenario 1

Das *Szenario 1* stellt den Speicherbetrieb außerhalb eines Pools dar, wobei die Besicherung des Speichers durch einen Dritten erfolgt. Für das Speicherstandmanagement werden die Möglichkeiten des *Fahrplanmanagement* und das *Aussetzen des Speichers aus der PRL* gewählt. Die beschriebenen Freiheitsgrade *Totband* und *Übererfüllung* sind deaktiviert. **Tabelle 3** zeigt die Lade- und Entladeleistungen und die dazugehörigen SOC-Schaltpunkte sowie veranschlagte Zeiten.

Tabelle 3 Parameter des Speicherstandmanagements im Szenario 1

Größe	Einheit	Wert
Entladeleistung bei Speicherstandkorrektur*	[MW]	7,4
SOC-Schaltpunkt Speicherstandkorrektur Entladen	[-]	0,7
Ladeleistung bei Speicherstandkorrektur*	[MW]	8,7
SOC-Schaltpunkt Speicherstandkorrektur Laden	[-]	0,3
SOC-Schaltpunkt Fahrplanverschiebung Verkauf	[-]	0,6
SOC-Schaltpunkt Fahrplanverschiebung Einkauf	[-]	0,4
Reaktionszeit (t_R)	[min]	5
Handelsschluss vor Lieferung ($t_{gate\ close}$)	[min]	15

*am Netzanschlusspunkt

Im *Szenario 1* werden die vermarktete PRL in einem Bereich von einem Angebot von 10 MW und keiner Reserve für Fahrplanverschiebung bis zu einem Angebot von 7,5 MW und 2,5 MW Reserve für Fahrplanverschiebung variiert. Die Schrittweite ist 0,5 MW (*Szenario 1A* bis *1F*).

3.3 Szenario 2

Der Betrieb des Speichers in einem Pool wird in *Szenario 2* simuliert. Im Gegensatz zu *Szenario 1* wird hier statt des kompletten Aussetzens aus der PRL der asynchrone Betrieb gewählt. Zusätzlich zur Variation der vermarkteten PRL-Leistung werden hier die Freiheitsgrade (*Totband* und *Übererfüllung*) und der Handelsschluss vor Lieferung variiert. Die **Tabelle 4** gibt einen Überblick über die gewählten Parametervariationen

Tabelle 4 Parameter des Speicherstandmanagements im Szenario 2

Größe	Einheit	Werte der Parametervariationen	
PRL-Angebot	[MW]	9,5	9
Lade- und Entladeleistung Fahrplanverschiebung*	[MW]	0,5	1
Handelsschluss vor Lieferung ($t_{gate\ close}$)	[min]	15	45
Übererfüllung		Ja	Nein
Totband		Ja	Nein

*am Netzanschlusspunkt

Die SOC-Schaltpunkte für den asynchronen Betrieb sind mit 0,75, bzw. 0,25 so gesetzt, dass zu jeder Zeit bei einem Ausfall der Kommunikation zwischen Poolbetreiber und Speicher eine 15-minütige Erbringung der maximalen PRL-Leistung gewährleistet ist. Die SOC-Schaltpunkte für die Fahrplanverschiebung und die Reaktionszeit bleiben wie im *Szenario 1* bei 0,6 und 0,4 bzw. 5 Minuten.

4 Ergebnisse der Simulation

4.1 Ergebnisse des Szenario 1

Die **Tabellen 5** und **6** zeigen die wesentlichen Simulationsergebnisse für das *Szenario 1* für das Jahr 2013. Es ist zu erkennen, dass ohne die Vorhaltung einer Reserve für die Fahrplanverschiebung (*Szenario 1A*) für 258 h im Jahr 2013 der Besicherungsfall aktiviert werden muss. Bereits durch eine Vorhaltung von 0,5 MW kann dieser Wert erheblich reduziert werden (auf 68,8 h). Im *Szenario 1F* ist kein Besicherungsfall mehr zu verzeichnen.

Durch die Wirkungsgradverluste des Speichers übersteigen die eingekauften Strommengen deutlich die verkauften Strommengen. Die insgesamt umgesetzte Energiemenge (hier ausgedrückt in Vollzyklen des Speichers) beträgt im *Szenario 1A* 408 Vollzyklen des Speichers und sinkt im *Szenario 1B* bereits deutlich auf 329 Vollzyklen ab. Die Ursache liegt zum einen bereits am geringeren PRL-Angebot. Zum anderen und dies stellt den Schwerpunkt der Reduzierung dar, wird durch die Fahrplanverschiebung der Speicher tendenziell in hohen PRL-Erbringungsphasen entlastet.

Tabelle 5 Ergebnisse der Simulation für Szenario 1A, 1B und 1C, Simulationsjahr 2013

Größe	Einheit	Szenario 1A	Szenario 1B	Szenario 1C
PRL-Angebot	[MW]	10	9,5	9
Lade- und Entladeleistung Fahrplanverschiebung*	[MW]	0	0,5	1
Aussetzen aus der PRL	[h]	258	68,8	8,8
Vollzyklen des Speichers	[-]	408	329	299
Handel Einkauf**	[GWh]	1,62	1,51	1,44
Handel Verkauf**	[GWh]	0,53	0,55	0,55
PRL–Einnahmen	[Mio. €]	1,50	1,46	1,39
Einnahmen, abzüglich Handel und Besicherungsfall	[Mio. €]	1,11	1,27	1,27

*am Netzanschlusspunkt; **nur Fahrplanverschiebung und Speicherstandkorrektur

Tabelle 6 Ergebnisse der Simulation für Szenario 1D, 1E und 1F, Simulationsjahr 2013

Größe	Einheit	Szenario 1D	Szenario 1E	Szenario 1F
PRL-Angebot	[MW]	8,5	8	7,5
Lade- und Entladeleistung Fahrplanverschiebung*	[MW]	1,5	2	2,5
Aussetzen aus der PRL	[h]	1	0,25	0
Vollzyklen des Speichers	[-]	297	294	286
Handel Einkauf**	[GWh]	1,39	1,33	1,26
Handel Verkauf**	[GWh]	0,52	0,49	0,45
PRL–Einnahmen	[Mio. €]	1,32	1,24	1,16
Einnahmen, abzüglich Handel und Besicherungsfall	[Mio. €]	1,21	1,13	1,06

*am Netzanschlusspunkt; **nur Fahrplanverschiebung und Speicherstandkorrektur

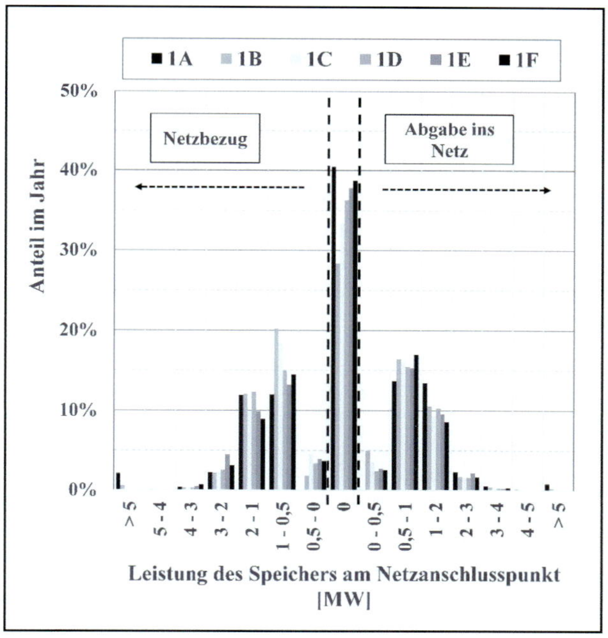

Bild 2 Verteilung der Leistungserbringung des Speichers am Netzanschlusspunkt im Jahr 2013 für Szenario 1

Die Verteilung der Leistungserbringung des Speichers ist in **Bild 2** dargestellt. In 28,3 bis 40,5 % der Zeit im Jahr 2013 wird keine Leistung durch den Speicher erbracht oder aufgenommen. Im *Szenario 1A* ist im Bereich zwischen 0 und 0,5 MW aufgrund des Totbandes und der fehlenden *Fahrplanverschiebung* keine Leistungserbringung zu erkennen, wohingegen in den anderen Szenarien ca. 5 bis 7 % der Zeit eine Leistungserbringung mit Leistungen < 5 % der Netzanschlussleistung in beide Richtungen erfolgt. Der Schwerpunkt der Leistungserbringung oder Leistungsaufnahme liegt im Bereich zwischen 0,5 und 2 MW (5 bis 20 % der Netzanschlussleistung).

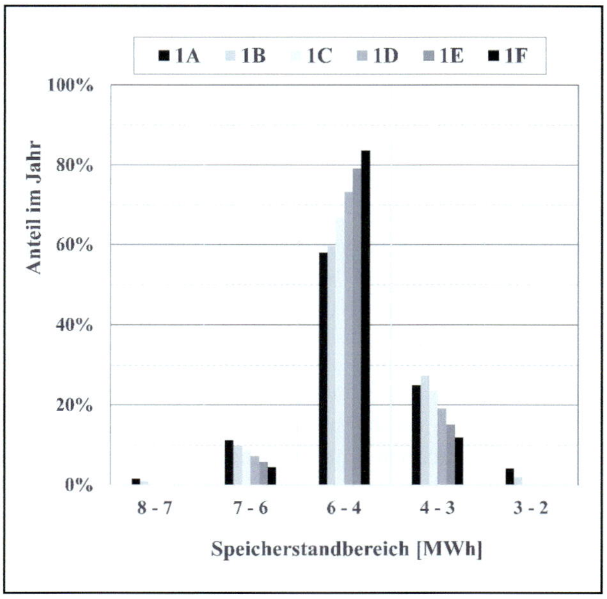

Bild 3 Verteilung der Speicherstände im Jahr 2013 für Szenario 1

Durch die Fahrplanverschiebung lässt sich, wie in **Bild 3** zu erkennen ist, ein deutlich engerer Speicherstandbereich realisieren. Sind im *Szenario 1A* noch unter 60 % aller Speicherstände im Bereich zwischen 4 bis 6 MWh, so steigt dieser Wert bis *Szenario 1F* auf über 80 %. Besonders hohe (größer 7 MWh) und besonders niedrige (kleiner 3 MWh) Speicherstände treten ab *Szenario 1C* nur sehr selten auf.

Auf der Basis der gewählten wirtschaftlichen Randbedingungen (**Tabelle 2**) ergeben sich zum einen die direkten Einnahmen aus der PRL-Vergütung, zum anderen die Einnahmen abzüglich der Kosten für den Strombezug und die Besicherung. Die ermittelten Einnahmen sollen hier lediglich zum Vergleich der einzelnen Szenarien untereinander herangezogen werden.

Es zeigt sich hier, dass unter den gegebenen Annahmen im *Szenario 1B und 1C* mit 1,27 Mio. € die höchsten Einnahmen erzielt werden können. Die um ca. 10 % niedrigeren Vollzyklen im *Szenario 1C* gegenüber *Szenario 1B*, sowie die geringere Belastung des Speichers im Hinblick auf Leistungserbringung und der engeren Speicherstandbandbreite, lassen den vorläufigen Schluss zu, dass das *Szenario 1C* für die gewählte Betriebsweise in Summe das beste Szenario darstellt. Jedoch ist eine weiterführende Betrachtung der Wirtschaftlichkeit unter Einbezug aller Randbedingungen erforderlich. Dies gilt insbesondere für den Einfluss des Betriebs auf die Alterung der Batteriezellen und auf die Betriebskosten des Speichers. Endgültige Aussagen sind somit erst am Ende des Forschungsprojektes zu erwarten, wenn erste Betriebserfahrungen und Messwerte des RRKW Feldheim vorliegen.

4.2 Ergebnisse des Szenario 2

Im *Szenario 2* wird der Betrieb des Speichers in einem Pool angenommen. Das Aussetzen des Speichers aus der PRL wird hier durch die Möglichkeit des asynchronen Betriebs ersetzt. Die **Tabelle 7** und **Tabelle 8** zeigen die Ergebnisse jeweils 3 ausgewählter Parametervariationen.

In den Szenarien mit einem PRL-Angebot von 9,5 MW betragen die asynchronen Betriebszeiten 215 *(Szenario 2A)* und 236 *(Szenario 2F)* Stunden im Jahr 2013. Hierbei liegt, begründet durch die Wirkungsgradverluste des Speichers, der Anteil des asynchronen PRL NEG Betriebs (Speicher wird durch PRL beladen) deutlich über dem des asynchronen PRL POS Betriebs (Verhältnis ca. 70:30). Mit einer Erhöhung der Lade- und Entladeleistung auf 1 MW und gleichzeitigen Reduzierung des PRL-Angebotes auf 9 MW ist ein deutlicher Abfall in den asynchronen Betriebszeiten zu erkennen (*Szenario 2B* und *2D*). Durch die erhöhte Leistung im Fahrplanmanagement steigen jedoch die Handelsmengen. In Summe reduzieren sich die Vollzyklen leicht um 2 %. Durch die Hinzunahme der Freiheitsgrade *Totband* und *Übererfüllung* (*Szenario 2C* und *2E*) können die Handelsmengen reduziert werden. Dies geht allerdings zu Lasten des Energieumsatzes, welcher deutlich um ca. 20 Vollzyklen steigt. Die Differenzen der Zeiten des asynchronen Betriebes ergeben sich aus den unterschiedlichen Zeiträumen, in denen der Asynchron-Betrieb aktiviert wird und der im jeweiligen Zeitraum erreichten Rückführung durch die PRL.

Tabelle 7 Ergebnisse der Simulation für Szenario 2A, 2B und 2C, Simulationsjahr 2013

Größe	Einheit	Szenario		
		2A	2B	2C
PRL-Angebot	[MW]	9,5	9	9
Lade- und Entladeleistung Fahrplanverschiebung*	[MW]	0,5	1	1
Handelsschluss vor Lieferung	[min]	15	15	15
Übererfüllung		N	N	J
Totband		N	N	J
Asynchronbetrieb PRL NEG	[h]	154,4	6,6	8,3
Asynchronbetrieb PRL POS	[h]	60,6	4,3	4,4
Vollzyklen des Speichers	[-]	303	296	315
Handel Einkauf**	[GWh]	1,33	1,43	1,20
Handel Verkauf**	[GWh]	0,48	0,55	0,48
PRL–Einnahmen	[Mio. €]	1,47	1,39	1,39
Einnahmen, abzüglich Handel und Besicherungsfall	[Mio. €]	1,15	1,27	1,28

*am Netzanschlusspunkt; **nur Fahrplanverschiebung

Tabelle 8 Ergebnisse der Simulation für Szenario 2D, 2E und 2F, Simulationsjahr 2013

Größe	Einheit	Szenario		
		2D	2E	2F
PRL-Angebot	[MW]	9	9	9,5
Lade- und Entladeleistung Fahrplanverschiebung*	[MW]	1	1	0,5
Handelsschluss vor Lieferung	[min]	45	45	45
Übererfüllung		N	J	N
Totband		N	J	N
Asynchronbetrieb PRL NEG	[h]	10,5	8,4	166,7
Asynchronbetrieb PRL POS	[h]	6,2	5,8	69,5
Vollzyklen des Speichers	[-]	293	314	303
Handel Einkauf**	[GWh]	1,44	1,21	1,33
Handel Verkauf**	[GWh]	0,56	0,50	0,49
PRL–Einnahmen	[Mio. €]	1,39	1,39	1,47
Einnahmen, abzüglich Handel und Besicherungsfall	[Mio. €]	1,27	1,28	1,13

*am Netzanschlusspunkt; **nur Fahrplanverschiebung

Die Verteilung der Leistungserbringung des Speichers (**Bild 4**) verdeutlicht die Auswirkungen der aktivierten Freiheitsgrade *Totband* und *Übererfüllung*. Die Stillstandszeit (P = 0 MW) wird durch den Freiheitsgrad *Totband* von 33 % (*Szenario 2B* und *2D*) auf 26 % (*Szenario 2C* und *2E*) reduziert, gleichzeitig steigt der im Totband aktivierte Netzbezug um 5 % und die Abgabe um 2 %. Der Freiheitsgrad *Übererfüllung* sorgt für eine leichte Verschiebung der Anteile aus dem 0,5 MW bis 1 MW Leistungsband auf das 1 MW bis 2 MW Leistungsband.

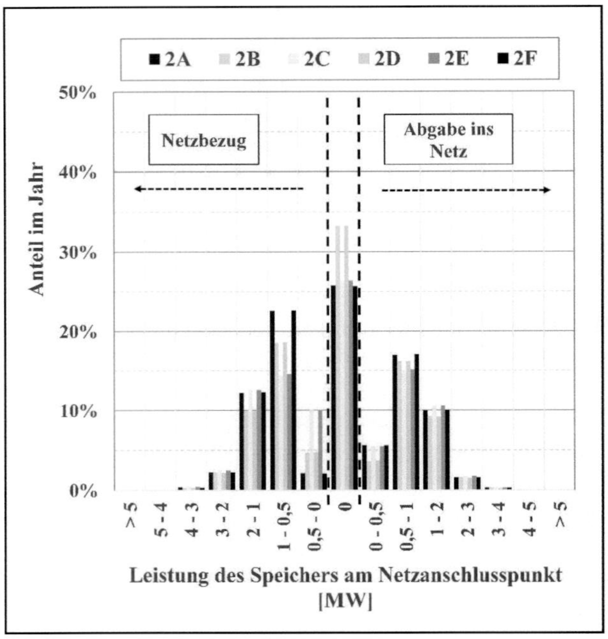

Bild 4 Verteilung der Leistungserbringung des Speichers am Netzanschlusspunkt im Jahr 2013 für Szenario 2

Mit dem Einführen der Freiheitsgrade ist, wie in **Bild 5** ersichtlich, eine Verbesserung der Verteilung des Speicherstands zu erreichbar (Vergleich *Szenario 2B* mit *2C* und *Szenario 2D* mit *2E*). Der Anteil des angestrebten idealen Speicherstandbereiches (4 bis 6 MWh) steigt von 66 % auf 72 % bzw. 60 % auf 67 %.

Ein Vergleich mit *Szenario 1* (**Bild 3**, *Szenario 1B* und *1C*) zeigt, dass der Bereich der hohen Speicherstände vor allem bei einer Reserveleistung von 0,5 MW deutlich zugenommen hat. Dies ist auf die langsamere Rückführung durch den asynchronen Betrieb und die geringe Leistungsreserve für das Fahrplanmanagement zurückzuführen.

Unter den gegebenen Randbedingungen können im *Szenario 2C* und *2E* die höchsten Einnahmen (1,28 Mio. €) erzielt werden. Jedoch sind der Energieumsatz, die Leistungsverteilung und die Speicherstsandverteilung, ähnlich wie bereits in *Szenario 1* beschrieben, in die wirtschaftlichen Berechnungen mit einzubeziehen.

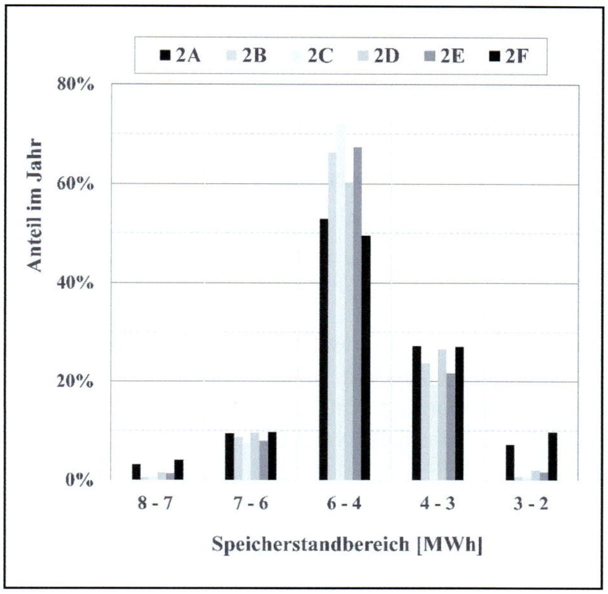

Bild 5 Verteilung der Speicherstände im Jahr 2013 für Szenario 2

5 Fazit und Ausblick

Durch das MATLAB/Simulink®-Modell ist es möglich, den Betrieb eines Batteriespeichers in der Primärenergieregelung detailliert zu untersuchen. Nachfolgend sind die wesentlichen Aussagen der berechneten Szenarien bei unterschiedlichen Betriebsstrategien zusammengefasst:

- Wird der Speicher außerhalb eines Pools betrieben (Szenario 1), kann durch die Bereitstellung einer Reserveleistung und reduziertem PRL-Angebot im Betrachtungsjahr 2013 der Besicherungsfall stark reduziert (lediglich 8,8 Stunden im Jahr bei 9 MW PRL-Angebot, 1 MW Reserveleistung und 10 MWh Speicherkapazität), bzw. komplett ausgeschlossen werden (ab 2,5 MW Reserveleistung).
- Durch die Aktivierung der Freiheitsgrade Übererfüllung und Totband ist es möglich die Handelsmengen um ca. 15 % zu reduzieren.
- Der Schwerpunkt der Leistungserbringung des Speichers bewegt sich mit 50 bis 60 % der Zeit in einem Bereich von 5 bis 20 % der Netzanschlussleistung. Die Ruhezeiten sind stark abhängig von der Totbandnutzung und nehmen ca. 25 % (Totband aktiviert) bis ca. 40 % der Zeit in Anspruch.
- Die Erhöhung der Zeit „Handelsschluss vor Lieferung“ zeigte im Poolbetrieb nur einen sehr geringen Einfluss.
- Unter den angegebenen wirtschaftlichen Randbedingungen konnten im Jahr 2013 ca. 140.000 € pro vermarktete MW erzielt werden.

Im Rahmen des „SDL-Batt“-Forschungsprojektes werden in Zukunft mit der Inbetriebnahme des RRKW Feldheim Informationen bezüglich des realen Speicherverhaltens gesammelt und ausgewertet. Im Weiteren folgen die Bestimmung des Wirkungsgrades und des Eigenstromverbrauchs des Speichers und dessen Nebenanlagen. Diese Daten werden anschließend in die Simulation eingepflegt. Ebenso werden die seitens des rechtlichen Rahmens bestehenden Unsicherheiten am realen Betriebsvorgang analysiert. Dies betrifft vor allem die Strommengen, welche durch den PRL-Abruf umgesetzt werden, aber auch die Kosten für den Strombezug und die Stromabgabe durch die Speicherstandkorrekturen.

Die weiteren Randbedingungen werden erfasst und in eine wirtschaftliche Gesamtbetrachtung überführt.

6 Literatur

[1] ENTSO-E Netzfrequenz 2013. http://www.50hertz.com/de/Maerkte/Regelenergie/Regelenergie-Downloadbereich. Zugegriffen: 30. Juli 2014

[2] Ausschreibungsübersicht. Anonymisierte Ergebnisliste für die wöchentliche Ausschreibung der Primärregelleistung 2012 bis 2014. https://www.regelleistung.net/ip/action/ausschreibung/public. Zugegriffen: 30. Juli 2014

[3] Anhang D1 zum Transmission Code 2003; Deutsche ÜNB 2003. https://www.regelleistung.net/ip/action/static/prequal. Zugegriffen: 30. Juli 2014

[4] Stromnetzzugangsverordnung (StromNZV) vom 25. Juli 2005 (BGBl. I S. 2243), die zuletzt durch Artikel 10 des Gesetzes vom 28. Juli 2011 (BGBl. I S. 1634) geändert worden ist.

[5] Eckpunkte und Freiheitsgrade bei Erbringung von Primärregelleistung. Deutsche ÜNB 2014. https://www.regelleistung.net/ip/action/static/prequal. Zugegriffen: 30. Juli 2014

Wirtschaftlichkeit von PV-Anlagenkonzepten zur Regelleistungserbringung bei veränderten Marktbedingungen

Julia Seidel, M. Sc., Björn Osterkamp, M. Sc., Prof. Dr.-Ing. Bernd Engel
Technische Universität Braunschweig, Institut für Hochspannungstechnik und Elektrische Energieanlagen – elenia
j.seidel@tu-braunschweig.de, Tel.: 0531 / 391 9720

Kurzfassung

Im Zuge der laufenden Dekarbonisierung und der damit einhergehenden Umstrukturierung des gesamten Energiesektors wird die Photovoltaik mittlerweile als tragende Säule der Energieversorgung angesehen. Jedoch übernehmen die Photovoltaik-Anlagen in Deutschland eine vergleichsweise geringe Systemverantwortung bezüglich der Systemdienstleistungen. Die betrachtete Systemdienstleistung ist hier die Frequenzhaltung und die dafür vorgehaltene Regelleistung. Um diesen stetigen Ausgleich zwischen Erzeugung und Verbrauch sicherzustellen, werden größtenteils konventionelle, thermische Kraftwerke genutzt. Zur Steigerung des Anteils von erneuerbaren Energien an der Stromversorgung müssen diese Systemdienstleistungen langfristig auch durch regenerative Stromerzeuger erbracht werden. Aktuell verhindern verschiedene Marktbarrieren die Teilnahme dieser Technologie an der Regelleistungserbringung. Eine Gegenüberstellung verschiedener PV-Anlagenkonzepte in Bezug auf die Bereitstellung von Regelleistung und deren ökonomische Bewertung zeigt jedoch, dass sich unter angepassten Marktbedingungen zusätzliche Erlösquellen für die Anlagenbetreiber erschließen lassen.

1 Einleitung

Die Bundesregierung verfolgt das Ziel, bis 2050 den Bruttostromverbrauch mit über 80 % aus regenerativen Energien zu decken [1]. Um diese und weitere klimapolitischen Ziele zu erreichen, sind massive Umstrukturierungen im Stromsystem nötig. Ein Großteil des heute regenerativ erzeugten Stroms wird durch fluktuierende Stromerzeuger wie Windkraft oder Photovoltaik (PV) bereitgestellt. Sie ersetzen nach und nach die konventionellen Großkraftwerke, welche mit ihren bereitgestellten Systemdienstleistungen maßgeblich an der Systemstabilität beteiligt sind. Um den Anteil der erneuerbaren Energien weiter steigern zu können, sind diese Systemdienstleistungen zukünftig vermehrt durch die zumeist dezentralen Anlagen aus erneuerbaren Energien zu erbringen.

2 Ziel des Forschungsvorhabens

Vor dem Hintergrund der größtenteils fehlenden Systemverantwortung von PV-Anlagen wird im Forschungsprojekt PV-Regel zusammen mit dem Institut für Hochspannungstechnik und Elektrische Energieanlagen (elenia) der TU Braunschweig, der SMA Solar Technology AG und der GEWI AG die Bereitstellung von Regelleistung mit Photovoltaik untersucht. Neben der technischen Umsetzung spielen vor allem die regulatorischen und marktgestalterischen Rahmenbedingungen eine große Rolle. Ziel des Projekts ist die technische Erprobung der Regelleistungsbereitstellung in einem Feldtest sowie die Erforschung von optimalen Marktbedingungen zur Regelleistungserbringung mit erneuerbaren Energiequellen. Hierdurch sollen Empfehlungen für Entscheidungsträger gegeben werden, die eine optimale Einbindung der erneuerbaren Energien in die Regelleistungsmärkte ermöglichen sollen.

3 Ausgangslage

Als eine der zentralen Systemdienstleistungen wird die Frequenzhaltung bislang hauptsächlich von konventionellen Kraftwerken übernommen. Mit der Bereitstellung von Regelleistung aus dezentralen Energieanlagen ist es technisch möglich, den Anteil der konventionellen Mindesterzeugung zu reduzieren und den Prozess der Dekarbonisierung voranzutreiben.

In [2] wurde bereits nachgewiesen, dass die Bereitstellung von Regelleistung mit PV aus technischer Sicht grundsätzlich möglich ist. Kritisch sind jedoch z. T. die Rahmenbedingungen des Regelleistungsmarktes zu werten. Bereits im Jahr 2011 hat die Bundesnetzagentur in ihren Beschlüssen *BK6-10-097*, *BK6-10-098* und *BK6-10-099* eine bessere Integration erneuerbarer Energien in den Regelleistungsmarkt gefordert [3]. Auch in [4] wird mit dem Ziel der Effizienzsteigerung ein diskriminierungsfreier Marktzugang für erneuerbare Energien gefordert. Eine wesentliche Herausforderung ist somit in den aktuellen regulatorischen Rahmenbedingungen zu sehen. Derzeit ist eine Teilnahme am Regelleistungsmarkt nur sehr begrenzt möglich bzw. nicht wirtschaftlich darstellbar. Im Folgenden wird daher die Bereitstellung von Regelleistung mit Photovoltaik-Anlagen aus wirtschaftlicher Perspektive betrachtet.

Im internationalen Vergleich wird deutlich, dass zwar eine Wirkleistungsregelung zur Frequenzhaltung i.d.R. verpflichtend, eine Teilnahme am Regelleistungsmarkt für fluktuierende Erzeuger jedoch häufig nicht möglich ist.

Als Ausnahmen sind an dieser Stelle Dänemark und Großbritannien zu nennen, obwohl auch hier einige Restriktionen zu beachten sind. In Dänemark ist ausschließlich eine Teilnahme am Energy-Only-Markt für MRL möglich. Die Bereitstellung von PRL und SRL aus fluktuierenden Erzeugern ist nicht erlaubt. Auch in Großbritannien ist die Teilnahme am Regelleistungsmarkt grundsätzlich möglich. Ökonomisch sinnvoll ist dies jedoch aufgrund fehlender Anreize nicht. [5] Bislang gab es weltweit noch keine PV-Anlage, die an einem Regelleistungsmarkt teilgenommen hat.

In Deutschland werden die Regelleistungsprodukte Primär-, Sekundär- und Minutenreserveleistung über eine gemeinsame Internetplattform der vier Übertragungsnetzbetreiber Amprion GmbH, TenneT GmbH, TransnetBW GmbH und 50Hertz GmbH auktioniert. Zur Teilnahme müssen die technischen Einheiten vorgeschriebene Präqualifikationsbedingungen erfüllen. Ein Überblick über die wesentlichen Marktbedingungen wird, geordnet nach Regelleistungsart, in **Tabelle 1** dargestellt.

Tabelle 1 Hauptmerkmale des Regelleistungsmarktes in Deutschland [6]

	Primärregelleistung (PRL)	*Sekundärregelleistung (SRL)*	*Minutenreserveleistung (MRL)*
Ausschreibungszeitraum	Wöchentlich	Wöchentlich	Werktäglich
Produktzeitscheiben	Eine Woche	Hochtarif (Mo.-Fr. 8-20 Uhr, kein Feiertag) / Niedertarif	Sechs 4-h-Blöcke pro Tag
Produktdifferenzierung	Keine	Pos./neg. SRL	Pos./neg. MRL
Mindestangebotsgröße	±1 MW	+5 MW oder -5 MW	+5 MW oder -5 MW
Pooling	Nur innerhalb einer Regelzone	Nur innerhalb einer Regelzone – zur Erreichung der Mindestangebotsgröße auch regelzonenübergreifend	Nur innerhalb einer Regelzone – zur Erreichung der Mindestangebotsgröße auch regelzonenübergreifend
Vergütung	Leistungspreis (Pay-as-bid)	Leistungs- und Arbeitspreis (Pay-as-bid)	Leistungs- und Arbeitspreis (Pay-as-bid)
Aktueller Bedarf (Juli 2015)	783 MW (+ und -) (DE, CH, NL)	2103 MW (-) und 2076 MW (+) (DE)	1782 MW (-) und 1513 MW (+) (DE)

Die Marktbedingungen sind historisch gewachsen und basieren auf der ausschließlichen Bereitstellung von Regelleistung durch konventionelle Kraftwerke. Diese Anlagen können eine vorgegebene Leistung garantieren und somit Regelleistung verbindlich über einen langen Zeitraum vorhalten. Das Verfahren gewährleistet den Übertragungsnetzbetreibern jederzeit die Bereitstellung von Regelleistung. Der Wandel im deutschen Energieversorgungssystem bringt jedoch auch Anpassungen zur Integration fluktuierender Energieerzeuger mit sich.

4 Untersuchungsgegenstand

Grundlage der Untersuchungen bilden verschiedene PV-Anlagenkonzepte. Neben einzelnen Anlagen hoher Leistungsklassen (> 20 MW) werden auch das Pooling aus mehreren, geographisch verteilten PV-Anlagen sowie die Kombination aus PV-Anlagen und Batteriespeichern betrachtet. Den Kern der Simulation bildet die ökonomische Bewertung dieser Konzepte. Berücksichtigt werden die möglichen Einnahmen durch die Vorhaltung von Regelleistung.

Die Analysen werden zunächst auf Basis der heutigen Marktbedingungen durchgeführt. Anschließend werden unter Beibehaltung der hohen Systemsicherheit sinnvolle Anpassungen vorgenommen, um bestehende Hindernisse abzubauen. Im Vordergrund stehen hier vor allem die Verkürzungen der Ausschreibungszeiträume und Produktzeitscheiben für PRL, SRL und MRL [4]. Bereits abgeschlossene Forschungsprojekte bestätigen die Potentialsteigerung für fluktuierende Erzeuger aufgrund dieser Maßnahmen [2][5]. Es werden zukünftige Potentiale abgeschätzt und Empfehlungen für Weiterentwicklungen der Marktbedingungen ausgesprochen.

4.1 Teilnahme unter aktuellen Marktbedingungen

Die o. g. PV-Anlagenkonzepte werden in **Tabelle 2** zunächst hinsichtlich äußerer Rahmenbedingungen zur Bereitstellung von Regelleistung qualitativ gegenübergestellt. Betrachtet werden Aufwendungen für Informations- und Kommunikationstechnik (IKT), die geographische Verteilung der Anlagen sowie die saisonalen Einflüsse in Bezug auf die Jahres- und Tageszeiten.

Tabelle 2 Qualitative Bewertung der PV-Anlagenkonzepte

	IKT-Aufwand	*Geographische Verteilung*	*Saisonale Bereitstellung*
Einzelne Anlagen (> 20 MW)	+	-	-
Pool aus PV-Anlagen	-	+	-
PV-Anlagen + Speicher	o	-	+

Die Bereitstellung von Regelleistung aus einzelnen PV-Anlagen einer hohen Leistungsklasse (> 20 MW) besitzt den Vorteil, dass der Installationsaufwand der IKT sehr gering ausfällt. Anlagen dieser Größenordnung sind in der Regel bereits mit umfangreicher Fernwirktechnik ausgestattet. Die geographische Verteilung ist aufgrund der Einzelanlage lokal begrenzt, was dazu führt, dass z. B. meteorologische Einflüsse wie Wolkenzug nicht durch andere PV-Anlagen ausgeglichen werden können. Die Bereitstellung von Regelleistung ist aufgrund der höheren solaren Einstrahlung hauptsächlich in den Sommermonaten, teils auch im Frühjahr und Herbst, möglich. Mit einem Pool aus mehreren PV-Anlagen lässt sich hingegen der Vorteil einer geographischen Verteilung nutzen. Dadurch lassen sich unterschiedliche Wetterlagen, insbesondere Wolkenzug oder Nebel, ausgleichen, was wiederum zu einer höheren Verfügbarkeit bzw. Zuverlässigkeit des Pools führt. Nachteilig ist hier die saisonale Bereitstellung in Bezug auf die Jahres- und Tageszeiten zu bewerten. Außerdem ist die Informations- und Kommunikationstechnik mit erhöhtem Aufwand verbunden. Die Kombination aus PV-Anlagen und Batteriespeichern ist ähnlich wie einzelne große PV-Anlagen zu bewerten. Der Aufwand für Informations- und Kommunikationstechnik ist hingegen geringer einzuschätzen als bei der PV-Pool-Lösung. Die geographische Verteilung lässt sich ebenso nur innerhalb eines Anlagenpools positiv bewerten. Der große Vorteil der Nutzung von Speichern besteht darin, dass die zeitliche Verfügbarkeit zur Regelleistungsbereitstellung gesteigert werden kann. Abhängig von der Dimensionierung und dem jeweiligen Speicherstand kann demnach auch zu Jahres- oder Tageszeiten mit geringer oder keiner Sonneneinstrahlung Regelleistung bereitgestellt werden. Für die PRL wurde in [7] bereits gezeigt, dass eine Kombination mit Batteriespeichern aus ökonomischer Sicht sinnvoller ist als die dauerhafte Abregelung von PV-Anlagen.

Aufgrund der aktuellen Marktbedingungen, insbesondere des wöchentlichen Ausschreibungszeitraumes und der Produktzeitscheiben, ist die Bereitstellung von Primär- und Sekundärregelleistung aus fluktuierenden Erzeugern zum jetzigen Zeitpunkt nicht möglich. Die Minutenreserveleistung bietet hingegen ein Potential zur Einbindung von PV-Anlagen in den heutigen Regelleistungsmarkt. Von den sechs 4-h-Zeitscheiben bleiben aufgrund der solaren Einstrahlung max. drei Zeitscheiben (8 bis 12 Uhr, 12 bis 16 Uhr und 16 bis 20 Uhr) übrig. Im Winter ist es höchstens möglich, die Zeitscheibe von 12 bis 16 Uhr zu bedienen. Im Sommer kann MRL häufig zusätzlich in den Zeitscheiben von 8 bis 12 sowie von 16 bis 20 Uhr angeboten werden. Unter Annahme einer perfekten Vorhersage der Einspeiseleistung können unter aktuellen Marktbedingungen die in **Abbildung 1** markierten Flächen als Regelenergie bereitgestellt werden. Es ist zu erkennen, dass große Teile der PV-Einspeiseleistung vor allem aufgrund der Zeitscheibenrestriktion für eine Netzregelung nicht zur Verfügung stehen.

Nach aktuellen regulatorischen Bedingungen erfolgen die Vorhaltung und der Abruf von Regelleistung über einen vorab gemeldeten Fahrplan. Dieser ist von der PV-Anlage einzuhalten. Dadurch entstehen bereits energetische Verluste durch die Abregelung von der maximal möglichen Einspeisung auf den gemeldeten Fahrplan. Dieser liegt aufgrund der Forderung der hohen Zuverlässigkeit deutlich unter der maximal möglichen Einspeisung. Weitere Abregelungen treten bei der tatsächlichen Bereitstellung negativer Regelleistung bzw. bei Vorhaltung positiver Regelleistung auf. Als Einnahmen können dem die Leistungspreise der Vorhaltung von Regelleistung gegenübergestellt werden. Bei einem tatsächlichen Abruf werden weitere Einnahmen durch die Arbeitspreise generiert. Die Unterscheidung in die einzelnen Anlagenkonzepte ist an dieser Stelle nur z. T. notwendig. Die Verwendung von

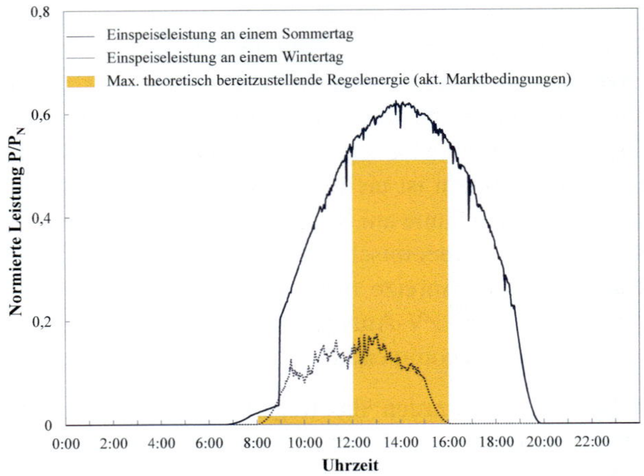

Abbildung 1 Theoretisch maximal mögliche Bereitstellung von MRL durch PV nach aktuellen Rahmenbedingungen

Speichern sowie das Pooling von Anlagen führen zu einer höheren Leistungsverfügbarkeit. Insbesondere zu der Kombination von PV und Speichern sind detailliertere Untersuchungen notwendig, in der die für die entsprechenden Anlagen sinnvollen Charakteristika berücksichtigt werden.

Zusammenfassend wird die Aussage getroffen, dass die Regelleistungsbereitstellung aus PV-Anlagen zu aktuellen Marktbedingungen entweder nicht möglich oder aus ökonomischen Gesichtspunkten nicht sinnvoll ist.

4.2 Teilnahme unter weiterentwickelten Marktbedingungen

Eine zusätzliche Dienstleistung durch eine Erzeugungseinheit wird nur dann erbracht, wenn eine wirtschaftliche Betriebsweise gegeben ist. Daher müssen aufgrund der eingangs erwähnten zurückgehenden konventionellen Erzeugung Alternativen geschaffen werden, um die Systemsicherheit auch in Zukunft zu gewährleisten. Eine Möglichkeit besteht darin, eine Dienstleistung wie die Netzregelung über eine Anschlussbedingung sicherzustellen. Allerdings stellt dies nicht die (volks-)wirtschaftlich sinnvollste Lösung dar. Eine Alternative besteht in der Teilnahme an der Netzregelung über den Regelleistungsmarkt. Eine mögliche Maßnahme ist dabei die Anpassung der Marktbedingungen, die in **Tabelle 3** dargestellt werden.

Tabelle 3 Angepasste Regelleistungsmarktbedingungen in Deutschland

	PRL	*SRL*	*MRL*
Ausschreibungs-zeitraum	täglich	täglich	täglich
Produktzeitscheiben	stündlich	stündlich	stündlich
Produktdifferenzie-rung	Pos./Neg. PRL	Pos./Neg. SRL	Pos./Neg. MRL
Mindestangebotsgröße	±1 MW	±1 MW	±1 MW

Erster Ansatzpunkt ist die Reduzierung der Vorlaufzeit bei der Regelleistungsauktion auf einen täglichen Rhythmus. Der entscheidende Vorteil besteht hierbei in einer deutlich verbesserten Prognosegüte von fluktuierenden erneuerbaren Energien. Weiterhin können von dieser Änderung auch konventionelle Kraftwerke profitieren, für die sich durchgängige, wöchentliche Betriebszeiten nicht rechnen. Außerdem stellen die aktuellen Produktzeitscheiben eine Marktbarriere für PV-Anlagen dar. Stündliche Produkte hingegen bieten ein erhöhtes Potential für die Bereitstellung von Regelleistung aus PV-Anlagen, da z. B. bei wechselnden meteorologischen Bedingungen auch für einen Teil der Stunden Angebote abgegeben werden können. Eine weitere Anpassung betrifft die Produktdifferenzierung der Primärregelleistung, um zukünftig separat positive oder negative PRL anbieten zu können. Die Mindestangebotsgröße wurde für alle Regelleistungsarten auf ±1 MW gesenkt. Dadurch entstehen günstigere Bedingungen für kleinere PV-Anlagen oder -Pools sowie bei schlechteren Wetterlagen.

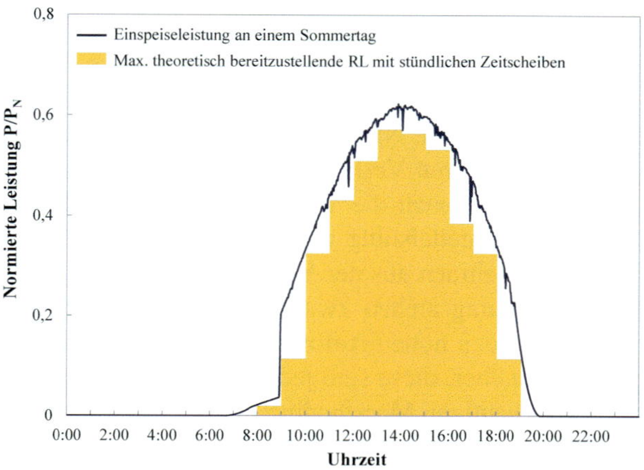

Abbildung 3 Theoretisch maximal mögliche Bereitstellung von Regelleistung durch PV im Sommer mit stündlichen Zeitscheiben

In **Abbildung 3** ist die Bereitstellung von Regelleistung bei angepassten Marktbedingungen zu sehen. Durch die Überführung auf stündliche Produktzeitscheiben kann der Anteil an Regelenergie bei perfekter Prognose im Vergleich zu den aktuellen Marktbedingungen deutlich gesteigert werden. Der tägliche Ausschreibungszeitraum erhöht zusätzlich die Prognosegüte für Sekundär- und Primärregelleistung. Auch an Wintertagen ist es nun eher möglich, Regelleistung zur Verfügung zu stellen.

Eine weitere wichtige Änderung betrifft das Nachweisverfahren zur Bereitstellung von Regelleistung. Aktuell muss die Regelleistung ausgehend von einem vorab gemeldeten Fahrplan erbracht werden. Die Leistungseinspeisung nach einem festen Fahrplan würde jedoch eine fortwährende Abregelung der PV-Anlagen und somit große Verluste mit sich bringen [5]. Alternativ soll in diesem Abschnitt die Regelleistung ausgehend von der maximal möglichen Einspeisung erbracht werden.

5 Ergebnisse

Die folgenden Simulationsergebnisse wurden bei Betrachtung eines Zeitraums von einem Jahr, Dezember 2013 bis November 2014, erzielt. Dadurch sind alle jahreszeitbedingten Wettereinflüsse berücksichtigt. Betrachtet wird ausschließlich die Bereitstellung von negativer Regelleistung. Diese stellt im Vergleich zur positiven Regelleistung die ökonomisch sinnvollere Variante dar, da ansonsten kontinuierliche Abregelungen notwendig wären. Dadurch können große Teile der zur Verfügung stehenden Energie nicht genutzt werden.

Durch das Einhalten der geforderten Zuverlässigkeit bei der Bereitstellung von Regelleistung kann in der Realität deutlich weniger Leistung zur Regelleistungsbereitstellung genutzt werden als hier angenommen. Diese Differenzen treten jedoch in beiden Betrachtungsweisen gleichermaßen auf. Außerdem entstehen unter aktuellen Marktbedingungen Verluste durch das Einhalten des vorab definierten Fahrplans. Da die weiterentwickelten Bedingungen die maximal mögliche Einspeisung als Richtwert nehmen, treten an dieser Stelle bei der Bereitstellung von negativer Regelleistung keine weiteren Verluste auf.

Die Anpassungen der Rahmenbedingungen des Regelleistungsmarktes führen dazu, dass Photovoltaik-Anlagen in einer weitaus größeren Anzahl an Stunden die Möglichkeit haben, Regelleistung anzubieten. Allein die stündlichen Produktzeitscheiben bei den drei Regelleistungsarten haben zur Folge, dass in mehr als doppelt so vielen Stunden des Jahres Regelleistung aus PV-Anlagen bereitgestellt werden kann. In dem betrachteten Zeitraum waren dies 1.656 Stunden unter aktuellen Marktbedingungen gegenüber 3.587 Stunden unter weiterentwickelten Marktbedingungen. Diese Werte beziehen sich auf PV-Anlagen an einem Standort mit einer installierten Leistung von 100 MW_p. Diese deutliche Steigerung bestätigt die Relevanz einer Anpassung der Märkte.

Abbildung 2 stellt die Anzahl der Stunden, in denen Regelleistung bereitgestellt werden kann, in dem betrachteten Zeitraum für aktuelle und angepasste Marktbedingungen dar. Es ist zu erkennen, dass die Änderungen das Potential für PV-Anlagen enorm steigern.

Abhängig von der Jahreszeit und den Wetterverhältnissen können die stündlichen Zeitscheiben zwischen 6 und

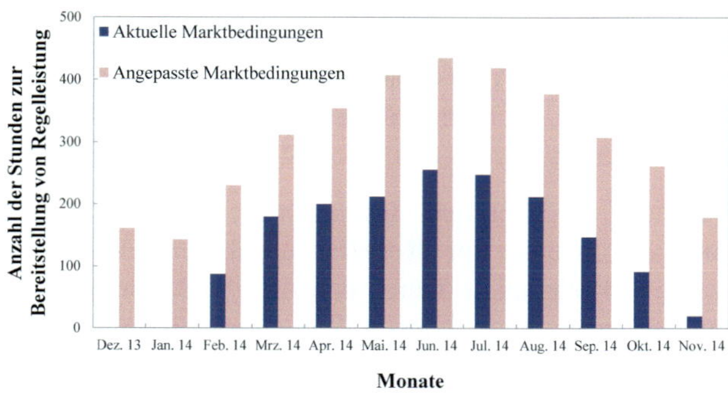

Abbildung 2 Verteilung der maximal möglichen Regelleistungsbereitstellung durch PV unter verschiedenen Marktbedingungen

21 Uhr abgedeckt werden. Im Winter beschränkt sich dieser Zeitraum auf die Mittagszeit (ca. 9-15 Uhr) – sofern eine Vorhaltung von Regelleistung aufgrund der insgesamt geringen solaren Einstrahlung möglich ist. In den Übergangsjahreszeiten vergrößert sich der Zeitraum, bis in den Monaten Mai bis August die meisten Zeitscheiben abgedeckt werden können. Insbesondere in den Sommermonaten kann eine nahezu 100-prozentige Verfügbarkeit zur Regelleistungsbereitstellung in den Stunden zwischen 9 und 19 Uhr gewährleistet werden. Im Vergleich zu den aktuellen Marktbedingungen kann, insbesondere in den Sommermonaten, bereits vor 8 Uhr sowie nach 16 Uhr Regelleistung vorgehalten werden.

Bei der Bereitstellung von Regelleistung werden unterschiedliche Erlöse abhängig von der jeweiligen Regelleistungsart erzielt. Aufgrund individueller Handelsstrategien werden keine Arbeitspreise, sondern lediglich Leistungspreise berücksichtigt. Diese werden für den o. a. Zeitraum der Onlineplattform www.regelleistung.net der vier Übertragungsnetzbetreiber entnommen [8]. Die Märkte für MRL und SRL bieten nach aktuellem Marktdesign die geringsten Profitmöglichkeiten, wohingegen durch Vorhaltung von Primärregelleistung die meisten Erlöse erzielt werden können. Durch die vorgeschlagene Weiterentwicklung der Marktbedingungen sind PV-Anlagen ebenfalls dazu geeignet, PRL bereitzustellen und können somit diese Einnahmen als zusätzlichen Erlös verbuchen.

Unter aktuellen Marktbedingungen kann lediglich MRL bereitgestellt werden. In dem betrachteten Zeitraum hätten dadurch Erlöse von etwa 120.000 Euro durch Leistungspreise erzielt werden können. Die Marktanpassungen führen unter denselben Randbedingungen zu Erlösen für die Vorhaltung von MRL i. H. v. 250.000 Euro. Dies entspricht etwa einer Verdoppelung der Einnahmen. Bei der Bereitstellung von SRL können ca. 200.000 Euro erzielt werden. Bei PRL hingegen wären sogar Erlöse i. H. v. etwa 900.000 Euro möglich. Mit Anpassung der Regelleistungsmarktbedingungen, welche den Rahmen für die Teilnahme von PV am Markt für PRL schaffen, kann demnach etwa das 7,5-fache an Erlösen erzielt werden.

Für einen Pool mit über Deutschland verteilten PV-Anlagen mit derselben Gesamtleistung von 100 MW_p hätten in demselben Zeitraum nach aktuellen Marktbedingungen ca. 140.000 Euro an Erlösen erzielt werden können. Nach den Anpassungen wären es für MRL bereits ca. 260.000 Euro gewesen. Mit der Vorhaltung von SRL wäre eine Steigerung auf etwa 210.000 Euro möglich gewesen. Die größten Erlöse hätten mit PRL i. H. v. 920.000 Euro erzielt werden können. Dies entspricht dem 6,5-fachen des Wertes unter aktuellen Marktbedingungen.

Demnach wird deutlich, dass auch die Auswahl der PV-Anlagenkonzepte zur Regelleistungsbereitstellung eine wichtige Rolle spielt. Durch den Ausgleich von Einstrahlungsschwankungen können Wetterereignisse wie Wolkenzug mithilfe eines Pools aus verteilten Anlagen eher ausgeglichen werden. Aufgrund des höheren Aufwands und der geringeren Zusatzerlöse sollte Pooling jedoch nur dann angewendet werden, wenn die Mindestangebotsgröße anderweitig nicht erreicht werden kann. Mit einem Speicher mit der für die jeweilige(n) PV-Anlage(n) passenden Kapazität ist es möglich, die Regelleistungsbereitstellung weiter zu steigern. Dies müsste im Einzelfall mit dem jeweiligen Aufwand abgewogen werden. Für eine genaue Bezifferung sind jedoch weiterführende Untersuchungen notwendig.

Eine weitere wesentliche Verbesserung konnte bei der Zuverlässigkeit der Vorhaltung erzielt werden. Während unter aktuellen Marktbedingungen bei dem lokalen Pool etwa 16 % der auf Basis der Day-Ahead-Prognose bereitgestellten Regelleistung unter den tatsächlich aufgetretenen Wetterbedingungen nicht vorgehalten werden konnten, waren dies unter weiterentwickelten Marktbedingungen nur noch lediglich 5 %. Hieran ist die Bedeutung der Verkleinerung der Produktzeitscheiben für die Regelleistungsbereitstellung zu erkennen.

Eine mögliche Umsetzung der angepassten Marktbedingungen stellt einen weiteren Schritt zur Übernahme von Systemverantwortung durch fluktuierende Erzeugungsanlagen dar und kann für das Energieversorgungssystem einen Mehrwert bilden. Nicht nur die Systemstabilität wird aufgrund der höheren Verfügbarkeit von PV-Anlagen unterstützt, sondern auch die Wirtschaftlichkeit der Bereitstellung von Regelleistung aus PV-Anlagen steigt deutlich. Den Einnahmen aus der Vorhaltung und Erbringung von Regelleistung stehen zwar noch immer Abregelungen, z. B. aus den hohen Anforderungen an die Zuverlässigkeit, gegenüber, diese sind jedoch wesentlich niedriger als noch bei heutigen Marktbedingungen.

6 Zusammenfassung und Ausblick

Zusammenfassend ist zu erkennen, dass die angeführten Änderungen der Regelleistungsmarktbedingungen zum einen die Teilnahme von PV-Anlagen am Regelleistungsmarkt ermöglichen und zum anderen auch die Wirtschaftlichkeit der Bereitstellung von Regelleistung erhöhen können. PV-Pools sind aufgrund der geographischen Verteilung in der Lage, besser auf wetterbedingte Schwankungen zu reagieren, während die Kombination aus PV-Anlagen und Speichern zu einer weitaus größeren Zeitspanne der Regelleistungsbereitstellung führt. Bei diesem Anlagenkonzept kann die abgeregelte Leistung gespeichert werden und dadurch in weiteren Zeitscheiben Regelleistung zur Verfügung gestellt werden.

Insbesondere die Weiterentwicklung der Marktbedingungen steigert das Potential der Regelleistungsbereitstellung aus PV-Anlagen. Regelleistung kann in jeder Art (PRL, SRL oder MRL) zur Verfügung gestellt werden, wobei die größten Erlöse im Bereich der Leistungspreise zum jetzigen Zeitpunkt durch die Vorhaltung von PRL erzielt werden können. Hier ist jedoch zu berücksichtigen, dass die Anforderungen der Übertragungsnetzbetreiber an die IKT technischer Einheiten, die PRL bereitstellen, deutlich höher sind. Weiterhin kann mit stündlichen Zeitscheiben eine deutlich größere Zeitspanne über den Tag verteilt abgerufen werden als unter aktuellen Rahmenbedingungen. Verluste treten bei beiden Betrachtungsweisen auf. Im Vergleich der Marktbedingungen bei der Bereitstellung von negativer Regelleistung sind ausschließlich diejenigen relevant, die bei aktuellen Rahmenbedingungen über die Einhaltung des Fahrplans auftreten.

Eine weitere Steigerung des Potentials zur Regelleistungsbereitstellung aus PV-Anlagen ist durch die Einführung eines Arbeitspreismarktes möglich. Für die Angebote können so die Intraday- anstelle der Day-Ahead-Prognosen verwendet werden. Dadurch kann besser auf wechselnde Wettereinflüsse reagiert und die Zuverlässigkeit der Bereitstellung erhöht werden.

Berücksichtigt werden muss bei dieser Simulation jedoch, dass bei Anpassung der Marktbedingungen auch der Markt selbst sich verändern wird. Ein Wandel der Erlösstruktur, sodass sich potentielle Einnahmen verschieben, ist naheliegend. Die „pay-as-bid"-Struktur, nach der im vorhandenen Markt gehandelt wird, kann daher nicht als fortwährend betrachtet werden. Eine Änderung kann etwa hin zu einer Bepreisung über Grenzkosten („pay-as-cleared") gehen, wie es in der überarbeiteten Version des *Network Code on Electricity Balancing* der ENTSO-E gefordert wird [9]. Dies würde weitere Untersuchungen zur Wirtschaftlichkeit der Bereitstellung von Regelleistung aus PV-Anlagen erfordern.

Weiterhin führt auch eine mögliche Umstellung des aktuellen statischen Verfahrens zur Ermittlung des Regelleistungsbedarfs auf ein adaptives Verfahren zu einer weiteren Effizienzsteigerung des Regelleistungsmarktes. Durch die Berücksichtigung von vortäglichen Prognosen des aktiven Kraftwerksparks können der Regelleistungsbedarf und somit die Kosten reduziert werden. [10]

Zu prüfen sind außerdem die Auswirkungen der genannten Anpassungen der Marktbedingungen auf die bisherigen Anbieter, insbesondere konventionelle Kraftwerke. Darunter fällt z. B. eine geringere Wirtschaftlichkeit, bedingt durch hohe An- und Abfahrkosten, bei einer Regelleistungsbereitstellung für lediglich eine Stunde. Denkbar wäre an dieser Stelle möglicherweise eine Flexibilität der Angebotslängen wie Blockangebote über mehrere Stunden. [4] Auf der anderen Seite kann die Bereitstellung von PRL und SRL durch konventionelle Kraftwerke flexibler gestaltet werden. So sind diese nicht mehr gezwungen, dauerhaft über eine Woche aktiv zu sein, sondern möglicherweise nur noch über Stunden oder Tage.

Welche Marktanpassungen in naher Zukunft umgesetzt werden, um fluktuierende Erzeuger in die Bereitstellung von Regelleistung einzubeziehen, ist zum jetzigen Zeitpunkt noch offen. Im Fokus der Empfehlungen liegt vor allem die Anpassung der Produktzeitscheiben sowie der Ausschreibungszeiträume für alle drei Regelleistungsarten.

7 Literatur

[1] Deutsche Bundesregierung, *„Energiekonzept für eine umweltschonende, zuverlässige und bezahlbare Energieversorgung"*, 2010

[2] Fraunhofer IWES, Siemens AG, Institut für Energieversorgung und Hochspannungstechnik, Fachgebiet Elektrische Energieversorgung der Universität Hannover, CUBE Engineering GmbH, „Kombikraftwerk 2 – Abschlussbericht", 2014

[3] Bundesnetzagentur für Elektrizität, Gas, Telekommunikation, Post und Eisenbahnen, Beschlusskammer 6, „BK6-10-097", „BK6-10-098", „BK6-10-099", Bonn, 2011

[4] Connect Energy Economics GmbH, „Leitstudie Strommarkt 2015", Berlin, 2015

[5] Fraunhofer IWES, „Regelenergie durch Windkraftanlagen - Abschlussbericht", Kassel, 2014

[6] Consentec GmbH, „Beschreibung von Regelleistungskonzepten und Regelleistungsmarkt", Aachen, 2014

[7] D. Unger, S. Laudahn, B. Engel, „Teilnahme von Photovoltaikanlagen an der Primärregelleistung", 13. Symposium Energieinnovation, Graz, 2014

[8] 50 Hertz Transmission GmbH, Amprion GmbH, TenneT TSO GmbH, TransnetBW GmbH, „regelleistung.net – Internetplattform zur Vergabe von Regelleistung"

[9] ENTSO-E – European Network of Transmission System Operators for Electricity, „ENTSO-E Network Code on Electricity Balancing, Version 3.0", Brüssel, 2014

[10] Deutsche Energie-Agentur GmbH (dena), Hannes Seidl, „Roadmap dena-Studie Systemdienstleistungen 2030", Berlin, 2014

Mobilitätanwendungen

Geschäftsmodelle für die Versorgung von Haushaltskunden mit Eigenerzeugung, Elektrofahrzeug und stationärem Speicher

M. Bolczek[1], S. Kippelt[1], C. Rehtanz[1]
[1]Institut für Energiesysteme, Energieeffizienz und Energiewirtschaft (ie[3])
Emil-Figge-Str. 70, 44227 Dortmund, Deutschland, malte.bolczek@tu-dortmund.de

Kurzfassung

Im Zuge der Energiewende ist bereits heute eine Vielzahl von Haushalten mit einer Eigenerzeugungsanlage, maßgeblich Photovoltaikanlagen, ausgestattet. Die konträre Entwicklung der Einspeisevergütung für Strom aus PV-Anlagen und dem Endkundenstrompreis für den Strombezug aus dem Netz der allgemeinen Versorgung seit Beginn der EEG-Förderung hat zuerst zur Netzparität und seit einigen Jahren sogar zu einer deutlichen Preisspreizung geführt, bei der der Vergütungssatz aktuell bei etwa der Hälfte des Strombezugspreises liegt. Die genannten Rahmenbedingungen haben Anreize geschaffen, neue Geschäftsmodelle für die Versorgung von Haushaltskunden, Steuerung von Verbrauchseinrichtungen und den Betrieb von Speichern im Haushaltsbereich zu etablieren. Treiber dieser Geschäftsmodelle ist im Wesentlichen die Optimierung der Erlöse aus der PV-Einspeisung und den Kosten für den Strombezug durch Maximierung des Eigenverbrauchs im Haushalt. Die Eigenverbrauchsmaximierung kann einerseits durch die zeitliche Verschiebung von energieintensiven Verbrauchseinrichtungen wie einem Elektrofahrzeug erfolgen, dessen Ladevorgang in Zeiten überschüssiger PV-Einspeisung verschoben wird. Andererseits kann ein stationärer Batteriespeicher die gewöhnlich auftretenden überschüssigen PV-Einspeisemengen aufnehmen, um in Zeitabschnitten mit einem Haushaltsbedarf, der die Einspeiseleistung der PV-Anlage übersteigt, den Strombezug aus dem Netz zu vermeiden.
In diesem Beitrag werden zwei Geschäftsmodelle in dem beschriebenen Umfeld hergeleitet und einer Wirtschaftlichkeitsanalyse unterzogen. Die dargestellten Untersuchungen finden im Forschungsprojekt „Smart-E“ statt, das durch das BMU gefördert wird.

1 Einleitung

Die Rahmenbedingungen für die Errichtung und den Betrieb von PV-Anlagen in Deutschland haben sich in den letzten Jahren drastisch geändert. Vor allem stark sinkende PV-Modulpreise haben den Gesetzgeber veranlasst, die Vergütungssätze im Rahmen des EEG für in das Energieversorgungsnetz eingespeisten PV-Strom entsprechend der gesunkenen Gestehungskosten anzupassen. Im Gegenzug haben sich die Endkundenpreise für den Strombezug aus dem Energieversorgungsnetz im gleichen Zeitraum durch gestiegene Umlagen deutlich erhöht. [1]

Der Vergleich der Entwicklungen zeigt, dass nach Jahren mit PV-Vergütungssätzen, die ein Vielfaches des Endkundenstrompreises darstellten, ab dem Jahr 2012 eine Umkehr dieses Verhältnisses eingetreten ist, die sich bis heute fortgesetzt hat. Dies hat einen Anreiz für PV-Anlagenbetreiber zur Folge, den selbsterzeugten Strom in der Kundenanlage selbst zu verbrauchen, anstatt diesen in das Energieversorgungsnetz einzuspeisen. Jede selbstverbrauchte Kilowattstunde, die sowohl die Einspeisemenge als auch den Bezug aus dem Netz verringert, erwirtschaftet somit die Differenz aus dem Endkundenstrompreis und dem PV-Vergütungssatz. Im Jahr 2012 bestand außerdem eine Übergangsregelung, die den Eigenverbrauch von PV-Strom sogar mittels einer speziellen Vergütung gefördert hat.

1.1 Messkonzepte für Haushalte mit PV-Anlage

Im Zuge der Entwicklungen der PV-Vergütungssätze haben sich entsprechende Messkonzepte ausgebildet. Dabei werden unterschiedliche Zählerarten (Ein-, Zweirichtungszähler) und eine verschiedene Anzahl von Zählern eingesetzt. Grundsätzlich basieren alle bestehenden Messkonzepte auf dem Standardlastprofilverfahren, bei dem in der Regel nur jährliche Verbrauchs- oder Einspeisemengen erfasst werden und die Einspeise- und Lastprofile auf Grundlage synthetischer Verfahren gebildet werden. Seit Novellierung der Stromnetzzugangsverordnung im Jahre 2013 steht für zukünftige Mess- und Bilanzierungskonzepte für Haushaltskunden außerdem die Zählerstandsgangmessung (ZSG) zu Verfügung. Die ZSG ermöglicht Lieferanten, durch tarifliche oder sonstige Anreize hervorgerufene Änderungen im Verbrauchsverhalten auch auf die Beschaffungsstruktur abzubilden, um die Beschaffungskosten zu senken.

Für Anlagen mit Vergütungssätzen oberhalb des Strombezugspreises wird das Konzept der *Volleinspeisung* des PV-Stroms verfolgt, bei dem die Einspeisung der gesamten erzeugten Strommenge über einen von der bestehenden Kundenanlage getrennten Hauptleitungsabzweig direkt in das Energieversorgungsnetz erfolgt. Dabei wird am Hauptleitungsabzweig ein separater Zweirichtungszähler zur Erfassung der eingespeisten und bezogenen Strommengen der PV-Anlage eingesetzt.

Für Anlagen, die nach der Übergangsregelung im Jahr 2012 errichtet wurden, wird auf das Messkonzept der

Überschusseinspeisung zurückgegriffen. Ein Zweirichtungszähler zwischen dem Energieversorgungsnetz und der Kundenanlage, an dem auch die PV-Anlage angeschlossen ist, erfasst dabei jeweils den Gesamtbezug des Haushalts und der PV-Anlage als auch den Überschuss der Einspeisung aus der Differenz zwischen der gesamten Einspeisung und dem Haushaltsbezug. Dieses Konzept bietet dem Anlagenbetreiber den erwähnten wirtschaftlichen Vorteil, der umso bedeutender ist, je größer die Preisspreizung zwischen Vergütungssatz und Strombezugspreis ausfällt. Der Vorteil hängt jedoch insbesondere davon ab, inwieweit eine zeitrichtige Überdeckung der PV-Einspeisung und des Haushaltsbezugs vorliegt.

1.2 Variable Verbraucher und stationäre Speicher

Die individuelle zeitrichtige Überdeckung (Eigenverbrauchsanteil) beim Haushaltskunden ist im Wesentlichen von dem spezifischen Verbrauchsverhalten des Kunden und der Einspeisecharakteristik der PV-Anlage abhängig, die auf der geografischen Ausrichtung der Anlage beruht. Der Eigenverbrauchsanteil kann einerseits durch Steuerung variabler Verbrauchseinrichtungen erhöht werden, indem diese zu Zeiten eingesetzt werden, in denen im ungesteuerten Fall eine Überschusseinspeisung vorliegt. Geeignete Verbrauchseinrichtungen für eine solche Steuerung sollten möglichst energieintensiv sein und eine hohe Flexibilität aufweisen ohne den Nutzerkomfort (zu stark) einzuschränken, um einen größtmöglichen Mehrwert zu generieren. In diesem Zusammenhang bieten sich vor allem Elektrofahrzeuge als steuerbare Verbrauchseinrichtung an, die bei gewöhnlicher Nutzung sowohl einen Energieverbrauch in der Größenordnung des gesamten Haushaltsbezugs als auch eine hohe Variabilität in Bezug auf das Verhältnis von benötigter Ladezeit und erwarteter Standzeit haben.

Des Weiteren kann der Eigenverbrauchsanteil durch den Einsatz eines stationären Batteriespeichers in der Kundenanlage erhöht werden. Der Batteriespeicher nimmt dann die Überschussmengen der PV-Anlage auf, die den momentanen Haushaltsbezug übersteigen. In Zeiten, in denen der Haushaltsbezug nicht durch die PV-Einspeisung gedeckt werden kann, gibt der Speicher die Energiemengenwieder innerhalb der Kundenanlage ab. Je nach Auslegung der installierten Leistung der PV-Anlage und der Kapazität des Batteriespeichers kann, abhängig vom individuellen Verbrauchsverhalten, sogar vollständige Autarkie des Haushalts erreicht werden. Sogenannte Solarstromspeicher werden in Verbindung mit der Errichtung einer PV-Anlage oder als kurzfristige Nachrüstung mittels des KfW-Förderprogramms *Erneuerbare Energien „Speicher"* mit 600 bzw. 660 Euro pro kWp der PV-Anlage bezuschusst. [2]

2 Geschäftsmodelle

In dem beschriebenen Umfeld bestehen für Energieversorgungsunternehmen und Energiedienstleister zahlreiche Möglichkeiten, innovative Geschäftsmodelle zu konzipieren. Die grundlegende Geschäftsidee besteht darin, einem Haushaltskunden ein Gesamtpaket von Produkten und Dienstleistungen anzubieten, das einerseits seine Stromkosten senkt und andererseits einen ideellen Mehrwert durch die persönliche Teilnahme an der Energiewende generiert. Dabei können auch Haushaltskunden mit bereits installierter Eigenerzeugungsanlage anvisiert werden. Kern der Geschäftsmodelle ist die Nutzung von Synergien zwischen dem Speicherbetrieb, der Laststeuerung, der Vermarktung der Überschusseinspeisungen und der Restbelieferung des Haushalts. Unterschiedliche Geschäftsmodellvarianten unterscheiden sich im Wesentlichen in der Ausprägung der in der Kundenanlage eingesetzten Betriebsmittel, dem Vergütungsmodell für die Überschusseinspeisung, der Art der Reststrombelieferung und dem übergeordneten Ansatz zur Einsparung von Kosten.

2.1 Eigenverbrauchsmaximierung durch Speichereinsatz

Als Referenz für die weitere Geschäftsmodellvariante wird die Kostenminimierung bzw. Erlösmaximierung für den Haushaltskunden betrachtet, bei der lediglich durch die Installation einer PV-Anlage und den Einsatz eines stationären Batteriespeichers eine Eigenverbrauchsmaximierung vorgenommen wird. Bei vorhandener PV-Anlage wird nur der Batteriespeicher nachgerüstet. Ein Elektromobil mit entsprechender Ladevorrichtung ist entweder auch schon vorhanden oder wird ebenfalls neu installiert. Dieses wird jedoch wie andere Verbraucher des Haushalts in dieser Variante nicht gesteuert.

Bei der Eigenverbrauchsmaximierung durch Speichereinsatz wird auf das Messkonzept der *Überschusseinspeisung* zurückgegriffen, das einen Zweirichtungszähler erfordert. Die reine Eigenverbrauchsmaximierung kann auf Basis des Standardlastprofilverfahrens (SLP) erfolgen, da keine Einspareffekte im Portfolio des Lieferanten erzielt werden sollen. Die Einspeisungsvermarktung der Überschussmengen wird damit folglich im Rahmen des EEG-Vergütungsmechanismus durchgeführt (siehe **Abb. 1**).

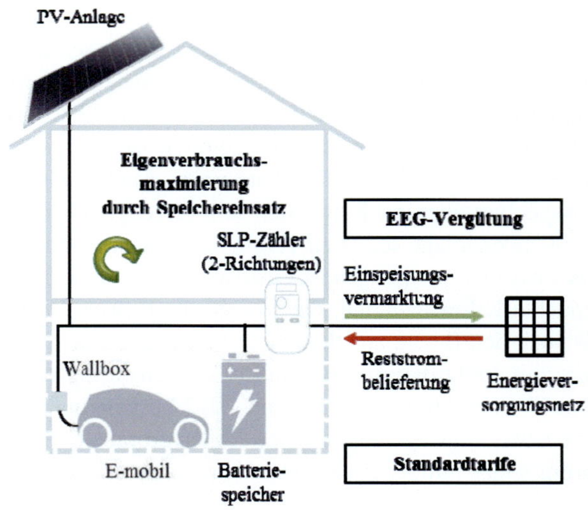

Abb. 1: Ausstattung und Merkmale des Geschäftsmodells zur Eigenverbrauchsmaximierung durch Speichereinsatz

Zur Reststrombelieferung des Kunden werden entsprechend dem SLP-Verfahren Standardtarife genutzt, die einen konstanten Preis über die gesamte Vertragslaufzeit beinhalten. Die Eigenverbrauchsmaximierung wird durch einen Batteriespeicher mit Wechselrichter und Batteriemanagementsystem vorgenommen, das bei auftretenden Überschusseinspeisungen den Batteriespeicher unverzüglich lädt. Im umgekehrten Fall wird bei auftretenden Netzbezügen der Speicher unverzüglich entladen.

2.2 Marktpreisorientierte Inhouse-Optimierung

Als Erweiterung zur ausschließlichen Eigenverbrauchsmaximierung könnte ein Mehrwert durch eine marktpreisorientierte Inhouse-Optimierung geschaffen werden. Ziel dieser Erweiterung ist die Einbindung der Beschaffungskonditionen des Lieferanten zur Optimierung des Einsatzes des Speichers und des Elektrofahrzeugs hinsichtlich der Maximierung des Eigenverbrauchs und der Strombeschaffung.

Zur Umsetzung des Optimierungsalgorithmus wird eine Optimierungshard- und -software benötigt, die eine informationstechnische Anbindung an ein Backend zur Übertragung der Preisinformationen besitzt. Die Erschließung von Beschaffungsportfolioeffekten erfordert eine ZSG zur viertelstündigen Erfassung des Last- und Einspeisegangs über einen Zweirichtungszähler am Netzanschlusspunkt. Dazu wird ein EDL-40-System bestehend aus elektronischem Zähler und IKT-Gateway/ Controller eingesetzt, das eine Tarifierung im Rahmen zeitvariabler Tarife außerhalb des Zählers zulässt (siehe **Abb. 2**). [3]

Auf Grundlage der zeitvariablen Tarife und der hinterlegten Prognose für die PV-Einspeisung und den nicht steuerbaren Verbrauch des Haushalts werden der Batteriespeicher und das Elektromobil kostenminimal eingesetzt. Da das Beschaffungsportfoliomanagement auf Stromprodukten in viertelstündiger Auflösung basiert, werden die zeitvariablen Tarife zur Reststrombelieferung in der gleichen Zeitgranularität gestaltet.

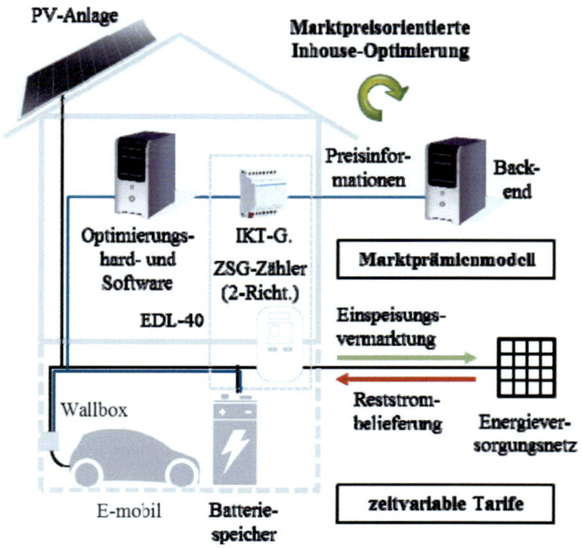

Abb. 2: Ausstattung und Merkmale des Geschäftsmodells zur marktpreisorientierten Inhouse-Optimierung

Die viertelstündige Messung der Einspeisung lässt außerdem eine Einspeisungsvermarktung im Marktprämienmodell zu, mit der eine höhere Vergütung der eingespeisten Strommengen erzielt werden kann.

3 Simulation der Geschäftsmodelle

Die Analyse der Geschäftsmodelle erfolgt anhand zweier unterschiedlicher Simulationen. Entsprechend der in den Geschäftsmodellen verwendeten Komponenten erfolgt die Analyse des Standardtarifs anhand einer Simulation des Speicherverhaltens bei einer Eigenverbrauchsmaximierung. Für das Geschäftsmodell Inhouse-Optimierung erfolgt die Simulation anhand einer optimierenden Einsatzplanung der verfügbaren Komponenten.

3.1 Eigenverbrauchsmaximierung

Die Simulation der Eigenverbrauchsmaximierung durch Speichereinsatz orientiert sich an den heutigen verfügbaren Systemen zur Eigenverbrauchsmaximierung. Dabei wird der Batteriespeicher zumeist vom Batteriemanagementsystem automatisch ge- und entladen. Dazu bedient sich das System an aktuellen Messwerten am Netzanschlusspunkt. Bei gemessener Überschusseinspeisung wird der Batteriespeicher mit der entsprechenden Leistung geladen, so dass weder eine Einspeisung noch ein Bezug aus dem Netz vorliegt. Dies geschieht so lange, bis der Batteriespeicher seine volle Kapazität erreicht hat. Misst das System einen Netzbezug, wird analog zur Überschusseinspeisung der Batteriespeicher entleert.

Die Modellierung der Einspeisung der PV-Anlage, des Haushaltsbedarfs und des Elektrofahrzeuges erfolgt über Standardlastprofile. Die Entwicklung von Standardlastprofilen für Elektrofahrzeuge basiert auf einem Modell aus [4]. Die Überlagerung der Standardlastprofile von Haushalt und Elektrofahrzeug mit dem Standardeinspeiseprofil von PV-Anlagen entspricht zwar nicht der spezifischen Überdeckung der tatsächlich auftretenden Last und Einspeisung, liefert jedoch eine hinreichend genaue Abschätzung für die gemittelte Überdeckung einer Vielzahl von Haushaltskunden mit PV-Anlage.

3.2 Inhouse-Optimierung

Die marktpreisorientierte Inhouse-Optimierung erfolgt anhand eines gemischt-ganzzahligen linearen Programmes (GGLP / MILP).

In den folgenden Abschnitten werden die für die Optimierung verwendeten Parameter, Entscheidungsvariablen und Nebenbedingungen vorgestellt. Hierbei bezeichnet T die Anzahl der zu optimierenden Perioden, t die aktuelle Periode und τ die Anzahl von Perioden innerhalb einer Stunde. Für die gewählte zeitliche Auflösung von 15-Minuten-Intervallen gilt $\tau = 4\ h^{-1}$.

3.2.1 Elektrofahrzeug

Für Elektrofahrzeuge besteht das Entscheidungsproblem aus der Suche nach der optimalen Ladeleistung in jedem Zeitpunkt. Dies erfolgt unter Berücksichtigung der individuellen Fahrzeugnutzung, des Fahrzeugstandortes sowie der maximal möglichen Ladeleistung und Speicherkapazität. Die Beladung des Fahrzeugs erfolgt mit variabler Leistung. Eine Rückspeisung aus dem Fahrzeugspeicher in das Stromnetz ist nicht vorgesehen. Feste Parameter für Elektrofahrzeuge sind:

- max. Ladeleistung P_{EV}^{max} in kW
- nutzbare Batteriekapazität E_{EV}^{max} in kWh
- Ladewirkungsgrad η_{EV}^{L}
- Speicherentladerate η_{EV}^{T}

Darüber hinaus existieren weitere, zeitveränderliche Anlagenparameter:

- Fahrzeugstandort $s_{EV}(t)$ in kW
- Ladebedarf $l_{EV}(t)$ in kWh
- Beginn einer Fahrt $b_{EV}(t)$
- Füllstand in t_0 E_{EV}^{0} in kWh

Der Fahrzeugstandort zeigt an, ob sich das Fahrzeug am zu optimierenden Objekt befindet ($s_{EV}(t) = 1$) oder nicht zur Verfügung steht ($s_{EV}(t) = 0$). Der Ladebedarf gibt an, welche Energieentnahme durch eine zurückliegende Fahr erfolgt. Der Beginn einer Fahrt wir durch den Parameter b_{EV} angezeigt. Das Entscheidungsproblem besteht anschließend aus den folgenden Entscheidungsvariablen:

- Ladeleistung $P_{EV}(t)$ in kW
- Speicherstand $E_{EV}(t)$ in kWh

Damit die Optimierung der Entscheidungsvariablen innerhalb der technischen Systemgrenzen erfolgt, werden die folgenden Nebenbedingungen aufgestellt: Eine Ladung des Fahrzeuges kann nur bei Anwesenheit des Fahrzeuges und nur bis zur maximalen Ladeleistung erfolgen:

$$0 \leq P_{EV}(t) \leq P_{EV}^{max} \cdot s_{EV}(t) \;\; \forall\, t \in T \tag{1}$$

Der Batteriespeicher kann nur bis zu seiner maximalen Kapazität geladen werden und muss bei Antritt einer Fahrt vollständig gefüllt sein:

$$E_{EV}^{max} \cdot b_{EV}(t) \leq E_{EV}(t) \leq E_{EV}^{max} \quad \forall\, t \in T \tag{2}$$

Der Speicherstand im Zeitpunkt t setzt sich aus dem Speicherstand der Vorperiode, der Entnahme durch Fahrten, dem Ladewirkungsgrad sowie der Selbstentladung zusammen:

$$E_{EV}(t) = \; E_{EV}(t-1) \cdot (\eta_{EV}^{T})^{-\tau} - l_{EV}(t) + \frac{P_{EV}(t)}{\eta_{EV}^{L}} \cdot \tau^{-1} \tag{3}$$

$$\forall\; t \in T \backslash \{1\}$$

Hiervon abweichend wird in der ersten Periode der Term $E_{EV}(t-1)$ gegen den Anfangsfüllstand E_{EV}^{0}ersetzt. Bedingt durch die geringe Speicherentladerate bei Elektrofahrzeugen gilt in der Simulation standardmäßig: $\eta_{EV}^{T} = 1$.

3.2.2 Batteriespeicher

Für die Betriebsoptimierung von Batteriespeichern mit mathematischen Programmen existiert eine Vielzahl von Ansätzen mit gemischt-ganzzahligen, linearen (z. B. [5]) oder quadratischen (z. B. [6]) Ansätzen. Zudem kommen naturanaloge Verfahren wie genetische Algorithmen (z. B. [7]) zum Einsatz.

Um eine geschlossene Lösung der Simulation zu ermöglichen, wird für Batteriespeicher ebenfalls ein mathematischer Programmansatz verwendet. Aufgrund der höheren Lösungsgeschwindigkeit und der Datenlage über (nichtlineare) Abhängigkeiten bei kommerziellen Energiespeichern wird ein gemischt-ganzzahliger, linearer Ansatz verwendet. Dieser verwendet die folgenden Parameter:

- max. Wirkleistung P_{BS}^{max} in kW
- max. Speicherkapazität E_{BS}^{max} in kWh
- Bereitschaftsenergieverlust η_{VS}^{B}
- Zykluswirkungsgrad η_{EV}^{Z}

Als zeitvarianter Parameter wird zudem der Speicherfüllstand zu Beginn der Simulation E_{BS}^{0} verwendet. Das Entscheidungsproblem besteht aus den Entscheidungsvariablen Lade- und Entladeleistung ($P_{BS}^{in}\; bzw.\, P_{BS}^{out}$), dem Speicherstand ($E_{BS}$) sowie einer Entscheidungsvariable Ladebetrieb ($\mathrm{L_{BS}}(t)$). Die binäre Entscheidungsvariable $\mathrm{L_{BS}}$ ist notwendig, um ein gleichzeitiges Ein- und Ausspeichern zu verhindern. Durch Kombination dieser Variablen mit der Begrenzung Speicherleistung kann dies gewährleistet werden:

$$0 \leq \; P_{BS}^{in}(t) \quad \leq P_{BS}^{max} \cdot \; \mathrm{L_{BS}}(t) \quad \forall\, t \in T \tag{4}$$

$$0 \leq \; P_{BS}^{out}(t) \; \leq P_{BS}^{max} \cdot (1 - \mathrm{L_{BS}}(t)) \quad \forall\, t \in T \tag{5}$$

Der Speicherstand wird in einer weiteren Nebenbedingung festgelegt. Diese Berücksichtigt die Selbstentladung des Speichers sowie den Zykluswirkungsgrad:

$$E_{BS}(t) = \; E_{BS}(t-1) \cdot (\eta_{BS}^{B})^{-\tau} \tag{6}$$

$$+ \left(P_{BS}^{in}(t) \cdot \eta_{BS}^{T} - P_{BS}^{out}(t)\right) \cdot \tau \quad \forall\, t \in T \backslash \{1\}$$

Für die erste Periode wird abweichend der Term $E_{BS}(t-1)$ gegen den Anfangsfüllstand E_{BS}^{0}ersetzt.

3.2.3 Elektrische Bilanz

Um die Außenwirkung des zu optimierenden Gebäudes abbilden zu können, müssen die elektrischen Leistungen von Elektrofahrzeug, Speicher und der Haushaltslast P_{Obj} bilanziert werden können. Die Haushaltslast P_{Obj} beinhaltet sowohl den Stromverbrauch des Haushaltes als auch den erzeugten PV-Strom. Im Falle von PV-Überschuss weist P_{Obj} entsprechend negative Werte auf. Für die elektrische Bilanz werden drei weitere Entscheidungsvariablen definiert, die die Außenwirkung des Objektes beschreiben:

- Leistung Strombezug $P_{Obj}^{in}(t)$ in kW
- Leistung Einspeisung $P_{Obj}^{out}(t)$ in kW
- Bezug $B_{Obj}(t)$ (binär)

Analog zur Abbildung von Batteriespeichern verhindert die binäre Entscheidungsvariable B_{Obj} den zeitgleichen Bezug und Einspeisung elektrischer Energie durch Kombination mit der Anschlussleistung P_{Obj}^{max} (Formel nicht angegeben, analog Formel 4 und 5). Die elektrische Bilanz ist gegeben durch:

$$P_{Obj}(t) + P_{BS}^{in}(t) + P_{Obj}^{out}(t) + P_{EV}(t) \tag{7}$$

$$= P_{Obj}^{in}(t) + P_{BS}^{out}(t) \quad \forall\, \mathrm{t} \in \mathrm{T}$$

3.2.4 Zielfunktionen

Die Zielfunktion bildet die Kosten- und Erlösstruktur des simulierten Geschäftsmodells ab. Für die marktpreisorientierte Inhouse-Optimierung werden hierbei die zeitvarianten Strombezugskosten als auch die Vergütung durch die Einspeisung von PV-Strom berücksichtigt.

$$\max \sum_{t=1}^{T} P_{Obj}^{out}(t) \cdot \mathrm{PR}_{\mathrm{Einsp}}(t) - P_{Obj}^{in}(t) \cdot \mathrm{PR}_{\mathrm{Bez}}(t) \qquad (8)$$

Hierbei bezeichnet $\mathrm{PR}_{\mathrm{Einsp}}$ die Vergütung für eingespeisten Strom. Im Standardfall entspricht dies der EEG-Einspeisevergütung. Im Falle einer Direktvermarktung können jedoch zeitvariante Preisvektoren angesetzt werden. $\mathrm{PR}_{\mathrm{Bez}}$ bezeichnet die Kosten des Strombezugs, die sich aus den Großhandelspreisen zzgl. der weiteren Preiskomponenten für Privatstromkunden ergeben.

3.2.5 Implementierung und Lösung

Das beschriebene mathematische Programm wird mit Hilfe der freien Toolbox *YALMIP* [8] in MATLAB implementiert. Die Toolbox erlaubt eine einfache und effiziente Formulierung des Optimierungsproblems und löst diese mit externen Solvern. Als Solver wurde IBM CPLEX gewählt und das Problem auf einem handelsüblichen PC gelöst.

Die Optimierung verwendet einen rollierenden Optimierungshorizont. Der Horizont beträgt 7 Tage, aus denen die ersten 5 als endgültige Lösung genutzt werden.

3.3 Simulation von Eingangsdaten

Um die Simulation der Geschäftsmodelle mit möglichst realitätsnahen Eingangsdaten durchführen zu können, werden verschiedene externe Modelle verwendet. Die folgenden Abschnitte fassen den Ursprung der in den Analysen verwendeten Eingangsdaten kurz zusammen.
Die Eingangsdaten der Fahrzeugnutzung sind einem Modell zur Bestimmung des Zustandes von Elektrofahrzeugen [4] entnommen. Hierbei wurden in Studien erhobene Daten zur Fahrzeugnutzung in Deutschland in ein stochastisches Modell zur Simulation des Fahrt- und Ladeverhaltens von Elektrofahrzeugen überführt.

Die verwendeten Daten zur Haushaltslast entstammen einem Feldversuch mit über hundert Smart-Metering-Pilotkunden. Die gemessenen Verbrauchsdaten wurden in ein stochastisches Simulationstool einzelner Haushaltslasten überführt [9].

Als PV-Einspeiseprofil wird ein Referenzlastgang genutzt, der aus Wettermodelldaten des DWD [10] generiert wird. Aus den hieraus enthaltenen Einstrahlungsdaten wird über ein physikalisches PV-Modell die lokale Erzeugung von PV-Strom simuliert.

Die zeitvariablen Tarife für den Haushaltskunden bilden die tatsächlichen Beschaffungskosten des Lieferanten ab. Die Beschaffung wird vereinfachend über zwei Standardprodukte der Elektrizitätswirtschaft simuliert. Dabei setzen sich die Strombeschaffungskosten zum einen aus den Kosten für Einzelstundenkontrakte am *Day-ahead*-Markt der EPEX Spot zusammen, die mit dem Mittelwert der vier viertelstündigen Leistungsmittelwerte des Netzbezugs innerhalb einer Stunde gewichtet werden. In jeder Viertelstunde wird die Abweichung des jeweiligen Leistungsmittelwertes der Viertelstunde zum Leistungsmittelwert der vollen Stunde mit dem regelzonenübergreifenden einheitlichen Bilanzausgleichsenergiepreis (reBAP) bewertet.

Die Berücksichtigung der Beschaffungskosten im Optimierungskalkül für den Haushaltskunden ermöglicht die Ladung des Elektrofahrzeugs und des Speichers zu Zeiten niedriger Preise der Einzelstundenkontrakte. Darüber hinaus wird gestattet, dass sich die viertelstündige Struktur des Netzbezugs gegen den reBAP optimiert.

4 Wirtschaftlichkeitsanalyse

Für die Wirtschaftlichkeitsanalyse werden alle wesentlichen Kosten- und Erlöspositionen ermittelt und eine Investitionsrechnung durchgeführt. Die Bewertung der Investition in den beiden vorgestellten Varianten wird anhand der Kapitalwertmethode vorgenommen. Der Kapitalwert wird anschließend annuisiert und auf die angenommene Jahresverbrauchsmenge bezogen, um eine Vergleichbarkeit mit dem ausschließlichen Netzbezug zu marktüblichen Konditionen in ct/kWh herzustellen. Die Kosten- und Erlöspositionen werden wie folgt angenommen:

Tabelle 1: Kosten und Erlöspositionen

Position	Annahmen/ Quelle
PV-Anlage+ Installation	1600 €/ kWp
Wallbox (Ladevorrichtung)	500 €
Optimierungshard- und Software	600 €
Wechselrichter	2800 €
Batteriespeicher	1650 € @ 3,2 kWh nutzbare Kapazität
KfW-Förderung	nach KfW-Programm [3]
Restwert Batterie + Wechselrichter	10 % der Investitionskosten nach Förderung
Installationskosten Batteriesystem	1000 €
EEG-Vergütung	13,14 ct/kWh (Errichtung Mai 2014, 10 kWp)
Strombeschaffung Großhandel	Mittelwerte Einzelstundenkontrakte EPEX Spot 2013; Mittelwerte reBAP 2013
Vertriebskosten/ Marge	2 ct/kWh
Netzentgelte	Grundpreis: 36,50 €/a Arbeitspreis: 4,61 ct/kWh
Umlagen (EEG, ...)	6,77 ct/kWh
Konzessionsabgabe	1,99 ct/kWh
Stromsteuer	2,05 ct/kWh
Messstellenbetrieb, Messung, Abrechnung (2-Richtungen)	33,86 €/a
Wartung/ sonstige Betriebskosten	1 % der Investitionskosten pro Jahr

Die Kapitalwertberechnung basiert auf einem Nutzungszeitraum von 20 Jahren und enthält einen internen Zinsfuß von 3 % p. a.

4.1 Ergebnisse der Eigenverbrauchsmaximierung

Die Investitionsrechnung für die Eigenverbrauchsmaximierung wird für einen Haushaltskunden mit einem Jahresverbrauch von 5000 kWh und einem Jahresverbrauch des Elektromobils von 2000 kWh durchgeführt. Es wird eine nutzbare Batteriekapazität von 3,2 kWh angenommen. Die Spitzenleistung der PV-Anlage wird von 2 kWp bis 8 kWp variiert, wobei die jährlichen Benutzungsstunden jeweils zu 800 h/a angenommen werden. Als Referenz für die Eigenverbrauchsmaximierung wird außerdem die verbrauchsbezogene Annuität der ausschließlichen Investition in eine PV-Anlage gegenübergestellt. **Abb. 3** zeigt die Ergebnisse der Investitionsrechnung.

Es wird deutlich, dass eine Investition in eine PV-Anlage im Vergleich zum ausschließlichen Netzbezug (zu 28 bis 30 ct/kWh, siehe [1]) lohnend ist, insofern dadurch ein Eigenverbrauch der erzeugten elektrischen Energie bewirkt werden kann. Mit zunehmender Spitzenleistung der PV-Anlage sinkt jedoch die Zunahme des Eigenverbrauchsanteils, so dass sich die verbrauchsbezogene Annuität 24 ct/kWh annähert.

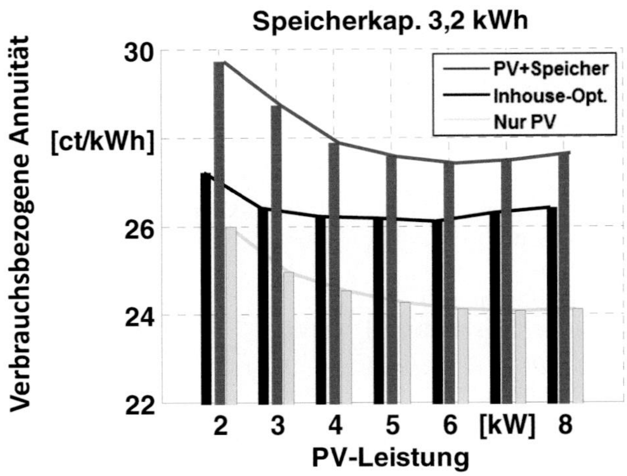

Abb. 3: Ergebnisse der Eigenverbrauchsmaximierung (dunkelgrau), der Inhouse-Optimierung (schwarz) sowie einer PV-Anlage ohne Speicher (hellgrau).

Die Ergebnisse der Eigenverbrauchsmaximierung weisen dagegen höhere verbrauchsbezogene Annuitäten über den gesamten Variationsbereich der Spitzenleistung der PV-Anlage auf und zeigen dabei weitgehend eine Parität zum ausschließlichen Netzbezug. Für das Verhältnis zwischen Jahresverbrauch des Haushalts und Einspeisemenge der PV-Anlage ergibt sich ein Kostenminimum bei einer Einspeisemenge, die dem Haushaltsbedarf entspricht.

4.2 Ergebnisse der Inhouse-Optimierung

Für die Inhouse-Optimierung werden zum Vergleich die gleichen Annahmen getroffen. Die Inhouse-Optimierung weist dabei geringe Gesamtkosten als die Eigenverbrauchsmaximierung auf, erreicht jedoch nicht die Wirtschaftlichkeit des ausschließlichen PV-Anlagenbetriebs (vgl. **Abb. 4**). Der Mehrwert gegenüber der Eigenverbrauchsmaximierung wird im Wesentlichen durch Ausgleichsenergieerlöse erzeugt. Auch die Inhouse-Optimierung zeigt ein Kostenminimum, das bei einer PV-Einspeisemenge auftritt, die dem Jahresverbrauch des Haushalts gleicht. Ähnliche Ergebnisse können für variierende Speichergrößen und Jahresverbrauchmengen der Haushaltskunden erzielt werden, auf eine Darstellung wird jedoch verzichtet. Hierbei zeigt sich bei größeren Speicherkapazitäten und verringerten Jahresverbräuchen einer verringerte Wirtschaftlichkeit.

4.3 Fazit

Anhand der Untersuchung kann gezeigt werden, dass sich die Strombezugskosten eines Haushaltes mit Elektrofahrzeug gegenüber üblichen Grundversorgungstarifen von Privatkunden durch alle untersuchten Maßnahmen verringern lassen. Hierbei zeigt sich jedoch, dass die Ergänzung einer PV-Anlage mit einem Batteriespeicher zu heutigen Anschaffungskosten nicht wirtschaftlich darstellbar ist. Die Mehrkosten gegenüber einer PV-Anlage ohne Speicher lassen sich durch den Einsatz einer Inhouse-Optimierung jedoch teils deutlich reduzieren.

5 Literatur

[1] Bundesnetzagentur: Monitoring Bericht 2014. 2014

[2] KfW-Programm Erneuerbare Energien "Speicher", Programmnummer 275 - Richtlinien zur Förderung von stationären und dezentralen Batteriespeichersystemen zur Nutzung in Verbindung mit Photovoltaikanlagen vom 01. Januar 2015.

[3] Forum Netztechnik/ Netzbetrieb im VDE (FNN): Lastenheft EDL – Elektronische Haushaltszähler – Funktionale Merkmale und Protokolle. V1.0, 2010.

[4] Rolink, J; Rehtanz, C.: „ Estimation of the availability of grid-connected electric vehicles by non-homogeneous semi-Markov processes", PowerTech, 2011 IEEE Trondheim, June 2011.

[5] Kanngießer, A.: „Entwicklung eines generischen Modells zur Einsatzoptimierung von Energiespeichern für die techno-ökonomische Bewertung stationärer Speicheranwendungen.“ Dissertation, TU Dortmund, 2013.

[6] Koller, M., Borsche, T.: “Defining a Degradation Cost Function for Optimal Control of a Battery Energy Storage System”, IEEE PowerTech, Grenoble, June 2013.

[7] Logenthiran, T.: “Intelligent Management of Distributed Storage Elements in a Smart Grid”, Power Electronics and Drive Systems (PEDS), 2011

[8] Löfberg, J.: YALMIP: A Toolbox for Modeling and Optimization in MATLAB, Proceedings of the CACSD Conference, S. 284-289, Taipei, 2004

[9] Belitz, H.-J.; Winter, S; Rehtanz, C.: „Load shifting of the households in the E-Energy project E-DeMa", IEEE PowerTech, Grenoble, June 2013.

[10] Deutscher Wetterdienst (DWD), „Regional Model COSMO-EU,“ www.dwd.de, 2015.

Fehlermöglichkeits- und Einflussanalyse (FMEA) eines Second-Life-Batteriespeichers bei einem Energieversorgungsunternehmen

C. Hindersmann[1], S. Gerhard[2]
[1] Vattenfall Europe GmbH, 22297 Hamburg, christoph.hindersmann@vattenfall.de
[2] Vattenfall Europe Innovation GmbH, 22297 Hamburg, sebastian.gerhard@vattenfall.de

Kurzfassung

Die Arbeit umfasst eine Fehlermöglichkeits- und -einflussanalyse (FMEA) eines Second-Life-Batteriespeichers aus Betreibersicht. Im Vorfeld werden das Problem des Ausbaus der erneuerbaren Energien und mögliche Lösungsansätze mit verschiedenen Speichern beschrieben. Die Ergebnisse der FMEA sind in verschiedenen Anlagenbereichen untergliedert und einzeln bewertet. Aus den Ergebnissen wurde eine Prüfung entwickelt, die für zwei existierende Second-Life-Batteriespeicher angepasst ist. Die Prüfung wird in bestimmten Zeitabständen durchgeführt und soll so dauerhaft die Entdeckungswahrscheinlichkeit von ausgewählten Fehlern verringern.

1 Einleitung

Mit der Zustimmung des Bundesrates zur aktuellen Erneuerbare-Energien-Gesetz (EEG) Novelle vom 27. Juni 2014 soll der Anteil des aus erneuerbaren Energien erzeugten Stroms am Bruttostromverbrauch stetig und kosteneffizient bis zum Jahr 2050 auf mindestens 80 % erhöht werden. [1]

Zur Erreichung dieses Ziels sowie der Gewährleistung einer stabilen Energieversorgung müssen aufgrund der fluktuierenden Einspeisung regenerativer Energiequellen mittel- und langfristig vermehrt Speicher eingesetzt werden. Im Gegensatz zu konventionellen Kraftwerken ist die Stromproduktion aus erneuerbaren Energien oft nicht konstant über einen längeren Zeitraum verfügbar und ihr Einspeiseprofil (Photovoltaik und Wind) volatil. Für die Speicherung gibt es eine Vielzahl von Ansätzen. [2]

Die Vattenfall Europe Innovation GmbH (VE Innovation) befasst sich mit einigen dieser Ansätze, wie bspw. dem Einsatz von Batteriespeichern zur Regelleistung, der Wasserstofftankstelle in der HafenCity in Hamburg sowie verschiedenen Elektromobilitätsprojekten in Hamburg und Berlin.

In diesem Kontext bildet ein stationärer Second-Life-Batteriespeicher ein Pilotprojekt mit dem Ziel die optimale Betriebsstrategie eines stationären Batteriespeichers zu entwickeln. Der Speicher hat eine Leistung von 50 kW und eine Kapazität von 120 kWh.

Um in der Wirtschaftlichkeitsanalyse Ergebnisse über Ausfallwahrscheinlichkeiten und Fehlerkosten mit einzubeziehen, wird eine „Fehlermöglichkeits- und Einflussanalyse“ (FMEA) verwendet.

Mittels der Durchführung der FMEA sollen Schwachstellen des Systems aufgedeckt und bewertet werden, sodass anhand eines Rankings entschieden werden kann, welchen möglichen Fehlern am schnellsten entgegengewirkt werden sollte, damit das Risiko im Betrieb des Batteriespeichers verringert werden kann.

2 Methode

Für die Ermittlung der potentiellen Fehler wird die 1949 veröffentlichte Methodik der FMEA verwendet. In der Vorbereitung wird der Batteriespeicher in Teilbereiche gegliedert und ein Template für die Auflistung der Fehler erstellt.

Das System des Batteriespeichers teilt sich wie folgt auf:

- Gebäude
- Leistungs- und Steuerungselektronik
- Batterieracks
- Schutzeinrichtung
- Use Cases (zu Deutsch: Anwendungsfälle)

Die Teilbereiche werden im Hauptteil der Analyse mit unterschiedlichen Fachkräften, die im besten Fall alle Kompetenzen für den Betrieb abdecken, nach und nach mit potentiellen Fehlern gefüllt. Es ist besonders darauf zu achten, dass die Fehler den richtigen Bereichen zugeteilt werden.

Die eigentliche Bewertung eines Fehlers teilt sich in folgende Schritte auf:

1. Benennung des Fehlers
2. Welche Folge hat der Fehler?
3. Welche Ursache hat der Fehler?
4. Gibt es momentan Gegenmaßnahmen?
5. Wie hoch ist die Bedeutung (**B**) des Fehlers? (1–10)
6. Wie hoch ist die Wahrscheinlichkeit, dass der Fehler auftritt (**A**)? (1–10)
7. Wie hoch ist die Wahrscheinlichkeit, dass der Fehler entdeckt (**E**) wird? (1–10)
8. Gibt es potentielle Gegenmaßnahmen, die die Bedeutung, die Auftretenswahrscheinlichkeit oder die Entdeckungswahrscheinlichkeit verbessern?
9. Wenn ja, wie verändern sich die **B**-, **A**- und **E**-Werte?

Das Produkt aus den drei Werten wird RPZ-Wert genannt und steht für die „Risikoprioritätszahl". Der RPZ-Wert sagt aus wie schwerwiegend die betrachteten Fehler relativ zueinander sind.

Die Bewertung der einzelnen Werte basiert auf eigenen Skalen, die für die Wahrscheinlichkeiten und Bedeutung erstellt sind:

Tabelle 1 Bedeutung der Fehlerfolge B

Wertung		Beschreibung
Sehr hoch	10 / 9	• Gefahr für Personen • Nicht Erfüllung gesetzlicher Vorschriften
Hoch	8 / 7	• Sicherheitsrisiko • Funktionseinschränkung wichtiger Teilsysteme • Nach Auslieferung notwendige Vorbeugemaßnahmen
Mäßig	6 / 5 / 4	• Einschränkung der Funktion
Gering	3 / 2	• Geringe Funktionseinschränkung • Keine Auslastung der Anlage
Sehr gering	1	• Sehr geringe Funktionseinschränkung

Tabelle 2 Bedeutung der Auftretenswahrscheinlichkeit A

Wertung		Beschreibung
Sehr hoch	10 / 9	• Sehr häufiges Auftreten
Hoch	8 / 7	• Fehlerursache tritt häufig auf
Mäßig	6 / 5 / 4	• Gelegentliches Auftreten des Fehlers
Gering	3 / 2	• Auftreten des Fehlers ist gering
Sehr gering	1	• Auftretens des Fehlers ist unwahrscheinlich

Tabelle 3 Bedeutung der Entdeckungswahrscheinlichkeit E

Wertung		Beschreibung
Sehr gering	10 / 9	• Entdecken des aufgetretenen Fehlers ist sehr gering • Nachweisverfahren sind sehr unsicher
Gering	8 / 7	• Entdecken des Fehlers ist gering und schwer nachweisbar
Mäßig	6 / 5 / 4	• Entdecken des Fehlers ist wahrscheinlich
Hoch	3 / 2	• Fehler ist sehr wahrscheinlich zu entdecken
Sehr hoch	1	• Aufgetretener Fehler ist sicher entdeckt

Nach dem Workshop werden in der Nachbereitung die Ergebnisse aufbereitet und Handlungsempfehlungen entwickelt.

3 Ergebnisse

Ergebnis des FMEA-Workshops ist, dass insgesamt 45 potentielle Fehler am Batteriespeicher identifiziert wurden.

Mittels Pareto-Analyse wurde entschiedenen, dass Fehler mit einem RPZ-Wert von größer-gleich 100 näher untersucht werden müssen. In den resultierenden neun Fällen wurde eine zukünftig einzuführende Gegenmaßnahme gefunden, die die RPZ-Zahl senkt. Die Grenze von einem RPZ-Wert größer gleich 100 war kein Hindernis weitere Gegenmaßnahmen zu finden. Insgesamt konnten für 16 Fehler Gegenmaßnahmen entwickelt werden.

Die Ergebnisse sind beispielhaft in den folgenden drei Tabellen zu sehen. Bei den Racks handelt es sich um Gestelle, in denen vier Batterien zusammen geschaltet werden.

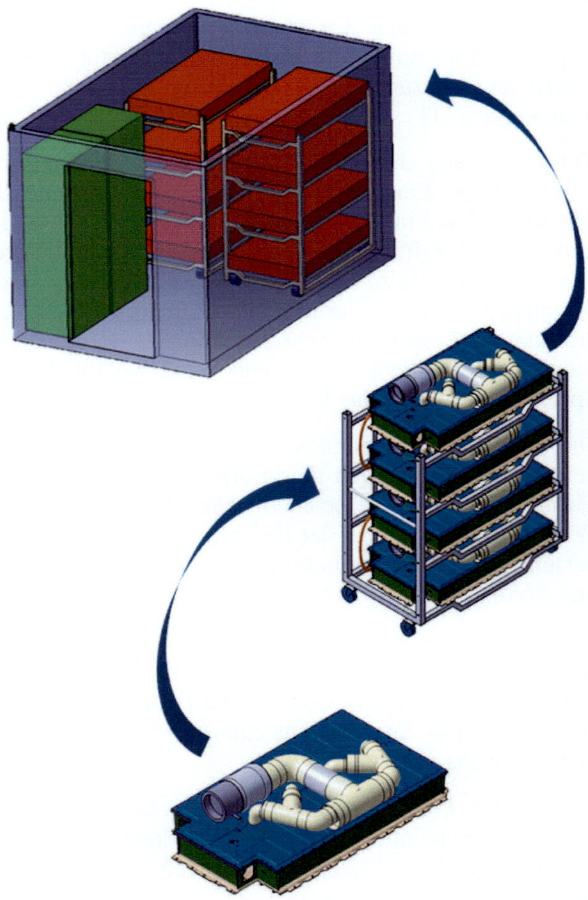

Abbildung 1 CAD-Modell des Batteriespeichers und seinen Komponenten

Zunächst wird der Fehler „Erdung an Racks unzureichend" aufgezeigt.

Tabelle 4 Fehler „Erdung an Racks unzureichend"

Fehler	Erdung an Racks unzureichend
Fehlerfolge	Berührungsschutz ist nicht gegeben
Ursache	Erdungskabel falsch ausgelegt oder defekt
Derzeitige Gegenmaßnahme (B,A,E,RPZ)	Erdung nach VDE 0100-540 (9,2,10,180)
Mögliche Gegenmaßnahme (B,A,E,RPZ)	Erdungswiderstand regelmäßig messen (WKP) (9,2,3,54)

Durch eine wiederkehrende Prüfung kann die Wahrscheinlichkeit des Entdeckens von 10 (sehr unwahrscheinlich) auf 2 (recht wahrscheinlich) minimiert werden, sodass der RPZ-Wert von 180 auf 54 gesenkt wird.

Der zweite beispielhafte Fehler ist „Fehler in csv-Datei". Die csv-Dateien werden verwendet um Betriebsparameter in die Steuerungssoftware zu laden.

Tabelle 5 Fehler „Fehler in csv-Datei"

Fehler	Fehler in csv-Datei
Fehlerfolge	Nicht planmäßige Fahrweise
Ursache	Tippfehler
Derzeitige Gegenmaßnahme (B,A,E,RPZ)	- (5,4,6,120)
Mögliche Gegenmaßnahme (B,A,E,RPZ)	Default-Werte, Integration von Plausibilisierungssoftware, Anleitung (5,4,1,20)

Um einem Tippfehler entgegen zu wirken, können neben Default-Werten, eine Anleitung oder eine Plausibilisierung integriert werden, die die Entdeckungswahrscheinlich erhöhen.

Tabelle 6 Fehler „Störung der Ethernet-Kommunikation"

Fehler	Störung der Ethernet-Kommunikation
Fehlerfolge	Keine Bedienung per Fernzugriff möglich, Stillstand der Anlage
Ursache	Defektes Kabel oder Netzwerkausfall
Derzeitige Gegenmaßnahme (B,A,E,RPZ)	Durch eine Querüberwachung wird bei Anlage bei einer Störung abgeschaltet (5,5,3,75)
Mögliche Gegenmaßnahme (B,A,E,RPZ)	- (-)

Ein Fehler in der Ethernet-Kommunikation hat zur Folge, dass die Anlage abgeschaltet wird, da sie permanent durch eine Querüberwachung kontrolliert wird. Aufgrund der relativ geringen Risikoprioritätszahl (75) ist eine mögliche Gegenmaßnahme nicht notwendig.

3.1 Wiederkehrende Prüfung

Die Auswertung der FMEA-Ergebnisse hat ergeben, dass eine wiederkehrende Prüfung (WKP) eine große Anzahl von Fehlern in ihrer Risikoprioritätszahl senkt.
Das Prüfprotokoll wurde auf Grundlage von DIN VDE 0105-100 – Betrieb von elektrischen Anlagen – Teil 100: Allgemeine Festlegungen entwickelt und teilt sich in drei Teile auf:

1. Sichtprüfung
2. Erprobung
3. Messung

Die Sichtprüfung und Erprobung der Anlage sollte alle sechs Monate wobei die Messung alle vier Jahre oder nach baulichen Änderungen durchgeführt werden sollte. Die Messung beinhaltet die Schutzleiter- und Isolationsmessung.

Folgende risikobelastete Fehler können mit der WKP gesenkt werden:

- Fehler an Steckverbindung
- Erdschluss am Batterierack
- Fehlerhafte Erdung
- Defekter RCD-Schutzschalter
- Defektes Bedienpanel

Darüber hinaus konnten weitere Punkte in die WKP eingebaut werden, die nicht direkt Ergebnis des FMEA-Workshops waren:

- Erprobung der Not-Aus-Einrichtung
- Erprobung der Temperatursensoren und programmtechnische Auslösung der Grenzwerte
- Erprobung der Thermostateinstellungen und Lüfter

Das Prüfprotokoll ist für zwei Second-Life Batteriespeicher angepasst worden und umfasst insgesamt 13 Prüfpunkte.

4 Fazit

Die FMEA-Methodik hat sich bewährt das System auf potentielle Fehler zu untersuchen. Auch wenn die Analyse normalerweise in der Projektentwicklung angewendet wird, konnten passende Gegenmaßnahmen inkl. des Prüfprotokolls entwickelt werden.

Hierbei ist zu beachten, dass die Maßnahmen möglichst direkt umgesetzt werden müssen, damit Aufgaben nicht unter den Tisch fallen und vergessen werden. Hierbei bietet sich an einen Verantwortlichen und eine Deadline zu setzten.

Für das Durchführen des Workshops bietet sich Teilnehmer auszuwählen, die unterschiedliche Fachkenntnisse der Anlage haben, da somit besonders viele Fehlerpotentiale erarbeitet werden. Ebenfalls ist wichtig, dass bei allen Teilnehmern das gleiche Anlagenverständnis und Detailtiefe bekannt ist, sodass homogene Ergebnisse erarbeitet werden können, die im besten Fall durch Gegenmaßnahmen effizient gelöst werden können.

Die Übertragbarkeit der Ergebnisse auf andere Batteriespeichersysteme ist schwierig, da die Infrastruktur des Batteriespeichers einmalig ist und viele Details angepasst werden müssten, wenn z. B. das Prüfprotokoll übertragen werden sollte.

5 Ausblick

Der Verband der Elektrotechnik (VDE) hat Anfang 2015 Anwendungsregeln für den sicheren Betrieb von stationären Batteriespeichern, die auf Basis von Lithium-Ionen arbeiten, veröffentlich und reagiert damit auf den stark wachsenden Markt der Lithium-Ionen-Batterien. [3]

Durch neue Anwendungsregeln soll der Betrieb von Batteriespeichern noch sicherer gestaltet werden. Hierdurch entstehen ebenfalls neue Herausforderungen für die Betreiber, die u. U. Prüfprotokolle anpassen (Grenzwerte, erlaubte Reaktionen im Fehlerfall, etc.) oder um Prüfungen erweitern müssen.

Die Erweiterungen sind nicht nur durch neue Regeln der Technik getrieben, sondern auch durch neue Technologien, die z. B. den Einblick in die Batterie mit einfachen Mitteln ermöglichen.

Hierbei können kabellose Sensoren verwendet werden, die sämtliche Zellen einer Großbatterie überprüfen. Da bei kabelgebundenen Sensoren häufig galvanische Isolationsprobleme auftreten, entwickelt das Forschungsteam „BATSEN“ einen Zellsensor, der kabellos Messdaten übermittelt. [4]

Durch diese Möglichkeit können Änderungen der einzelnen Zellen schnell und zuverlässig ermittelt werden und bei Bedarf Batterien, bevor sie defekt werden, ausgetauscht werden.

Darüber hinaus ist es möglich die Messergebnisse, die in den Prüfungen gesammelt werden, auszuwerten. Hierfür können Langzeitwerte entwickelt werden, die die Gerätesysteme zeigen.

Beispielhaft kann der Widerstand von Schutzleitern ausgewertet und eine stetige Verschlechterung aufgedeckt werden.

6 Literaturverzeichnis

[1] BMWI: *EEG-Gesetz*. – Aktualisierungsdatum: 2015-05-18

[2] DR. ANDREA FELDMUELLER: *Wie flexibel ist der heutige konventionelle Kraftwerkspark aus Herstellersicht?* URL http://www.effiziente-energiesysteme.de/fileadmin/user_upload/PDF-Dokumente/Veranstaltungen/Workshop_Retrofit/3_SIEMENS_Feldmueller.pdf

[3] FRAUENHOFER ISI: *Technologie-Roadmap Lithium-Ionen-Batterien 2013*. URL http://www.isi.fraunhofer.de/isi-wAssets/docs/t/de/publikationen/PRM-LIB2030.pdf

[4] RIEMSCHNEIDER K.-R., SCHNEIDER M: *Drahtlose Sensoren in den Zellen von Fahrzeug-Batterien*. URL http://www.haw-hamburg.de/department-informations-und-elektrotechnik/forschung/forschungsschwerpunkt-adys/batsen.html. – Aktualisierungsdatum: 2015-05-20

Technische Herausforderungen und Möglichkeiten zur Steuerung des Ladeverhaltens von Elektrofahrzeugen durch intelligente Ladestationen auf Basis der DIN IEC-61851

Gerrit Schlömer, Constantin Reese, Hannes von Appen, Lutz Hofmann
Leibniz Universität Hannover, Institut für Energieversorgung und Hochspannungstechnik, 30159 Hannover

Kurzfassung

In vielen Studien und Aufsätzen wird die Verwendung von Elektrofahrzeugen als Energiespeicher für die Energiewende diskutiert. Dabei kommen viele Studien (z. B.: [5]) zu dem Ergebnis, dass sich durch eine intelligente Steuerung des Ladeverhaltens die Kosten der Energiewende, sei es durch vermiedenen Netzausbau oder durch Speicherung andernfalls abgeregelter Energie, deutlich reduzieren lassen. Im Rahmen des Projektes „Demand-Response – Das Auto als aktiver Speicher und virtuelles Kraftwerk" aus dem Förderprogramm Schaufenster Elektromobilität wird im Rahmen eines Konsortiums aus Energieversorger (enercity, Stadtwerke Hannover AG), Elektrotechnikern (Institut für Energieversorgung und Hochspannungstechnik, Leibniz Universität Hannover) und Sozialwissenschaftlern (Institut für Transportation Design, HBK Braunschweig) erforscht, ob für dieses Vorhaben eine entsprechende Nutzerakzeptanz und -bereitschaft vorhanden ist und wie eine technische Lösung dazu aussehen könnte. Um fahrzeugherstellerübergreifend dieses Ziel untersuchen zu können, wurde eine Ladestation gemäß IEC 61851 entwickelt, in Kleinserie gebaut und bei den Projektteilnehmern im privaten und gewerblichen Umfeld installiert. Das Feld der Projektteilnehmer besteht aus Personen und Unternehmen, die sich aus eigenen finanziellen Mitteln ein Elektrofahrzeug zugelegt haben.
Als Herausforderung hat sich bei der Durchführung des Feldtests gezeigt, dass die IEC 61851 eigentlich nur ein Low-Level-Protokoll ist, in dem nur das sofortige Laden exakt geregelt ist. Schon bei leichten Verzögerungen, beispielsweise durch die momentan eingesetzten Nutzerauthentifizierungsprozesse an öffentlichen Ladesäulen, kann es zu Problemen bei der Ladung kommen, da die Fahrzeughersteller die Norm unterschiedlich umgesetzt haben. Im Extremfall führt dies dazu, dass einige Fahrzeuge an öffentlichen Ladesäulen gar nicht oder nur unter Beachtung einer festen Vorgehensweise laden können. In diesem Beitrag wird zunächst der entsprechende Ladestandard erläutert, im Anschluss auf die fahrzeugspezifischen Eigenschaften eingegangen und Lösungen für den Versuchsaufbau aufgezeigt.

1 Einleitung und Zielstellung

Im Rahmen des Forschungsprojektes „Demand-Response – Das Auto als aktiver Speicher und virtuelles Kraftwerk" aus dem Förderprogramm Schaufenster Elektromobilität wird erforscht, ob es möglich ist Elektrofahrzeuge als flexibele Speicher in ein virtuelles Kraftwerk einzubinden. Dass es technisch möglich ist, haben viele Projekte – oftmals mit angepassten Fahrzeugen – bereits bewiesen. In diesem Feldversuch soll im Gegensatz dazu untersucht werden, in wie weit dieses auch mit den heute vorhandenen Fahrzeugen möglich ist und wie hoch die Nutzerakzeptanz für verschiedene Energieprodukte ist.

Das gesamte Projekt ist in vier Projektphasen aufgeteilt. Mit jeder Projektphase gibt es für die Projektteilnehmer eine zusätzliche Ladeprogrammwahlmöglichkeit, in dem der Nutzer dem Energieversorger eine größere Freiheit in der Steuerung des Ladens einräumt. In der ersten Phase des Projektes fand die Referenzphase statt. In dieser Phase gibt es lediglich die übliche sofort-laden Variante. Die Phase dient der Gewöhnung der Teilnehmer an die neue Ladestation und dem Aufzeichnen der üblichen Ladegewohnheiten. Ferner soll der Projektteilnehmer Vertrauen in die Ladestation entwickeln, dass er sich auf die Ladestation verlassen kann.

In den weiteren Projektphasen haben die Nutzer die Möglichkeit ihr Fahrzeug nicht sofort laden zu lassen, sondern sich an Ladeprogrammen zu beteiligen, die dem Energieversorger einen immer höheren Freiheitsgrad bei der Steuerung der Ladezeiten gibt. In jeder Versuchsphase kommt ein Programm hinzu, sodass die Nutzer in der letzten Versuchsphase zwischen dem sofortigen Laden und drei gesteuerten Ladeprogrammen wählen können. Das Auswählen der gesteuerten Ladeprogramme wird jeden Tag mit einer Prämie belohnt. Die Prämienhöhe steigt mit dem Freiheitsgrad des Energieversorgers und wird in den unterschiedlichen Projektphasen jeweils neu angepasst.

In diesem Beitrag stehen zunächst die Erfahrungen im Vordergrund, die bei der Umsetzung der E-KFZ Ansteuerung gemacht wurden. Während das sofortige Laden der Fahrzeuge mit der heutigen Technik in der Regel kein Problem mehr darstellt, verhalten sich die Fahrzeuge bei einer zeitgesteuerten Ladung vollkommen unterschiedlich. Im Forschungsprojekt mit 40 Ladestationen nehmen 19 verschiedene E-KFZ-Typen teil, wodurch eine hohe Bandbreite aktuell verfügbarer Elektrofahrzeuge erreicht wird. Durch fünf verschiedene Verhaltensweisen der Ladestation, die im Folgenden erläutert werden, ist es möglich, fast alle Fahrzeuge zeitgesteuert zu laden. Spricht man allerdings ein Fahrzeug mit der falschen Ansteuerungstechnik an, kann es sein, dass das Fahrzeug in einen

Fehlermodus fällt und bis zum erneuten Anschluss an die Ladestation nicht mehr geladen werden kann.

2 Anschluss von E-KFZ an Ladestationen

Als einheitliche Norm für Steckertypen und Lademodi für Elektrofahrzeuge (E-KFZ) existiert die Norm IEC 62196. Sie ist auf keine physische Steckertechnik bezogen, verweist aber in Teil 2 auf die IEC 60309 (CEE-Steckverbindungen) und übernimmt die Definition des Signalpins aus der IEC 61851.

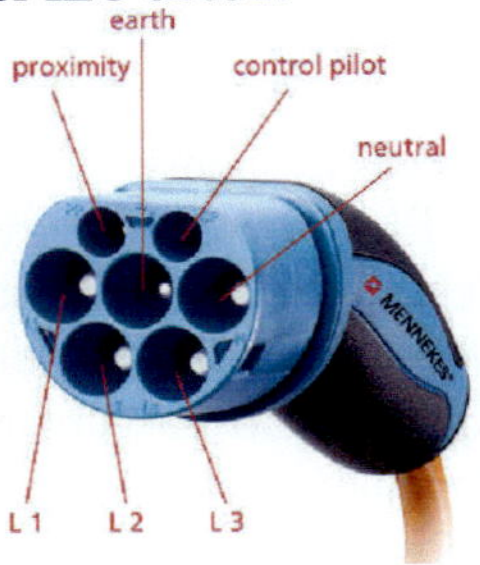

Bild 1: Darstellung des Typ2-Steckers (Firma Mennekes)

Die Typ2-Steckverbindung (siehe **Bild 1**) gemäß der IEC 62196-2 ist in der Lage, in der Baugröße einer 16-A-CEE-Drehstromsteckdose dreiphasig Ströme bis zu 63 A zu übertragen. Da von dieser Steckdose aufgrund der hohen möglichen Absicherung eine potenzielle Gefahr im öffentlichen Raum ausgeht, darf keine Einschaltung der Steckdosen erfolgen, solange nicht sichergestellt ist, dass ein E-KFZ auf der Gegenseite angeschlossen ist. Daher sind zwei Signalkontakte in die Steckverbindung integriert worden, mit deren Hilfe die Leistungskontakte erst zugeschaltet werden, wenn eine sichere Fahrzeugverbindung besteht. Der erste dieser Kontakte ist der „Proximity Pilot" (PP). Dieser ist auf Fahrzeug- und Ladestationsseite mit einem Widerstand gegen den Erdleiter (Earth / PE) verbunden. Der Wert des Widerstands steht stellvertretend für den Kabelquerschnitt und gibt somit an, wie groß die Stromtragfähigkeit des Kabels ist. **Tabelle 1** zeigt mögliche Werte des Widerstandes.

Tabelle 1: Widerstandstabelle Proximity Pilot gemäß IEC 61851

Widerstand PP-PE	1500 Ω	680 Ω	220 Ω	100 Ω
max. Ladestrom	13 A	20 A	32 A	63 A
Querschnitt	1,5 mm²	2,5 mm²	6 mm²	16 mm²

Da die beiden kabelseitigen Steckverbinder mechanisch nicht kompatibel sind, ist eine Verlängerung der Anschlussleitung durch ein zweites Verbindungskabel sowohl mit einem gleichwertigen als auch einem anderen Kabelquerschnitt nicht möglich. Dies stellt eine zusätzliche Sicherheitsfunktion dar.

Der zweite Kommunikationskontakt ist der „Control Pilot" (CP). Dieser Kontakt wird mit einem pulsweitenmodulierten Signal (PWM-Signal) belegt, welches die maximale Strombelastbarkeit der Ladestation codiert. Als Rückkanal wird der Schutzleiter verwendet. Auf diesem Weg erfolgt gleichzeitig auch eine ständige Überprüfung des Schutzleiters auf Durchgängigkeit (siehe **Bild 2**). Durch unterschiedliche Widerstandskombinationen auf der Fahrzeugseite (siehe **Tabelle 2**) wird der Ladestation der Zustand des Fahrzeuges mitgeteilt. Die Diode im Schaltkreis sorgt dafür, dass der am Messpunkt eintretende Spannungsabfall des PWM-Signals nur auf einer Halbwelle zu beobachten ist und verhindert somit ein Einschalten der Leistungsschütze beispielsweise bei einem widerstandsbehafteten Fehler zwischen PP und PE. Ein Widerstandswert kleiner als 880 Ω signalisiert der Ladestation eine Ladebereitschaft des Fahrzeuges, sodass die Station das Leistungsschütz zuschalten kann.

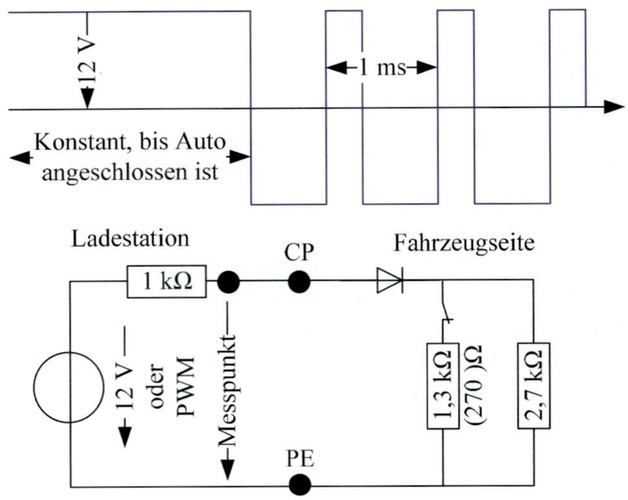

Bild 2: Darstellung des PWM-Signals und des Proximity-Signalkreises mit Autosimulation

Tabelle 2: Kommunikationswiderstände gemäß IEC 61851 für die Fahrzeugseite

Widerstand CP-PE	offen	2700 Ω	880 Ω	240 Ω
Status	Standby	Vehilcle detected	ready	ready with ventilation

Durch Einstellung des Tastverhältnis der PWM kann die Ladestation dem E-KFZ die an der Ladestation maximal mögliche Ladeleistung mitteilen. Die Berechnung zur Höhe des möglichen Ladestromes kann **Tabelle 3** entnommen werden.

Tabelle 3: Tastverhältnis PWM-Signal gemäß IEC 61851

Tastverhältnis	Bedeutung
< 8 %	keine Ladung erlaubt
> 8 % – ≤ 10 %	Ladung mit 6 A
> 10 % – ≤ 85 %	Ladung mit Tastverhältnis · 0,6 A
> 85 % – ≤ 96 %	Ladung mit (Tastverhältnis-64 %) · 2,5 A
> 96 % – ≤ 97 %	Ladung mit 80 A
> 97 %	keine Ladung erlaubt

In der zukünftigen Norm ISO/IEC 15118 wird das Pilotsignal durch ein höher frequentes Signal überlagert, welches weitaus mehr Informationen über den Ladevorgang bidirektional übertragen kann.

3 Realisierung des zeitgesteuerten Ladens

Im Projektablauf sollen sowohl E-KFZ, die mit einer herkömmlichen Schuko-Steckdose als auch Fahrzeuge, die über eine Typ2-Steckdose laden, berücksichtigt werden. Durch die beschriebene Sicherheitsphilosophie beim Laden von E-KFZ über die Typ2-Steckdose ist das zeitlich gesteuerte Laden der E-KFZ deutlich aufwendiger, als dass lediglich das Schütz der Typ2-Steckdose eingeschaltet werden muss.

Wie in Kapitel 2 geschildert, definiert die Norm IEC 61851, nach der die E-KFZ des Feldversuchs gebaut sind, nicht das Verhalten der Automobile für ein zeitgesteuertes Laden mit Ansteuerung durch die Ladestation. Aus diesem Grund reagieren die E-KFZ grundsätzlich unterschiedlich, wie im Folgenden erläutert wird.

Durch eine Untersuchung diverser Fahrzeugtypen konnten allerdings viele Gemeinsamkeiten im Fahrzeugverhalten herausgefunden werden, sodass mithilfe von vier unterschiedlichen Ansteuerungen fast alle Fahrzeuge des Testfeldes erfolgreich zeitgesteuert geladen werden konnten. Da es sich bei der Kommunikation durch das PWM-Signal lediglich um eine rudimentäre Kommunikation handelt, ist es problematisch, wenn die Ladestation im Voraus nicht ermitteln kann, was für ein E-KFZ angeschlossen ist und somit nicht weiß, auf welche Weise die Ladekommunikation stattfinden muss. Dadurch, dass die Ladestationen innerhalb des Forschungsvorhabens allerdings bei Privat- und Gewerbekunden installiert und somit einem festen Auto zugeordnet sind, ist ein individuelles Ladestationsverhalten für das zeitgesteuerte Laden tolerierbar. Weitere notwendige Informationen, wie die verfügbare Batteriekapazität und der Ladezustand, werden durch Nutzereingabe an der Ladestation abgefragt. In den folgenden Abschnitten werden die unterschiedlichen Lösungen der Fahrzeugansteuerungen vorgestellt, die im Projekt Anwendung finden.

3.1 Schuko-Fahrzeuge

Alle E-KFZ, die innerhalb des Projektes über die Schuko-Steckdose geladen werden, beginnen den Ladevorgang selbstständig, sobald eine Spannung an der Steckdose anliegt.

Aus Sicherheitsgründen erfolgt in der Norm eine Begrenzung des Ladestroms auf 13 A an der Schuko-Steckdose. Dies ist erforderlich, weil durch die Bauart des Schuko-Stecksystems die Kontakte des Steckers nur wenig vor mechanischer Beschädigung geschützt sind und durch leichte Beschädigungen oder auch Verschmutzungen hohe Übergangswiderstände entstehen können. Ferner findet keine automatische Überprüfung statt, ob die Steckverbindung bis zum Anschlag gesteckt wurde. Durch die lange Belastungszeit bei hoher Ladeleistung kann dies zu starker Erwärmung und auch zu Bränden führen. Innerhalb des Projektes hat sich gezeigt, dass mehrere E-KFZ sich nicht an die Norm halten und die Schuko-Steckdose über mehrere Stunden mit 16 A belasten, teils sogar noch etwas mehr, da auf eine konstante Wirkleistung von 3,6 kW geregelt wird. Unter Einbeziehung der Blindleistung führt dies zu deutlich höheren Strömen, gerade wenn die Netzinnenwiderstände an der Grenze des zulässigen Bereichs liegen.

Es zeigt sich, dass die Ladung von E-KFZ über Schuko-Steckverbinder eine problematische Lösung ist, auch wenn sich die Ansteuerung und somit das zeitgesteuerte Laden in diesem Fall besonders einfach gestalten lässt.

3.2 Typ2 - Variante A

Eine Gruppe von E-KFZ weist die Eigenschaft auf, dass unmittelbar nach Beginn der PWM-Kommunikation eine Kontrolle der Netzspannung vorgenommen wird. Liegt diese nicht innerhalb von wenigen Sekunden an, fällt das Auto in einen Fehlermodus, der nur durch erneutes Kabelstecken verlassen werden kann. Aus diesem Grund erfolgt der Beginn der Ladekommunikation erst unmittelbar bevor die Ladung beginnen soll. Ein weiteres Problem dieser E-KFZ ist es, dass eine Unterbrechung der Ladung nicht stattfinden darf, da ein erneutes Starten des Ladevorgangs auf Grund des Fehlermodus nicht möglich ist. Diese Eigenschaft kann dazu führen, dass das E-KFZ nicht weiter geladen werden kann, auch wenn es nur zu einer kurzen Versorgungsunterbrechung der Ladestation kommt. Einige dieser E-KFZ weisen ferner die Eigenschaft auf, dass eine Variation des Ladestromes nur oberhalb von 16 A möglich ist.

Ein gesteuertes Laden dieser Fahrzeuge ist möglich, wenn beachtet wird, dass die Kommunikation erst bei Ladebeginn gestartet wird, die Ladung nicht unter 16 A stattfindet und der Ladevorgang nicht vorzeitig beendet werden kann.

3.3 Typ2-Steckdose - Variante B

Eine Sonderstellung nehmen die E-KFZ eines einzelnen Herstellers ein. Sie verhalten sich ähnlich wie die E-KFZ der Variante A, allerdings mit dem wesentlichen Unterschied, dass sie sich bis zu dreimal wecken lassen. Hierdurch ist eine Unterbrechung der Ladung bis zu zweimal möglich. Nach dem dritten Weckversuch fällt auch diese Fahrzeuggruppe in einen Fehlermodus.

Ein gesteuertes Laden ist möglich, wenn beachtet wird, dass die Kommunikation erst bei Ladebeginn gestartet wird und der Ladevorgang zweimal vorzeitig beendet werden kann.

3.4 Typ2-Steckdose - Variante C

Eine weitere Gruppe von E-KFZ weist schon beim normalen Laden an vielen Ladesäulen Probleme auf. Dies liegt daran, dass die Fahrzeuge gleich nach dem Kabeleinsteckvorgang das PWM-Signal auf dem CP-Kontakt erwarten. Steckt der Fahrer zuerst das Kabel am Auto und dann erst in der Ladestation ein, fällt diese Gruppe sofort in einen Standby-Modus.

Durch ein kurzzeitiges Anlegen einer Gleichspannung von 12 V auf dem CP-Kontakt lässt sich die Fahrzeuggruppe allerdings beliebig oft wecken und der Ladevorgang kann erneut gestartet werden. Bei diesen Fahrzeugen lässt sich ein gesteuertes Laden ohne Probleme realisieren.

3.5 Typ2-Steckdose - Variante D

Die letzte Fahrzeuggruppe weist keine Möglichkeit auf, eine zeitverzögerte Ladung zu realisieren. Sofern nicht innerhalb von einer Minute nach dem Abschließen des Fahrzeuges eine 230 V Spannung an der Typ2-Steckdose anliegt, fallen die Fahrzeuge in einen Standby-Modus, aus dem sie von der Ladestation nicht wieder geweckt werden können. In diesem Fall ist keine direkte Steuerung des Ladeverhaltens möglich, es kann lediglich eine Steuerung des Ladevorgangs durch eine Variation der Ladestromstärke erfolgen.

4 Beeinflussung der Ladeleistung

In Kapitel 2 ist bereits die Möglichkeit beschrieben, den Ladevorgang durch die Beeinflussung des PWM-Tastverhältnises zu steuern. Durch die in **Tabelle 2** angegebenen Werte lässt sich für viele Fahrzeuge der Ladestrom flexibel zwischen 6 A und 63 A einstellen, sofern Netzanschluss, Ladestation, Kabel und Fahrzeug dazu in der Lage sind. Problematisch ist aber, dass nicht alle Fahrzeuge in der Lage sind, diesen Stellbereich auszunutzen.

Unter den Fahrzeugen der Testgruppe befinden sich zum einen Fahrzeuge, die nicht unter 16 A bzw. 10 A betrieben werden können. Ist der Ladestation nicht bekannt, welche Regel Anwendung finden darf, fällt das Auto bei Unterschreitung des Mindestwertes in einen Fehlermodus, der sich in der Regel nur durch ein Entfernen und Neueinstecken des Anschlusskabels wieder zurücksetzen lässt.

5 Weitere Ergebnisse des Forschungsprojektes

In diesem Kapitel sollen kurz weitere Ergebnisse des Forschungsprojektes dargestellt werden. Zunächst wird der natürliche Lastgang der E-KFZ gezeigt, im Anschluss unter Eingriff des Energieversorgers, sofern die Kunden dem zustimmten.

5.1 Ungesteuerte Ladung der E-KFZ

Die erste 18-wöchige Projektphase, in der die Projektteilnehmer wie gewohnt und ohne finanzielle Anreise geladen haben, lieferte zunächst Referenzwerte für die anschließenden Projektphasen. Der mittlere Lastgang der E-KFZ aus dem Projekt ist in **Bild 3** dargestellt. Es zeigen sich ganz deutliche Unterschiede im Ladeverhalten zwischen den dreißig privaten und den zehn gewerblichen Fahrzeugen. Während die gewerblichen Fahrzeuge überwiegend werktags zwischen 8:00 Uhr und 16:00 Uhr geladen werden, laden die privaten Fahrzeuge überwiegend nachdem die Projektteilnehmer zu Hause angekommen sind. Deutlich ist dieses an der Lastspitze gegen 19:00 Uhr zu sehen. Auch wenn die durchschnittliche Ladeleistung für die dreißig privaten E-KFZ in diesem Zeitraum mit 13 kW sehr gering ist, treten dennoch an einzelnen Tagen Spitzen mit bis zu 40 kW auf. Dieser Wert muss zum einen mit der theoretisch möglichen Gesamtladeleistung von 200 kW und zum anderen mit der von den Projektteilnehmern tatsächlich gewählten Ladeleistung von etwa 140 kW gewichtet werden. Die geringere Ladeleistung ergibt sich, da viele Projektteilnehmer an ihrer Ladestation die maximale Ladeleistung begrenzen, um die Erwärmung und somit den Verschleiß der Akkumulatoren zu reduzieren. Einige Teilnehmer streben durch Anpassung der Ladeleistung ebenfalls eine Optimierung des Eigenverbrauchs ihrer PV-Anlagen an.

Zum Vergleich ist in **Bild 3** zusätzlich das H0-Lastprofil für Haushaltslasten darstellt. Deutlich ist zu sehen, dass es, wenn in einigen Jahren die von der Bundesregierung gewünschten 1 Mio. Elektrofahrzeuge auf deutschen Straßen unterwegs sind, ohne eine gezielte Steuerung der E-KFZ zu noch deutlicheren Lastspitzen im Verteilnetz kommen wird.

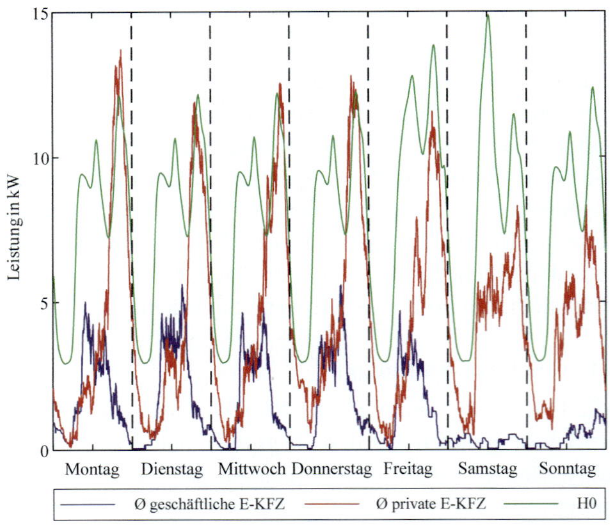

Bild 3: Lastgang des ungesteuerten Ladeverhaltens von privaten und geschäftlichen E-KFZ im Vergleich zum H0-Lastprofil während einer Woche

5.2 Zeitgesteuerte Ladung der E-KFZ

In der zweiten Projektphase hatten die Nutzer die Entscheidungsmöglichkeit, entweder das Fahrzeug wie gewohnt sofort zu laden oder zu fest vorgegebenen Ladezeiten durch den Energieversorger laden zu lassen. Diese Zeiten wurden so gelegt, dass in der Regel entweder die Sonne scheint (11:00–15:00 Uhr) oder die Strom gering ist (22:00–06:00 Uhr). In der Versuchsphase konnte gezeigt werden, dass die Nutzer eine hohe

Akzeptanz gegenüber den fest vorgegebenen Ladezeiten haben. Durch die hohe Nutzerakzeptanz und die einfache Zeitschaltung der Fahrzeuge kommt es zu besonders starken Lastspitzen der Fahrzeuge um 11:00 Uhr und 22:00 Uhr (siehe **Bild 4**). Es ist geplant, die Elektrofahrzeuge in der letzten Versuchsphase in das Virtuelle Kraftwerk einzubinden, um die optimierte Steuerung der Ladevorgänge entsprechend der Anforderungen des Regelenergiemarktes zu untersuchen. Dann wird sich eine Glättung des Lastgangs einstellen.

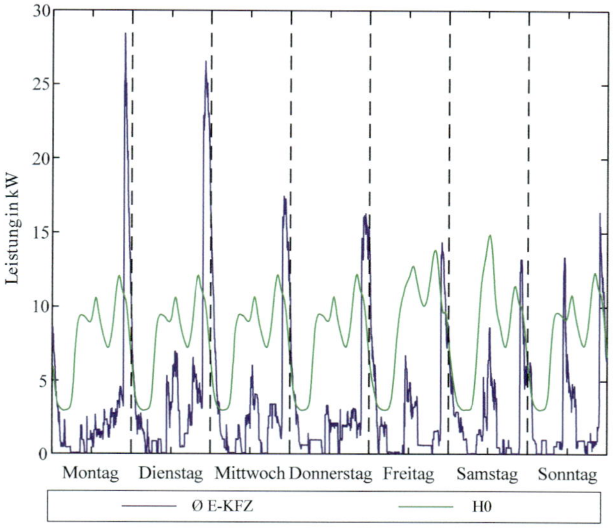

Bild 4: Durchschnittlicher Lastgang des gesteuerten Ladeverhaltens von E-KFZ im Vergleich zum H0-Lastprofil während einer Woche

In der dritten Projektphase werden die Ladezeiten jeweils am Vortag aus den prognostizierten Einspeiseleistungen von Wind- und PV-Strom abgeleitet und den Nutzer ab 15:30 Uhr bekannt gegeben. Ab einer bestimmten prognostizierten Einspeiseleistung werden die E-KFZ zum Laden freigeschaltet. Den Projektteilnehmern wird zur Sicherstellung der Ladung allerdings eine Mindestladezeit von 3 Stunden in der Nacht garantiert. Auch dieses noch flexiblere Lademodell wird von den Nutzern sehr gut angenommen, endgültige Ergebnisse liegen noch nicht vor.

In der letzten Projektphase wird ein Lademodell getestet, welches noch höhere Flexibilität von den Nutzern erfordert. Hierbei sollen Restriktionen erprobt werden, die die Bereitschaft der Fahrzeugnutzer zur Teilnahme am Regelenergiemarkt zeigen soll. Die Nutzer können beim Anschluss des E-KFZ an der Ladestation oder in der Handy-App einen Zeitpunkt in der Zukunft sowie den zu diesem Zeitpunkt gewünschten Ladestand des Akkus eingeben (z.B. am Folgetag um 7:00 Uhr mit 50 % Ladestand). Bis zu diesem Zeitpunkt steht es dem Energieversorger offen, wann das Fahrzeug geladen wird. Eine Entladung sollte theoretisch auch möglich sein, kann aber derzeit auf Grund der fehlenden technischen Voraussetzungen der Fahrzeuge noch nicht umgesetzt werden.

5.3 Verfügbarkeit der Fahrzeuge

Für die Einplanung der Fahrzeuge im virtuellen Kraftwerk ist es essenziell zu wissen, mit welcher Wahrscheinlichkeit sich die Fahrzeuge an der Ladestation befinden. Durch den Betrieb der Ladestation besteht in diesem Projekt die Möglichkeit, relativ genau zu erfassen, zu welchen Zeiten die Fahrzeuge verfügbar sind. In **Bild 5** ist dargestellt, zu welchen Zeiten die Fahrzeuge an der Ladestation verfügbar sind. Da auf Grund der am Anfang beschriebenen Problematiken nicht immer exakt festgestellt werden kann, ob ein Fahrzeug an der Ladestation angeschlossen ist, wird davon ausgegangen, dass ein E-KFZ zwischen dem Start und dem Ende eines Ladeprogramms vorhanden ist.

Die Auswertung der Datenreihe zeigt, dass wie erwartet die Fahrzeuge überwiegend nachts an der Ladestation stehen. Deutlich ist ebenfalls zu sehen, dass etwa ein Viertel der Fahrzeuge immer, allerdings äußerst selten mehr als 50 % der Fahrzeuge, verfügbar sind. Die niedrige Rate ist auch damit zu erklären, dass die Reichweite der Fahrzeuge für die meisten Wegstrecken mehr als ausreichend ist und die Nutzer daher nur jeden zweiten oder dritten Tag laden. Die höhere Verfügbarkeit am Wochenende liegt zum einen an den gewerblichen Fahrzeugen, die am Wochenende nicht genutzt werden, und zum anderen, dass am Wochenende auch die privaten Elektrofahrzeuge weniger gefahren werden.

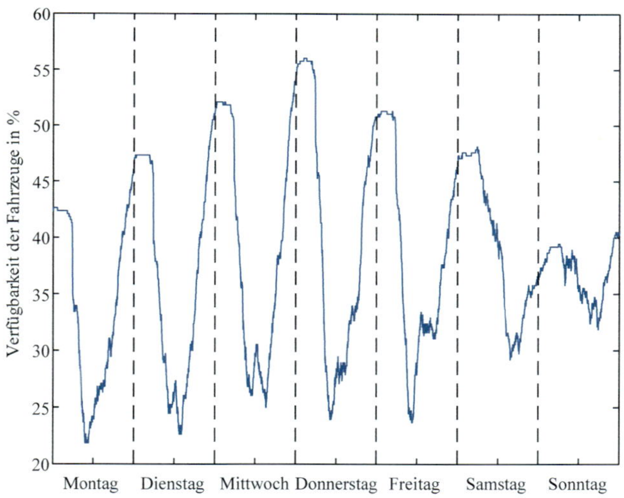

Bild 5: Durchschnittliche Verfügbarkeit der E-KFZ während einer Woche

6 Zusammenfassung

Der Einsatz von E-KFZ als Speicher für die Energiewende ist noch mit einigen technischen Herausforderungen verbunden. Während andere Forschungsprojekte wie [4] oder ähnliche mit speziell für diesen Zweck angepassten E-KFZ und Ladestationen arbeiten, können in dem diesem Beitrag zugrunde liegenden Feldversuch bereits nahezu alle im Feld vorkommenden Fahrzeuge zeitvariabel geladen werden, auch wenn dafür ein individuelles, fahrzeuggruppenspezifisches, Verhalten der Ladestationen erforderlich ist. Aufgrund der beschriebenen Problematiken ist es im Moment erforderlich, dass die Ladestation weiß, welcher Fahrzeugtyp angeschlossen ist. Dieses kann anhand des Ladeverhaltens des Fahrzeugs entweder automatisch erkannt oder durch Nutzereingabe im Vorfeld angegeben werden. Eine gezielte Entladung der Fahrzeuge ist zum heutigen Zeitpunkt nicht möglich.

In Zukunft werden viele E-KFZ mit der Norm ISO/IEC 15118 kompatibel sein. Diese definiert ein High-Level-Protokoll, welches für die Ladevorgangssteuerung und Abrechnung genutzt werden soll, womit sich viele der in diesem Beitrag geschilderten Herausforderungen lösen könnten.

7 Literatur

[1] IEC 61851: Elektrische Ausrüstung von Elektro-Straßenfahrzeugen - Konduktive Ladesysteme für Elektrofahrzeuge, Teil 1, 2012

[2] IEC 62196: Stecker, Steckdosen, Fahrzeugkupplungen und Fahrzeugstecker - Konduktives Laden von Elektrofahrzeugen, Teil 1-2, 2012

[3] IEC 60309: Stecker, Steckdosen und Kupplungen für industrielle Anwendungen, Teil 1-2, 2013

[4] Degner, T. et. al.: Intelligente Netzanbindung von Elektrofahrzeugen zur Erbringung von Systemdienstleistungen, Internationaler ETG-Kongress 2013, 05.-06.11.2013, Berlin, Deutschland

[5] Götz, A.: Innovative und effektive Integration von Elektrofahrzeugen ins Niederspannungsnetz, 13. Symposium Energieinnovation, 12.-14.02.2014, Graz, Österreich

Induktive Ladung mit hoher Leistung in schwachen Netzausläufern

Dipl.-Ing. Jonas Wussow, Prof. Dr-Ing. Bernd Engel, TU Braunschweig
Institut für Hochspannungstechnik und Elektrische Energieanlagen – elenia
Schleinitzstraße 23, 38106 Braunschweig
j.wussow@tu-braunschweig.de, Tel.: 0531/391-7707

Kurzfassung

Im Rahmen des Forschungsprojektes *emil* wurden in Braunschweig batteriebetriebene Elektrobusse in den öffentlichen Personennahverkehr (ÖPNV) integriert. Die Ladung dieser Elektrobusse erfolgt induktiv mit einer Ladeleistung von 200 kW. Dazu wurden bislang vier induktive Ladestationen installiert, deren Versorgung aus dem Straßenbahnnetz oder aus einer Kundenstation erfolgt. Zukünftig besteht die Möglichkeit, dass auch Strecken elektrifiziert werden, deren Endhaltestellen sich in suburbanen oder gar ländlichen Gebieten befinden. In diesen Bereichen müssen die Ladestationen in schwachen Netzausläufern integriert werden, ohne einen negativen Einfluss auf das Netz auszuüben. Durch die Verwendung von Stichleitungen von den Ortsnetzstationen zu den Ladestationen und durch die Bereitstellung von Blindleistung zur Spannungshaltung mit Hilfe eines aktiven Gleichrichters (Active-Front-End) können induktive Ladestationen mit hoher Ladeleistung zur Schnellladung in diesen Gebieten realisiert werden.

1 Einleitung und Motivation

Mit dem *Nationalen Entwicklungsplan Elektromobilität* wurden durch die Bundesregierung die Ziele formuliert, die Forschung, die Entwicklung, die Marktvorbereitung und die Markteinführung von batterieelektrisch betriebenen Fahrzeugen in Deutschland voranzubringen [1].

Im Jahr 2013 wurden 28,4 Mrd. Personenkilometer von ca. 40.000 Stadt- und Überlandbussen erbracht [2]. Dies verdeutlicht die Notwendigkeit, dass nicht nur Fahrzeuge des Individualverkehrs, sondern auch Fahrzeuge des ÖPNV bei der Thematik Elektromobilität betrachtet werden müssen. Im Gegensatz zu Oberleitungsbussen und schienengebundenem Verkehr, bei denen eine kontinuierliche Energieversorgung mit Oberleitungen erfolgt, sind batteriebetriebene Elektrobusse bislang nicht marktüblich. Die begrenzte Reichweite und die fehlende Möglichkeit zum schnellen Laden der Batterie waren bisher Gründe gegen die Elektrifizierung des ÖPNV. Zurzeit wird jedoch in Braunschweig die Möglichkeit einer induktiven Schnellladung mit 200 kW im Rahmen eines vom *Bundesministerium für Verkehr und digitale Infrastruktur* geförderten Forschungsprojektes[1] untersucht [3].

1.1 Die Forschungsprojekte emil und emilia

Bei den Forschungsprojekten *emil* (*Elektromobilität mittels induktiver Ladung*, offiziell *Primove*) und dem Folgeprojekt *emilia* (*Elektromobilität mittels induktiver Ladung im Automobil*, offiziell *InduktivLaden*) untersucht ein Konsortium bestehend aus der *Braunschweiger Verkehrs GmbH*, *BS|Energy*, drei Instituten der *Technischen Universität Braunschweig* und im Falle von emil zusätzlich noch *Bombardier* die Möglichkeiten zur induktiven Ladung von batteriebetriebenen Elektrobussen.

Ziel des Projektes emil ist die Überprüfung aller Facetten eines Linienbetriebs mit Elektrobussen, die stationär induktiv geladen werden. Für das Projekt wurde die Ringlinie *M19* ausgewählt (**Bild 1**). Diese zwölf Kilometer lange Linie wird wochentags im Zehn-Minutentakt befahren und besitzt eine hohe Auslastung, wodurch eine umfassende öffentliche Wahrnehmung gegeben ist.

Seit März 2014 verkehrt auf dieser Linie bereits der erste Elektrobus vom Typ *Urbino 12 electric* der Firma *Solaris*.

Das *Institut für Hochspannungstechnik und Elektrische Energieanlagen – elenia* hat im Rahmen des Projektes Netzanschlussmöglichkeiten untersucht, um eine wirtschaftliches und technisches Optimum zu finden.

Bild 1 Linienplan der Ringlinie M19 [4]

Beim Folgeprojekt emilia wird darüber hinaus eine gemeinsame Nutzung der induktiven Ladeinfrastruktur von Fahrzeugen des ÖPNV und des Individualverkehrs untersucht.

Die Schwerpunkte des elenias liegen auf der simulativen Untersuchung von Netzanbindungen und der Möglichkeit von induktivem Hochleistungsladen in schwachen Netzausläufern, wie sie in suburbanen und ländlichen Gebieten vorkommen.

1.2 Möglichkeiten zur Versorgung der Ladestation

Die Elektrobusse werden statisch (d. h. im Stillstand) geladen. Die benötigte Scheinleistung am Netzanschlusspunkt beträgt 260 kVA. Mögliche Ladezeiten variieren dabei zwischen 30 Sekunden und mehreren Minuten und sind u. a. davon abhängig, ob es sich um eine End- oder Zwischenhaltestelle handelt. Für die Versorgung der Ladestationen existieren verschiedene Möglichkeiten, die im Rahmen des Projektes emil ermittelt worden sind (siehe auch **Bild 2**) [5]:

a) Direkte Integration in das Niederspannungsnetz
b) Anbindung mit separater Leitung an das Niederspannungsnetz (Stichleitung), ggf. Verstärkung des vorhandenen Ortsnetztransformators
c) Anschluss an das Mittelspannungsnetz und Erstellung einer Kundenstation
   i. ausschließlich für die Ladestation
   ii. gemeinsame Nutzung mit anderen Verbrauchern
d) Anbindung an ein 600 V DC-Unterwerk
   i. vorhandenes Straßenbahnunterwerk
   ii. Erstellung eines neuen DC-Unterwerks ausschließlich für die Ladestation
e) Anbindung an eine vorhandene 600 V DC-Oberleitung

Die Vor- und Nachteile der einzelnen Versorgungsmöglichkeiten werden in [5] thematisiert.

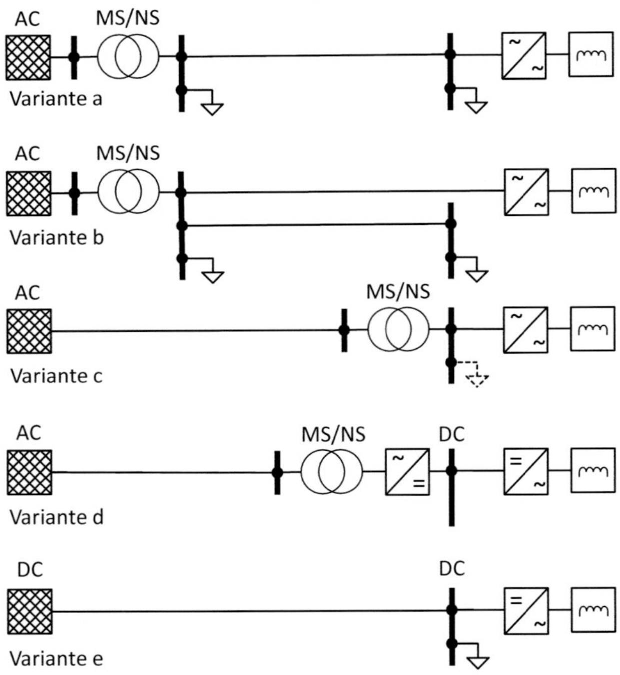

Bild 2: Ermittelte Möglichkeiten zur Versorgung der Ladestationen [5]

1.3 Versorgung der innerstädtischen Ringlinie

Für die Versorgung der Elektrobusse wurden im Rahmen des Projektes emil bislang drei Ladestandorte mit vier Ladestationen realisiert. Eine weitere Ladestation auf der Ringlinie befindet sich in Planung.

Unter Berücksichtigung der jeweiligen Gegebenheiten im Bereich der Standorte wurden die Netzanbindungen der Ladestationen festgelegt. Bei den Ladestation auf der Ringstrecke erfolgt die Versorgung über das Straßenbahnnetz (Variante d-i und Variante e) Die Ladestation im Busdepot wird über eine Kundenstation aus dem Niederspannungsnetz versorgt (Variante c-ii).

2 Netzanschlussherausforderungen bei weiteren Linien

Im Rahmen des Projektes emilia wurden Buslinien der Braunschweiger Verkehrs GmbH analysiert, die über ein Potential zur Elektrifizierung verfügen. Die Endstationen weiterer Linien liegen in suburbanen Gebieten. Mit wenigen Ausnahmen ist daher bei diesen Haltestellen das Straßenbahnnetz nicht in unmittelbarer Nähe. Die Versorgungsvarianten d und e sind in den meisten Gebieten somit nicht umsetzbar. Nachteile der Variante c sind hohe Kosten und ein hoher Platzbedarf. Zudem ließe sich eine solche Kundenstation, anders als beim Busdepot, nur für die Versorgung der Ladestation nutzen (Variante c-i), weil keine weiteren Verbraucher des Betreibers in unmittelbarer Nähe sind. Deshalb kann diese Möglichkeit der Versorgung ausgeschlossen werden. Die Netzanbindung für die induktive Ladung kann in solchen Fällen nur direkt im Niederspannungsnetz (Variante a) oder gegebenenfalls über eine Stichleitung zur jeweiligen Ortsnetzstation (Variante b) erfolgen.

Ein weiteres Problem stellen die Niederspannungsnetze dar, bei denen es sich um schwache Netzausläufer handelt. Die Netze sind durch eine geringe Nennleistung der Ortsnetztransformatoren und hohe Entfernungen zum Umspannwerk charakterisiert.

Die Lagen der potentiellen Ladestationen innerhalb der Netze können in drei Kategorien eingeteilt werden, die jeweils einer bestimmten Versorgungsvariante entsprechen:

- Potentielle Ladestation in unmittelbarer Nähe zur Ortsnetzstation (Variante b)
- Potentielle Ladestation innerhalb des Niederspannungsnetzes mit Distanz zur Ortsnetzstation (Variante a)
- Potentielle Ladestation in unmittelbarer Nähe zum Straßenbahnnetz (Variante d oder e)

Acht der untersuchten Netze entsprechen der Variante a. Bei dieser Versorgungsmöglichkeit ist der Einfluss der Ladestation auf das Netz am größten. Niederspannungsnetze sind gewöhnlich nicht für eine punktuelle Leistungsabnahme von 220 kW ausgelegt. Dadurch können sowohl große Distanzen zwischen Ortsnetzstation und Ladestation vorhanden sein als auch sind die Leiterquerschnitte sehr gering. Deshalb wurden die Netze dieser Variante mit Hilfe des Netzsimulationsprogramms *PowerFactory* untersucht. Die Netze unterscheiden sich in der Bemessungsleistung des Ortsnetztransformators, der Leitungslänge, der Anzahl der Verbraucher und der Anzahl und Größe dezentraler Erzeugungseinheiten.

2.1 Versorgung von vorstädtischen Endhaltestellen

Eine der potentiellen Strecken stellt die Linie 433 dar, deren Endhaltestellen in vorstädtischen Gebieten liegen. Diese Strecke ist knapp 21 km lang und wird werktags halbstündlich bedient. Sie bietet den Vorteil, dass sie auch über die Station Hamburger Straße verkehrt und somit die Möglichkeit zur Zwischenladung an einer bereits vorhandenen Ladestation besteht. Der Streckenverlauf ist in **Bild 3** dargestellt.

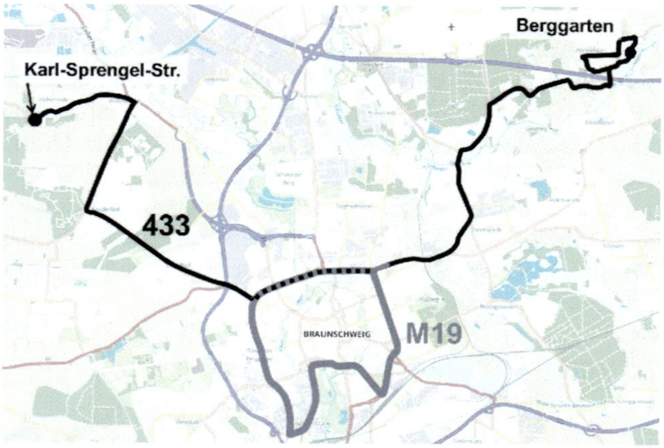

Bild 3: Streckenverlauf der Linie 433 Grafik aus [6], Linienverlauf nach [7]

Im Bereich der Endstation Berggarten (Hondelage) befindet sich eine Ortsnetzstation in unmittelbarer Nähe. Deshalb kann hier die Versorgung direkt über eine kurze Stichleitung zum Ortsnetztransformator erfolgen, ohne ins übrige Niederspannungsnetz einzugreifen (Variante b). Die Endstation Karl-Sprengel-Straße ist am Rand eines Wohngebietes. Die Integration erfolgt zunächst direkt im Niederspannungsnetz (Variante a). Die Leitungslänge zwischen der Ladestation und dem Ortsnetztransformator beträgt etwa 450 m. Die Leitungen weisen vor allem zum Ende des Wohngebietes nur sehr geringe Querschnitte auf, weil über diese Leitung nur wenige Haushalte versorgt werden.

Bei den Simulationen wurden den einzelnen Verbrauchern synthetische Lastprofile zugeordnet. **Bild 4** stellt die simulierten Spannungsverläufe für den Zeitraum 6:00 Uhr bis 19:00 Uhr dar. Die Spannung auf der Unterspannungsseite des Ortsnetztransformators variiert im Normalbetrieb (ohne Ladung) zwischen 395,5 V und 398 V. Der Spannungsverlauf am Ende des Wohngebietes, an der auch die Ladestation integriert werden könnte, variiert zwischen 392 V und 398 V.

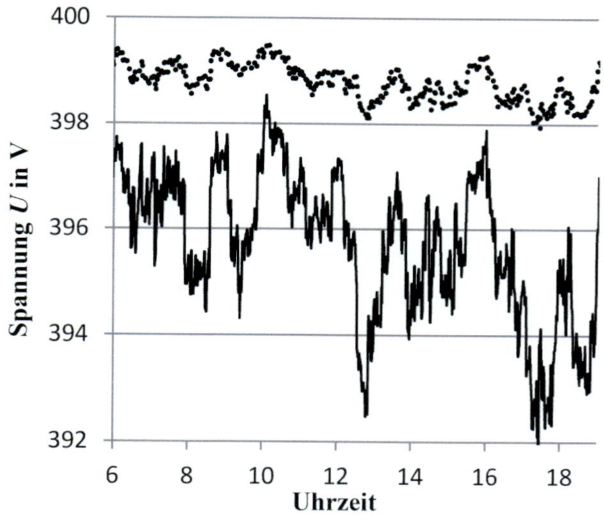

Bild 4: Spannung am Ortsnetztransformator, unterspannungsseitig (gepunktet), Spannung am Ende des Wohngebietes (durchgängig), ohne Ladung

Bei der Simulation mit integrierter Ladestation wurde angenommen, dass zu jeder vollen Stunde der Bus für sechs Minuten mit 200 kW geladen wird. Aufgrund des Eigenbedarfs der Leistungselektronik und aus thermischen Gründen wurde ein Verbraucher mit einer Gesamtleistungsaufnahme von 220 kW in die Simulation integriert. Die Ergebnisse sind für einen Zeitraum von 13 Stunden mit zwölf Ladungen in **Bild 5** dargestellt.

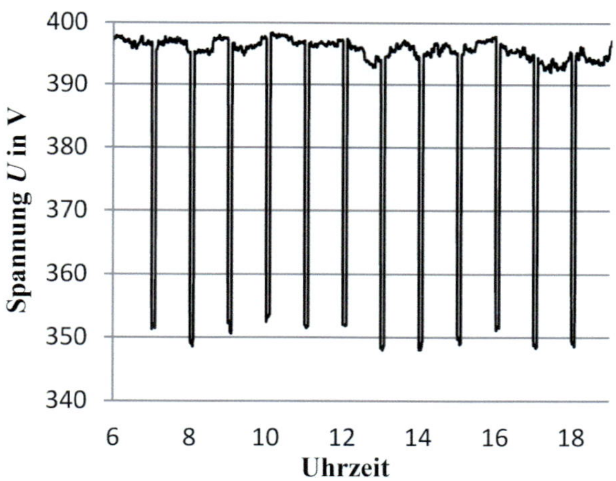

Bild 5: Spannungsverlauf an der Ladestation am Ende des Wohngebiets mit Ladevorgängen zu jeder vollen Stunde

Die Simulation zeigt, dass während des Ladevorgangs bei einer Versorgung direkt aus dem Niederspannungsnetz am Ende des Wohngebietes die Spannung auf Werte von unter 360 V fällt. Somit ist das erlaubte Spannungsband verletzt [8]. Eine derartige Versorgung der Ladestation ist demnach nicht möglich. Zusätzlich werden die Leitungen im Wohngebiet beim Ladevorgang mit ca. 220 % der maximal zulässigen Leistung betrieben, wodurch thermische Probleme auftreten und die Lebensdauer stark vermindert wird.

3 Lösungsansatz

Die Ergebnisse aus dem vorherigen Kapitel zeigen, dass eine Integration der Ladestation im betrachteten Niederspannungsnetz (Variante a) ohne weitere Maßnahmen auf Grund des zu hohen Spannungsfalls und der Überlastung der Zuleitung nicht möglich ist. Das Ziel ist, im Bereich dieser Endhaltestelle die induktive Hochleistungsladung zu ermöglichen, ohne dass negative Einflüsse auf das Netz und die Betriebsmittel auftreten.

Ein Lösungsansatz zur Spannungsstabilisierung stellt die Bereitstellung von Blindleistung am Netzanschlusspunkt der Ladestation dar. Für die leistungselektronische Spannungsregelung existieren bereits mehrere Verfahren für das Niederspannungsnetz [9]. Das Verfahren der Blindleistungsbereitstellung durch den Wechselrichter wird bei Photovoltaik-Wechselrichtern angewendet. In Analogie dazu stellt die Bereitstellung von Blindleistung durch den Umrichter eine Möglichkeit zur Integration der induktiven Ladestationen mit hoher Leistung dar. Dafür ist die Verwendung von aktiven Gleichrichtern (*Active Front End,AFE*) notwendig. Die Topologie der Ladestation ist in **Bild 6** dargestellt.

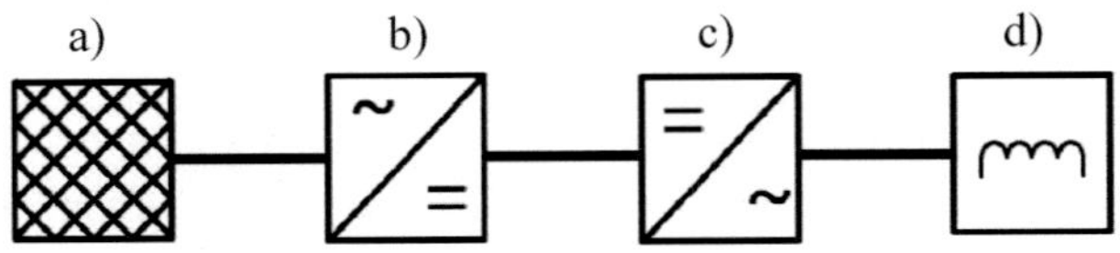

Bild 6: Topologie der Ladestation

Erklärung zur Topologie:

a) Niederspannungsnetz:
Versorgung der Ladestationen
b) Aktiver Gleichrichter/ Acitve-Front-End:
Gleichrichtung der Netzspannung
Bereitstellung von Blindleistung
c) Wechselrichter:
Umwandlung der Gleichspannung in hochfrequente Wechselspannung (20 kHz)
d) Primärspule:
Versorgung des Busses (Sekundärspule ist am Bus)

Als Ansatz für die Blindleistungsbereitstellung wird dem Gleichrichter der Ladestation ein fester Leistungsfaktor $\cos\varphi$ zugewiesen. Durch die Variation des Leistungsfaktors kann die Höhe der Blindleistung festgelegt werden, die am Netzanschlusspunkt der Ladestation zur Verfügung gestellt wird. Zur Verringerung des Spannungsfalls muss der Umrichter spannungshebend wirken. Der Leistungsfaktor muss daher im Bereich $\cos\varphi < 1$ (kap) liegen, um einen Beitrag zur Netzstabilität erbringen zu können.
Bei dem veränderten Leistungsfaktor wird weiterhin eine Wirkleistung von 220 kW von der Ladestation aufgenommen. Bei einem Leistungsfaktor von $\cos\varphi = 0{,}95$ (kap) beträgt die Scheinleistung 232 kVA. Daraus resultiert ein größerer Effektivwert des Stroms und somit eine höhere thermische Belastung der Betriebsmittel, als im Fall ohne Blindleistungsbereitstellung. Weil die Leitungen im Wohngebiet auch ohne Blindleistungsbereitstellung überlastet sind, ist die Versorgung der Ladestation so nicht möglich.

Damit auch die thermische Überlastung der Zuleitung verringert werden kann, bietet sich die Versorgung über eine etwa 300 m lange Stichleitung vom Ortsnetztransformator zur Ladestation (Variante b) an. Die Stichleitung müsste extra für die Versorgung verlegt werden. Dadurch ist eine unmittelbare Entkopplung mit dem übrigen Netz möglich. Ein erheblicher Spannungsfall innerhalb des Netzes aufgrund eines hohen Ladestroms ist somit ausgeschlossen.

Auch bei dieser Versorgung über die Stichleitung wird der Leistungsfaktor $\cos\varphi$ auf 0,95 (kap) festgelegt. In **Bild 7** ist der simulativ ermittelte Verlauf der Spannung am Ende des Wohngebietes dargestellt. In der Grafik werden die Versorgung über eine Stichleitung (Variante b) mit Blindleistungsbereitstellung als spannungsstabilisierende Maßnahme und die Versorgung durch das Wohngebiet (Variante a) ohne weitere Maßnahmen miteinander verglichen.

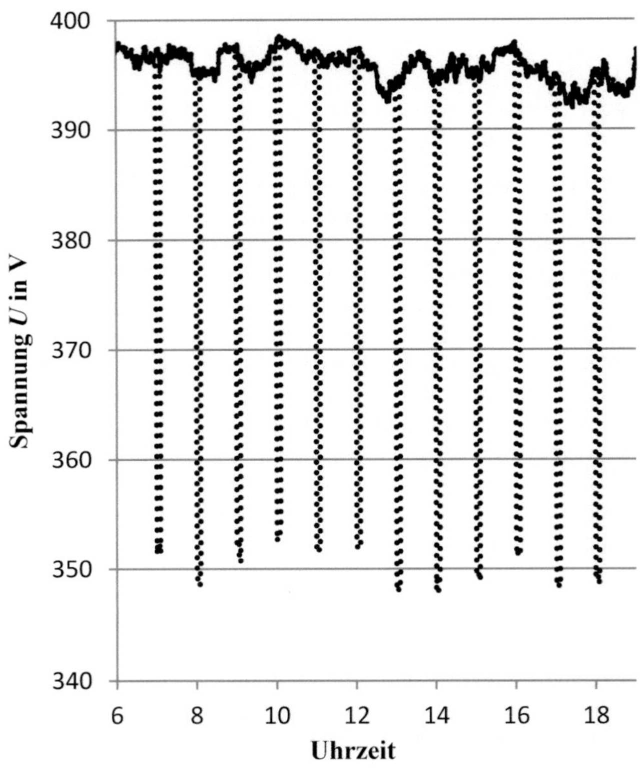

Bild 7: Spannung am Ende des Wohngebietes *mit* spannungsstabilisierenden Maßnahmen (durchgängig), Spannung am Ende des Wohngebietes *ohne* spannungsstabilisierende Maßnahmen (gestrichelt)

Die Ergebnisse der Simulation zeigen, dass bei der Verwendung einer Stichleitung und durch die Bereitstellung von kapazitiver Blindleistung durch den Gleichrichter der negative Einfluss der Ladung auf die Spannung innerhalb des Netzes kompensiert werden kann.

Während des Ladevorgangs bricht die Spannung mithilfe der spannungsstabilisierenden Maßnahmen nicht mehr zusammen. Der Spannungsverlauf ähnelt dem Verlauf für den Fall, dass keine Ladung erfolgt (vergleiche **Bild 4**).

Infolge der Blindleistungsbereitstellung kommt es während des Ladevorgangs bei dem verwendeten Leistungsfaktor nur zu minimalen Spannungserhöhungen an der Ortsnetzstation.

3.1 Herausforderungen

Ein Lösungsansatz für die Kompensation der negativen Einflüsse auf das Niederspannungsnetz wurde bereits vorgestellt. Aufgrund der Länge der Stichleitung und der Ladeleistung von etwa 232 kVA bestehen die Gefahren einer zu niedrigen Spannung an der Ladestation und einer Überlastung der Stichleitung.

Für die Stichleitung wurden Kabel vom Typ NAYY genutzt. Dabei wurden vier unterschiedliche Querschnitte betrachtet: 120 mm, 150 mm, 185 mm und 240 mm. Die Auswahl des Querschnitts hat keinen Einfluss auf die Spannung im Netz, sondern nur auf die Spannung an der Ladestation und der Auslastung der Stichleitung.

In **Tabelle 1** ist die maximale Auslastung in Bezug auf den jeweiligen Nennstrom der einzelnen Querschnitte und die minimale Spannung an der Ladestation während der Ladevorgänge beim betrachteten Szenario aufgeführt.

Tabelle 1 Spannung an der Ladestation und Auslastung der Stichleitung in Abhängigkeit des Kabeldurchmessers

Kabeldurch-Messer [mm]	Spannung an Ladestation [V]	Auslastung [%]
120	353	164
150	364	142
185	372	122
240	380	103

Die Ergebnisse zeigen, dass selbst bei einem Querschnitt von 240 mm die Auslastung beim Ladevorgang über 100 % beträgt. Das Spannungsband wird in diesem Szenario nur bei der Verwendung eines Kabels mit einem Querschnitt von 120 mm verletzt. Bei der Verwendung einer Stichleistung hat sowohl der Querschnitt als auch die Länge einen Einfluss auf die Spannung an der Ladestation. Die Ladeleistung ist festgesetzt. Daher ist bei geringerer Spannung an der Ladestation der Ladestrom und somit die Auslastung der Stichleitung folglich größer, damit die geforderte Ladeleistung bereitgestellt werden kann. Weil die Länge einer Stichleitung nicht verändert werden kann, muss die Auswahl des Querschnitts so erfolgen, dass sowohl der Spannungsfall über die Stichleitung als auch deren Auslastung möglichst gering, aber die Auswahl gleichzeitig noch weiterhin wirtschaftlich ist.

Die thermische Zeitkonstante liegt oberhalb der Ladedauer von sechs Minuten (150 mm und 185 mm [10]). Daher ist bislang unklar, welchen Einfluss die kurzzeitigen, sich wiederholenden elektrischen Überlastungen auf das thermische Verhalten des Kabels haben. Dieses Verhalten ist das entscheidende Kriterium für die Auswahl des Querschnittes der Stichleitung.

4 Fazit

Eine Integration von induktiven Ladestationen mit hohen Anschlussleistungen (hier 220 kW) ist in Niederspannungsnetzen mit schwachen Netzausläufern nicht ohne geeignete Maßnahmen möglich. Werden diese nicht ergriffen, kann das Spannungsband verletzt und die Betriebsmittel überlastet werden.

Mithilfe einer Stichleitung zwischen der Ladestation und der Ortsnetzstation sowie der Bereitstellung von Blindleistung durch den aktiven Gleichrichter, Active-Front-End, (kapazitiver Betrieb) können die Spannungseinbrüche innerhalb des Niederspannungsnetzes kompensiert werden. Durch diese Maßnahmen entspricht die Spannung während des Ladevorgangs annähernd der Spannung im Normalbetrieb ohne Ladung. Durch die Verwendung der Stichleitung werden thermische Probleme innerhalb des Netzes vermieden. Eine derartige Versorgung stellt somit eine kostengünstige Möglichkeit des induktiven Hochleistungsladens in schwachen Netzausläufern dar. Damit ist der Einsatz von batteriebetriebenen Elektrobussen auch in suburbanen und ländlichen Gebieten aus netztechnischer Sicht denkbar.

5 Ausblick

Bei der Versorgung über die Stichleitung wird das Kabel aufgrund der hohen Ladeleistung elektrisch überlastet. In einem nächsten Schritt sollen die kurzzeitigen, wiederkehrenden Überlastungen und der Einfluss auf das thermische Verhalten der Kabel untersucht werden.

Unterschiedliche Verhältnisse zwischen Ladedauer (Überlastung) und Zeit ohne Ladung (Leerlauf) sollen dabei analysiert werden. Mithilfe der Ergebnisse soll eine Aussage getroffen werden können, ob die kurzzeitigen Überlastungen aus thermischer Sicht vertretbar sind und somit größere Querschnitte vermieden werden können.

6 Literatur

[1] Nationaler Entwicklungsplan Elektromobilität der Bundesregierung, 2009

[2] Verband Deutscher Verkehrsunternehmen (VDV): ÖPNV | Schienengüterverkehr Daten und Fakten 2012/2013, Köln, VDV, 2013, http://www.vdv.de/statistik-personenverkehr.aspx

[3] Braunschweiger Verkehrs GmbH: Braunschweig fährt jetzt induktiv! (Infoflyer), http://www.verkehr-bs.de/fileadmin/user_upload/downloads/Emil/A5_emil_Flyer_WEB.pdf; 17.02.2015

[4] Braunschweiger Verkehrs GmbH, http://www.verkehr-bs.de/unternehmen/ forschungs-projekt-emil/technik.html; 13.02.2015

[5] Soyck, F.; Laudahn, S., Engel, B.: Aktive Spannungsstützung beim Laden von Elektrofahrzeugen mit großer Leistung, ETG-Kongress, Berlin, 2013

[6] Openstreetmaps, **www.openstreetmap.de**

[7] Braunschweiger Verkehrs GmbH: Liniennetzplan kartografisch, http://www.verkehr-bs.de/fileadmin/user_upload/downloads/Liniennetzplan/Liniennetzplan_ab_28.11.14/Gesamtnetz_Verkehrs-GmbH_Stand_November_2014.pdf

[8] DIN EN 50160:2011-02: Merkmale der Spannung in öffentlichen Elektrizitätsversorgungsnetzen

[9] Kruschel, W.; da Costa, J. P.; Sahan, B.; Zacharias, P.; Bülo, T.; Engel, B.: Dezentrale Spannungsregelung in intelligenten Niederspannungsnetzen – Leistungselektronische Komponenten zur Erhöhung der Aufnahmekapazität für erneuerbare Erzeugungseinheiten. ETG-Kongress, Würzburg, 2011

[10] Kerber, Georg: Aufnahmefähigkeit von Niederspannungsverteilnetzen für die Einspeisung aus Photovoltaikkleinanlagen. Dissertation, Technische Universität München, 2011

---

[1] Gefördert vom Bundesministerium für Verkehr und digitale Infrastruktur aufgrund eines Beschlusses des Deutschen Bundestages vertreten durch den TÜV Rheinland. FKZ: 16SNI006C

Hybridantriebskonzepte auf Schiffen

Prof. Dr.-Ing. Thorsten Völker, Hochschule Bremen, Bremen, Deutschland, (thorsten.voelker@hs-bremen.de)
Komplexlabor Elektrische Energietechnik, Am Neustadtswall 30, 28199 Bremen

Kurzfassung

Die Hybridantriebstechnik ist nicht nur im Bereich der Elektrofahrzeuge ein aktuelles Thema. Auch im Bereich von Schiffen gibt es zurzeit an mehreren Stellen Überlegungen die bewährten diesel-mechanischen Antriebe durch neuartige Antriebskonzepte zu ersetzen.
In diesem Paper wird die Realisierbarkeit von Hybridantriebskonzepten auf unterschiedlichen Schiffstypen untersucht. Das Paper zeigt, dass es durchaus Einsparpotentiale durch den Einsatz von Hybridantriebskonzepten auf Schiffen gibt. In einigen Fällen können dadurch sogar die wesentlich höheren Investitionen durch Einsparung von Treibstoff kompensiert werden. Jedoch gibt es Einschränkungen im Vergleich zu Elektrofahrzeugen.

Abstract

The hybrid technology is not only a current issue in the field of electric vehicles. Also in the area of ships there are some projects to replace the proven diesel mechanical drives by innovative drive concepts.
In this paper the feasibility of hybrid propulsion concepts is examined for different types of vessels. The paper shows that there are definitely opportunities to save energy by using hybrid drive systems on ships. In some cases this can even return the much higher investment by savings in fuel. But there are some restrictions compared to electric vehicles.

1 Einleitung

In letzter Zeit wird für Schiffe immer häufiger in hybrides Antriebskonzept in Betracht gezogen. Die Auslöser sind zum einen der ständig steigende Preis für fossile Treibstoffe und zum anderen eine strengere Abgasnorm in sogenannten Sulphur-Emission-Control-Areas (SECAs) ab dem 1. Januar 2015 [9]. Danach dürfen in vielen Küstenregionen nur noch Kraftstoffe mit einem Schwefelgehalt von 0,1 % verwendet werden, was zusätzliche Kosten für den Bezug von Treibstoff bedeutet.

Der größte Unterschied beim Einsatz von Elektroantrieben auf Schiffen im Vergleich zu Elektroantrieben in PKWs ist die Rückgewinnung von Bremsenergie. Der fehlende direkte Kraftschluss zwischen der Schiffsschraube und dem Wasser erschwert die Rückgewinnung von Energie während des Bremsvorgangs bei Schiffen. Dies ergibt bei Schiffen wesentlich geringere Einsparpotentiale durch den Einsatz hybrider Antriebskonzepte im Vergleich zu Straßenfahrzeugen.

In diesem Paper wird die Realisierbarkeit von Hybridantriebskonzepten auf unterschiedlichen Schiffstypen überprüft.

Für Schiffe, die den überwiegenden Teil der Fahrt mit konstanter Geschwindigkeit zurücklegen, wie z. B. Containerschiffe, ergeben hybride Konzepte keinen Sinn, da ein diesel-mechanischer Antrieb für diesen Betrieb optimiert werden kann und dann immer einen niedrigeren Verbrauch hat als ein diesel-elektrisches oder hybrides System. Daher werden in diesem Paper mögliche zukünftige Hybridantriebskonzepte für die speziellen Antriebskonzepte eines Hafenschleppers und einer Motorfähre untersucht. Die untersuchten Konzepte bestehen aus einem Verbrennungsmotor, einem Generator, den elektrischen Antrieben und Akkus als elektrischer Energiespeicher.

Es erfolgt zunächst die Auswertung der Ist-Zustände der untersuchen Schiffe und die Erstellung eines beispielhaften Fahrprofils basierend auf Messungen vor Ort.

Schwerpunkt bei den weiteren Untersuchungen war der Treibstoffverbrauch der unterschiedlichen Antriebskonzepte und die daraus berechneten Kosten bezogen auf einen Fahrzyklus. Es werden unterschiedliche Hybridantriebskonzepte untersucht und mit dem klassischen dieselmechanischen Antrieb verglichen.

Die möglichen Einsparpotentiale liegen in der Optimierung des Betriebspunktes des Dieselmotors bezüglich Drehzahl und Leistung. Ein damit verbundener günstiger spezifischer Brennstoffverbrauch bietet dabei die Möglichkeit einer Energieeinsparung.

Dies erfordert auch ein Umdenken bei den Herstellern von Verbrennungsmotoren, da die Motoren dann nicht auf einer konstanten Drehzahl betrieben werden, sondern bei einer geforderten Antriebsleistung nun mit der optimalen Drehzahl betrieben werden können. Dazu wird jedoch eine zweidimensionale Kennlinie für den spezifischen Treibstoffverbrauch in Abhängigkeit von Drehzahl und Leistung benötigt.

Das Paper zeigt, dass es durchaus Einsparpotentiale durch den Einsatz von Hybridantriebskonzepten auf Schiffen gibt. In einigen Fällen können dadurch sogar die wesentlich höheren Investitionen durch Einsparung von Treibstoff kompensiert werden. Durch den in Zukunft wohl weiter steigenden Treibstoffpreis und den wahrscheinlich sinkenden Preis für Akkus werden solche hybriden Antriebskonzepte in Zukunft noch attraktiver.

2 Hafenschlepper

Zunächst wird ein hybrides Antriebskonzept für einen Hafenschlepper untersucht.

2.1 Beschreibung Hafenschlepper

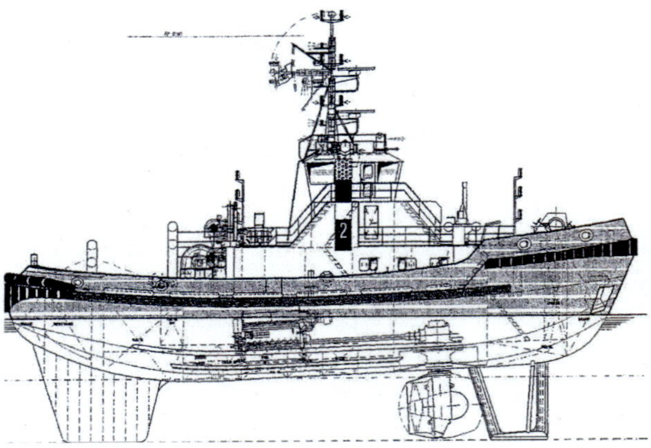

Bild 1 Traktorschlepper mit Ruderpropellern [Bugsier]

Der untersuchte Hafenschlepper ist ein Traktorschlepper. Das heißt er besitzt zwei schwenkbare Ruderpropeller im vorderen Teil der Schiffes (siehe **Bild 1**). Die Hauptaufgabe eines Hafenschleppers ist das Begleiten bzw. Schleppen von großen Schiffen wie z. B. Containerschiffen in den Hafenbereich. Ein beispielhaftes Lastprofil eines typischen Schleppereinsatzes ist im folgenden Kapitel gezeigt.

2.2 Lastprofile Hafenschlepper

Ein beispielhaftes Lastprofil ist in **Bild 2** zu sehen.
Dieses Lastprofil entspricht einem realen Schleppeinsatz.

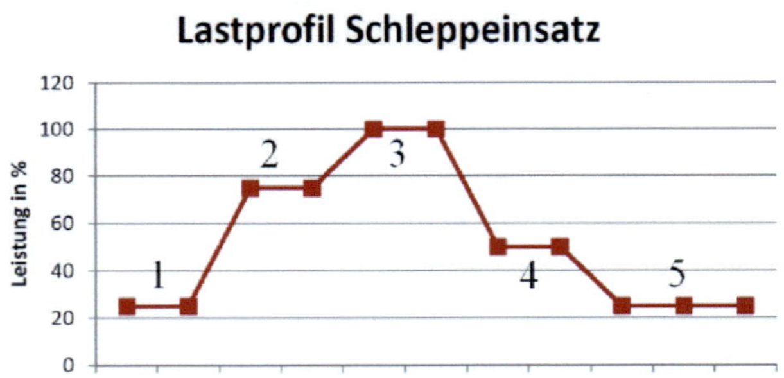

Bild 2 Lastprofil Schleppeinsatz eines Hafenschleppers

Lastprofil von Bild 2:
1: Anfahrt (25 % Last),
2: Drehen (75 % Last),
3 Manöver (85 %–100 % Last),
4: Assistenz (50 % Last),
5: Rückfahrt (25 % Last).

Die prozentualen zeitlichen Last-Anteile sind im folgenden Diagramm in **Bild 3** dargestellt.

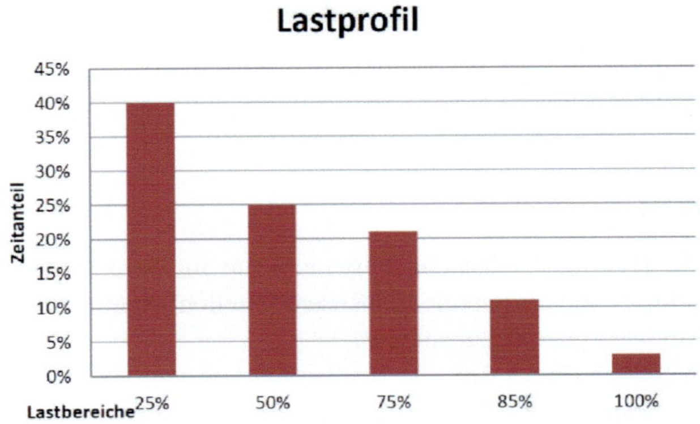

Bild 3 Lastprofil Schleppeinsatz eines Hafenschleppers

Wie dem Lastprofil aus **Bild 3** zu entnehmen ist, sind beim Hafenschlepper überwiegend Zeiten in Teillast vorhanden. Gerade dieser hohe Anteil an Teillast ermöglicht Einsparpotentiale durch den Einsatz eines hybriden Antriebssystems. Im folgenden Kapitel werden mögliche Antriebskonzepte dargestellt. Bei der Verwendung von mehr als einem Verbrennungsmotor lassen sich im Teillastbetrieb gegebenenfalls einzelne Verbrennungsmotoren abschalten oder die verbleibenden Motoren in einem verbesserten Wirkungsgrad betreiben.

2.3 Antriebskonzepte

Im Folgenden werden mögliche Konzepte für einen Hybrid-Betrieb von diesel-elektrischen Antrieb mit Akkus vorgestellt. **Bild 4** zeigt das untersuchte hybride Antriebskonzept. Dabei wird die Energie über zwei Synchrongeneratoren, die von Dieselmotoren angetrieben werden, in eine Gleichstromschiene gespeist.

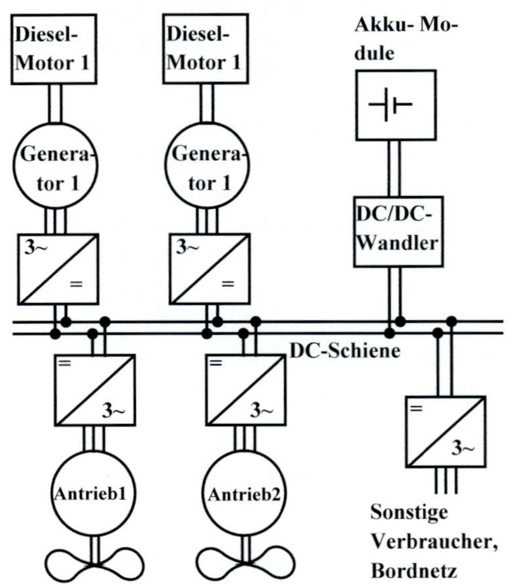

Bild 4 Hybrides Antriebskonzept

An diese Gleichstromschiene sind nun zum einen die Antriebsmaschinen angeschlossen, die über jeweils einen Wechselrichter angesteuert werden. Außerdem sind Akku-Module an die Gleichstromschiene angeschlossen.

Über einen DC/DC-Wandler kann nun Energie in den Akkus gespeichert werden, oder aus den Akkus in die Gleichstromschiene gespeist werden.

Das Einsparpotential eines hybriden Antriebssystems auf Schiffen liegt nun darin, dass der verwendete Dieselmotor in Betriebs-Bereichen betrieben werden kann, in denen der spezifische Brennstoffverbrauch geringer ist.

Außerdem können unter Umständen einer oder beide Diesel Generatoren abgeschaltet werden, und die gespeicherte Energie kann genutzt werden. In **Bild 5** ist der spezifische Brennstoffverbrauch eines Dieselmotors der Marke Wärtsilä 6 L 26 (2040 kW) zu sehen.

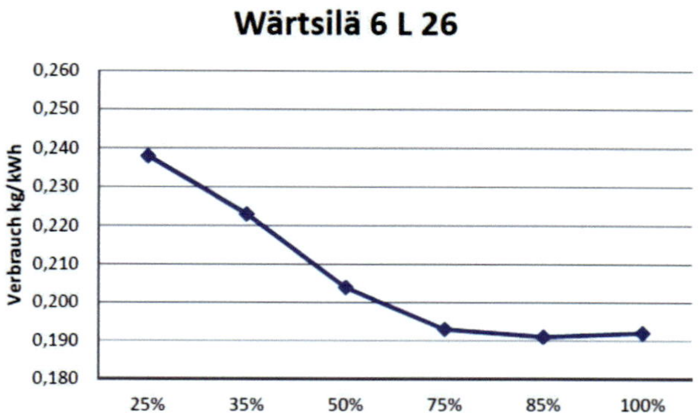

Bild 5 Spezifischer Brennstoffverbrauch

Es ist zu erkennen, dass der spezifische Brennstoffverbrauch im Teillastbereich im Vergleich zu Volllast um ca. 25 % steigt. Die Grafik in **Bild 5** gilt bei konstanter Drehzahl. Bei variabler Drehzahl, die bei dem in **Bild 4** beschriebenen Konzept möglich ist, können die Unterschiede im spezifischen Brennstoffverbrauch noch größer sein.

2.3.1 Auslegung der Komponenten

Für eine realistische Abschätzung werden die einzelnen Komponenten am Beispiel eines realen Schleppers ausgelegt.

2.3.1.1 Antriebsmotoren und Nebenverbraucher

Für die erforderliche Zugkraft muss der Schlepper mit zwei Antrieben der Leistung je 2400 kW ausgestattet sein. Die Leistung der Nebenverbraucher beläuft sich auf ca. 200 kW. Das macht eine maximale Leistung von 5000 kW.

2.3.1.2 Dieselmotoren und Akkus

Da der Lastbereich von 25 % im Betrieb häufig vorkommt, wird die Leistung der Akkus auf diesen Wert abgestimmt. Diese Leistung soll von den Akkus für eine Dauer von einer Stunde bereitgestellt werden können. Daher haben die Akkus eine Leistung von ca. 1250 kW und eine Energie von 1250 kWh.

Diese Dieselmotoren müssen daher mindestens 75 % der maximalen Leistung aufbringen können, damit im Extremfall die volle Zugkraft bereit steht.

Für die Motoren wurden 2 Motoren der Wärtsilä 6 L 26 mit einer Leistung von 2040 kW ausgewählt.
Für die Spitzenleistung wird zusätzlich der Energiespeicher genutzt.

Für die einzelnen Komponenten wurden die folgenden Wirkungsgrade angenommen:

Tabelle 1: Wirkungsgrade [10], [11]

	Verluste	Wirkungsgrad
Generator	3,5%	96,5%
Batterie laden	4%	96%
Batterie entladen	4%	96%
IGBT	1,5%	98,5%

2.3.1.3 Investitionskosten

Die Investitionskosten für die unterschiedlichen Antriebskonzepte sind in **Tabelle 2** im Vergleich zu den Kosten des diesel-mechanischen Konzepts dargestellt [10].

Bei diesen Vergleichen wird von 2500 Betriebsstunden (bh) jährlich ausgegangen. Die Kraftstoffkosten werden mit 0,78 Euro/kg angenommen. Für die Kosten pro kWh Strom wurden ein Industriepreis von 9,11 Cent angenommen (Preisauskunft Industriebetriebe Bremen).

Tabelle 2 Vergleich der Betriebskosten

	Diesel-elektrisch	Hybrid	Hybrid mit Landstrom
Investitionskosten (€) im Vergleich zu diesel-mech.	+2.000.000	+2.823.400	2.823.400
Investitionskosten (€) im Vergleich zu diesel-elektr.	-	+823.400	+823.400
Kosten (€) pro Betriebsstunde	413	406	379
Jährliche Betriebskosten (€)	1.033.500	1.015.950	947.500
Ersparnis zu diesel-mech. (Amortisation)	29.250 (68 Jahre)	46.800 (60 Jahre)	115.250 (24,5 Jahre)
Ersparnis zu diesel-elektr. (Amortisation)	-	17.550 (47 Jahre)	68.450 (12 Jahre)

In **Tabelle 2** ist der Vergleich der unterschiedlichen Antriebskonzepte zum einem gegenüber dem dieselmechanischen Konzept sowie dem diesel-elektrischen Konzept zu sehen.

Durch die Reduzierung des spezifischen Treibstoffverbrauchs der Dieselmotoren ergibt sich gegenüber dem diesel-mechanischen Konzept bei 2500 Betriebsstunden im Jahr eine jährliche Ersparnis von 29.250 € beim dieselelektrischen Konzept und 46.800 € ohne Landstromladung bzw. 115.250 € mit Landstromladung bei den hybriden Konzepten. Bei der Nutzung von Landstrom zur Ladung der Akkus ist die Ersparnis also doppelt so groß. Die Amortisationszeiten im Vergleich zum diesel-mechanischen Konzept liegen weit über der zu erwartenden Lebensdauer der Systeme.

Dagegen fällt der Vergleich zum reinen diesel-elektrischen besser aus. Hier ist die Amortisationszeit des hybriden Konzeptes bei 47 Jahren und beim hybriden Konzept mit Landstromnutzung bei 12 Jahren. Es lässt sich damit zusammenfassen, dass der Einsatz eines hybriden Antriebssystems bei einem Schlepper nur in Frage kommt, wenn ohnehin ein diesel-elektrisches System geplant ist, oder bereits vorliegt. Eine realistische Amortisationszeit ist hier nur mit Nutzung der Landstromversorgung zur Ladung der Akkus möglich.

3 Motorfähre

Als weiterer Schiffstyp wird in diesem Paper ein hybrides Antriebskonzept für eine Motorfähre untersucht.

3.1 Beschreibung Motorfähre

Die hier untersuchte Motorfähre NORDENHAM (kurz: MF NORDENHAM) ist eine Doppelendfähre im Besitz der Weserfähre GmbH, Bremerhaven und verkehrt als Binnenschiff auf der Weser (**Bild 6**).

Bild 6 Motorfähre Nordenham [Quelle: Jürgen Winkler]

Die Motorfähre wird zurzeit über zwei Voith-Schneider-Propeller angetrieben, die von zwei Dieselmotoren direkt angetrieben werden.

Die MF NORDENHAM verkehrt auf der Weser zwischen der Stadt Bremerhaven und dem Stadtteil Blexen der Stadt Nordenham (**Bild 7**). Die Überfahrt inklusive Liegezeit dauert jeweils ca. 20 min. Die Entfernung beträgt ca. 3 km.

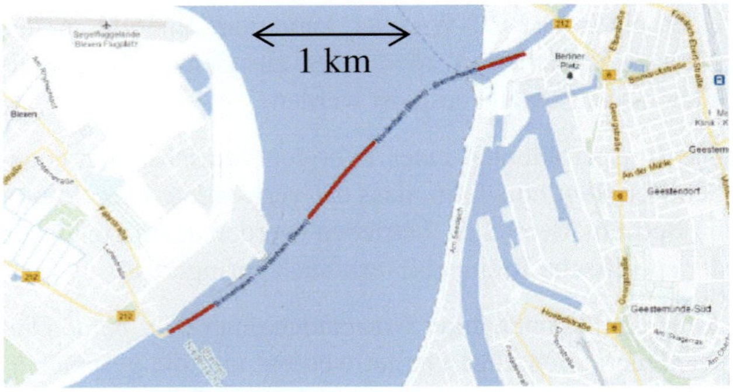

Bild 7 Fährverbindung der Motorfähre Nordenham

In der folgenden **Tabelle 3** ist der Leistungsbedarf für die Weserfähre aufgeführt. Die zwei Voith-Schneider-Propeller benötigen jeweils eine Antriebsleistung von 470 kW und für die Verbraucher im Bordnetz sind ca. 50 kW erforderlich.

Tabelle 3 Leistungsbedarf

Verbraucher	Leistungsbedarf
Antriebe	2 x 470 kW
Bordnetz	50 kW
Summe	990 kW

3.2 Antriebskonzepte

Im Folgenden werden vier unterschiedlich Antriebsvarianten untersucht. Alle Konzepte beinhalten einen Gasmotor als Antrieb und Akkus als zusätzlichen Energiespeicher.

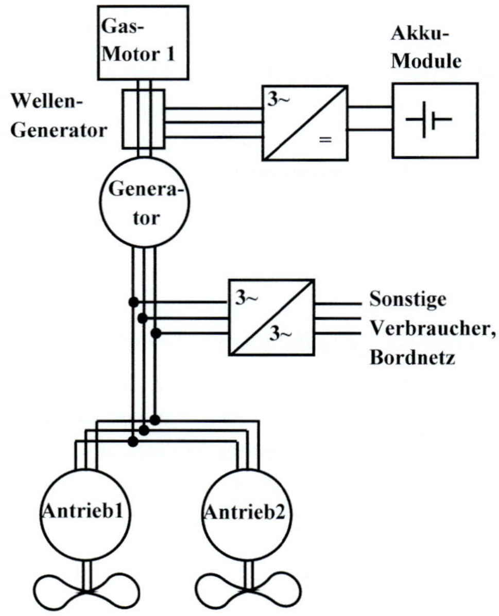

Bild 8 Antriebsvarianten Konzept 1 und 2

Die Konzepte unterscheiden sich darin, dass beim Konzept 1 und 2 (**Bild 8**) ein Drehstromnetz als Verteilungsnetz an Bord vorliegt und bei Variante 3 und 4 (**Bild 9**) ein Gleichstromnetz. Außerdem unterscheidet sich die Ankopplung der Akkus. Bei den Konzepten 1 und 2 erfolgt die Ankopplung durch einen zusätzlichen Antrieb, der an die mechanische Welle zwischen dem Gasmotor und dem Synchrongenerator angekoppelt ist. Bei den

Konzepten 3 und 4 erfolgt die Ankopplung der Akkus über einen DC/DC-Wandler direkt an die Gleichstromschiene.

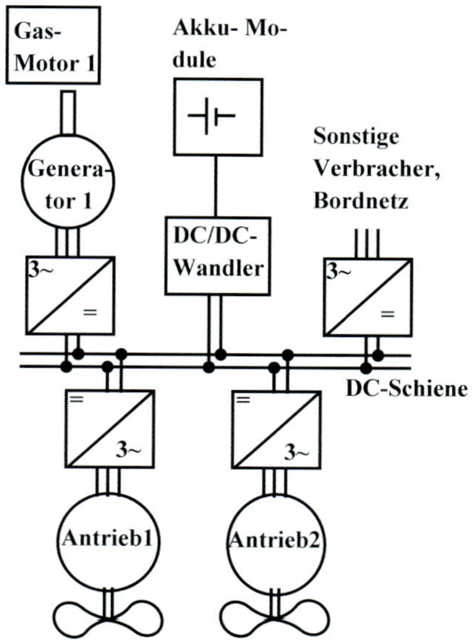

Bild 9 Antriebsvarianten Konzept 3 und 4

In der folgenden Tabelle 4 sind die Leistungen von Gasmotor und Akku-System sowie die Kapazität der Akkus der unterschiedlichen Antriebsvarianten zu sehen.

Tabelle 4 Leistungen der einzelnen Antriebsvarianten

	Variante 1	Variante 2	Variante 3	Variante 4
$P_{Gasmotor}$	850 kW	1200 kW	850 kW	1200 kW
P_{Akku}	83 kW	1040 kW	97 kW	981 kW
E_{Akku}	35 kWh	420 kWh	40 kWh	400 kWh

3.3 Lastprofil Motorfähre

Um ein Antriebskonzept entsprechend des Leistungsbedarfs der MF NORDENHAM auslegen zu können, ist es notwendig das aktuelle Fahrverhalten auszuwerten.
Zunächst wird ein typisches Lastprofil für eine Überfahrt erstellt, die insgesamt ca. 20 Minuten dauert.

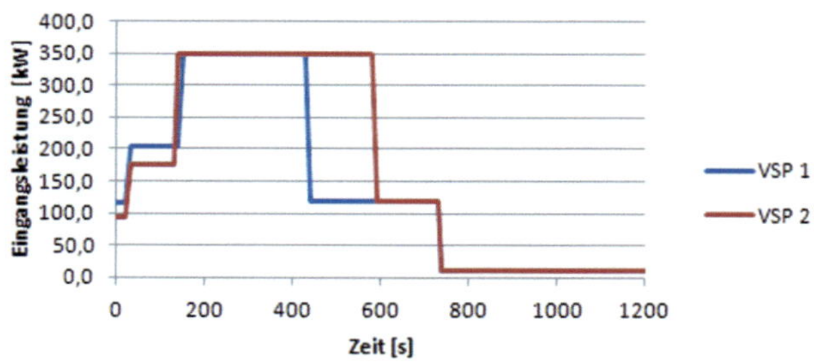

Bild 10a Eingangsleistung [kW] der Voith-Schneider-Propeller (VSP) Überfahrt Bremerhaven-Blexen

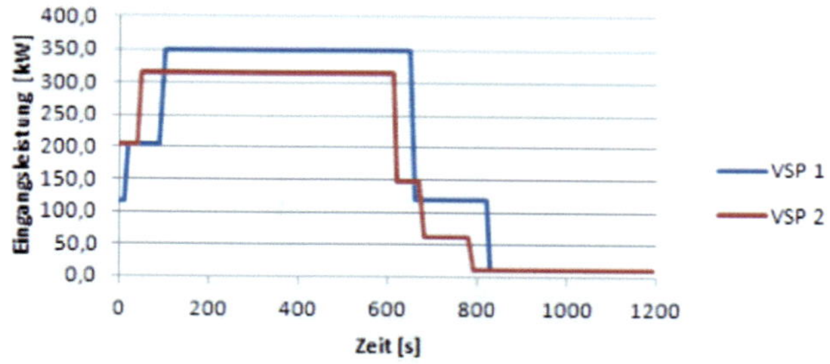

Bild 10b Eingangsleistung [kW] der Voith-Schneider-Propeller (VPS) Überfahrt Blexen-Bremerhaven

Die **Bilder 10a** und **10b** zeigen typische Lastprofile für eine Überfahrt von Bremerhaven nach Blexen und zurück. Zunächst wird jeweils mit Teillast in der Nähe des Anlegers manövriert. Nach 2–3 min wird zur Überfahrt auf volle Leistung erhöht.

Nach ca. 10 min wird die Leistung beim Annähern an den Anleger reduziert. Während die Fähre am Anleger liegt, ist die Leistung auf ca. 10 % reduziert.
Die prozentualen Anteile der Lastzustände sind im **Bild 11** zu erkennen.

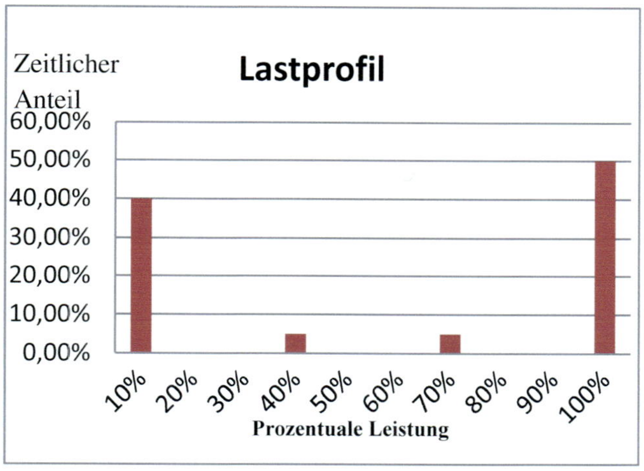

Bild 11 Lastprofil von einem der Dieselmotoren der MF NORDENHAM

Im folgenden **Bild 12** sind die Verbrauchskosten der unterschiedlichen hybriden Antriebskonzepte im Vergleich zum mechanischen System zu sehen, der als Ist-Zustand beschrieben ist.

Ohne Betrachtung der höheren Investitionskosten zeigt sich hier bereits, dass keine der betrachteten hybriden Antriebsvarianten bei dem hier untersuchten Lastprofil eine Ersparnis mit sich bringt.

Es ist sogar so, dass bei allen vier Varianten mit einem Mehrverbrauch zu rechnen ist, der gerade bei den Varianten mit einer großen Akku-Kapazität bei ca. 20 % liegt.

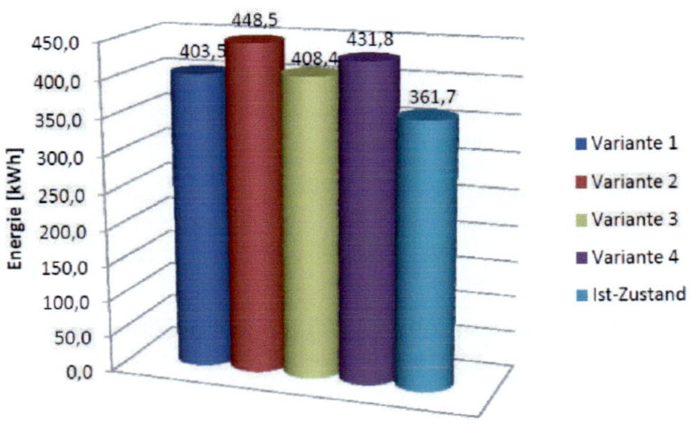

Bild 12 Vergleich der Antriebsvarianten

Die Hauptursache für den höheren Energiebedarf der hybriden Antriebskonzepte im Vergleich zum mechanischen System ist der hohe Volllastanteil des Fahrprofils der Motorfähre. Wenn der Verbrennungsmotor hier bei optimalem spezifischen Verbrauch betrieben wird und die gesamte Leistung bei den hybriden Konzepten zusätzliche Verluste bei Generator, Elektromotor und ggf. bei Gleichrichter und Wechselrichter verursacht, kann sich unter Volllast ein solches System nicht rechnen. Eine Ladung der Akkus mit Landstrom ich bei den kurzen Liegezeiten nicht möglich, so dass auch hierdurch keine Verbesserung möglich ist.

4 Zusammenfassung

Zusammenfassend lässt sich sagen, dass der Einsatz von hybriden Antriebskonzepten auf Schiffen unter bestimmten Umständen Sinn ergeben kann.

Im Vergleich zu Elektrofahrzeugen kann bei Schiffen keine nennenswerte Energie bei einem Bremsvorgang zurückgewonnen werden.

Ein hybrides Antriebskonzept bietet lediglich die Möglichkeit durch die Anpassung von Leistung und Drehzahl des Verbrennungsmotors einen geringeren spezifischen Treibstoffverbrauch zu erzielen. Dies kann besonders im Teillastbereich große Auswirkungen haben.

In diesem Paper wurden unterschiedliche Antriebskonzepte für einen Hafenschlepper und eine Motorfähre untersucht.

Für den untersuchten Hafenschlepper sind die Investitionskosten eines hybriden Antriebssystems etwa 820.000 € höher im Vergleich zu einem diesel-elektrischen System. Diese Kosten können während der Lebensdauer des Systems nur dann eingespart werden, wenn die Akkus nach jedem Einsatz über Landstrom zum günstigen Industriepreis geladen werden können. Dazu muss jedoch genügend Zeit zwischen den Einsätzen sein und eine ausreichende Leistung des Landanschlusses vorliegen.

Der zweite Schiffstyp der untersucht wurde, war eine Motorfähre. Diese fährt alle 20 min zwischen Bremerhaven und Blexen/Nordenham. Verglichen mit dem Hafenschlepper hat die Motorfähre einen wesentlich höheren Volllastanteil. Das ist der Grund, dass sich für die Motorfähre ein hybrides Antriebskonzept während der Lebensdauer des Systems nicht rentiert. Ohne auf die Investitionskosten einzugehen lässt sich sagen, dass die Verbrauchskosten in allen betrachteten Konzepten 10 %–20 % über den Kosten des diesel-elektrischen Konzepts liegen. Somit ergibt ein hybrides Antriebskonzept für diesen Einsatzfall aus finanzieller Sicht keinen Sinn.

5 Literatur

[1] Völker, Thorsten: VDE-ETG-Kongress 2009 "Leistungselektronik in Netzen", Düsseldorf, „Regelung unsymmetrisch belasteter Inselnetze durch unabhängige Wechselrichter im Parallelbetrieb"

[2] Göttsche, D: „Untersuchung und Vergleich verschiedener Antriebskonzepte eines Schleppers", Bachelor-Thesis, Hochschule Bremerhaven, Oktober 2012

[3] Nikolai, C: „Auswertung des Fahrverhaltens der Motorfähre NORDENHAM und wirkungsgradorientierte Untersuchung von möglichen zukünftigen Hybridantriebskonzepten", Bachelor-Thesis, Hochschule Bremerhaven, September 2012

[4] Völker T., Raffel H., Orlik B.: Parallel Operation of thyristor- and IGBT-based HVDC", EPE 2007, Aalborg

[5] Völker T., Orlik B.: „Zwischenkreisspannungsregelung eines rückspeisefähigen Frequenzumrichters für unsymmetrische Netzspannungen", SPS/IPC/DRIVES 2008, Nürnberg, Germany

[6] Jerzy Herdzik: "Problems of propulsion systems and main engines choice for offshore support vessels", Scientific Journals Zeszyty Naukowe, 2013, 36(108) z. 2 pp. 45–50 2013, ISSN 1733-8670

[7] Bas Kwasieckyj: „Hybrid propulsion systems"; Master Thesis, TU-Delft, March 2013

[8] Simon Sortland: "Hybrid propulsion system for anchor handling tug supply vessels"; WÄRTSILÄ TECHNICAL JOURNAL 01.2008

[9] European Maritime Safety Agency: " The 0.1% sulphur in fuel requirement as from 1 January 2015 in SECAs"; Technical Report 13 December, 2010

[10] Leipziger Institut für Energie GmbH: „Wirtschaftlichkeit Batteriespeicher", Januar 2014

[11] Sahan, Benjamin: „Wechselrichtersysteme mit Stromzwischenkreis zur Netzanbindung von Photovoltaik-Generatoren", Dissertation, Uni-Kassel, 2010

Brennstoffzellen und thermische Systeme

Simulation of Electromagnetically and Thermally Controlled Ionic Flow in a Fuel Cell

M. Stiemer[1], A. Lücken[2], T. T. Do[2], D. Schulz[2]
[1]Institute for the Theory of Electrical Engineering, Holstenhofweg 85, 22043 Hamburg, stiemer@hsu-hh.de
[2]Institute for Electrical Power Systems, Holstenhofweg 85, 22043 Hamburg

Abstract

A major issue related to the use of fuel cells to convert electrical energy in chemical energy in modern power supply concepts are their bad dynamical properties. To overcome these problems, it seems promising to introduce a suitable mechanism to control the ionic flow inside the fuel cell. The purpose of this work is to estimate the potential of certain approaches to controlling the ionic flow inside the fuel cell via magnetic and temperature fields. To this end, mathematical models combining a description of the ionic movement in a hydrogen fuel cell with a model for the effects of an additional magnetic or temperature field, respectively, are proposed. Further the implementation of these models in the context of the finite element method combined with other simulation techniques is discussed, such as, e.g., a molecular dynamic model. Finally, some preliminary results are presented.

1 Introduction

Fuel cells possess a huge potential to act as energy converter transforming electrical into chemical energy in the context of regenerative energy concepts. Hence, they may help to solve the crucial problem of managing the natural variation of the supply of regenerative power [1,2]. An overview over typical system variants may be found in [1,2]. However, before fuel cells can extensively be employed as an economical energy converter in a large range of applications still scientific research is required. Due to their bad dynamical properties, fuel cells often have to be oversized to meet the requirements of changing electrical loads, implying a low efficiency factor. Moreover, for the same reason, contemporary fuel cells need to be switched off at a quite early stage in case of high electric loads to avoid damage. Further, changing load demands may reduce a cell's life span by deterioration of the cell membrane as a consequence of thermal stress invoked by an excessive or unbalanced flow of hydrogen ions via the membrane. All these issues indicate that a control of the ionic flow over the cell membrane is commendable to increase the cell's dynamical response to load changes, to avoid damage in case of high electrical loads, and to increase the cell's life span. Even during normal operation, i.e., under a moderate load, such a control of ion movement seems to be favorable to adapt the ionic flow to the electrical properties of the membrane. In case of an ideal, i.e., spatially homogeneous membrane, the current caused by ionic motion in the cell should be controlled in such a way that the membrane is homogeneously loaded, i.e., such that the current-density is spatially homogeneous. However, the membrane usually possesses material inhomogeneities [3], which are partially due to fabrication and partially due to progressive deterioration during operation. These could also be compensated by a magnetic adaptation of the ionic flow through the membrane.

The idea of controlling the ionic flux in a fuel cell's membrane has firstly been presented in [4]. In this patent specification, several ways of implementing the idea of a control of the ionic flow in a fuel cell are proposed, particularly with the help of magnetic fields exerting a Lorentz force on the ions moving through the fuel cell. It is further proposed that the employed magnetic fields are generated via an electric current through conducts adequately installed within the fuel cell. Also connected with furnishing a fuel cell's membrane with electrical conducts is the ability to control the ionic movement by inducing a spatial variation of the membrane's temperature. Depending very sensitively on the temperature, the membrane's conductivity can quickly be raised locally by a short current pulse to direct the ions along the heated region. In fact, our analysis will show that under typical situation the temperature effect definitively exceeds the influence caused by the magnetic field triggered by the control current. Combined with a suitable control of the triggering currents through the incorporated conducts, it is aimed at generating an optimum ion flow through the membrane yielding a high efficiency factor.

To estimate the range of this new idea's applicability as well as to identify suitable implementation parameters and control mechanisms, the availability of a simulation of the ionic flow through a membrane is favorable. At best, such a simulation would comprise the whole coupled system consisting of the fuel cell and the devices providing the additional thermal and magnetic fields. Hence, the purpose of this work is to present a mathematical model for this coupled problem and its numerical implementation, within the finite element method and some extended techniques. While the idea of a thermal and magnetic control of the ionic flow inside a fuel cell basically applies to all kind of cells, we restrict the modeling to fuel cells consisting of two porous media acting as anode and cathode, respectively, separated by a polymer membrane.

The paper is organized as follows: In the following Section 2, first the here considered types of standard fuel cells are described and concept variants are presented. Then, in Section 3, a mathematical model for the ionic flow through the membrane is outlined and extended by the influence of additional control currents. Section 4 is devoted to an estimate of the size of the different physical mechanisms the flow control relies on. It turns out that thermal effects dominate over magnetic effects. Particularly a thermal interference seems to allows for a large range control of the ionic flow. In Section 5 the numerical implementation of the model is described and first results are shown. The paper ends with conclusions and an outlook in Section 6.

2 Concept of Electromagnetically Controlled Ionic Flow

2.1 Fuel Cell Setup

We consider hydrogen fuel cells, i.e., cells that convert the chemical energy stored in molecular hydrogen and molecular oxygen (fuel) into electrical energy and vice versa. The considered type consists of two electrodes (anode and cathode) separated by a membrane acting as an electrolyte. **Fig. 1** shows a typical setup. Anode and cathode consist of porous media. After passing a fuel distributer (gas diffuser), molecular hydrogen streams through the anode's pores and is ionized in its active layer at carbon particles with embedded platinum catalyst nanoparticles.

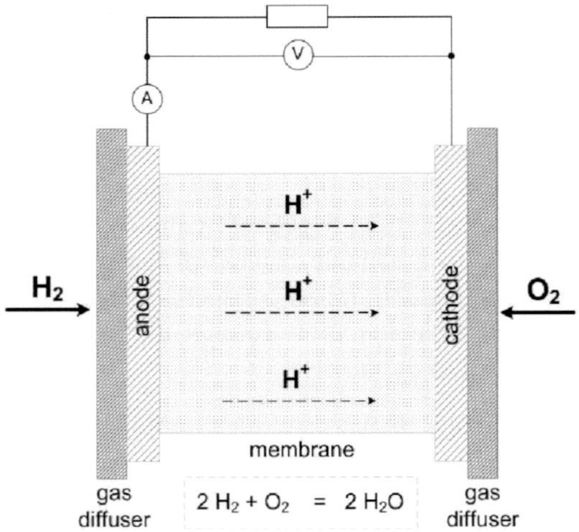

Fig. 1 Typical fuel cell setup

The carbon particles receive the electrons, while the protons remain in the vaporized phase and enter the membrane. The thickness of the anode's active layer amounts to approximately 10 μm, while its upstream diffusion layer possesses a thickness of approximately 100 - 200 μm. As electrolyte we consider a polymer membrane, which can be penetrated by the hydrogen ions. The proton's path, velocity, and the corresponding losses depend on the spatial distribution of the membrane conductivity [5] and the interface currents through the cathode's and anode's active layers. After passing through the membrane the protons reach the cathode region, whose pores are streamed by molecular oxygen, which is supplied by a further gas diffuser. The oxygen is ionized at the cathode's active layer, and the O^- ions react with 2 H^+ ions to water. The size of the cathode's active layer and diffusion layer equals the corresponding dimensions of the anode.

2.2 Control of the Ionic Flow

The idea of fuel cell control proposed in [4] is to change the proton paths and velocity during their transit through the membrane by Lorentz forces conveyed by magnetic fields, which are triggered by line currents through additional conducts integrated in the membrane. Connected with this approach is also a local heating of the membrane due to Ohmic losses, which can also be employed for a control of the ions' paths, since it causes a significant rise of the electrical conductivity of the heated area. In fact, we will see that this effect dominates over magnetic forces.

Depending on the direction in which the protons are to be deviated, different geometric implementations are possible. In case of a thermal control, good conducting regions attract protons. Particularly spatial variations of the membrane's conductivity due to fabrication or deterioration may be thermally compensated. As far as the magnetic fields are considered, different ways of interaction with the protons occur, particularly via the Lorentz force and the magnetic moment of protons. Lorentz forces deviate a moving proton always in a direction perpendicular to its velocity vector and in or against the direction of the conducts depending on the direction of the triggering currents. Hence, with conducts traversing the membrane from the anode to the cathode or vice versa, the proton flow can be focused or dispersed. Conducts leading in a direction perpendicular to the electrodes (see **Fig. 2**), allow for a selective deviation of the proton paths in the corresponding direction. Grids with different orientation can be combined to obtain a control of the ionic motion in all spatial directions.

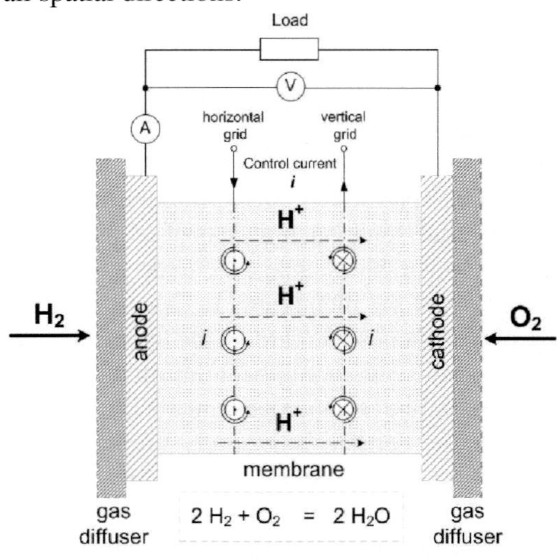

Fig. 2 Fuel cell with additional conducts providing a control current

3 Mathematical Model

3.1 Fuel Cell

First we sketch a model for a fuel cell without additional conducts for the controlling currents. For more than 20 years different models have been proposed and used for simulation, see, e.g., [5-7]. Grujicic and Chittajallu [5], e.g., formulated for the anode, the cathode, and the membrane three systems of differential equations that are coupled via the interchange current flowing through the active layers of the electrodes. The size of these active layers is neglected, and the interchange currents are computed by the analytical solution of a diffusion-reaction problem in a spherical porous agglomerate particle [8,9]. However, this model prescribes a steady state of the coupled system, and hence is not suited for modeling a transient control process. Such a model is presented, e.g., in [10], where moreover one set of equation has been employed to model anode, cathode, and membrane. This also offers the advantage that no interfacial coupling conditions are required. However, jumps in the coefficients of the system have to be included at interfaces between different functional parts. The system proposed in [10] consists of equations reflecting the principles of mass, momentum, species, and charge conservation:

$$\begin{aligned} \frac{\partial(\rho\epsilon)}{\partial t} + \nabla\cdot(\epsilon\rho\boldsymbol{u}) &= 0 \\ \frac{\partial(\epsilon\rho\boldsymbol{u})}{\partial t} + \nabla\cdot(\epsilon\rho\boldsymbol{u}\boldsymbol{u}) &= -\epsilon\nabla p + \nabla\cdot\left(\epsilon\mu^{\mathrm{eff}}\nabla\boldsymbol{u}\right) + S_{\mathrm{u}} \\ \frac{\partial(\epsilon X_{\mathrm{k}})}{\partial t} + \nabla\cdot(\epsilon\boldsymbol{u}X_{\mathrm{k}}) &= \nabla\cdot\left(D_{\mathrm{k}}^{\mathrm{eff}}\nabla X_{\mathrm{k}}\right) + S_{\mathrm{k}} \\ \nabla\cdot\left(\sigma_{\mathrm{e}}^{\mathrm{eff}}\nabla\Phi_{\mathrm{e}}\right) + S_{\Phi} &= 0 \end{aligned} \tag{1}$$

with $\boldsymbol{u}, \rho, X_{\mathrm{k}}$, and Φ_{e} denoting the fluid velocity vector, pressure, mole fraction of the chemical species k, and the phase potential of the electrolyte membrane, respectively. According to [10], the parameter ϵ describes the backing layer porosity. The diffusion coefficients of species k and the ionic conductivity of the membrane phase

$$\begin{aligned} D_{\mathrm{k}}^{\mathrm{eff}} &= \epsilon_{\mathrm{m}}^{1.5} D_{\mathrm{k}} \\ \sigma_{\mathrm{k}}^{\mathrm{eff}} &= \epsilon_{\mathrm{m}}^{1.5} \sigma_{\mathrm{k}} \end{aligned} \tag{2}$$

are effective values modified via Bruggman correlation to account for porosity and tortuosity in porous electrodes, catalyst layers, and the membrane [10]. Here, ϵ_{m} represents the volume fraction of the membrane phase. Further,

$$X_{\mathrm{k}} = \begin{cases} \dfrac{c_{\mathrm{k}}^{\mathrm{g}}}{c_{\mathrm{tot}}} & \text{in the gas phase} \\ \dfrac{c_{\mathrm{k}}^{\mathrm{e}}}{c_{\mathrm{tot}}} & \text{in the electrolyte} \end{cases} \tag{3}$$

according to the definitions from [10], where c_{k} is the molar concentration of species k, superscript g refers to the gas phase, and superscript e to the electrolyte phase, respectively. Additionally, the jumps in the coefficients at interfaces between the different model regions have to be considered as described in [10]. Moreover, the source terms S_{k}, S_{u}, and S_{Φ} are specified there in such a way that the *current sources* in the catalytic layers are correctly incorporated. The latter is modeled via a nonlinear, analytic expression.

The diffusivities D_{k} for the individual species depend on the temperature and the pressure [10]. Similarly, the membrane conductivity σ_{e} is a function of temperature T and the water content $\lambda>1$ in the membrane:

$$\sigma_{\mathrm{e}}(T) = (0.5139\,\lambda - 0.326)\exp\left[1268\left(\frac{1}{303} - \frac{1\mathrm{K}}{T}\right)\right] \tag{4}$$

Depending on λ, different ionic transport mechanisms are active. This results in a large variation of the corresponding electrical conductivity. The dependence of these values on the temperature and on the water content is displayed in **Fig. 3**.

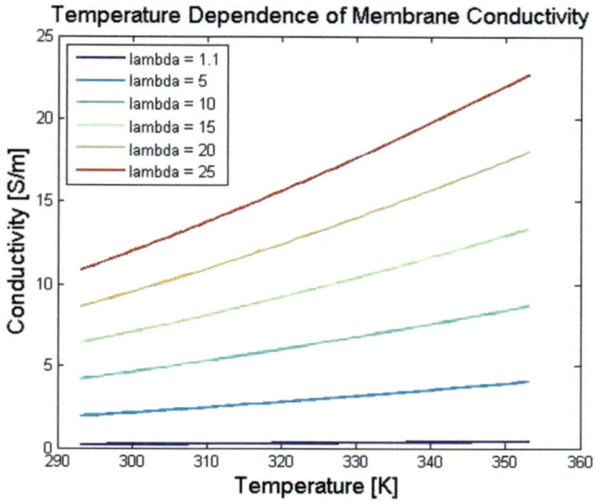

Fig. 3 Dependence of the membrane conductivity on the temperature for different levels of water content

3.2 Modeling of the Ionic Flow Control

In the next step, we augment the above model by the influence of an external magnetic or thermal field. If the magnetic field with flux density $\boldsymbol{B}$, which is assumed not to be influenced by the changes inflicted upon the ionic movement, is evoked by line currents I_i, $i = 1, \dots, n$, flowing along the contours C_i, the external flux density is given by

$$\boldsymbol{B}(\boldsymbol{r}) = \sum_{i=1}^{n} \boldsymbol{B}_i(\boldsymbol{r}) \tag{5}$$

with

$$\boldsymbol{B}_i(\boldsymbol{r}) = -\frac{\mu_0 I_i}{4\pi}\oint_{C_i} \frac{(\boldsymbol{r}'-\boldsymbol{r})\times d\boldsymbol{r}'}{|\boldsymbol{r}'-\boldsymbol{r}|^3} \tag{6}$$

according to the Biot-Savart-law for line currents, where μ_0 denotes the magnetic field constant, $\boldsymbol{r}'$ runs over all points in the contour C_i and $\boldsymbol{r}$ is the point of interest. Via the Lorentz force $\boldsymbol{F} = q\,\boldsymbol{v}\times\boldsymbol{B}$ an ion with charge q and mass m is deviated from its original path. Its total acceleration due to Lorentz forces and the electric membrane potential amounts to

$$\frac{\partial \boldsymbol{v}}{\partial t} = \frac{q}{m}(\nabla\phi_e + \boldsymbol{v} \times \boldsymbol{B}) \tag{7}$$

with the ratio q/m well known for protons. The corresponding magnetic dragging force has to be accounted for in the impulse balance. Below we will argue, that the magnetic moment of the protons does not need to be considered, since it is too small. Hence, we omit a discussion of its influence here.

If the heat production via Joule heating shall be considered, the dissipated energy has to be interpreted as source term for the heating. The corresponding power density amounts to $\frac{1}{\sigma_e(T)} j^2$, where j denotes the current density imposed on the membrane due to the conducts. Then, the corresponding temperature rise in the membrane can be computed with the help of the heat equation

$$\varrho_m c_m \frac{\partial T}{\partial t} = \nabla \cdot (\lambda_m \nabla T) \tag{8}$$

with ϱ_m, c_m, and λ_m being the membrane's mass density, specific heat capacity, and heat conductivity, respectively. The coupling between the heat equation and the membrane potential equation can be carried out in a loose way, i.e., in any time step an update of the temperature is computed and then, the altered temperature is employed to compute the resulting spatial distribution of the membrane's electrical conductivity.

4 Estimate on the Size of Different Effects

While it will turn out that the electric conducts employed for the control of the ion flow have a significant influence via the corresponding temperature rise, it is not clear, whether the magnetic fields may be sufficiently strong to have a significant impact. To get an idea of the size of the temperature effect, assume, for example, that through a conduct with a radius of 10 μm and a length of 2 mm a current of 100 μA is applied with a voltage of 1 V. Under adiabatic conditions, which we assume to obtain a simple estimate, and a value $\varrho_m c_m = 1.65 \times 10^6$ Ws/m^3 taken from [11], we obtain a temperature rise of 96.5 K within 1 s. For a sufficiently hydrated membrane this leads to more than a doubling of the local conductivity, as can be seen from Fig. 3. An increased conductivity implies a proportionally increased proton flux, and hence, this technique can be employed to guide the proton flux by an induced spatial variation of the conductivity.

To estimate the influence of the magnetic fields generated by the conducts, we have to consider both an influence via Lorentz forces and via the magnetic moment of a proton. To estimate the averaged size of the Lorentz forces, we compute the drift velocity of the protons. In fact, the protons move much faster than their drift velocity, but most of their movement is an undirected stochastic movement according to a Rayleigh distribution, and, hence, does not contribute to the averaged Lorentz force. With the number $C = 1200$ mol / m^3 of protons per unit volume [12] and the Faraday constant $F = 96485.33289$ As / mol, one obtains for the charge density

$$\varrho_e = CF \tag{9}$$

and from the relation between proton drift velocity $\boldsymbol{v}_{\text{drift}}$ and current density

$$j = \sigma_e \frac{U}{d} = \varrho_e v_{\text{drift}} \tag{10}$$

with cell voltage U, membrane thickness d, and electrical conductivity σ_e , one obtains

$$v_{\text{drift}} = \frac{\sigma_e U}{dCF} \approx \left(10^{-6}\ \frac{m^2}{Vs}\right) \frac{U}{d} \tag{11}$$

Even for a magnetic field with a strength of $B = 1$ T, which is much too high for our purposes, the ratio between magnetic force and electrical driving force would consequently amount to

$$\frac{f_l}{f_e} \approx 10^{-6} \tag{12}$$

As a consequence, the average Lorentz force only allows for short range control. However, as mentioned before, the absolute value of the particle velocity is much higher. We will discuss below, if this can be exploited for our purposes. Similarly to the analysis of the averaged Lorentz force, the magnetic dipole moment of protons, 1.6726×10^{-27} Tm, is too small to lead to significant effects if magnetic fields are employed possessing inhomogeneous magnetic flux densities with realistic field gradients (cf. [13]).

5 Finite Element Simulation

5.1 Simulation methods

As usual in the context of the finite element method, the field equations displayed in the preceding sections are transformed in the so called weak form and then discretized by a Galerkin method based on approximation by continuous functions whose restriction to the elements of a tetrahedral mesh of the fuel cell are linear [14]. In contrast to preceding work [5-7], a full 3D simulation is required to consider the derivation of the ion traces by the implemented line currents. Moreover, since the system is not in a steady state when control is applied, the transient simulation has to be carried out in a couple of time steps. To this end, an overall time stepping algorithm for parabolic systems (trapezoidal rule) is used. The anode, cathode, and membrane regions are meshed independently, and the coupling between these regions via the exchange currents is implemented by an iterative procedure. To this purpose, the non-linear expressions for i_a and i_c are linearized and a Newton-type iteration is applied. The same is done for the heat equation, which is only loosely coupled to the remaining system, i.e., it is computed in any time step with the Joule heating as source term. With the determined temperature field, the spatial distribution of electrical conductivity is then computed, and, finally, these data are used to compute the next time step with the fuel cell equations. The electric control currents are treated independently from the rest of the system, i.e., they are pure input quantities, and no feedback possibly influencing them is considered. Hence, the magnetic flux density triggered by the line currents can be computed by a quadrature formula applied to the Biot-Savart integral representation.

5.2 Randomly distributed velocity

Knowing that the drift velocity of the protons is too small to be employed for large range control via Lorentz forces, we developed a molecular dynamical simulation to test if we can find a way to exploit the stochastic movement of the protons which takes place at much higher velocity.

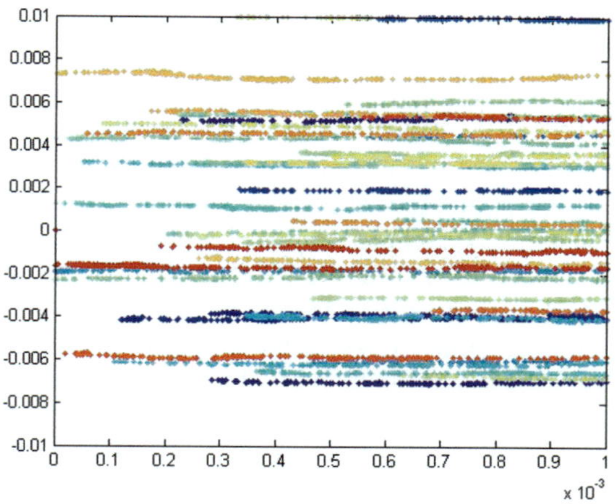

Fig. 4 Traces of 40 randomly chosen protons through the membrane

To this end, a finite number of protons is considered and their electric potentials according to their current states are computed. With these data together with the boundary conditions and external electric and magnetic fields, equations of motion are solved for the particles and their traces are studied. **Fig. 4** shows the result of a first trial with the too small number of only 40 protons. It becomes obvious, that the applied Lorentz forces only lead to a small deviation from their paths. However, it is too early to draw any conclusions from these preliminary numerical experiments.

5.3 Heat effects

For the time being the implementation of the numerical simulation is still work in progress. However, we present below some preliminary applications to give an idea, how the simulation will work. In the examples presented below, the correct coupling between electrode and membrane currents had not been implemented, but instead, typical values for the exchange currents in the electrodes' active layers have been inserted, i.e., the constant values $i_a = 10^5$ A/m^2 and $i_c = 1$ A/m^2. The membrane's conductivity as displayed in **Fig. 5** has been computed as explained in Section 3 from a temperature field previously computed with the heat equation (**Fig. 7**), assuming that the membrane is heated by two electrical conducts. The interface to the surrounding area has been assumed to be on a constant temperature level. Finally, the conductivity values displayed in Fig. 5 have been employed as data for the determination of the electric membrane potential as presented in **Fig. 6**. The black arrows indicate the electric field vectors, while the colors have been set according to the potential value. As mentioned before, the proton inflow and outflow conditions have only been guessed and not determined via a coupled finite element computation. The latter is a matter of current work.

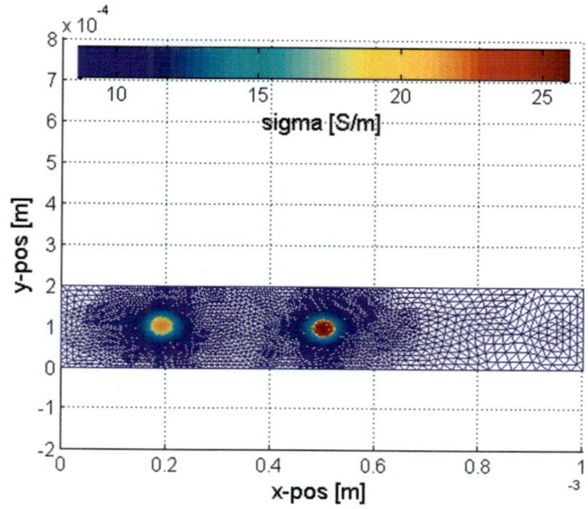

Fig. 5 Spatial distribution of conductivity in a well hydrated membrane resulting from the temperature distribution displayed in Fig. 7

The effect of the magnetic fields due to the control currents is not displayed here. In any case it would be much smaller than the influence of the temperature rise. If a long range control is possible via Lorentz forces, cannot definitively be judged from a finite element simulation, since a finite element simulation can only account for the averaged particle velocity. The Rayleigh distribution of the particle velocity has to be considered within a molecular dynamic framework as presented in Section 5.2.

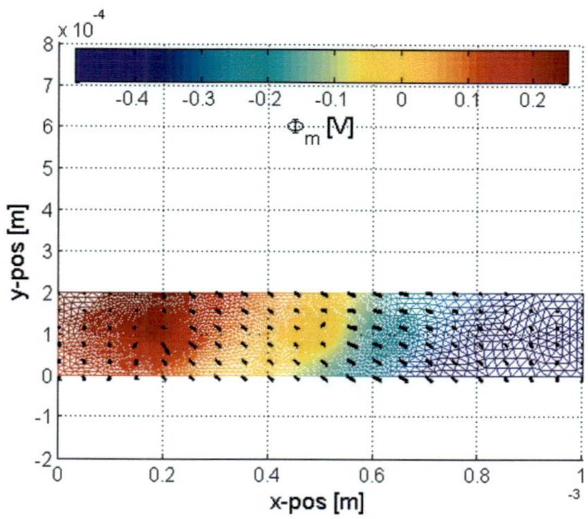

Fig. 6 Membrane potential computed with the conductivity distribution displayed in Fig. 5 and a guess for the electric inflow and outflow conditions

The results presented here are only preliminary. As far as the simulation of the whole system is completed, precise results on thermal and magnetic effects introduced by conducts through a membrane will be presented.

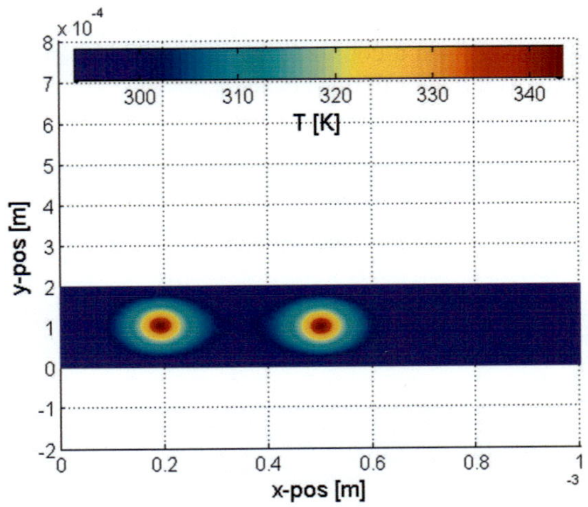

Fig. 7 Temperature field generated in a fuel cell's membrane by two electrical conducts ($I = 1$ mA). The boundary of the displayed region is held at constant temperature

6 Conclusions and Outlook

Electric line currents may allow for a control of the ionic flow through the membrane of a fuel cell. Here, the dominant mechanism to interfere the ionic flow is the sensitivity of the electrical membrane conductivity towards temperature variations, particularly on a high hydration level. The resulting influence on the proton movement in the membrane seems to be sufficiently strong. Magnetic effects may additionally play a minor role. For a large range control via magnetic forces, exploitation of the stochastically distributed velocity of the protons would be required, since the drift velocity is too small to allow for a significant aberration from a particle's original path. To further study the size of magnetic and thermal effects and to validate the feasibility of a control of the ionic movement through a fuel cell's membrane by this method, a mathematical model of a fuel cell has been extended both by the effect of a static magnetic fields conveyed via electrical conducts in the membrane and by Joule heating. The finite element implementation of the coupled model represents work in progress. In addition, molecular dynamic studies are carried out, since questions related to the velocity profile of protons in the membrane cannot be treated with a field theoretic approach. As soon as the finite element implementation is available, it will be employed for a systematic study showing to what extend the ionic flow can be controlled by thermal and magnetic fields. In addition, the simulation results will be compared to experimental data and, hence, validated. Suitable experiments are currently developed. In the long term, simulation is intended to be employed for the development of suitable designs of electromagnetically controlled fuel cells. Such cells are hoped to possess a better response to sudden load changes (avoiding oversized fuel cells), reduced cell deterioration, and a higher efficiency factor.

7 References

[1] U.S. Department of Energy: Fuel Cell Handbook. 7th ed. (2004)

[2] Heuck, K., Dettmann, K.-D., Schulz, D.: Elektrische Energieversorgung, 8th Edition, Vieweg, Wiesbaden (2010)

[3] Aleksandrova, E., Hink, S., Hiesgen, R., Roduner, E.: Spatial distribution and dynamics of proton conductivity in fuel cell membranes: potential and limitations of electrochemical atomic force microscopy measurements. J. Phys. Condens. Matter 23 (2011)

[4] Schulz, D.: Brennstoffzellenmembraneinheit, steuerbare Brennstoffzelle und Hochdruckelektrolysezelle, 15.12. 2011, Patent DE 10 2011 088 613

[5] Grujicic, M., Chittajallu, K. M.: Design and optimization of polymer electrolyte membrane (PEM) fuel cells. Appl. Surface Sci. 227, 56-72 (2004)

[6] Obayopo, S. O., Bello-Ochende, T., Meyer, J. P.: Performance enhancement of a PEM fuel cell through rectant gas channel and gas diffusion layer optimisation. University of Pretoria

[7] Kakac, S., Pramuanjaroenkij, A., Xiang Yang Zhou: A review of numerical modeling of solid oxide fuel cells. Int. J. Hydrogen Energy 32, 761-786 (2007)

[8] Fogler, H. S.: Elements of Chemical Reaction Engineering. 3rd ed., Prentice Hall Englewood Cliffs (1999)

[9] Bird, R. B., Stewart, W. E., Lightfoot, E. N.: Transport Phenomena. Wiley, New Nork (1960)

[10] Sukkee Um, Wang, C.-Y. , Chen, K. S.: Computational Fluid Dynamics Modeling of Proton Exchange Membrane Fuel Cells. J. Electrochem. Soc. 147 (12). 4485-4493 (2000)

[11] Chen Yun Wang, Ken Chen, Mishler, J., Sung Chan Cho, Cordobes Adroher, X.: A review of polymer electrolyte membrane fuel cells: Technology, applications, and needs on fundamental research. US Department of Energy Publications, US Department of Energy (2011)

[12] Baschuk, J. J., Xianguo Li: Modelling of polymer electrolyte membrane fuel cells with variable degrees of water flooding. J. Power Sources 86, 181-196 (2000)

[13] Kinouchi, Y., Tanimoto, S., Ushita, T., Sato, K. Yamaguchi, H., Miyamoto, H.: Effects of Static Magnetic Fields on Diffusion in Solutions. Bioelectromagnetics 9, 159-166 (1988)

[14] Schwarz, H. R.: Methode der finiten Elemente. Teubner (1991)

Verwertung von wasserstoffreichen Gasgemischen in PEM-Brennstoffzellensystemen

M. Steinberger, R. Öchsner
Fraunhofer-Institut für Integrierte Systeme und Bauelementetechnologie IISB, Schottkystraße 10, 91058 Erlangen, michael.steinberger@iisb.fraunhofer.de

Kurzfassung

Wasserstoff wird bisher fast ausschließlich als Ausgangsstoff für industrielle Produktionsprozesse verwendet und vor Ort in großtechnischen Prozessen größtenteils aus fossilen Energieträgern hergestellt. Zunehmend nimmt Wasserstoff auch die Rolle eines mehrdimensionalen Energievektors ein, zum Beispiel als Energiespeicher oder als Treibstoff im Mobilitätssektor. Bei nachhaltigen, regenerativen Energiesystemen muss daher versucht werden, alternative Wasserstoffquellen zu erschließen, im Speziellen die Herstellung und Speicherung von „grünem" Wasserstoff durch Elektrolyse von Wasser mittels erneuerbaren Energien. Zudem sind Potentiale durch Wiederverwertung von Wasserstoff als Prozessmedium zu nutzen, zum Beispiel durch die energetische Verwertung von Prozessabfällen oder -nebenprodukten. Die PEM-Brennstoffzellentechnologie bietet aufgrund ihres hohen Wirkungsgrades und ihrer Skalierbarkeit ein hohes Potential für die Nutzung von Wasserstoff vor allem für dezentrale Energiesysteme. Je nach Wasserstoffquelle können unterschiedliche Kontaminationen und Fremdgasanteile im Wasserstoff vorhanden sein. In dieser Arbeit wird gezeigt, dass unkritische Fremdgasanteile prinzipiell in Brennstoffzellensystemen verwertet werden können. Die Wirkungsgradeinbußen sind deutlich, wenn anstatt der Anodengasrezirkulation auf Flow-Through-Betrieb umgestellt werden muss. Die obere Grenze für den Rezirkulationsbetrieb liegt bei etwa 2 % Fremdgasanteil im einströmenden Werkstoff.

1 Problematik und Fragestellung

In der Wasserstoffinfrastruktur gibt es unterschiedliche Spezifikationen, welche die Wasserstoffreinheit kategorisieren. Die beiden Normen SAE J2719 und ISO 14687-2 definieren die Qualität von gasförmigen Wasserstoff für PEM-Brennstoffzellenfahrzeuge [01] und werden oft bei Standardbrennstoffzellensystemen als erforderliche Wasserstoffreinheit angegeben. Die beiden Normen schreiben einen Wasserstoffstoffmengenanteil größer als 99,99 % vor und spezifizieren zusätzlich die Grenzwerte für einzelne Fremdstoffe. Die darin zulässigen Verunreinigungen haben laut Molter [02] keinen Einfluss auf PEM-Brennstoffzellen. Für die Versorgung mit Wasserstoff ist es wichtig, dass die komplette Wasserstoffinfrastruktur immer die gleiche Reinheitsklasse verwendet. Das heißt, alle beteiligten Prozess-, Lager-, Transport und Abfüllschritte einer Versorgungskette haben die gleiche Reinheitsanforderung. Oft wird Reinstwasserstoff mit der Qualität 5.0 (99,999 % Wasserstoff) als Standardreinheit verwendet. Die Qualität dieses häufig durch Erdgas-Dampfreformierung hergestellten Wasserstoffs wird durch nachgeschaltete Reinigungsstufen gewährleistet [03] und erfüllt die Reinheitsanforderungen der beiden Normen. Die hohe Reinheitsanforderung verursacht einen erheblichen Filteraufwand, der im großtechnischen Maßstab Standard ist. Für kleine, meist lokale Systeme zur Wasserstofferzeugung oder -freisetzung ist der Filteraufwand zur Erreichung der Standardreinheit deutlich höher und schwieriger zu realisieren. Sowohl finanziell, als auch energetisch und hinsichtlich des Platzbedarfs ist die Filterung bei kleinen Wasserstoffsystemen eine große Herausforderung. Wenn der lokal erzeugte oder freigesetzte Wasserstoff direkt vor Ort verbraucht wird und nicht in die allgemeine Wasserstoffinfrastruktur einfließt, so ist die hohe Reinheitsanforderung kritisch zu hinterfragen. Je toleranter die Wasserstoffkomponenten in solch einem lokalen Wasserstoffsystem gegenüber Fremdstoffen sind, desto geringer wird der Filteraufwand. In diesem Beitrag werden zwei unterschiedliche Arten von Wasserstoffqualitäten betrachtet und die Auswirkung auf PEM-Brennstoffzellensysteme untersucht. Dies ist zum einen Wasserstoff mit Kontaminationen, wie er zum Beispiel bei der Freisetzung von Wasserstoff aus flüssigen, organischen Wasserstoff-Trägerstoffen auftritt. Hier liegt ein Wasserstoffstoffmengenanteil von über 99 % vor. Zum anderen wird Wasserstoff mit Fremdgasanteilen untersucht, was beispielsweise bei wasserstoffhaltigen Prozessabgasen zutrifft. Der Wasserstoffstoffmengenanteil liegt hier oft zwischen 40 - 98 %. Die Unterscheidung zwischen Kontamination und Fremdgas wird in Kapitel 3 beschrieben. Zudem benötigen Standardbrennstoffzellensysteme laut Datenblatt oft einen Wasserstoffeingangsdruck von 5 - 7 bar absolut. Ist die Wasserstoffquelle nicht in der Lage den Druck bereitzustellen, wird eine Komprimierung des Gases erforderlich. Dies ist ebenfalls mit einem hohen energetischen und finanziellen Aufwand verbunden. Aus diesem Grund wird in diesem Beitrag analysiert, welche Mindestreinheit für PEM-Brennstoffzellensysteme erforderlich ist und woher die Anforderungen an den hohen Eingangsdruck für PEM-Brennstoffzellensysteme resultieren. Bezüglich des Einflusses von Kontaminationen auf eine Brennstoffzelle oder einen Brennstoffzellenstack, vor allem auf den Katalysator, gibt es in der Literatur zahlreiche Untersuchungen (siehe Kapitel 3.1). Jedoch ist zur Beantwortung der zuvor erläuterten Fragestellung die

Betrachtung des gesamten Brennstoffzellensystems erforderlich, welche in diesem Beitrag durchgeführt wird.

2 Aufbau von PEM-Brennstoffzellensystemen

Die kleinste Einheit eines Brennstoffzellensystems bildet die Brennstoffzelle. Diese besteht aus einer Anoden- und Kathodenbipolarplatte, der Gasdiffusionsschicht (GDL) und der Membran-Elektroden-Einheit (MEA). Aufgrund der niedrigen Ausgangsspannung einer Brennstoffzelle, werden mehrere Zellen in einer elektrischen Reihenschaltung zum Brennstoffzellenstack verbunden. Die Medienversorgung der Zellen mit Wasserstoff, Sauerstoff und Kühlwasser erfolgt dabei parallel. Neben dem Stack als Herzstück besteht ein Brennstoffzellensystem noch aus dem Anodenkreis, dem Kathodenkreis, dem Kühlkreis und der Leistungselektronik. Die Kathode wird dabei im sogenannten Flow-Through-Modus betrieben. Dabei wird dem Stack mittels des Kathodengebläses Luftsauerstoff eingeblasen. Der Luftsauerstoff wird stark überstöchiometrisch mit einem Stöchiometriefaktor $\lambda_{O2} = 2 - 4$ zugeführt. Der restliche Luftsauerstoff und die restlichen Luftbestandteile verlassen den Stack am Kathodenausgang und befördern dabei das entstandene Produktwasser aus dem Stack. Die Betriebsweise des Anodenkreises wird in Kapitel 2.1 näher erläutert. Bei der Leistungselektronik handelt es sich entweder um einen DC/DC-Wandler oder einen DC/AC-Wandler, wobei die niedrige Ausgangsspannung und damit der hohe Strom des Stacks eine Herausforderung darstellen. Der Kühlkreis führt die im Stack aufgrund der Überspannungen entstehende Verlustleistung aus dem Stack ab. Ein vereinfachtes Schema eines Brennstoffzellensystems ist in **Bild 1** skizziert.

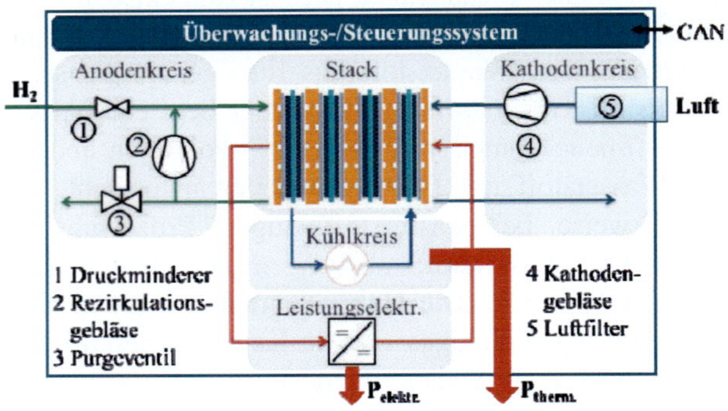

Bild 1: schematischer Aufbau eines PEM-Brennstoffzellensystems

Zur Charakterisierung des Brennstoffzellensystems und zur Bewertung der Kontaminationseinflüsse werden im Folgenden die charakteristischen Wirkungsgrade eines Brennstoffzellensystems definiert. Für eine PEM-Brennstoffzelle mit Betriebstemperaturen unter 100°C liegt das Produktwasser in flüssiger Form vor. Die ideale Leerlaufspannung beträgt für eine Zelle:

$$U_{\text{rev}}^0 = -\frac{\Delta G_0^0}{z \cdot F} = 1{,}23V \tag{1}$$

Hierbei beschreibt F die Faraday-Konstante und z die Zahl der beteiligten Elektronen pro Reaktion, wobei für die Reaktion der PEM-Brennstoffzelle $z = 2$ gilt. ΔG_0^0 ist die freie Gibb'sche Reaktionsenthalpie und beträgt unter Standardbedingungen -237,13 kJ/mol für flüssiges Produktwasser. Aufgrund von Überspannungen in der Zelle ist die tatsächlich verfügbare Ausgangsspannung jedoch deutlich kleiner als die ideale Leerlaufspannung. Der Spannungswirkungsgrad beschreibt die gemessene Zellspannung bezogen auf die ideale Leerlaufspannung und ist abhängig von der Stromdichte.

$$\eta_{\text{uz}} = \frac{U_{\text{Zelle}}}{U_{\text{rev}}^0} \tag{2}$$

Ebenso kann der Spannungswirkungsgrad auf den Stack mit n Zellen bezogen werden. Aufgrund der elektrischen Reihenschaltung der Zellen wird dieser wie folgt definiert:

$$\eta_{\text{u}} = \frac{U_{\text{Stack}}}{U_{\text{rev}}^0 \cdot n} \tag{3}$$

Der kombinierte Brennstoff-Strom-Wirkungsgrad beschreibt, mit welcher Effizienz der einfließende Wasserstoff für die Reaktion genutzt wird und hängt maßgeblich von der Betriebsweise des Anodenkreises ab.

$$\eta_{\text{H2}} = \frac{I_{\text{Zelle}} \cdot n}{z \cdot F \cdot \dot{n}_{\text{H2,in}}} \tag{4}$$

Dabei ist I_{Zelle} der Laststrom der Zelle bzw. des Stacks und $\dot{n}_{\text{H2,in}}$ der Stoffmengenfluss an Wasserstoff, welcher in das Brennstoffzellensystem hineinfließt. Der Gesamtwirkungsgrad η_{ges} ist die Multiplikation aller Teilwirkungsgrade und wird in diesem Fall auf den Heizwert von Wasserstoff $\Delta H_{\text{U,H2}}^0$ mit -241,98 kJ/mol bezogen. Soll der Gesamtwirkungsgrad auf den Brennwert bezogen werden, so ist im Nenner des ersten Quotienten der Brennwert anstatt des Heizwertes einzusetzen.

$$\eta_{\text{ges}} = \frac{P_{\text{elektr.}}}{\Delta H_{\text{U,H2}}^0 \cdot \dot{n}_{\text{H2,in}}} = \frac{\Delta G_0^0}{\Delta H_{\text{U,H2}}^0} \cdot \eta_{\text{u}} \cdot \eta_{\text{H2}} \cdot \eta_{\text{sys}} \tag{5}$$

Der Systemwirkungsgrad η_{sys} beschreibt den Energieverbrauch der Peripheriekomponenten wie Kathodengebläse, Rezirkulationsgebläse, Anodengasbefeuchter und die Verluste des leistungselektronischen Wandlers.

2.1 Betriebsweise des Anodenkreises

Der Anodenkreis hat die Aufgabe, den Stack mit ausreichend Wasserstoff bei definierter Temperatur und relativer Feuchte zu versorgen. Um eine Unterversorgung zu vermeiden, wird der Stack auch anodenseitig überstöchiometrisch mit dem Stöchiometriefaktor $\lambda_{\text{H2}} = 1{,}2 - 1{,}6$ betrieben. Das heißt, es muss der Anode mehr Wasserstoff zugeführt werden als durch die Reaktion verbraucht wird. Dies kann durch den Flow-Through-Betrieb oder den Rezirkulationsbetrieb erreicht werden. Beim Flow-Through-Betrieb geht der überschüssige Wasserstoff am Anodenausgang verloren, was in einem schlechten Brennstoff-Strom-Wirkungsgrad resultiert. Dieser liegt üblicherweise zwischen 60 - 80 %. Zudem muss eine Befeuchtung und Temperierung des zugeführten Wasserstoffs erfolgen.

Der Rezirkulationsbetrieb führt den überschüssigen Wasserstoff am Anodenausgang zum Anodeneingang mit Hilfe eines Rezirkulationsgebläses zurück (siehe **Bild 1**). Um Fremdgas- und Wasseransammlungen auf der Anode aufgrund von Diffusionsvorgängen durch die Membran zu unterbinden, wird der Anoden-Rezirkulationskreis durch Öffnen des Purgeventils zyklisch mit frischem Anodengas durchspült. Der Brennstoff-Strom-Wirkungsgrad beträgt mit dieser Betriebsart 90 - 99,9 % und ist abhängig von der Anzahl und Dauer der Purgevorgänge. Der erzielte Brennstoff-Strom-Wirkungsgrad liegt damit deutlich über dem im Flow-Through-Betrieb. Durch die Rezirkulation wird gleichzeitig auch eine Temperierung und Befeuchtung des frisch zufließenden Wasserstoffs erreicht, was als selbstbefeuchtender Betrieb bezeichnet wird. Aufgrund des höheren Wirkungsgrades und der einfacheren Anodengaskonditionierung wird bei den meisten Brennstoffzellensystemen die Anode im Rezirkulationsmodus betrieben.

2.2 Druckanforderungen im Anodenkreis

Die Stackanode arbeitet in beiden Betriebsmodi mit variablem Druck. Der Anodendruck wird dabei so geregelt, dass der Differenzdruck über der Membran zwischen Anode und Kathode minimal ist. Bei drucklosen Kathodensystemen wie in **Bild 1** gezeigt, wird die Stackanode meist im Bereich einiger hundert Millibar betrieben. Zur Druckanpassung befindet sich am Anodeneingang ein entsprechend geregelter Druckminderer (siehe **Bild 1**). Die Anforderung von Brennstoffzellensystemen, der Eingangsdruck solle mindestens 5 – 7 bar betragen, ist nicht mit den Stackanforderungen zu erklären. Die Ursachen sind auf Systemebene zu finden. Brennstoffzellensysteme benötigen häufig für die Prozedur zum Herunterfahren oder zum Frostschutz eine Restmenge an Wasserstoff. Diese wird durch ein kleines, im Anodenkreis integriertes Reservoir bereitgestellt. Je höher der minimale Eingangsdruck ist, desto kleiner kann das Volumen des Reservoirs ausgelegt werden. Zum anderen können bei höherem Eingangsdruck die Anschlussleitungen und -armaturen aufgrund des geringeren Volumenstroms kleiner dimensioniert werden. Aus technischer Sicht ist es aber durchaus möglich, ein Brennstoffzellensystem mit niedrigem Eingangsdruck zu betreiben. Dafür müssen das Reservoir, die Leitungen, die Armaturen und vor allem der Druckminderer entsprechend ausgelegt werden. Durch diese Anpassung können Brennstoffzellensysteme auch durch Wasserstoffquellen mit einigen hundert Millibar Wasserstoffausgangsdruck ohne Komprimierung versorgt werden.

2.3 Aufbau und Vermessung des PEM-Testsystems

Zur praktischen Bestimmung der Wirkungsgrade für ein mit reinem Wasserstoff betriebenes Brennstoffzellensystem wird das am Institut vorhandene Brennstoffzellensystem mit 8 kW elektrischer Leistung vermessen. Das System wird in Rezirkulation und mit reinem Wasserstoff bei unterschiedlichen Lastpunkten betrieben. In **Bild 2** ist der Verlauf der charakteristischen Wirkungsgrade über die Zeit aufgetragen. Deutlich zu erkennen sind die kurzzeitigen Einbrüche des Brennstoff-Strom-Wirkungsgrades η_{H2}, was auf die Purgevorgänge zurückzuschließen ist. Die über einen konstanten Lastpunkt gemittelten Wirkungsgrade sind im **Bild 3** dargestellt. Der Brennstoff-Strom-Wirkungsgrad liegt je nach Lastpunkt zwischen 96 % und 99,5 %. Zusätzlich ist der Wirkungsgradverlauf für den Flow-Through-Betrieb anhand der Messdaten errechnet und in **Bild 3** ergänzt. Dabei ist eine Anodenstöchiometrie von $\lambda_{H2} = 1{,}4$ angenommen, was in einem Brennstoff-Strom-Wirkungsgrad von gerade noch 72 % resultiert. Es wird angenommen, dass die Peripherieleistung in etwa gleich bleibt. Gegenüber dem Rezirkulationsbetrieb entfällt zwar die Leistung des Rezirkulationsgebläses, jedoch ist eine Befeuchtung und Temperierung erforderlich. Der Gesamtwirkungsgrad im Flow-Through-Betrieb liegt etwa 16 % unterhalb dem des Rezirkulationsbetriebes, so dass hinsichtlich Gesamtwirkungsgrad immer der Rezirkulationsbetrieb vorzuziehen ist.

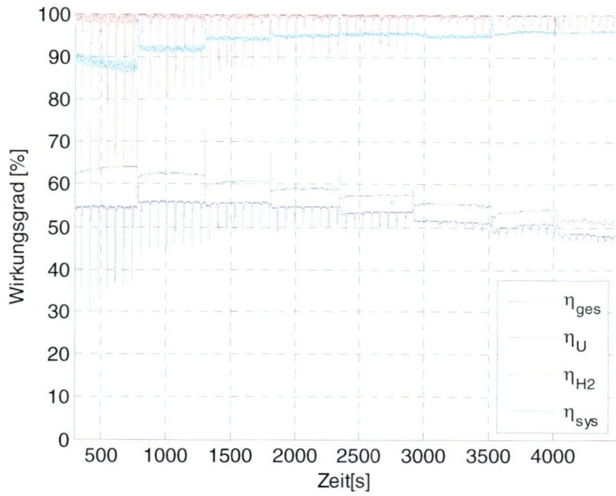

Bild 2: Charakteristische Wirkungsgradverläufe über der Zeit für ein 8kW PEM-Brennstoffzellensystem bei steigenden Lastpunkten

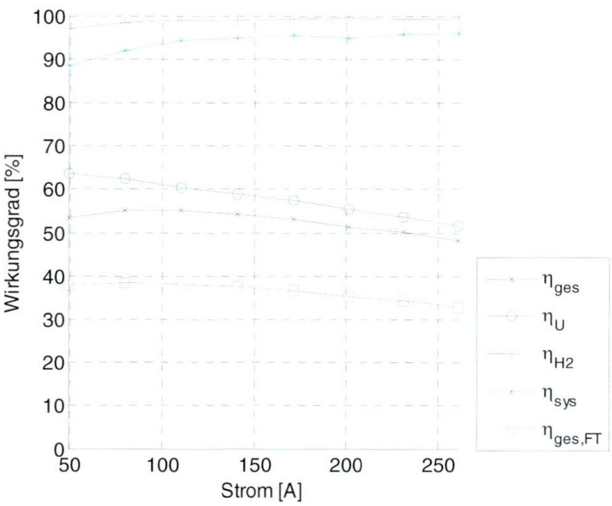

Bild 3: Gemittelte Wirkungsgradverläufe aufgetragen über dem Laststrom und berechneter Wirkungsgrad $\eta_{ges,FT}$ für Flow-Through-Betrieb

3 Einflussanalyse von Kontaminationen und Fremdgasen auf PEM-Brennstoffzellensysteme

Die Unterscheidung zwischen Kontaminationen und Fremdgasen geschieht anhand der Konzentrationen, in welcher ein Fremdstoff im Gasgemisch vorliegt. Ist der Fremdstoffstoffmengenanteil im Bereich von ppm, so wird im Folgenden von Kontaminationen gesprochen. Stoffmengenanteile im Prozentbereich werden als Fremdgas bezeichnet. Bevor die Einflussanalyse von unterschiedlichen Fremdstoffen auf PEM-Brennstoffzellensysteme erfolgen kann, muss eine Aufgliederung in zwei Auswirkungsebenen durchgeführt werden. Dies sind die Zell-/Stackebene und die System-/Betriebsführungsebene.

3.1 Fremdstoffeinflüsse auf Zellebene

In der Literatur gibt es zahlreiche Untersuchungen zu Auswirkungen unterschiedlicher Stoffe auf PEM-Brennstoffzellen. Meist werden Kontaminationen betrachtet. In [02] wurde gezeigt, dass die 5-fachen nach ISO 14687-2 zulässigen Kontaminationen bereits große Auswirkungen auf die Zelle haben. Kontaminationen verkürzen meist die Lebensdauer, verringern die Systemeffizienz und verursachen aufgrund höherer Platinkatalysatorbeladung höhere Kosten [04]. Verschiedene Arbeitsgruppen, meist in den USA, untersuchen daher, welche Auswirkungen Kontaminationen haben und nach welchen Mechanismen diese funktionieren. Als Quelle für die Kontaminationen werden Systembauteile wie Dichtungen und die Wasserstofferzeugung betrachtet. Daher steht die Untersuchung bestimmter Fremdstoffgruppen im Vordergrund. Als Kontaminationsmechanismus wird vorwiegend unterschieden zwischen Einfluss auf den Platinkatalysator und Einfluss auf die Protonenleitfähigkeit der Membran. Bei den Untersuchungen werden fast ausschließlich einzelne Kontaminationen betrachtet. Der Effekt von multiplen Kontaminationen ist daher weitgehend unbekannt [05]. Für diesen Beitrag wurden ca. 60 Literaturquellen zu Verunreinigungen analysiert und zusammengefasst. In **Tabelle 1** wird eine Übersicht gegeben, welche Kontaminationen in der Literatur bisher untersucht wurden. Es ist vermerkt, ob die Kontamination auf der Anoden- oder Kathodenseite der Zelle zugegeben wurde. Zudem sind die getesteten Konzentrationen und die daraus resultierenden Grenzen angegeben. Die Grenze wird jeweils so gewählt, dass bei der Untersuchung kein Effekt durch die Kontamination festgestellt werden kann. Bei manchen Stoffen wurde über den gesamten getesteten Bereich ein Effekt festgestellt. Die Grenzen liegen dann je nach Stärke des Effekts unterhalb oder weit unterhalb der kleinsten getesteten Konzentration. Die Analyse zeigt einige Verunreinigungen, die bereits in geringen Konzentrationen einen Effekt auf die Zelle haben. Hier kann unterschieden werden in reversible und irreversible Effekte. Beim reversiblen Einfluss kann sich die Zelle wieder erholen und die ursprüngliche Performance erreichen, sobald die Verunreinigung gestoppt wird. Während der Erholungsphase wird Reinstwasserstoff verwendet und dieser Vorgang kann einige Stunden dauern. Irreversible Schäden an der Zelle können nicht regeneriert werden. Angaben über reversible und irreversible Effekte sind in den Literaturquellen teils sehr unterschiedlich. Oft sind kleine Mengen von Kontaminationen reversibel, bei höheren Konzentrationen wirken die gleichen Stoffe jedoch irreversibel. Teilweise werden irreversible Schäden aber auch erst nach mehreren hundert Betriebsstunden sichtbar (wie beispielsweise in [06] beschrieben). Die meisten Kontaminationstests werden nur über wenige Stunden durchgeführt – meist 5–30h [07] –, so dass über die Langzeiteffekte nur sehr wenig Datenmaterial verfügbar ist. Deshalb sind nicht nur Stoffe mit irreversiblen Effekten, sondern auch mit reversiblen Auswirkungen auf ein Minimum zu begrenzen [05]. Diese kritischen Stoffe verhindern die Anwendung von PEM-Brennstoffzellensystemen und werden daher im Weiteren nicht betrachtet. Unkritische Stoffe hingegen, wie beispielsweise Stickstoff, die keinen Effekt auf den Katalysator oder die Membran ausüben, gefährden die Brennstoffzelle weder als Kontamination noch als Fremdgas. In [08] wird dies mittels eines Stickstoff-Wasserstoffgemischs untersucht. Unkritische Fremdgasanteile haben jedoch einen Einfluss auf die Zellspannung und damit auf die Zellebene. Die Zellspannung U_{Zelle} setzt sich aus der Nernstspannung U_{n} als ideale Leerlaufspannung (siehe Gleichung(1)) und den Überspannungen zusammen. Dazu gehören die ohmsche Überspannung U_{ohm}, die Aktivierungsüberspannung U_{act}, die Konzentrationsüberspannung U_{conc} und das Mischpotential U_{int}.

$$U_{\text{zelle}} = U_{\text{n}} - U_{\text{int}} - U_{\text{act}} - U_{\text{ohm}} - U_{\text{conc}} \qquad (6)$$

Die Nernstspannung ist abhängig vom Wasserstoffpartialdruck auf der Anodenseite und damit direkt abhängig von der Konzentration von Fremdgasen in der Anode. Mit Gleichung (7) kann der Spannungsabfall der Nernstspannung für ein Gasgemisch mit einem Wasserstoffstoffmengenanteil von x_{H2} gegenüber reinem Wasserstoff berechnet werden. Dabei ist R die allgemeine Gaskonstante, F die Faradaykonstante und T die Zelltemperatur in Kelvin.

$$\Delta U_{\text{n}} = \frac{R \cdot T}{2 \cdot F} \cdot ln(x_{\text{H2}}) \qquad (7)$$

In **Bild 4** ist der errechnete Spannungsfall in Abhängigkeit des Wasserstoffstoffmengenanteils für Standardbedingungen dargestellt. Zusätzlich sind aus [08] verfügbare Messwerte für den Spannungsfall bei unterschiedlichen Fremdgaskonzentrationen eingezeichnet.

Aufgrund des Unterschiedes zwischen berechneten und gemessenen Werten liegt die Schlussfolgerung nahe, dass mindestens eine der Überspannungen ebenfalls von der Wasserstoffkonzentration abhängig ist. In der Literatur gibt es dazu Quellen, dass dies die Aktivierungsüberspannung ist [09]. Zudem wird ersichtlich, dass der maximale Spannungsfall zwischen 100 % und 25 % Wasserstoffstoffmengenanteil lediglich 30 mV beträgt. Dies verschlechtert den Spannungswirkungsgrad lediglich um 2,4 %. Für den in Kapitel 2.3 untersuchten Gesamtwirkungsgrad bedeutet dies einen Verlust von etwa 2 - 2,2 %. Die Auswirkung unkritischer Fremdgase auf die Zellebene ist somit sehr gering und steht einem Betrieb von PEM-Brennstoffzellen mit Gasgemischen nicht im Wege.

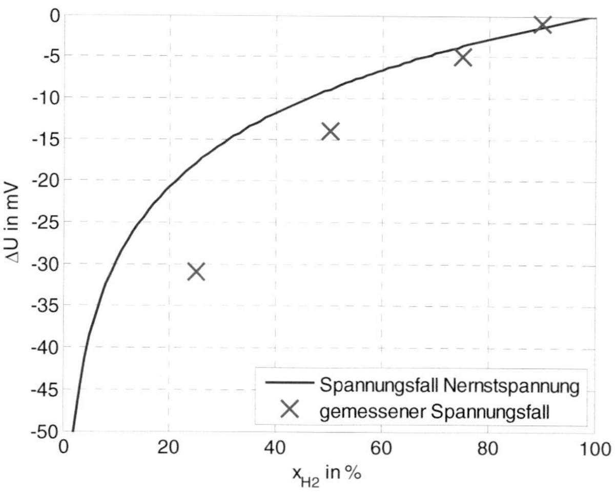

Bild 4: Abfall der Zellspannung abhängig vom Stoffmengenanteil Wasserstoff an der Anode durch Nernstspannung berechnet und Messwerte aus der Literatur

3.2 Fremdstoffeinflüsse auf Systemebene

Auf der Systemebene werden ebenfalls nur unkritische Stoffe betrachtet, da kritische Stoffe, wie beschrieben, ohnehin den Brennstoffzellenbetrieb nicht ermöglichen. Dies wird in [07] beschrieben, wonach bereits 1ppb eines irreversiblen Fremdstoffes im Rezirkulationsbetrieb nach 10.000 h zu einer Katalysatoraktivität von nur noch 25 % gegenüber der ursprünglichen Aktivität führen kann. In [10] wird der Einfluss von Stickstoff auf Brennstoffzellensysteme im Dead-End-Betriebsmodus untersucht, wobei ein Betrieb grundsätzlich möglich erscheint. Im Folgenden wird die Auswirkung auf die beiden Betriebsmodi Flow-Through und Rezirkulation des Anodenkreises getrennt analysiert.

3.2.1 Flow-Through

In diesem Betriebsmodus gibt es nahezu keine Auswirkungen aufgrund von Kontaminationen und nur geringe durch Fremdgase. Die absolute, durch den Stack fließende Stoffmenge erhöht sich, da die Anodengasstöchiometrie λ_{H2} gewahrt werden muss. Je nach Beschaffenheit der Gaskanäle auf den Bipolarplatten ist die maximal durchströmende Stoffmenge begrenzt. Vor allem wenn Fremdgase mit hoher Dichte beigefügt sind, reduziert sich damit der maximale Stoffmengenfluss und damit auch die maximale Stromdichte und respektive die maximale elektrische Leistung des Brennstoffzellensystems. Zusätzlich entsteht ein erhöhter Aufwand bei der Anodengaskonditionierung, da auch das Fremdgas befeuchtet und temperiert werden muss. Dies reduziert den Systemwirkungsgrad. Der Brennstoff-Strom-Wirkungsgrad wird durch die Fremdgasanteile nicht beeinflusst, so dass der Gesamtwirkungsgrad des Brennstoffzellensystems aufgrund des niedrigeren Spannungs- und Systemwirkungsgrades nur geringfügig sinkt gegenüber Flow-Through-Betrieb mit reinem Wasserstoff.

3.2.2 Rezirkulation

Im Rezirkulationsbetrieb haben bereits Kontaminationen Auswirkungen, da sich diese durch die Rezirkulation anreichern können. Bei Fremdgasanteilen ist dieser Effekt nochmal viel deutlicher. Diese reichern sich im Rezirkulationskreis schnell an und führen dazu, dass die Anodengasstöchiometrie λ_{H2} nicht mehr gewahrt wird und damit eine Wasserstoffunterversorgung eintritt. Durch häufigere und längere Purgevorgänge kann dies teilweise kompensiert werden, wodurch jedoch auch der Wasserhaushalt der Zelle und der Brennstoff-Strom-Wirkungsgrad beeinflusst werden. Im Grenzfall, wenn die Purgedauer gegen die Purgezykluszeit geht, wird aus dem Rezirkulationsbetrieb letztlich ein Flow-Through-Betrieb, wodurch die Vorteile der Rezirkulation verloren gehen. Zusätzlich erhöht sich der Energieaufwand für die Rezirkulation, da auch die Fremdgasanteile rezirkuliert werden. Dadurch ist für jeden Fremdgasanteil eine eigene Rezirkulationsstrategie inklusive Purgestrategie erforderlich. Erste Simulationsergebnisse zeigen, dass für Fremdgasanteile über 2 % ein Rezirkulationsbetrieb nicht mehr zielführend ist bzw. der Flow-Through-Modus angewendet werden sollte.

4 Zusammenfassung und Ausblick

In dieser Veröffentlichung wird gezeigt, dass PEM-Brennstoffzellensysteme prinzipiell auch mit wasserstoffreichen Gasgemischen auf niedrigem Druckniveau betrieben werden können. Die Anforderung des hohen Eingangsdruckes resultiert aus dem Systemdesign und kann technisch durch eine entsprechende Auslegung der Anodengaszuführung beseitigt werden. Der minimale Eingangsdruck liegt dann im Bereich einiger hundert Millibar Überdruck. Für Gasgemische ist die Unterscheidung in kritische und unkritische Fremdstoffe wichtig. Kritische Stoffe können bereits als Kontamination eine reversible oder irreversible Schädigung des Stacks hervorrufen und schließen daher die Anwendung in einer Brennstoffzelle aus. Kontaminationen mit unkritischen Stoffen verursachen auf der Zellebene hingegen keine Auswirkung. Selbst Fremdgasanteile bis 75 % von unkritischen Stoffen verringern lediglich den Spannungswirkungsgrad um maximal 2,5 %, was nahezu keine Auswirkungen hat. Der Effekt auf der Systemebene ist deutlich größer. Für Gasgemische mit einem Wasserstoffstoffmengenanteil von unter 98 % ist voraussichtlich nur der Flow-Through-Betrieb zielführend. Gegenüber dem Betrieb von Brennstoffzellensystemen in Rezirkulation mit reinem Wasserstoff bedeutet dies eine Gesamtwirkungsgradeinbuße von 16 % und eine aufwendige Konditionierung des Anodengaszustroms. In weiteren Untersuchungen ist daher zu klären, ob der Rezirkulationsbetrieb auch für Gasgemische unter 98 % Wasserstoffstoffmengenanteil, zum Beispiel durch eine Mischung von stetigem und diskontinuierlichem Purgen, angepasst werden kann.

Danksagung
Die Forschungen wurden im Rahmen der vom Bayerischen Staatsministerium für Wirtschaft und Medien, Energie und Technologie geförderten Projekte SEEDs und LZE durchgeführt.

5 Literatur

[01] ISO/TS 14687-2: Hydrogen fuel - Product specification - Part 2: Proton exchange membrane (PEM) fuel cell applications for road vehicles, ISO (2008)

[02] Molter, T.M.: Effects of Impurities on Fuel Cell Performance and Durability. DOE Hydrogen and Fuel Cell Program, FY 2011 Annual Progress Report (2011)

[03] Bolle, F., Schröder, M., et.al.: Einsatz der Wasserstofftechnologie in der Abwasserbeseitigung - Phase I, Band I: Kompendium Wasserstoff, Aachen, Aktenzeichen IV-7 – 042 600 003 G (2012)

[04] Garzon, F.H., Rockward, T., Mukundan, R., Kienitz, B., Chlistunoff, J., Brosha, E.L., Sansiñena, J.: Effect of Fuel and Air Impurities on PEM Fuel Cell Performance. DOE Hydrog. Progr., FY 2010 Progress Report (2010)

[05] Goodwin, Jr. J.G.: Effects of Impurities on Fuel Cell Performance and Durability. DOE Hydrog. Progr. , Final Report, DOE award No.: DE-FG36-07GO17011, Clemson University (2011)

[06] Zamel, N., Li, X.: Effect of contaminants on polymer electrolyte membrane fuel cells. Pro. in Energy and Combustion Science 37, 292-329 (2011)

[07] Verhage, A.J.L., Coolegem, J.F., Mulder, M.J.J., Yildirim, M. H., de Bruijn, F.A.: 30,000 h operation of a 70 kW stationary PEM fuel cell system using hydrogen from a chlorine factory. Int, J. of Hydrog. Energy 38, 4714-4724 (2013)

[08] Smolinka, T.: Untersuchungen an einer mit Reformat betriebenen PEM-Brennstoffzelle, Dissertation, Universität Ulm (2005)

[09] Amphlett, C., Baumert, R. M., Mann, R. F., Peppley, B. A., Roberge,P. R.: Performance Modeling of the Ballard Mark IV Solid Polymer Electrolyte Fuel Cell, J. Electrochem. Soc., Vol. 142, No. 1 (1995)

[10] Zhu, W.H., Payne, R.U., Tatarchuk , B.J.: Critical flow rate of anode fuel exhaust in a PEM fuel cell system. J. of Power Sources 156, 512–519 (2006)

[11] Moore, J.M., Adcock, P.L., Lakeman, J.B., Mepsted, G.O.: The effects of battlefield contaminants on PEMFC performance. J. of Power Sources 85, 254-260 (2000)

[12] Mohtadi, R., Lee, W.-k., van Zee, J.W.: Assessing durability of cathodes exposed to common air impurities. J. of Power Source 138, 216-255 (2004)

[13] Cheng, X., Shi, Z., Glass, N., Zhang, L,. Zhang, J., Song, D., Liu, Z., Wang, H., Shen, J.: A review of PEM hydrogen fuel cell contamination: Impacts, mechanisms and mitigation. J. of Power Sources 165, 739-756 (2007)

[14] Jing, F., Hou, M., Shi, W., Fu, J., Yu, H., Ming, P., Yi, B.: The effect of ambient contamination on PEMFC performance. J. of Power Sources 166, 172-176 (2007)

[15] Li, H., Zhnag, J., Shi, Z., Song, D., Fatih, K., Wu, S., Wang, H., Zhang, J., Jia, N., Wessel, S., Abouatallah, R., Joos, N.: PEM Fuel Cell Contamination: Effects of Operating Conditions on Toluene-Induced Cathode Degradation. J. Electrochem. Soc. 156, B252-B257 (2009)

[16] Uribe, F., Brosha, E., Garzon, F., Mikkola, M., Pivovar, B., Rockward, T., Valerio, J., Wilson, M.: Effect of Fuel and Air Impurities on PEM Fuel Cell Performance. DOE Hydrog. Progr., FY 2005 Progress Report (2005)

[17] Li, H., Wang, H., Qian, W., Zhang, S., Wessel, S., Cheng, T.T.H., Shen, J., Wu, S.: Chloride contamination effects on proton exchange membrane fuel cell performance and durability. J. of Power Sources 196, 6249-6255 (2011)

[18] Garzon, F.H., Lopes, T., Rockward, T., Sansiñena, J., Kienitz, B., Mukundan, R., W.: The Impact of Impurities on Long-Term PEMFC Performance. ESC Transactions 25 (1), 1575-1583 (2009)

[19] Yuan, X.-Z., Li, H., Yu, Y., Jiang, M., Qian, W., Zhang, S., Wang, H., Wessel, S., Cheng, T. T. H.: Diagnosis of contamination introduced by ammonia at the cathode in a polymer electrolyte membrane fuel cell. Int. J. of Hydrog. Energy 37, 12464-12473 (2012)

[20] Li, H., Zhang, S., Qian, W., Yu, Y., Yuan, X.-Z., Wang, H., Jiang, M., Wessel, S., Cheng, T.T.H.: Impacts of operating conditions on the effects of chloride contamination on PEM fuel cell performance and durability. J. of Power Sources 218, 375-382 (2012)

[21] Kortsdottir, K., Lindström, R.W., Lindbergh, G.: The influence of ethene impurities in the gas feed of a PEM fuel cell. Int. J. of Hydrog. Energy 38, 497-509 (2013)

[22] Faber, P., Drewnick, F., Piske, J., Kurz, T., Borrmann, S.: Effects of atmospheric aerosol on the performance of environmentally sustainable passive air-breathing PEM fuel cells. Int. J. of Hydrog. Energy 37, 17203-17208 (2012)

[23] Rockward, T., Brosha, E.L., Chlistunoff, J., Fernandez, A., Springer, T., Uribe, F., Davey, J., Urdampilleta, I., Kienitz, B., Zawodzinski, T., Lopes, T.: Effect of Fuel and Air Impurities on PEM Fuel Cell Performance. 2end Int. Workshop on Degrad. Issues of Fuel Cells (2011)

[24] Molter, T.M.: Effects of Impurities on Fuel Cell Performance and Durability. DOE Hydrog. and Fuel Cell Program, University of Conneticut (2011)

[25] Biesdorf, J., Zamel, N., Kurz, T.: Influence of air contaminants on planar, self-breathing hydrogen PEM fuel cells in an outdoor environment. J. of Power Sources, doi: 10.1016/j.jpowsour.2013.08.064 (2013)

[26] Goodwin, J.G., Zhang, J., Hongsirikarn, K.: Effects of Fuel and Air Impurities on PEM Fuel Cell Performance. DOE Hydrog. Progr. Anu. Merit Review and Peer Evaluation (2011)

[27] Molter, T. M.: Project Title: Effects of Impurities on Fuel Cell Performance and Durability, DOE Hydrog. Progr. , Final Report, DOE award No.: DE-FG36-07GO17020, University of Connecticut (2011)

Tabelle 1 Literaturübersicht von Kontaminationsanalysen für PEM-Brennstoffzellensysteme

	Anode/ Kathode	Getestete Konzentrationen	Grenzen	Literaturquelle
Inertgase (N2, Ar, H2)	A	50…80 %	k. E.	[05]
Schwefeldioxid SO2	A & K	0,5…5 ppm	<0,1...0,5 ppm	[16], [18], [04], [23], [11], [12], [13], [14], [06], [25]
Schwefelwasserstoff H2S	A & K	0,01…5 ppm	0,05 ppm	[18], [04], [23], [13], [06], [12]
Kohlenstoffmonoxid CO	A & K	0,1…250 ppm	ca. 1 ppm	[16], [26], [05], [13], [06], [02], [11], [25]
Kohlenstoffdioxid CO_2	A	5…500 ppm (25 %)	k. E. (g. E.)	[05], [16], [06]
Ammoniak NH_3	A & K	1…500 ppm	< 1 ppm	[05], [16], [18], [19]
Stickstoffdioxid NO_2	A & K	0,4…5 ppm	ca. 0,4 ppm	[18], [04], [23], [11], [12], [13], [14], [06], [25]
Ethen, Ethylen C_2H_4	A	0,01…5 %	k. E.	[02], [05], [21], [24], [27]
Ethan C2H6	A	0,01…5 %	k. E.	[27]
Propan C3H8	A & K	60…100 ppm	g. E.	[05], [23], [11]
Propylene C3H6	A	25 ppm	k. E.	[23]
Butane C4H10	A	100 ppm	g. E.	[05]
Methan CH4	A	0,01…5 %	k. E.	[27]
Pentane C5H12	A	100 pm	g. E.	[05]
Benzene C6H6	A & K	10…1000 ppm	ca. 10 ppm	[02], [11], [27]
Hexane C6H14	A	100 ppm	g. E.	[05]
Heptane C7H16	A	100 ppm	g. E.	[05]
Toluene C7H8	A	1…50 ppm	1 ppm	[02], [15], [27]
Methanol CH4O	A	500…2500 ppm	500 ppm	[27]
Aceton C3H6O	A	100…250 ppm	< 100 ppm	[27]
Formaldehyd CH2O	A	1…20 ppm	1 ppm	[27]
Tetrahydrofuran (THF) C4H8O	A	0,5…5 ppm	<0,5 ppm	[05], [26]
Tetrachloroethylene, Perchloroethylene (PCE) C2Cl4	A	0,05…400 ppm	<0,05 ppm	[05], [26], [27]
Cyanwasserstoff (Blausäure) HCN	K	1780 ppm	<< 1780 ppm	[11]
Chlorcyan CClN	K	780 ppm	<< 780 ppm	[11]
Senfgas C4H8Cl2S	K	13 ppm	<< 13 ppm	[11]
Sarin C4H10FO2P	K	170 ppm	<< 170 ppm	[11]
Cu2+, Fe3+, Na+, Ni2+	A & K		s. E.	[05], [06]
Chlorwasserstoff HCl	A & K	0,1…20 ppm	0,1…0,5 ppm	[17], [20]
Natriumcarbonat Na_2CO_3	K			[07]
Soot (Ruß)	K	100ppm	k.E.	[22]
Ammoniumsulfat $(NH_4)2SO_4$	K	100ppm	< 100ppm	[22]
Oxalsäure $C_2H_2O_4$	K	100ppm	< 100ppm	[22]
Acetaldehyd CH3CHO	A	30…250ppm	<30ppm *	[02], [05], [24], [27]
Ameisensäure HCOOH	A	2…500ppm	2…5ppm *	[05], [24], [27]
Chlormethan (Methylchlorid) CH3Cl	A	1…19ppm	k.E.	[24], [27]
Diboran B2H6	A	50ppm	<50ppm	[05]
Natriumchlorid NaCl	A & K			[16], [18], [07], [22]

* widersprüchliche Aussage der Literatur; k. E. = kein Effekt; g. E. = geringer Effekt; s. E. = starker Effekt

Modellprädiktive Leistungsregelung eines Brennstoffzellensystems mit Wechselrichter in einem Microgrid

M. Schultze, C. Hähnel und J. Horn,
Helmut-Schmidt-Universität / Universität der Bundeswehr Hamburg,
Fakultät für Elektrotechnik, Professur für Regelungstechnik, Holstenhofweg 85, 22043 Hamburg,
martin.schultze@hsu-hh.de, christian.haehnel@hsu-hh.de, joachim.horn@hsu-hh.de

Kurzfassung

Neben dem Aufbau und der Integration erneuerbarer Energien rücken zunehmend Vorgänge und Systeme in den Blickpunkt, die Schwankungen witterungsabhängiger Energiewandler bei nicht ausreichender Leistungsabgabe ausgleichen. Eine Lösung zur Bereitstellung dieser Energie kann die Brennstoffzelle sein. Ihre Energiewandlung ist hocheffizient und zeitlich unabhängig. Setzt man ein solches Brennstoffzellensystem beispielsweise in einem Inselnetz zur Unterstützung oder dem Ausgleich von Leistungsschwankungen oder -ausfällen ein, so müssen verschiedene Bedingungen beim Betrieb beachtet werden. Die Modellprädiktive Regelung stellt hier ein Regelungsverfahren dar, welches diese Nebenbedingungen direkt in den Entwurfsprozess einbindet und berücksichtigt. Zur Kopplung des Brennstoffzellensystems mit dem Netz wird ein Wechselrichter genutzt. Dieser komplettiert das System zur Energiebereitstellung. Relevante Modelle der beteiligten Komponenten werden ausgeführt. Abschließend werden experimentelle Ergebnisse zur Leistungsregelung anhand einer Polymer-Elektrolyt-Membran-Brennstoffzelle mit bis zu 4,4 kW elektrischer Leistung gezeigt.

1 Einleitung

Die heutige Landschaft der Energieversorgung in der Bundesrepublik Deutschland umfasst einen Anteil Erneuerbarer Energien (EE) von 25,3 % des Bruttostromverbrauchs (Stand 2013) [1]. Dieser soll bis zum Jahr 2025 auf 40–45 % und auf mindestens 80 % bis zum Jahr 2050 ausgebaut werden [2]. Bedingt durch die Fluktuation in Wind und Sonnenscheindauer muss der Ausfall von Erzeugern der EE kompensiert werden. Für ein Inselnetz (Microgrid) bieten sich zur Kompensation der Leistungseinbußen Polymer-Elektrolyt-Membran-(PEM)-Brennstoffzellen an. Sie verstromen den zu Überschusszeiten erzeugten und gespeicherten Wasserstoff (siehe **Bild 1**). Die PEM-Brennstoffzellen liefern einen Gleichstrom (engl. direct current, dc) und werden daher über einen Wechselrichter an das Wechselstromnetz (engl. alternating current, ac) des Inselnetzes angekoppelt.

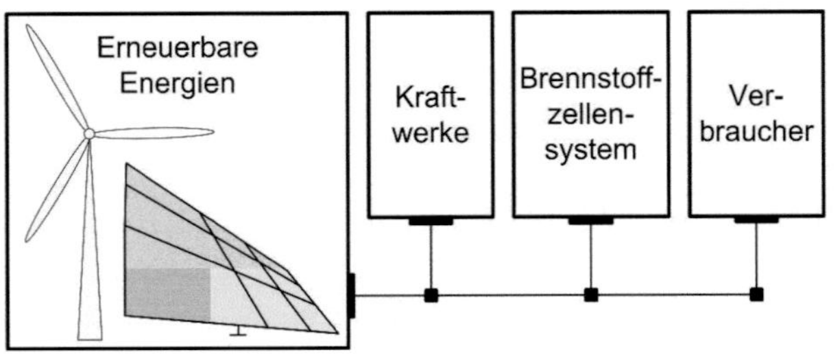

Bild 1 Elektrische Energieversorgung eines Inselnetzes mit Erneuerbaren Energien und Brennstoffzellen

Beim Betrieb von PEM-Brennstoffzellen ist eine ausreichende Zufuhr von Sauerstoff und Brennstoff zu gewährleisten, um eine Unterversorgung zu vermeiden. Diese kann sonst zu Schädigungen am Brennstoffzellenstack führen [3]. Eine Unterversorgung tritt bei schnellen Lastwechseln auf, wenn die Dynamik der Stoffzufuhr nicht ausreicht, um Sauerstoff oder Brennstoff in der benötigten Menge nachzuführen. Eine Begrenzung des Lastanstiegs wird durch eine Begrenzung des Gradienten der Stackstromstärke der Brennstoffzelle vorgenommen und vermindert so eine Unterversorgung. Bei der Kopplung des Brennstoffzellenstacks mit einem Wechselrichter ist die Stackstromstärke nunmehr ein interner Zustand und kann nicht direkt vorgegeben werden. Das Verfahren der Modellprädiktiven Regelung (MPR) ermöglicht es, systeminterne Zustände und damit auch den Stackstromstärke-Gradienten zu beschränken. Modellprädiktive Regelungen wurden bereits für Brennstoffzellensysteme mit Gleichspannungswandlung [4] entwickelt. Für ein PEM-Brennstoffzellensystem mit Wechselrichter ist dieser Ansatz neuartig. Dieser Beitrag stellt für ein Microgrid mit EE, Kraftwerk und Verbrauchern eine Nichtlineare Modellprädiktive Regelung der vom Brennstoffzellensystem mit Wechselrichter ins Netz eingespeisten elektrischen Leistung vor.

2 Modellierung

Das Brennstoffzellensystem mit Brennstoffzellenstack, geregelter Wasserstoff- und Luftzufuhr und Wechselrichter ist schematisch in **Bild 2** dargestellt.

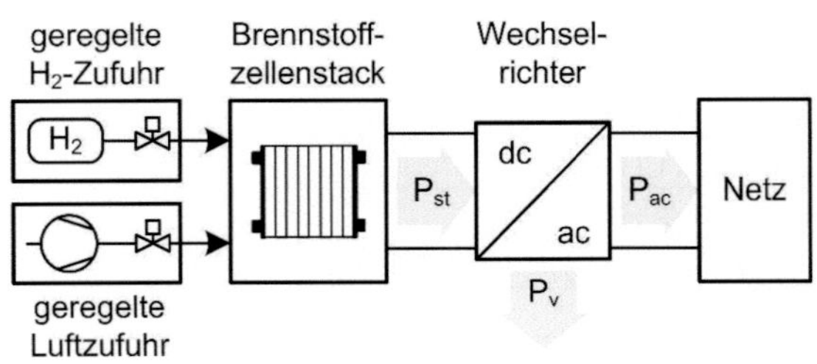

Bild 2 Schematischer Aufbau des Brennstoffzellensystems mit Wechselrichter zum Netzanschluss

Es besteht aus dem PEM-Brennstoffzellenstack mit einer geregelten Wasserstoff (H_2)- und Luftzufuhr und dem Wechselrichter. Im Folgenden wird die Modellierung des Systems aufgezeigt.

2.1 Modell des PEM-Brennstoffzellenstacks

Das Modell des PEM-Brennstoffzellenstacks beschreibt die wesentliche Dynamik des Systems und beinhaltet die geregelte Zufuhr des Luftmassenstroms W_{mfc}, den Kathodendruck p_{st} im Stack und die Stackstromstärke I_{st}. Das Modell ist vom nichtlinearen Brennstoffzellenmodell [5] abgeleitet. Die Dynamiken werden linear als Systeme erster Ordnung mit den Zeitkonstanten τ_{mfc}, τ_p und τ_I angenähert. Die Sollwerte für W_{mfc} und p_{st} sind Funktionen von I_{st} mit $W_{mfc,soll} = W_{mfc,soll}(I_{st})$ und $p_{st,soll} = p_{st,soll}(I_{st})$. Die vom Wechselrichter aufgeprägte Stromstärke ist $I_{dc,in}$.

$$\frac{dW_{mfc}}{dt} = \frac{1}{\tau_{mfc}}\left(W_{mfc,soll} - W_{mfc}\right) \tag{1}$$

$$\frac{dp_{st}}{dt} = \frac{1}{\tau_p}\left(p_{st,soll} - p_{st}\right) \tag{2}$$

$$\frac{dI_{st}}{dt} = \frac{1}{\tau_I}\left(I_{dc,in} - I_{st}\right) \tag{3}$$

Luft besteht zu 21 % (vol.) aus Sauerstoff (O_2) und 79 % (vol.) aus Stickstoff (N_2). Der O_2-Partialdruck p_{O2} im Brennstoffzellenstack ergibt sich aus der stationären Bilanz der Stoffströme von N_{O2} für O_2 und N_{N2} für N_2. Die Stoffströme sind in (5) und (6) aufgezeigt. Die chemische Reaktion entzieht W_{mfc} einen Anteil O_2 proportional zu I_{st} und der Zellenanzahl n_{zellen} im Stack. M_{O2} und M_{N2} sind die Molaren Massen, F ist die Faraday-Konstante und x_{O2} ist der Massenanteil von O_2 in Luft mit $x_{O2} = 0{,}23$.

$$p_{O2} = \left(\frac{N_{O2}}{N_{O2} + N_{N2}}\right)p_{st} \tag{4}$$

$$N_{O2} = \frac{1}{M_{O2}} x_{O2} W_{mfc} - \frac{I_{st}}{4F} n_{zellen} \tag{5}$$

$$N_{N2} = \frac{1}{M_{N2}}\left(1 - x_{O2}\right) W_{mfc} \tag{6}$$

Eine Einführung zur Modellierung der Zellspannung gibt [6]. Sie setzt sich aus der reversiblen Zellspannung U_{rev}, Aktivierungsverlusten U_{akt} und ohmschen Verlusten zusammen. Konzentrationsverluste werden aufgrund des limitierten Betriebsbereichs vernachlässigt. Die Zellspannung hängt von I_{st}, der Stacktemperatur T_{st}, von p_{O2} und dem H_2-Partialdruck p_{H2} ab. Die Anodendruckregelung motiviert die Annahme eines konstanten p_{H2}. Die Spannung U_{akt} wird nach [7] modelliert.

$$\begin{aligned} U_{rev} &= 1{,}229 - 0{,}85 \cdot 10^{-3}\left(T_{st} - 298{,}15\right) \\ &\quad + 4{,}3 \cdot 10^{-5} T_{st}\left(\ln\left(p_{H2}/p_0\right) + 0{,}5 \cdot \ln\left(p_{O2}/p_0\right)\right) \end{aligned} \tag{7}$$

$$\begin{aligned} U_{akt} &= \zeta_1 + \zeta_2 T_{st} + \zeta_3 T_{st} \ln\left(p_{O2} e^{498/T_{st}} / \left(5{,}08 \cdot 10^6\right)\right) \\ &\quad + \zeta_4 T_{st} \ln\left(I_{st}\right) \end{aligned} \tag{8}$$

Die sehr dünne Polymermembran wird als ausreichend befeuchtet angenommen und besitzt somit einen konstanten ohmschen Membranwiderstand R_{ohm}. Die Stackspannung U_{st} ergibt sich als Summe aller Zellspannungen im Stack, wobei identische Zellen angenommen werden.

$$U_{st} = n_{zellen}\left(U_{rev} - U_{akt} - R_{ohm} I_{st}\right) \tag{9}$$

Die Parameter $\zeta_1, \ldots, \zeta_4$ und R_{ohm} wurden durch Minimierung der Fehlerquadrate aus experimentellen Daten identifiziert. Die vom Brennstoffzellenstack bereitgestellte elektrische Leistung P_{st} ergibt sich wie folgt.

$$P_{st} = U_{st} I_{st} \tag{10}$$

2.2 Modell des Wechselrichters

Das Modell des Wechselrichters ergibt sich aus der Leistungsbilanz und der Verlustleistung P_v nach [8]. P_v wird in Abhängigkeit von der abgegebenen Leistung P_{ac} modelliert. Hierbei ist $p_{ac} = P_{ac}/P_{nom}$ die auf die nominelle Leistung von $P_{nom} = 5$ kW normierte Leistung. Die Parameter a_0, a_1 und a_2 ergeben sich nach [9] mit dem in [8] verwendeten Wechselrichter. Der Wirkungsgrad η_{inv} des Wechselrichters ist in **Bild 3** über P_{ac} aufgetragen. Die vom Wechselrichter eingeprägte Stromstärke ist $I_{dc,in}$.

$$P_v = \left(a_0 + a_1\left(p_{ac}\right) + a_2\left(p_{ac}\right)^2\right)P_{nom} \tag{11}$$

$$\eta_{inv} = P_{ac}/\left(P_{ac} + P_v\right) \tag{12}$$

$$I_{dc,in} = \left(P_{ac}/\eta_{inv}\right)/U_{st} \tag{13}$$

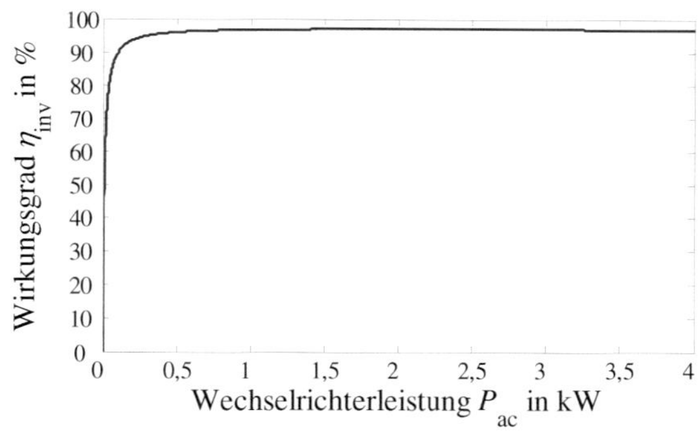

Bild 3 Wirkungsgrad des Wechselrichters

2.3 Gesamtsystemmodell

Das Gesamtsystemmodell hat die Zustände $\underline{x} = [x_1\ x_2\ x_3]^T$ mit $\underline{x} = [W_{mfc}\ p_{st}\ I_{st}]^T$ und die fünf Ausgangsgrößen $\underline{y} = [y_1\ y_2\ y_3\ y_4\ y_5]^T$ mit $\underline{y} = [W_{mfc}\ p_{st}\ I_{st}\ U_{st}\ P_{st}]^T$. Zur schnellen Bereitstellung elektrischer Leistung würde das Gesamtsystem mit einem schnellen elektrischen Speicher gekoppelt werden. Ein Powermanagement führt dabei die Leistungsflüsse zwischen Brennstoffzellensystem und Speicher durch. Das Managementsystem ist hier als ideal angenommen, so dass es die vom Brennstoffzellensystem bereitzustellende Leistung P_{ac} unmittelbar vorgibt. Die Stellgröße $u = P_{ac}$ ist die bereitzustellende elektrische Leistung. Das Gesamtsystemmodell lautet wie folgt.

$$\begin{aligned} \dot{\underline{x}} &= \underline{f}\left(\underline{x}, u\right) \\ \underline{y} &= \underline{h}\left(\underline{x}, u\right) \end{aligned} \tag{14}$$

3 Modellprädiktive Regelung der elektrischen Leistung

In der Modellprädiktiven Regelung (MPR) wird ein Optimal-Steuerungsproblem mit endlichem Zeithorizont gelöst. Die Lösung ist eine optimale Steuerfolge, deren erstes Element angewendet wird. Der aktuelle Systemzustand, der durch Messung oder Schätzung ermittelt wird, dient als Ausgangszustand. Nach jedem Taktschritt verschiebt sich der Zeithorizont um einen Schritt und der Vorgang wird wiederholt. Die Vorteile der MPR sind die Einbeziehung von Limitierungen der Zustands- und Stellgrößen als Nebenbedingungen. Nichtlineare Nebenbedingungen oder ein nichtlineares Systemmodell führen dabei zur Nichtlinearen Modellprädiktiven Regelung (NMPR) [10]. Eine Einführung zur NMPR ist in [11] gegeben. Das Optimierungsproblem einer NMPR ist in der Regel nicht analytisch lösbar, so dass numerische Verfahren eingesetzt werden. Ein gängiges Verfahren zur Lösung des Optimierungsproblems ist die Sequentiellen Quadratische Programmierung (SQP).

Die im Folgenden vorgestellte NMPR soll die vom Gesamtsystem aus Brennstoffzellenstack und Wechselrichter bereitzustellende elektrische Leistung P_{ac} derart vorgeben, dass dabei die folgenden Nebenbedingungen auf die Stackstromstärke I_{st} als Zustandsgröße x_3, auf die Stackspannung U_{st} als Ausgangsgröße y_4 und auf P_{ac} eingehalten werden. Die NMPR ist somit als ein Bereich des Powermanagements zu betrachten.

$$\begin{aligned} P_{min} &\leq P_{ac} \leq P_{max} \\ I_{min} &\leq I_{st} \leq I_{max} \\ -\Delta I &\leq \frac{d}{dt} I_{st} \leq +\Delta I \\ U_{min} &\leq U_{st} \leq U_{max} \end{aligned} \tag{15}$$

3.1 Prädiktion der Anfangsbedingung

Die Berechnung der Stellgröße erfolgt innerhalb des aktuellen Taktintervalls $[t\ t+T_s]$. Das Ergebnis wird zum nächsten Zeitschritt $t+T_s$ angewendet. Die NMPR benötigt hierzu den aktuellen Systemzustand $\underline{x}$. Um eine Verzögerung von einer Taktperiode T_s zu vermeiden, wird anhand des aktuell anliegenden Stelleingangs u der Systemzustand $\underline{x}(k=0)=\underline{x}(t+T_s)$ zum Zeitpunkt $t+T_s$ prädiziert und an die NMPR übergeben. Die Prädiktion erfolgt durch numerische Zeitintegration anhand eines Runge-Kutta-Verfahrens vierter Ordnung. Die Prädiktion ist bei Echtzeitanwendungen für stabiles Verhalten notwendig, wenn die Rechenzeiten in Relation zu den Abtastzeiten relativ groß sind.

3.2 Herleitung des Optimierungsproblems

Für das Optimierungsproblem erfolgt eine Diskretisierung des kontinuierlichen Systemmodells (14) über N Zeitschritte mit der Abtastzeit T_s durch numerische Zeitintegration anhand eines Runge-Kutta-Verfahrens vierter Ordnung. Das führt auf das zeitdiskrete Gesamtsystemmodell (16). Hierbei bezeichnet $\underline{x}(k) = \underline{x}(t+(k+1)T_s)$.

$$\begin{aligned} \underline{x}(k+1) &= \underline{F}(\underline{x}(k), u(k)) \\ \underline{y}(k) &= \underline{H}(\underline{x}(k), u(k)) \end{aligned} \tag{16}$$

Die N Stellgrößen $u(k)$, die zur Zeitintegration vorgegeben werden, sind $\underline{u}=[u(0), u(1), \dots u(k), \dots u(N\text{-}1)]^T$. Sie sind zugleich die N Optimierungsvariablen des Minimierungsproblems. Zusätzlich zur Anfangsbedingung $\underline{x}(k=0)$ liefert die Diskretisierung N weitere Zustandsvektoren $\underline{x}(k)=[x_1(k), x_2(k), x_3(k)]$ mit je drei Zuständen, so dass insgesamt $N+1$ Zustandsvektoren vorliegen. Weiterhin liefert die Diskretisierung $N+1$ Ausgangsvektoren $\underline{y}(k) = [y_1(k), y_2(k), y_3(k), y_4(k), y_5(k)]$ mit je fünf Ausgängen. Der Prädiktionshorizont beträgt $N{\bullet}T_s$.

Ziel ist die Folgeregelung der vom Wechselrichter abgegebenen Leistung P_{ac} entsprechend der Sollwertvorgabe $P_{ac,soll}$. Zusätzlich sollen sprunghafte Änderungen der eingeprägten Stackstromstärke I_{st} zur Schonung der Brennstoffzelle vermieden werden. Diese Anforderungen fließen in das Optimierungskriterium $J(\underline{u})$ ein. Zur Berechnung der Änderung der Stackstromstärke von Schritt $k=0$ zum Schritt $k=1$ wird die im vorherigen Taktschritt eingeprägte Stackstromstärke $x_3(k=0)$ herangezogen. Sie liegt als Messgröße in der Anfangsbedingung $\underline{x}(k=0)$ vor.

$$J(\underline{u}) = \sum_{k=0}^{N-1} q_1 \left(u(k) - P_{ac,soll}\right)^2 + \sum_{k=1}^{N-1} q_2 \left(x_3(k) - x_3(k-1)\right)^2 \tag{17}$$

Die in (15) aufgeführten Beschränkungen fließen als Nebenbedingung ein. Der Gradient der Stackstromstärke I_{st} (Zustand x_3) wird durch den Differenzenquotienten approximiert. Das Optimierungsproblem zur Minimierung der Gütefunktion $J(\underline{u})$ unterliegt der Systemdynamik und den Nebenbedingungen und lautet wie folgt.

$$\begin{aligned} &\min_{\underline{u}} J(\underline{u}) && \qquad (18) \\ &\underline{x}(k+1) = \underline{F}(\underline{x}(k), u(k)) && \\ &P_{min} \leq u(i) \leq P_{max} && (i = 0, 1, \dots, N\text{-}1) \\ &I_{min} \leq x_3(i) \leq I_{max} && (i = 1, 2, \dots, N) \\ &-\Delta I \leq \frac{x_3(i) - x_3(i-1)}{T_s} \leq +\Delta I && (i = 1, 2, \dots, N) \\ &U_{min} \leq y_4(i) \leq U_{max} && (i = 1, 2, \dots, N) \end{aligned}$$

Das Minimierungsproblem (18) wird durch das SQP-Verfahren *solnp* [12] in MATLAB (The MathWorks®) gelöst. Für eine schnelle Konvergenz wird die Startlösung des SQP-Verfahrens aus dem Lösungsvektor des letzten Taktschritts als $[u(1), \dots u(N\text{-}1), u(N\text{-}1)]$ konstruiert. Die Lösung des Optimierungsproblems (18) ist der Stellgrößenvektor $\underline{u}^*$. Von diesem wird der erste Eintrag $u^*(0)$ angewendet.

4 Simulation

Die NMPR der abgegebenen elektrischen Leistung P_{ac} wurde am Modell des Brennstoffzellensystems mit Wechselrichter implementiert und in MATLAB simuliert. Ziel ist die Bereitstellung der Sollleistung $P_{ac,soll}$ bei gleichzeitiger Einhaltung der in (15) gegebenen Nebenbedingungen. Die Nebenbedingungen sind in den einzelnen Abbildungen als Limitierung dargestellt. Die Sollleistung wird als stufenförmiges Profil mit $P_{ac,soll}$= 1,2 kW, 2 kW und 2,8 kW vorgegeben. Es wurden $\Delta I = 5$ A/s gesetzt und der Brennstoffzellenstack bei $T_{st} = 323$ K betrieben. Die Länge des Prädiktionshorizonts nimmt signifikant Einfluss auf die Rechenzeit. Hier wurde $N = 25$ gewählt und somit ein Prädiktionshorizont von 2,5 s gesetzt.

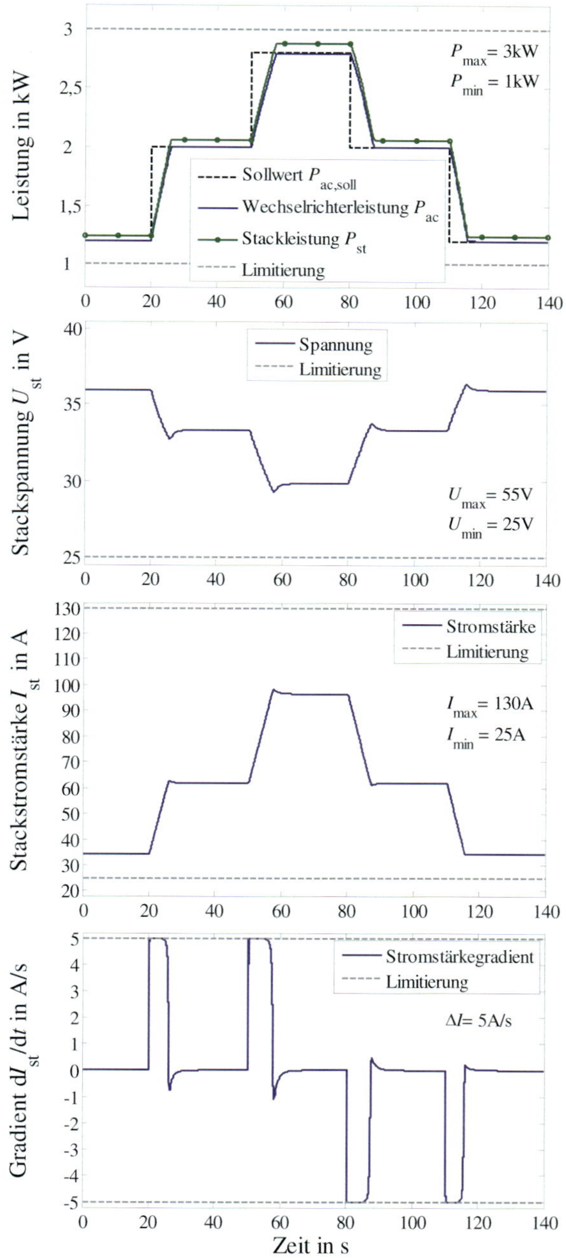

Bild 4 Simulationsergebnisse für einen Prädiktionshorizont von 2,5s (N = 25 bei einer Taktrate T_s = 0,1s)

Bild 4 zeigt die Sollwerte und die Simulationsergebnisse. Die Berechnung des Stellgrößenvektors $\underline{u}^*$ ist in weniger als 60 ms abgeschlossen und ermöglicht unter Nutzung des beschrieben Prädiktionsschrittes eine echtzeitfähige NMPR. Die Simulationsergebnisse zeigen, dass die Beschränkungen auf die Stackstromstärke und ihren Gradient, die Stackspannung und die abgegebene Leistung eingehalten werden. Der Wirkungsgrad des Wechselrichters bewirkt, dass die vom Brennstoffzellenstack bereitgestellte Leistung P_{st} größer als die vom Wechselrichter eingespeiste Leistung P_{ac} ist.

5 Brennstoffzellenstack und experimentelle Ergebnisse

Neben den zuvor aufgeführten Simulationsergebnissen werden im Folgenden der verwendete PEM-Brennstoffzellenstack, der Brennstoffzellenprüfstand und die Versuchsbedingungen vorgestellt. Auszüge aus den experimentellen Ergebnissen werden dargestellt, erläutert und mit den Simulationsergebnissen verglichen.

5.1 Brennstoffzellenstack PM200

Das Kernstück des Brennstoffzellensystems bei den durchgeführten Experimenten ist vom Typ „PM200“ der Fa. Proton Motor Fuel Cell GmbH. Der Brennstoffzellenstack besteht aus 48 Einzelzellen und ist der Lage, im Überlastbereich bis zu 4,4 kW elektrische Leistung bereitzustellen. Der genutzte Stack benötigt keine gesonderte Befeuchtung der Reaktionsgase und kann in mobilen, stationären und maritimen Anwendungen eingesetzt werden. Grundsätzlich wird der Stack mit trockener, komprimierter Umgebungsluft und hochreinem Wasserstoff betrieben. Anodenseitig wird der nicht verwendete Wasserstoff entfeuchtet und über eine Rezirkulationspumpe am Stackeingang erneut zur Verfügung gestellt. Diese Maßnahme erhöht die Effizienz, da der Volumenstrom innerhalb des Stacks angehoben wird. Mit dieser Maßnahme wird zusätzlich sichergestellt, dass eingelagertes Produktwasser leichter den Stack verlässt und die Reaktion nicht beeinflusst. **Bild 5** zeigt den verwendeten Brennstoffzellenstack in seiner Prüfumgebung.

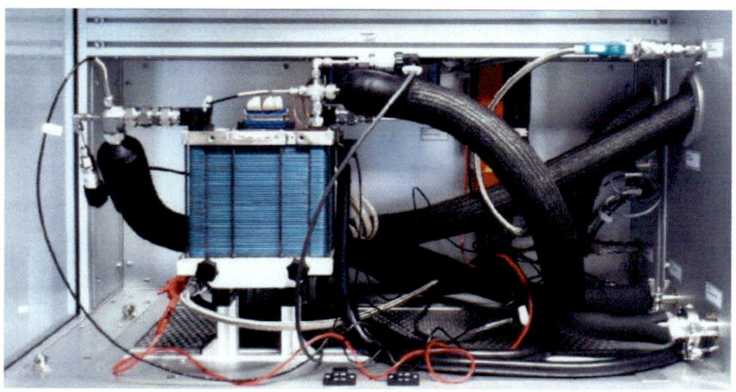

Bild 5 Brennstoffzellenstack PM 200 (48 Zellen) im Prüflingsraum des Brennstoffzellenprüfstandes

Der verwendete Brennstoffzellenstack kann kontinuierlich in einem Bereich von 25 A bis 130 A betrieben werden. Stromstärken unter 20 A werden nur für den Anfahrvorgang durchlaufen. Die verbleibenden Stromstärken von 130 A bis 150 A sollen maximal für 5 % der Gesamtbetriebsdauer genutzt werden. Wie in der Simulation bereits berücksichtigt soll ein Lastwechsel mit nur $\Delta I = 5$ A/s durchgeführt werden. Eine Druckbeaufschlagung der Anoden- und Kathodenseite ist bis zu 1,7 bar möglich. Der vorgesehene Temperaturbereich ist zwischen 48 °C und 60 °C. Effiziente und sichere Arbeitspunkte für alle Drücke, Temperaturen, Temperaturdifferenzen, Stöchiometrien und Rezirkulationsvolumen ergeben sich immer aus der aktuellen Stackstromstärke.

5.2 Experimentelle Ergebnisse

Für die Durchführung der Experimente wurden die gleichen Sollleistungsverläufe und Begrenzungen wie in der Simulation gewählt.

Für die Bereitstellung der Reaktionsgase sowie der anoden- und kathodenseitigen Druckregelung sind unterlagerte Regler eingesetzt. Die Überwachung der Stackeingangstemperatur, der Temperaturdifferenz über Ein- und Ausgang sowie des anodenseitigen Rezirkulationsvolumenstroms werden über unterlagerte Regler realisiert.

Die Abbildung des Wechselrichters erfolgt als Modell in Form einer Simulation, da dieser nicht Teil des Prüfstandes ist. Die dargestellte Ausgangsleistung des Wechselrichters ergibt sich somit aus seinem vom Arbeitspunkt abhängigen Wirkungsgrad (12).

Die Berechnung des Stellgrößenvektors $\underline{u}^*$ ist bei der Durchführung der Experimente in weniger als 65 ms abgeschlossen und ermöglicht somit auch hier eine echtzeitfähige NMPR. Die etwas höhere Rechenzeit folgt aus der Messung der Drücke, des Massenstroms und der Temperaturdaten und deren Abweichungen, was einen Einfluss auf die Prädiktion und zugrundeliegenden Modelle und somit auf die Lösung des Optimierungsproblems hat. Die experimentellen Ergebnisse sind in **Bild 6** dargestellt. Die Sollausgangsleistung des Wechselrichters wird mit $P_{ac,soll\,l} = 1{,}2$ kW, 2 kW und 2,8 kW vorgegeben. Es wurden $\Delta I = 5$ A/s gesetzt und der Brennstoffzellenstack bei $T_{st} = 323$ K betrieben. Die Länge des Prädiktionshorizonts nimmt signifikant Einfluss auf die Rechenzeit. Hier wurde $N = 25$ gewählt und somit ein Prädiktionshorizont von 2,5 s gesetzt. Der in **Bild 6** dargestellte Gradient der Stackstromstärke wurde durch numerische Approximation anhand des Differenzenquotienten $(I_{st}(k+1)\text{-}I_{st}(k))/T_s$ gebildet. Geringe Überschreitungen der Grenzen von 5 A/s und -5 A/s resultieren aus der Abweichung zwischen dem Regelungsmodell der NMPR und der realen Strecke.

Die NMPR regelt die am Wechselrichter abgegebene Leistung mit einer Abweichung von weniger als 1 % sehr genau ein. Die geringe Abweichung zwischen Soll- und Istleistung $P_{ac,soll}$ und P_{ac} resultiert aus der Abweichung zwischen dem Modell und der realen Strecke Brennstoffzelle. Weiterhin hält die NMPR der am Wechselrichter bereitgestellten Leistung die in (15) vorgegebenen Begrenzungen für Ausgangsleistung, Stackspannung, Stackstromstärke und Stackstromstärkegradient ein.

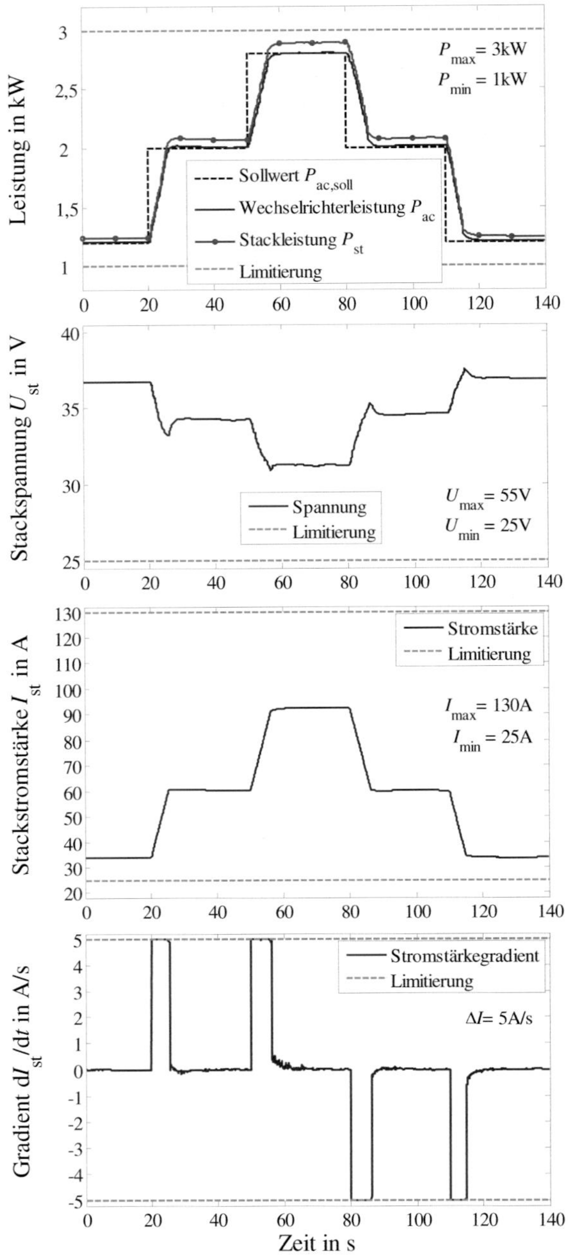

Bild 6 Experimentelle Ergebnisse für einen Prädiktionshorizont von 2,5 s (N = 25 bei einer Taktrate T_s = 0,1 s)

6 Zusammenfassung und Ausblick

Gezeigt wurde eine mögliche Anwendung von Brennstoffzellen zur exakten Bereitstellung von elektrischer Energie unter Beachtung von Nebenbedingungen und dem nichtlinearen Verhalten der Brennstoffzelle sowie des Spannungswandler. Das Verfahren der Modellprädiktiven Regelung wurde hierzu angewendet. Die Modellierung des Gesamtsystems aus Brennstoffzelle und Wechselrichter wurde vorgestellt und Ergebnisse der Modellprädiktiven Regelung an der Simulation gezeigt. Experimentelle Ergebnisse der Modellprädiktiven Regelung zur Bereitstellung der Wechselrichterleistung wurden präsentiert. In zukünftigen Arbeiten wird das Verhalten der Leistungsregelung bei Vorliegen von Abweichungen zwischen Strecke und Modell untersucht.

7 Modellparameter

Die in Abschnitt 2. Modellierung aufgeführten Zustände und verwendeten Modellparameter sind in nachstehender **Tabelle 1** aufgeführt.

Tabelle 1 Zustände und Modellparameter

Größe	Beschreibung	Einheit
W_{mfc}	Luftmassenstrom	kg/s
$W_{mfc,soll}$	Luftmassenstromsollwert $W_{mfc,soll} = (1{,}7 \cdot 10^{-3} I_{st}^2 + 0{,}9203 I_{st} + 24{,}3189)/(4{,}6417 \cdot 10^4)$	kg/s
p_{st}	Kathodendruck	Pa
$p_{st,soll}$	Sollwert des Kathodendrucks $p_{st,soll} = 1{,}2 \cdot 10^5 + 1{,}25 \cdot 10^3 (I_{st} - 30)$ $1{,}013 \cdot 10^5 \leq p_{st,soll} \leq 1{,}7 \cdot 10^5$	Pa
I_{st}	Stackstromstärke	A
$I_{dc,in}$	vom Wechselrichter eingeprägte Stromstärke	A
p_{O2}	Sauerstoffpartialdruck	Pa
p_{H2}	Wasserstoffpartialdruck	$1{,}4 \cdot 10^5$ Pa
p_0	Standarddruck	$1{,}013 \cdot 10^5$ Pa
N_{O2}	Sauerstoff-Stoffstrom	mol/s
N_{N2}	Stickstoff-Stoffstrom	mol/s
U_{rev}	Reversible Zellspannung	V
U_{akt}	Aktivierungsverluste	V
U_{ohm}	Ohmsche Verluste	V
U_{st}	Stackpannung	V
T_{st}	Stacktemperatur	K
P_{st}	Stackleistung (el.)	W
P_{ac}	Wechselrichterleistung (el.)	W
P_v	Verlustleistung Wechselrichter	W
P_{nom}	Nominelle Leistung	5000 W
η_{inv}	Wirkungsgrad	-
τ_{mfc}	Zeitkonstante Modell der Luftzufuhr	1,80 s
τ_p	Zeitkonstante Kathodendruckmodell	0,50 s
τ_I	Zeitkonstante Modell der Stackstromstärke	0,01 s
x_{O2}	Luftsauerstoffmassenanteil	0,23
M_{O2}	Molare Masse Sauerstoff	$3{,}1998 \cdot 10^{-2}$ kg/mol
M_{N2}	Molare Masse Stickstoff	$2{,}8013 \cdot 10^{-2}$ kg/mol
F	Faraday-Konstante	$9{,}64853 \cdot 10^4$ C/mol
n_{zellen}	Zellenanzahl im Stack	48
ζ_1	Parameter Stackspannungsmodell	1,0768
ζ_2		$-3{,}2885 \cdot 10^{-3}$
ζ_3		$-2{,}2012 \cdot 10^{-4}$
ζ_4		$6{,}1705 \cdot 10^{-5}$
R_{ohm}	Ohmscher Widerstand	$1{,}8142 \cdot 10^{-3}$
a_0	Parameter Wechselrichterverlustleistungsmodell	$1{,}85 \cdot 10^{-3}$
a_1		$2{,}22 \cdot 10^{-2}$
a_2		$1{,}01 \cdot 10^{-2}$

8 Literatur

[1] Bundesministerium für Wirtschaft und Energie (Hrsg.): Zeitreihen zur Entwicklung der erneuerbaren Energien in Deutschland, Stand 02/2014

[2] Gesetz für den Ausbau erneuerbarer Energien (Erneuerbare-Energien-Gesetz-EEG 2014), Nichtamtliche Lesefassung des EEG in der ab 1. August 2014 geltenden Fassung

[3] Borup, R., Meyers, J., Pivovar, B., Kim, Y.S., Mukundan, R. et al., Scientific aspects of polymer electrolyte fuel cell durability and degradation, Chemical Reviews, 107 Nr. 10, S. 3904–3951 (2007)

[4] Niemeyer, J.: Modellprädiktive Regelung eines PEM-Brennstoffzellensystems, Dissertation, Universität Karlsruhe, Karlsruhe (2009)

[5] Hähnel, C., Aul, V. und Horn, J., Power Control for Efficient Operation of a PEM Fuel Cell System by Nonlinear Model Predictive Control , 1st Conference on Modelling, Identification and Control of Nonlinear Systems (MICNON), St. Petersburg, Russland (2015)

[6] O'Hayre, R., Cha, S.-W., Colella, W. und Prinz, F.B., Fuel Cell Fundamentals, John Wiley & Sons, Inc., Hoboken, NJ. (2009)

[7] Amphlett, J. C., Baumert, R. M., Mann, R. F., Peppley, B. A. et al., Performance Modeling of the Ballard Mark IV Solid Polymer Electrolyte Fuel Cell I. Mechanistic Model Development, Journal of the Electrochemical Society, 142 Nr. 1, S. 1–8 (1995)

[8] Thomas, M. C.: Optimierte Betriebsführung von Niederspannungsnetzen mit einem hohen Anteil an dezentraler Erzeugung. Dissertation, ETH Zürich (2007)

[9] Schmidt, H. und Sauer, D. U.: Wechselrichter-Wirkungsgrade - Praxisgerechte Modellierung und Abschätzung. Fraunhofer-Institut für Solare Energiesysteme. Sonnenenergie Bd. 4, 43–47 (1996)

[10] Allgöwer, F., Findeisen, R. und Nagy, Z. K., Nonlinear Model Predictive Control: From Theory to Application, Journal of Chinese Institute of Chemical Engineers, 35 Nr. 3, S. 299-315 (2004)

[11] Grüne, L. und Pannek, J., Nonlinear Model Predictive Control: Theory and Algorithms, Springer, London (2011)

[12] Y. Ye: SOLNP user's guide - A Nonlinear Optimization Program in MATLAB. Dept. of Management Sciences, University of Iowa (1989)

Die Druckaufgeladene Dampfwirbelschicht-Trocknung von Braunkohlen (DDWT)

M. Merzsch[1], J. Posselt[1], P. Wiedemann[1], H.-J. Krautz[1]
[1] Brandenburgische Technische Universität Cottbus - Senftenberg, Lehrstuhl Kraftwerkstechnik, Universitätsstraße 22 (MZG), 03046 Cottbus, matthias.merzsch@b-tu.de

Kurzfassung

Aufgrund der Vorrangregelung gemäß Erneuerbare-Energien-Gesetz (EEG) bei der Einspeisung Erneuerbarer Energien in die Energieversorgungsnetze, werden konventionelle Kraftwerke bereits heute vor große Herausforderungen hinsichtlich flexibler Betriebsweisen gestellt. Das Ziel der Bundesregierung, den Anteil von Strom aus Erneuerbaren Energien auf 40 bis 45 Prozent im Jahre 2025 anzuheben, erhöht diese Anforderungen zunehmend.
Die Sicherung einer weiterhin stabilen und kostengünstigen Energiebereitstellung erfordert für die Energieunternehmen und Netzbetreiber anlagen- und verfahrenstechnische Maßnahmen zur flexiblen Anpassung an die neuen Lastverhältnisse. Neben intelligenten Lastregelungen, liegt ein großes Innovationspotential für konventionelle, braunkohlebasierte Kraftwerke in der effizienten Vortrocknung des wasserreichen Brennstoffes Rohbraunkohle (RBK) mittels Druckaufgeladener Dampfwirbelschicht-Trocknung (DDWT).
Nach einem kurzen Aufzeigen einiger Auswirkungen der Energiewende für konventionelle Kraftwerke sowie dem Übertragungsnetz sollen in diesem Beitrag die Grundlagen des druckaufgeladenen Trocknungsverfahren, die Projektentwicklung am Lehrstuhl Kraftwerkstechnik, ergänzt mit einigen konkreten Forschungsergebnissen, sowie ein kurzer Ausblick dargelegt werden.

1 Herausforderungen der Energiewende für konventionelle Kraftwerke

Bestehende konventionelle Kraftwerke müssen aufgrund der neuen Anforderungen an die Bereitstellung der versorgungssichernden und stabilitätsgewährleistenden Residuallast bereits heute zeitweise eine diffizile Betriebsweise durchfahren. Als Beleg hierfür sei genannt, dass in der Regelzone des Netzbetreibers 50Hertz eine Wind- und Photovoltaik- (PV) Leistung von insgesamt 20,8 GW (Stand: 31. Juli 2014, Quelle EEG-Anlagenstammdaten von 50Hertz) installiert ist, die maximale Regelzonenlast jedoch lediglich 17,0 GW beträgt. Des Öfteren übersteigt dabei die eingespeiste Leistung aus Erneuerbaren Energien die Regelzonenlast. Als Beispiel sind in **Bild 1** der Lastgang der Regelzone von 50Hertz, die darin eingespeiste Windleistung und weitere wichtige charakteristische Kennzahlen der Netzbelastung, wie Windgradienten sowie Eingriffe nach § 13 EnGW für den Dezember 2014 dargestellt. Dabei sind Windgradienten von mehr als 1,5 GW/h auszumachen, die größten Stundensprünge belaufen sich laut 50Hertz bereits auf über 2 GW/h. Diese werden in absehbarer Zeit aufgrund des Zuwachses an erneuerbarer Erzeugungsleistung weiter steigen.

In **Bild 1** ist weiterhin zu erkennen, dass diese hohen Gradienten gerade in Zeiten hoher Windeinspeisung, d. h. geringer Last der konventionellen Kraftwerke auftreten. In solchen Lastfällen müssen konventionelle Blöcke mitunter bereits in den Stillstand versetzt worden sein und stehen damit für die Bereitstellung und dem Ausgleich dieser Regelleistungsanforderung nicht zur Verfügung, was die Lage verschärft und die Notwendigkeit der Absenkung technischer Mindestlasten intensiviert. Konventionelle Kraftwerke werden auch weiterhin benötigt, davon zeugen die auch länger währenden Zeiten in denen Windflauten herrschen.

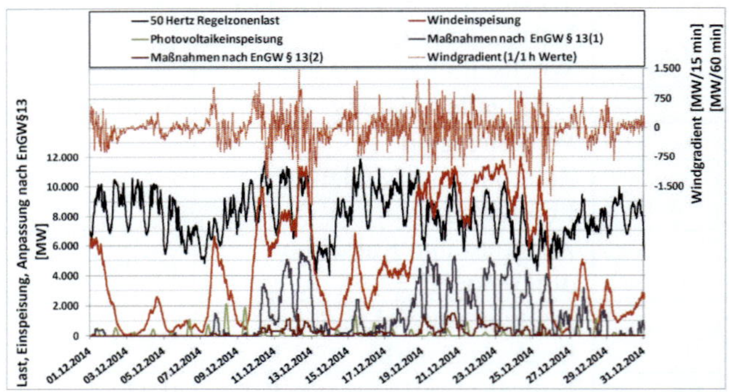

Bild 1 Windenergieeinspeisung und Anpassungen nach §13 EnGW in der Regelzone von 50 Hertz [Rohdaten: www.50hertz.com]

2 Grundlagen des DDWT-Verfahrens

Stand der Technik bei Kraftwerksneubauten zur Braunkohleverstromung ist die Trocknung der grubenfeuchten Rohkohle mittels Schlagradmühlen direkt vor dem Einblasen in den Kessel. Hierbei werden über 900 °C heiße Rauchgase mit hoher Exergie aus dem Kessel als Trocknungs- und Fördermedium benutzt. Das Kohle-Wasserdampf-Rauchgasgemisch befindet sich beim Eintritt in den Kessel auf einem Temperaturniveau von rd. 160 °C [1].

Lausitzer Braunkohle weist einen Wassergehalt von etwa 56 % auf. Sämtliches Kohlewasser ist dampfförmiger Bestandteil des Gasstromes in der Rauchgasstrecke und verringert als Wärmesenke die Brennkammertemperaturen. Zwar ergibt sich hieraus eine niedrigere Flächenbelastung des Kessels, aber auch größeres Bauvolumen auf dem gesamten Rauchgasweg, angefangen vom Kessel bis hin zum Kühlturm. Die Kohlebrüden verlassen das Kraftwerk gasförmig über die Kühltürme, die Verdampfungsenthalpie von über 300 t/h Wasserdampf bei einem 670-MW-Block bleibt ungenutzt.

Die Trocknung von Braunkohle in Dampfwirbelschichtanlagen wird seit Ende der 70er-Jahre [2,3] mit dem Ziel der Wirkungsgradverbesserung von Kohlekraftwerken untersucht. Die aktuellen Entwicklungsstände an den sich in Betrieb befindenden atmosphärischen (RWE) [4] und druckaufgeladenen (BTU CS, Vattenfall) Dampfwirbelschicht-Trocknungsanlagen [5,6] zeigen die großtechnische Einsatzreife des Verfahrens, sodass es in bestehende bzw. zukünftige Kraftwerksblöcke Einzug finden kann.
Die externe Trocknung der Rohkohle bietet die Chance, die latente Wärme der Kohlebrüden in den Kraftwerksprozess zurückzuführen. Desweiteren kann die Kohletrocknung mit Heizdampf bei 160 °C und damit deutlich geringerem Temperatur- und somit auch Exergieniveau erfolgen. Diese beiden Sachverhalte führen dazu, dass der Wirkungsgrad um ca. 5 %-Punkte bei einer Vollstromtrocknung (Neubaublock) bzw. um ca. 1,5–2%-Punkte bei einer Teilstromtrocknung (Retrofit bestehender Kraftwerke) gesteigert werden kann.

Neben dem reinen Wirkungsgradgewinn wird das Verfahren der DDWT auch für die Gewährleistung der Netzstabilität immer essentieller. Im Bereich der Lastflexibilität von konventionellen Kraftwerken führt die DDWT zu einer Erweiterung des Lastregelbereiches durch den möglichen Einsatz eines Trockenbraunkohle- (TBK) Stützfeuersystems, ausgedrückt durch eine Reduzierung der technischen Mindestlast. Der Regelbereich des Blockes wird nach unten ggf. nicht mehr vom Kessel sondern von anderen Anlagenkomponenten limitiert. Hierdurch kann die Anzahl kosten- und verschleißintensiver An- und Abfahrvorgänge verringert werden. Der Block kann durchgängiger am Netz bleiben und so auch in Zeiten hoher EE-Einspeisung Regelleistung bereitstellen, was während An- und Abfahrvorgängen nicht möglich ist. Weiterhin wird das Lastwechselverhalten durch die Heizwert-Steigerung des Brennstoffes begünstigt, deutlich höhere Lastwechselgradienten werden gestattet. Die Möglichkeit der TBK-(Zwischen-) Speicherung kann zusätzlich zur Verbesserung der Blockflexibilität beitragen. Ein Vergleich von Kennziffern der Lastflexibilität zwischen einem modernen RBK- und einem TBK-befeuerten Braunkohleblock zeigt **Tabelle 1**.

Tabelle 1 Vergleich der Lastflexibilität von Braunkohlekraftwerken, Daten entnommen von [7]

	Lippendorf	**TBK-Block**[1]
Max. Kapazität (MW)	• 920	• 675
Min. Kapazität (MW)	• 400	• 200
Minimallast (%/P_N)	• 43	• 30
Laständerungsgeschwindigkeit (MW/min)	+/- 28	+/- 34
Laständerungsgradient (% P_N/min)	+/- 3	+/- 5

Das kraftwerksintegrierte Verfahrensprinzip der Vortrocknung von Braunkohle in einer Dampfwirbelschicht veranschaulicht **Bild 2**. Lausitzer Kohle wird dabei von rund 56 % auf den Zielwassergehalt von 12 % getrocknet. Die Beheizung erfolgt indirekt über dicht gepackte, in die Wirbelschicht integrierte Heizrohrbündel, in der Regel durch kondensierenden Anzapfdampf aus dem Turbosatz. Zu den Vorteilen der Wirbelschicht zählen die sehr gute Stoffdurchmischung, quasi-Isothermie und hohe erzielbare Wärmeübergangskoeffizienten. Der Trockner wird in nur leicht überhitzter, reiner Dampfatmosphäre betrieben, wodurch eine isotherme Kondensation und damit vollständige energetische Nutzung des verdampften Kohlewassers erst ermöglicht wird. Bei herkömmlichen Trocknungsverfahren entfällt diese Option. Das bei der Trocknung verdampfte Kohlewasser (Brüden) verlässt nach Entstaubung den Trockner zur weiteren energetischen Nutzung, ein konstanter Massenstrom verbleibt als Fluidisierungsgas im System.

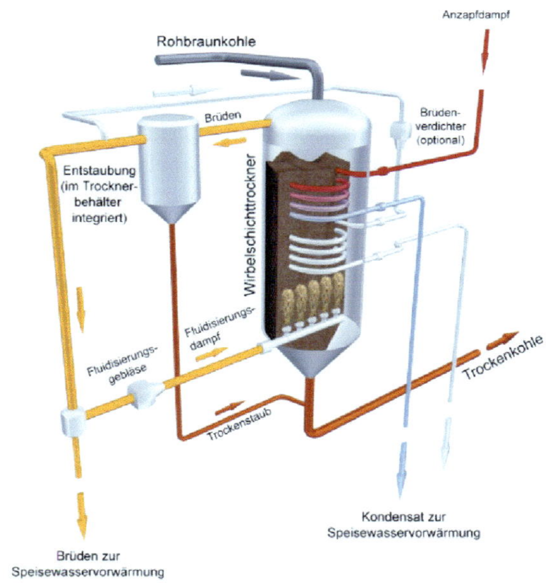

Bild 2 DDWT-Verfahrensprinzip (eigene Darstellung)

Die Überhitzung der Dampfatmosphäre im Trockner ist gleichzeitig die Stellgröße für den Endwassergehalt der Trockenbraunkohle. Das Kohlewasser wird generell in zwei Arten aufgeteilt. Zum einen gibt es die nicht bzw.

[1] Umsetzungsreifes Konzept.

nur geringfügig gebundene Oberflächenfeuchte, zum anderen die innere Feuchte, welche stärkere Bindung an die Kohle aufweist. Wird die Kohle bis in den Bereich dieser stärker werdenden Bindungen getrocknet, ist nicht nur eine höhere Überhitzung erforderlich, sondern auch ein erhöhter Energieaufwand, um die Bindungsenthalpie der inneren Feuchte aufzubringen. Aktuelle Verfahren trocknen Braunkohle auf einen Wassergehalt von bis zu 12 %.

3 Entwicklung des DDWT-Verfahrens

Seit 2002 wird das Verfahren zur Druckaufgeladenen Dampfwirbelschicht-Trocknung (DDWT) von Braunkohlen an der BTU Cottbus-Senftenberg als Forschungsverbundvorhaben mit zahlreichen Partnern aus der Industrie entwickelt. In diesem Rahmen wurde die erste kontinuierlich arbeitende DDWT-Anlage im Technikum der BTU mit einem RBK Durchsatz von 500 kg/h erfolgreich in Betrieb genommen.

Im Jahr 2006 begann das Folgevorhaben mit dem Titel „Konzeptentwicklung für eine großtechnische Versuchsanlage zur druckaufgeladenen Dampfwirbelschicht-Trocknung von Braunkohlen“ (DDWT-GTVA) ebenfalls als Forschungsverbundvorhaben. Das Konsortium bildete sich aus Unternehmen, die jeweils über einschlägige Spezialkenntnisse für die einzelnen Bereiche der verfahrensübergreifenden Technologie verfügten (Rohkohleförder- und Rohkohleaufbereitungstechnologie, TBK-Verbrennung, Kesselauslegung, Anlagenbau, thermodynamische Gesamtprozessbetrachtung).

Parallel zum Vorhaben DDWT-GTVA wurde 2007/08 eine Demonstrationsanlage (RBK Durchsatz 10 t/h) im Industriepark Schwarze Pumpe durch die Vattenfall Europe Generation AG errichtet und in Betrieb genommen, um einen wesentlichen Scale-Up-Schritt in Richtung Kraftwerksmaßstab zu gehen. Die **Tabelle 2** stellt die Eckdaten der beiden DDWT-Anlagen gegenüber.

Tabelle 2 Allgemeine Daten der Versuchs- und Pilotanlage zur DDWT

	BTU	Vattenfall
Durchsatz RBK (t/h)	0,1...0,5	2...10
Durchsatz TBK (t/h)	0,05...0,35	1...6
Trocknersystemdruck (bar a.)	1,1...6,5	1,3...6,0
Wassergehalt RBK (Ma.-%)	40...60	55...60
Wassergehalt TBK (Ma.-%)	5...30	6...18
Max. Heizdampfdruck (bar a.)	22	16
Gesamtanlagenhöhe (m)	12	36
Grundfläche der Anlage (m)	31,5	192
Errichtung	bis 02/02	10/07 bis 07/08
Inbetriebnahme	03/02	08/08

Im März 2011 startete das zweite Folgeprojekt zur Weiterentwicklung der DDWT an der BTU CS. Ziel dieses Projektes sind die vertiefte Grundlagenforschung und die verfahrenstechnische Optimierung der Dampfwirbelschicht-Trocknung. Des Weiteren sind Betrachtungen zur bevorstehenden Integration von DDWT-Anlagen in den Kraftwerkskomplex wesentlicher Bestandteil des Vorhabens.

4 Ausgewählte Ergebnisse

4.1 Grundlagenuntersuchung zum Desorptionsverhalten von Braunkohlen

Bei der Trocknung in Wasserdampfatmosphäre ist die Überhitzung der Wirbelschicht über die Siedetemperatur entscheidend für den Restwassergehalt der Trockenbraunkohle. Mit steigender Überhitzung sinkt der Restwassergehalt. Die Kenntnis dieser funktionalen Abhängigkeit (Desorptionsverhalten) ist für die Einstellung der gewünschten Produkteigenschaften von entscheidender Bedeutung.

Zur Charakterisierung des Desorptionsverhaltens von Braunkohlen steht eine Magnetschwebewaage (MSW) als gravimetrisches Messsystem zur Verfügung. Diese erfasst durch berührungslose Kraftübertragung (mittels Magnetfeld) die Masse(-änderungen) der Probe. Dabei lässt sich die im DDWT-Trockner herrschende Atmosphäre (Medium, Druck und Temperatur) exakt nachbilden. Das Fließbild der Waage ist in **Bild 3** und deren technische Daten in **Tabelle 3** dargestellt.

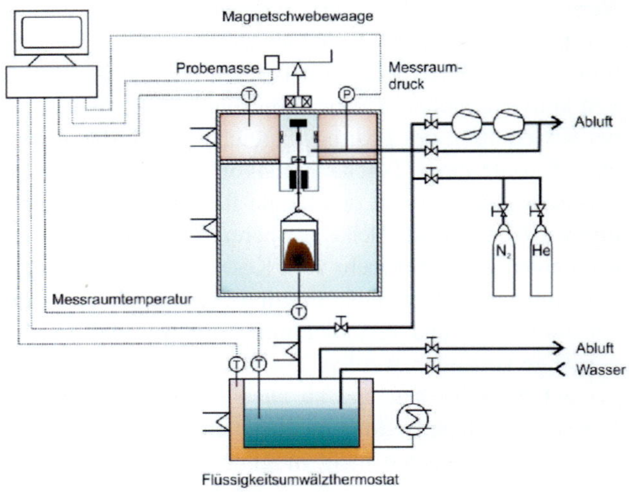

Bild 3 Fließbild der Magnetschwebewaage (MSW) [8]

Neben dem mit Laborpräzision ermittelten Desorptionsverhalten wird auch die Trocknungskinetik erfasst. Vergleiche unterschiedlicher Kohlequalitäten und Korngrößen sollen das theoretische Verständnis für ablaufende Wärme- und Stofftransportphänomene vervollständigen und gleichfalls die Auslegung von DDWT-Anlagen präzisieren.

Tabelle 3 Technische Daten der MSW [8]

Technische Daten der Apparatur	
Messlast (g)	0–25
Auflösung (mg)	0,01
Reproduzierbarkeit (mg)	< ± 0,02
Relativer Fehler (% v. Messwert)	< ± 0,002
Zulässiger Druck (bar)	150
El. Begleitheizung (°C)	bis 200
Versuchsparameter	
Temperatur (°C)	100 bis 200
Druck (bar)	1 bis 6

Bild 4 zeigt die gefunden Abhängigkeiten des Restwassergehaltes von der Wasserdampfüberhitzung sowie dem Systemdruck für Lausitzer Kraftwerkskohle.

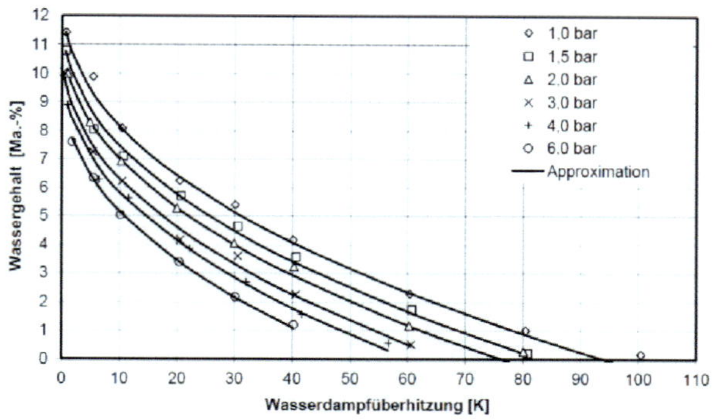

Bild 4 Desorptionsisobaren Lausitzer Kesselkohle [8]

Aufbauend auf den ermittelten Desorptionsisobaren erfolgte die semianalytische Ermittlung der Bindungskräfte des Wassers an die Feststoffsubstanz der Kohle, um die energetische Auslegung und Bilanzierung von Trocknern zu vervollkommnen.

4.2 Wärmeübergangsmessung am Einzelrohr

Entscheidend für die Dimensionierung des Trockners ist der erzielbare Wärmeübergangskoeffizient zwischen Heizfläche und Braunkohlewirbelschicht.

Um zu erfassen, wie sich dieser in Abhängigkeit von Produktfeuchte, Fluidisierung und Partikelgröße ändert, wurden Kaltmodelle in verschiedenen Größenordnungen entwickelt.

Mit Hilfe von speziell entwickelten Wärmeübergangsmesssonden werden die notwendigen Informationen zur Ermittlung des Wärmeübergangskoeffizienten erfasst. Den prinzipiellen Aufbau der Sonde sowie deren Installation in einem Kaltmodell zeigt **Bild 5**.

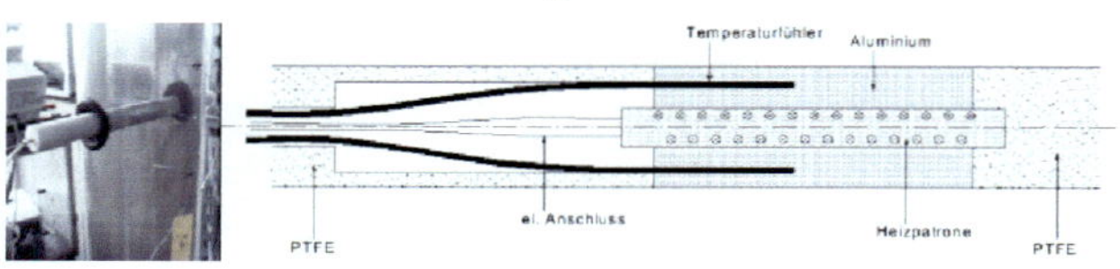

Bild 5 Schematischer Aufbau der Messsonde (rechts), Installation im Kaltmodell (links)

Die Heizsonde besteht aus Aluminium mit sehr guter Wärmeleitfähigkeit, getragen von Teflon-Elementen mit vergleichsweise sehr geringer Wärmeleitfähigkeit, um eine definierte Fläche für die Wärmeübertragung zu schaffen. Im Kern des Aluminiumkörpers sitzt eine mit Gleichspannung betriebene Heizpatrone, deren regelbare Leistung erfasst wird. Über Temperaturfühler im Aluminiumkörper wird die Oberflächentemperatur der Heizsonde erfasst.

In **Bild 6** sind beispielhafter Weise die gefundenen Abhängigkeiten des Wärmeübergangskoeffizienten von der Leerrohrgeschwindigkeit und dem Wassergehalt der Braunkohle mit einer staubbrennertauglichen Körnung von 0–2 mm dargestellt. Mit steigender Leerrohgeschwindigkeit wird aus dem Zustand des durchströmten Festbettes ab einer Minimalfluidisierungsgeschwindigkeit von ca. 0,05–0,10 m/s der Wirbelschichtzustand mit sehr hohen Wärmeübergangskoeffizienten erreicht. Dieses Verhalten ist für Wirbelschichten typisch. Die maximalsten Wärmeübergangskoeffizienten werden bei einem Wassergehalt von 35 % erzielt. Mit abnehmendem Wassergehalt sinken die Wärmeübergangskoeffizienten aufgrund schlechter werdender thermopysikalischer Eigenschaften der Kohle, ausgedrückt durch die Verringerung der volumetrischen Wärmekapazität ($\rho * c_p$) und der Wärmeleitfähigkeit λ. Die Partikel können bei ihren Wandkontakten immer weniger und immer langsamer Wärme aufnehmen. Entgegen der Theorie ist für Wassergehalte größer 35 % ein Wiederabfall des Wärmeüberganges zu verzeichnen, welcher aus den zunehmend schlechter werdenden Fluidisierungseigenschaften resultiert. Wassergehalte über 35 % äußern sich ausschließlich in der Zunahme des Oberflächenwassers, wodurch sich die innerpartikulären Haft- und Anziehungskräfte zwischen den einzelnen Kohlepartikeln verstärken. Es kommt zur Agglomeration und somit zur Verkleinerung der relativen Wärmeaustauschfläche.

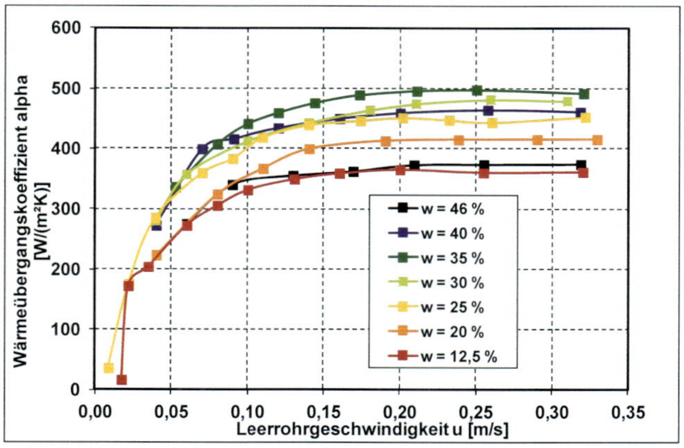

Bild 6 Erzielte Wärmeübergangskoeffzienten abhängig von der Fluidisierung und vom Wassergehalt einer 0–2 mm Braunkohlekörnung des Lausitzer Reviers [9]

4.3 Optimierung der Wärmetauschergeometrie

Neben den spezifischen Kohleeigenschaften kommt der Geometrie der Wärmeübertragereinheit entscheidender Einfluss zu. Damit der Wirbelschicht-Trockner so kompakt wie möglich ausgelegt werden kann, sollte die Wärmeübertragungsfläche geringstmögliche Raumvolumen einnehmen, wodurch den Rohteilungsverhältnissen sowie den Rohrdurchmesser Beachtung geschenkt werden muss. Aus geometrischen Gesichtspunkten sind minimale Rohrdurchmesser und Teilungsverhältnisse anzustreben, da somit die größte Wärmeübertragungsfläche pro Raumvolumen der Wirbelschicht erzielt werden kann (höchste Packungsdichten, m²/m³). Je geringer jedoch der freie Abstand zwischen den Rohren wird, desto mehr wird die Hydrodynamik der Wirbelschicht gestört, die Wirbelschicht kann sich nicht mehr frei ausbilden. Die freie Partikelbewegung wird eingeschränkt und das Hauptmerkmal des Wirbelschichtverfahrens, die vollständige Fluidisierung und Durchmischung wird mehr und mehr limitiert. Aufgrund der verringerten Partikelgeschwindigkeiten und -austauschraten verschlechtert sich so auch der Wärmeübergangskoeffizient, wodurch als Resultat die wärmeübertragende Fläche wieder erhöht werden muss. Im Ergebnis steht hierbei die gegenseitige Beeinflussung von Packungsdichte und Wärmeübergangs- und damit Wärmedurchgangskoeffizienten, wodurch ein klassisches Optimierungsproblem entsteht.

Im Rahmen der Untersuchungen von Lechner [10] wurde der Einfluss der Wärmetauschergeometrie auf das Wärmeübergangsverhalten an einem Wirbelschicht-Kaltmodell mit einer quadratischen Grundfläche von 39 x 39 cm durchgeführt. Fluidisiert wurde mit Stickstoff im atmosphärischem Betrieb. Als Wirbelgut kam eine Brikettkohle aus dem Lausitzer Revier mit staubbrennertauglicher Körnung zum Einsatz. Der Wassergehalt der Braunkohle wurde aus Gründen des besseren Handlings im Vorfeld auf ca. 32 % reduziert. Der Schwerpunkt der Untersuchung lag auf versetzter Rohranordnung mit Rohrdurchmessern von 12–35 mm sowie Rohrteilungen zwischen 1,5–4,0 (horizontal) und 1,0–3,5 (vertikal). Die betrachteten Rohranordnungen zeigt **Bild 7**.

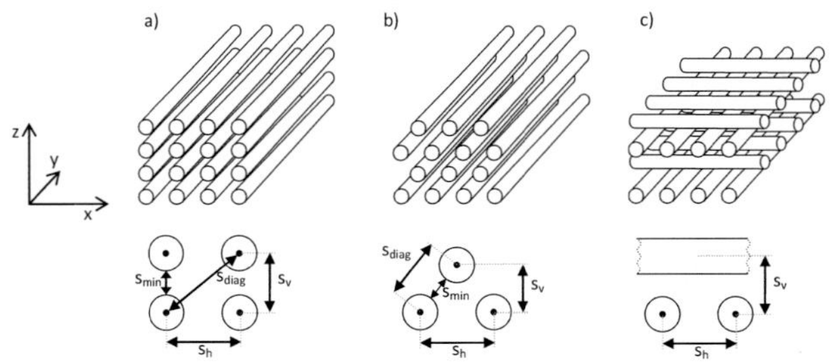

Bild 7 Betrachtete Rohranordnungen: a) fluchtend, b) versetzt, c) gekreuzt [10]

Als Ergebnisse werden in **Bild 8** die in Rohrbündelmitte max. erzielten Wärmeübergangskoeffizienten ($alpha_{max}$-Werte), in Abhängigkeit der Packungsdichte für die drei untersuchten Rohrdurchmesser vorgestellt. Als Referenzmessung sind die erzielten Wärmeübergangskoeffizienten mit lediglich einen, in die Wirbelschicht installierten Wärmetauscherrohr (Einzelrohrmessung) aufgeführt, welche stellvertretend für eine ungestörte Wirbelschichtausbildung sind. Mit kleineren Rohrdurchmessern werden durch die geometrisch bedingt, höhere Partikelaustauschrate am Heizelement, höchste Wärmeübergangskoeffizienten erzielt. Der Abfall der $alpha_{max}$-Werte zu höheren Packungsdichten hin erfolgt im untersuchten Bereich nahezu linear. Die größten hier untersuchten Packungsdichten sind jedoch für die Konzeptionierung von Wirbelschicht-Trocknungsanlagen nicht zu empfehlen, da diese gerade im oberen Trocknerbereich, mit partiell höchsten Anteil feuchter, kohäsiver Kohle zu Betriebsinstabilitäten führen könnten.

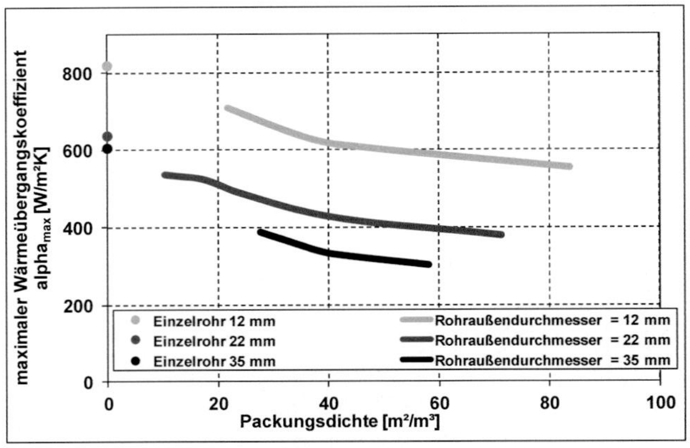

Bild 8 Wärmeübergang abhängig von der Heizflächendichte für unterschiedliche Rohrdurchmesser [10]

4.4 Modellierung und CFD-Simulationen der Dampfwirbelschicht-Trocknung

Die Ergebnisse der experimentellen Untersuchungen an der vorhandenen Technikumsanlage zeigten, dass einige für die Fragestellung des Scale-up und der Verfahrensoptimierung wichtige fluiddynamische Details der Wirbelschicht, wie Geschwindigkeitsfelder und Partikelkonzentrationen, nicht oder nur mit sehr hohem messtechnischen Aufwand ermittelt werden können. Aufgrund dieser Problematik wurden die Forschungsaktivitäten seit 2008 auf den Bereich der numerischen Strömungssimulation erweitert. Ziel ist es hierbei, ein Modell zu entwickeln, welches die komplexen und untereinander stark gekoppelten Phänomene der Hydrodynamik sowie der Wärmeübertragung und des Stoffübergangs im Trockner adäquat abbilden kann. Dabei kommt das Two Fluid Model (TFM) zum Einsatz, welches dem Euler-Euler-Modellierungsansatz zuzuordnen ist.

Zur experimentellen Validierung der durchgeführten Simulationen wurde ein 2D-Kaltmodell aufgebaut. Den schematischen Versuchsaufbau zeigt **Bild 9**. Diese Anlage ermöglicht die experimentelle Untersuchung der Eigenschaften der Wirbelschicht für eine große Bandbreite von Partikelgrößen und Fluidisierungszuständen sowie für verschiedene, unterschiedlich dicht gepackte Rohrbündelgeometrien.

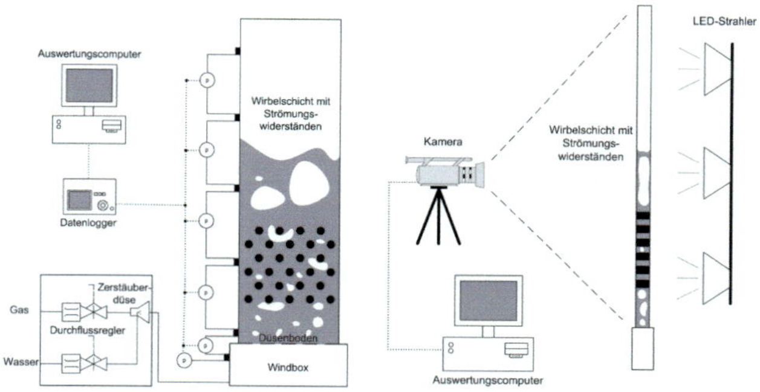

Bild 9 Schema des Kaltmodells - Frontale (links) und seitliche (rechts) Ansicht [11]

Zur Ermittlung des Verhaltens der aufsteigenden Blasen in der Wirbelschicht wurde eine optische Messmethode, die so genannte Digital Image Analysis (DIA), realisiert. Das dabei von hinten durch die transparenten Wände der Versuchsanlage hindurch scheinende Licht lässt das Wirbelbett dunkel und die aufsteigenden Blasen hell erscheinen (siehe auch **Bild 10**, links). Die mit Hilfe einer Digitalkamera von vorn aufgenommene Videosequenz kann anschließend mittels eines Auswerteprogramms analysiert werden. Dazu wird jedes Einzelbild zunächst in ein Graustufenbild umgewandelt. Über eine Auswertung des sich ergebenden Intensitätshistogramms kann der zur Unterscheidung von Wirbelbett und Gasblase notwendige Helligkeitsgrenzwert ermittelt und das entsprechende Binärbild erzeugt werden. Dieses ermöglicht nun die Detektierung der Blasen und die Berechnung der Blasengröße und Aufstiegsgeschwindigkeit.

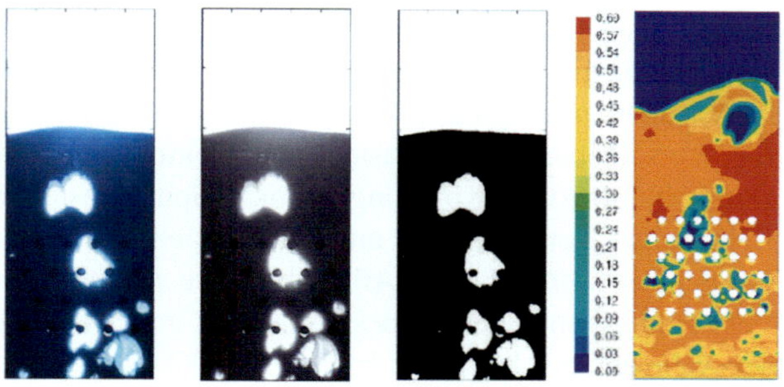

Bild 10 V. l. n. r.: Original, Graustufenbild und Binärbild der automatisierten Bildanalyse, Simulationsergebnis [11]

5 Zusammenfassung und Ausblick

Dieser Beitrag soll zum Einen verdeutlichen, dass die Kohlevortrocknung mittels DDWT für die notwendige Flexibilisierung der konventionellen Kraftwerke, als auch für die Effizienzsteigerung und Absenkung der CO_2-Emissionen eine aussichtsreiche Brückentechnologie im Energiemix mit Erneuerbaren Energien darstellt. Zum Anderen soll durch die Darstellung einiger Untersuchungsschwerpunkte die Komplexität des Verfahrens aufgezeigt werden. Die Trocknung und die damit zusammenhängenden Wärme- und Stofftransportphänomene in reiner Dampfatmosphäre mit indirekter Beheizung im kompliziert zu beschreibenden Wirbelschichtsystem, sowie die dabei zum Einsatz kommende stark polydisperse Braunkohleschüttung bestehend aus nicht sphärischen kapillarporösen Partikeln, welche einer stärken Änderung ihrer thermophysikalischen Eigenschaften während des Trocknungsprozesses unterliegen, zeugt von der verfahrensübergreifenden Prozessstruktur.

In Zukunft sollen weitere F&E-Vorhaben zum Thema DDWT, auch verstärkt im Bereich der Grundlagenforschung, folgen.

6 Literatur

[1] Kallmeyer, D., Wicke, W.: Entwicklungsschritte zum trockenbraunkohlebefeuerten Kraftwerk. VBG Kraftwerkstechnik. 77 (4), 300-303 (1997)

[2] Potter, O.E., Keough, A.J.: Cheaper Power from High Moisture Brown Coals - Part I: System Description and Preliminary Costing. Part II: Feasibility and Development. Journal of the Institute of Energy. 52 (3), 143-149, (1979)

[3] Wolf, B., Trocknung - Prozeßstufe der energetischen und stofflichen Nutzung fossiler Brennstoffe. VDI Berichte Nr. 690, VDI Verlag Düsseldorf (1988)

[4] Klutz, H.-J.: Theoretische und experimentelle Untersuchungen zur Entwicklung eines Wirbelschichttrockners für Braunkohle. Dissertation, Universität Hamburg-Harburg, (2008)

[5] Merzsch, M. et al.: Druckaufgeladene Dampfwirbelschicht-Trocknung (DDWT) von Braunkohlen: Grundlagenuntersuchungen zur verfahrenstechnischen Optimierung der Dampfwirbelschicht-Trocknung von Braunkohlen. In: Beckmann, M, Hurtado A. (Hrsg.) Kraftwerkstechnik 4, Neuruppin:, S. 923934. TK Verlag Karl Thomé-Kozmiensky, Neuruppin, (2012)

[6] Höhne, O. et al.: Die DDWT-Technologie für hocheffiziente und flexible Braunkohlekraftwerke. In: Beckmann, M, Hurtado A. (Hrsg.) Kraftwerkstechnik 5, Neuruppin:, S. 211221. TK Verlag Karl Thomé-Kozmiensky, Neuruppin, (2013)

[7] Jentsch, N.; Kahlert, J.: Herausforderungen der Energiewende in Brandenburg - Welchen Beitrag kann die Braunkohle leisten?, Vortrag zum 14. Brandenburger Energietag, (06.09.2012)

[8] Höhne O.: Grundlagenuntersuchungen zum Einfluss des Systemdrucks auf die Trocknung von Braunkohle in Wasserdampf, Dissertation, BTU Cottbus, (2012)

[9] Merzsch, M.; Lechner, S.; Krautz, H.J.: Heat-transfer from single horizontal tubes in fluidized beds: Influence of tube diameter, moisture and diameter-definition by Geldart C fines content. Powder Technology, (2012)

[10] Lechner, S.: Untersuchungen zur Berechnung und Optimierung des Wärmeübergangs bei der Dampfwirbelschicht-Trocknung am Beispiel von Lausitzer Braunkohlen, Dissertation, BTU Cottbus, (2012)

[11] Schreiber, M.: Modellierung von Hydrodynamik und Wärmeübertragung in blasenbildenden Wirbelschichten zur Trocknung von Braunkohle, Dissertation, BTU Cottbus, (2013)

Entwicklung von Betriebsstrategien zur Verbesserung der Effizienz von Kälteversorgungssystemen mit thermischen Speichern

P. Puls, S. Linhardt, R. Öchsner
Fraunhofer Institut für Integrierte Systeme und Bauelementetechnologie (IISB), Schottkystraße 10, 91058 Erlangen, philipp.puls@iisb.fraunhofer.de

Kurzfassung

Kälteversorgungssysteme verursachen einen großen Teil der Jahresenergiekosten von Gebäuden mit Kühlbedarf. Oft sind Kälteanlagen für die wenigen Tage im Jahr mit hohen Außentemperaturen ausgelegt und arbeiten den Rest der Zeit im Teillastbereich. Durch individuell angepasste Regelstrategien und die Möglichkeit der Lastverschiebung über thermische Speicher kann der Betrieb von Kältesystemen flexibilisiert und ein signifikanter Anteil der Energiekosten von Versorgungskomponenten eingespart werden. Eine vorausschauende Verschiebung thermischer Lasten ermöglicht zudem die Reduktion elektrischer Energiebezüge zu Spitzenlastzeiten und die Nutzung von günstigeren Tarifen, wie z. B. Nachtstromtarifen.

Am Beispiel des Fraunhofer IISB konnte im Rahmen von Simulationen eine Einsparung der elektrischen Energiekosten für die Kälteversorgung von über 18 % durch Integration eines Kältespeichers vorausgesagt werden. Das Einsparpotenzial beruht auf einer Verschiebung der Kälteerzeugung von der Tages- in die Nachtzeit und kann durch weitere Effizienzmaßnahmen, wie z. B. die Nutzung von freier Kühlung, weiter angehoben werden. Die Untersuchungen zeigen, dass die Entwicklung einer individuell angepassten Betriebsstrategie des Gesamtsystems maßgebend für den wirtschaftlichen Erfolg eines solchen Versorgungskonzeptes ist. Obwohl thermische Speicher im privaten Bereich eine weitverbreitete Technologie darstellen und die Kosten pro Kilowattstunde im Vergleich zu elektrischen Speichern gering ausfallen [1], wird in der Praxis selten Gebrauch von größeren Tankspeichersystemen gemacht. Ziel der Arbeiten am IISB ist nicht nur die Simulation innovativer Betriebsstrategien für thermische Speicher, sondern auch die anschließende technische Umsetzung der Konzepte als Forschungsdemonstrator mit Leuchtturmwirkung.

1 Effizienzpotenziale in Kälteversorgungssystemen

Kälteversorgungsanlagen sind auch in mitteleuropäischen Breiten ein energieintensiver Baustein thermischer Versorgungssysteme. Ein Großteil der bereitgestellten Kälteenergie im nicht-privaten Sektor wird für Klimatisierungszwecke, industrielle Kühlprozesse sowie für Lebensmittelkühlung in der Industrie und in Supermärkten eingesetzt. Im Jahr 2009 wurden für die Kälteversorgung etwa 14 % des jährlichen elektrischen Strombedarfs in Deutschland aufgewendet [2]. Hier besteht ein großes Energieeinsparpotenzial, da Kälteanlagen häufig für heiße Tage ausgelegt sind und den Rest des Jahres ineffizient arbeiten. Gängige Effizienzmaßnahmen wie Sanierung im Bestand und bessere Dämmung der Gebäude und Kälteverteilungssysteme bringen aber wenig Erfolg, sofern nicht auch die Kältemaschinen durch adäquate, effiziente Anlagen ersetzt werden.

Am Fraunhofer-Institut für Integrierte Systeme und Bauelementetechnologie werden weitere Maßnahmen erarbeitet, um die Energieeffizienz von bestehenden und geplanten Kältesystemen zu erhöhen. Ein vielversprechender Ansatz ist ein optimiertes Lastmanagement zur Verschiebung der Kältebereitstellung in die Nachtstunden mit Hilfe von thermischen Speichern. Durch die gezielte zeitliche Entkopplung von Kältebereitstellung und -nutzung können verbrauchsgebundene Betriebskosten der Anlagen um etwa 9 bis 18 % reduziert werden. Ein weiterer Vorteil dieser Strategie ist die Flexibilisierung der thermischen Infrastruktur und damit verbunden auch die Möglichkeit zur Verschiebung von Lasten auf elektrischer Seite. Durch die flexible Kopplung können Spitzenlasten im elektrischen Netz reduziert und Nachtstromtarife effektiver genutzt werden. Sind Arbeitsbereiche der einzelnen Versorgungskomponenten bekannt, z. B. aus Simulation der Anlagen oder Messwert-basierten Abhängigkeiten, so lässt sich über gezielte Lastverschiebungsstrategien auch eine Beeinflussung der elektrischen Leistungsaufnahme des Gebäudes („*peak shaving*") umsetzen. Dadurch ergibt sich bei großen Kälteanlagen oder -verbünden auch die Möglichkeit der Teilnahme am Markt für Sekundär- und Minutenreserve. Voraussetzung für solche Ansätze ist eine genaue Kenntnis des Lastverschiebungspotenzials und der Entwurf individueller Betriebsstrategien.

2 Aufbau von Kältesystemen

Ein Kältenetzwerk besteht aus Kälteversorgern und Kälteverbrauchern, welche bei wasserführenden Systemen typischerweise über Rohrleitungssysteme miteinander gekoppelt sind. Daneben werden Elemente der Sicherheitstechnik, Stromversorgung sowie passive und aktive Komponenten der Mess-, Steuer- und Regelungstechnik zum Betrieb eines Kältenetzes benötigt [3]. Die Kälteenergie wird über ein Kälteträgermedium im sogenannten Vorlauf zu den Verbrauchern transportiert, wo es die Abwärme aufnehmen kann und anschließend im Rücklauf wieder auf den Kälteversorger überträgt. Der Transport des Kälteträgermediums erfolgt üblicherweise mit Hilfe von Netzpumpen, welche hinsichtlich ihrer hydraulischen Leistungen idealerweise an das Bezugsverhalten der Verbraucher angepasst oder danach geregelt werden. Die gesamte Kälteleistung $\dot{Q}_K$ eines geschlossenen Systems ist über den Volumenstrom $\dot{V}$, die Dichte ρ_W und die isobare Wärmekapazität $c_{p,W}$ des Wärmeträgermediums (z. B. Wasser) bei bekannter Vor- und Rücklauftemperatur T definiert [4]:

$$\dot{Q}_K = \rho_W \cdot c_{p,W} \cdot \dot{V}_W \cdot \left(T_{Vorlauf} - T_{Rücklauf}\right) \quad (1)$$

Im Verdampfer der Kältemaschine in **Bild 1** wird der eintretende Kaltwasservolumenstrom vom Rücklauftemperaturniveau auf die Vorlauftemperatur des Netzes herabgekühlt. Dies geschieht in einem linksläufigen thermodynamischen Kreisprozess, der speziell bei elektrisch angetriebenen Maschinentypen unter Einsatz von technischer Arbeit einen Wärmestrom von einem niederen Temperaturniveau auf ein höheres transportiert und an die Umgebung abführt [5].

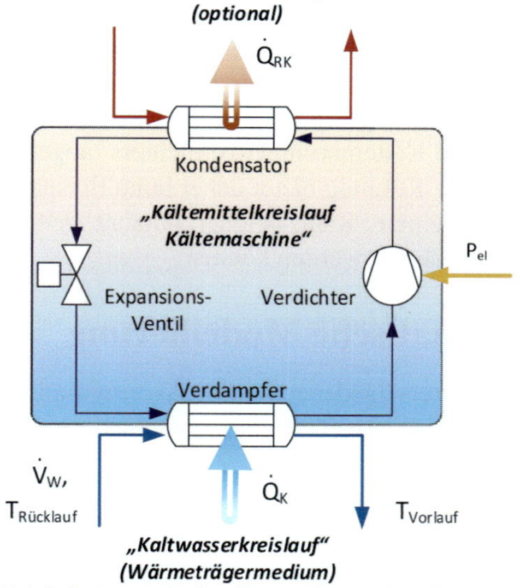

Bild 1 Schematische Darstellung der Energieströme einer Kältemaschine mit mechanischem Verdichter

Innerhalb des Kältemaschinenkreislaufs wird ein tiefsiedendes Kältemittel geführt, welches in mehreren Teilprozessen unter Zustandsänderung Wärmeenergie aus dem Verdampfer aufnimmt und sie im Kondensator wieder abgibt. Die flüssige Kältemittelphase wird vor dem Eintritt in den Verdampfer über ein Expansionsventil entspannt und kann den auf niedrigem Temperaturniveau angebotenen Wärmestrom des Kältenetzes in einem Siedeprozess aufnehmen. Damit der im Verdampfer aufgenommene Wärmestrom $\dot{Q}_K$ an die Umgebung abgeführt werden kann, wird der trockene Kältemitteldampf mit einem Verdichter komprimiert und in einen höheren Temperaturzustand überführt. Die Kondensation des Kältemitteldampfes kann somit bei Temperaturen deutlich oberhalb der Verdampfungstemperatur erfolgen. Im Kondensator muss zusätzlich zur Verdampfer-Leistung auch die Verdichtungsarbeit $P_{el,Verd}$ des Kältemittelverdichters abgeführt werden, weshalb der Wärmestrom auf der Kondensatorseite größer ist als auf der Verdampferseite. Der aus dem System abzuführende Wärmstrom entspricht der notwendigen Rückkühlleistung $\dot{Q}_{RK}$ der Kältemaschine und ist definiert über [4]:

$$\dot{Q}_{RK} = \dot{Q}_K + P_{el,Verd} \quad (2)$$

In der Praxis werden unterschiedliche Konzepte verfolgt, den Wärmestrom aus dem Kondensator an die Umgebung abzuführen. Bei kurzen Wegen oder Außenaufstellung der Kältemaschine bietet sich eine direkte Rückkühlung des Kältemittels über einen Wärmeübertrager an der Außenluft an. Bei größeren Entfernungen zwischen Kältemaschine und Außenbereich wird der Wärmestrom, wie im **Bild 2** dargestellt, vom Kondensator zunächst auf einen sog. Kühlmittelkreislauf übertragen, ehe er über das Rückkühlwerk an die Umgebung abgeführt wird. Dies kann je nach Kühlturmbauart entweder direkt oder indirekt über trockene und feuchte Kühlung geschehen [3]. In der Regel besteht das Wärmeträgermedium in geschlossenen Rückkühlkreisläufen aus einem Wasser-Glykol-Gemisch, um ein Einfrieren der Leitungen im Winter zu vermeiden.

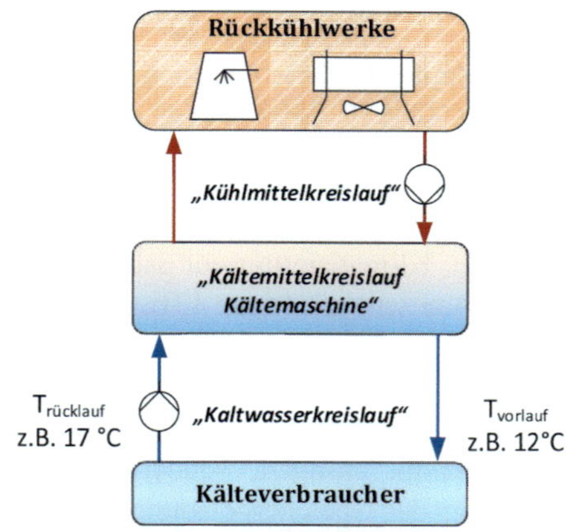

Bild 2 Schematische Darstellung der Einbindung einer flüssigkeitsgekühlten Kältemaschine in die Kälteinfrastruktur

3 Effizienzpotenzial durch Lastverschiebung

Das Ziel der simulationsbasierten Untersuchungen ist die energetische Charakterisierung des gesamten Kälteversorgungssystems mit allen relevanten Versorgungskomponenten in Abhängigkeit von realen Randbedingungen. Dazu gehören gemessene Verläufe von Kältebezügen und die zeitlich korrelierten Daten der Außentemperatur und Luftfeuchtigkeit. Für einen energetisch optimierten Betrieb der Kälteinfrastruktur hat es sich in konkreten Untersuchungen an den Kälteanlagen des IISB nicht als ausreichend erwiesen, allein die Effizienz der Kältemaschinen zu berücksichtigen. Vielmehr besitzen einzelne Versorgungskomponenten voneinander teilweise unabhängige Wirkungsgrade, die nicht mit dem besten Arbeitspunkt des Kälteversorgers korrelieren. So arbeitet beispielsweise die Rückkühltechnik der Kälteanlagen bei niedrigerer Außentemperatur in der Nachtzeit deutlich effizienter. **Bild 3** stellt das Verhältnis aus bereitgestellter Kälteleistung zur aufgenommen elektrischen Leistung (engl. *„coefficient of performance"*, COP) der Kältemaschine gegenüber dem Gesamtsystem dar. Die höheren COP-Werte der Kältetechnik bei 10 °C gegenüber 25 °C sind auf den effizienteren Betrieb der Rückkühlwerke bei niedrigeren Temperaturen in der Nacht zurückzuführen.

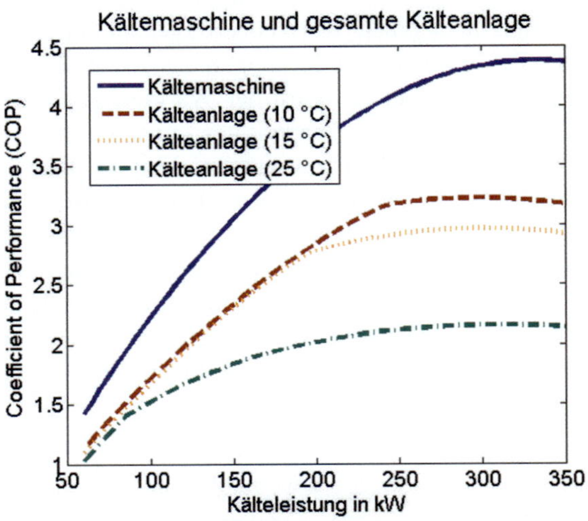

Bild 3 COP-Verlauf der Kältemaschine im Vergleich zur gesamten Kühltechnik in Abhängigkeit geforderter Kälteleistung und verschiedenen Außentemperaturen

Zur Nachtzeit fahren gerade nicht-drehzahlgeregelte Kältemaschinen mit Klimatisierungsaufgabe im Teillastbereich und arbeiten mit niedriger Effizienz. Über einen Kältespeicher lässt sich dieses Missverhältnis teilweise ausgleichen, da die Verschiebung eines Teils der Kältebereitstellung in die Nacht nicht nur die Effizienzsteigerung der Rückkühltechnik ausnutzt, sondern über die Leistungsanforderung der Speicherbeladung auch die Kältemaschinen stärker auslastet. Wird keine Abschaltung der Kältemaschinen am Tag zugelassen, so scheint die erhaltene Zunahme der Gesamteffizienz zunächst nicht beständig, da der Speicher im Anschluss an die Beladephase wieder entladen werden muss. Eine Entladung ist notwendig, um die gespeicherte Kälteenergie zu nutzen und den Speicher für die nächste Nachtperiode zu regenerieren. Die Kälteleistung der Kältemaschine wird daher zur Spitzenlastperiode am Tag reduziert, was bei gängigen Maschinentypen zunächst mit Effizienzeinbußen verbunden ist. **Bild 3** zeigt, dass die untersuchte Schraubenkompressions-Kältemaschine gerade im Niedriglastbereich die höchste Steigung des COP-Verlaufes aufweist. Dagegen nimmt die Steigung des Wirkungsgrades der gesamten Kältetechnik durch den Einfluss der Rückkühlwerke oberhalb einer kritischen Kälteleistung von etwa 80 kW_{th} (25 °C Außentemperatur) am Tag deutlich ab. Somit ist die Effizienzerhöhung bei geringer Steigerung der Auslastung im nächtlichen Niedriglastbereich insgesamt größer als die Effizienzabnahme, welche mit einer Reduktion der Kälteleistung im Spitzenlastbereich am Tag auftritt.

Das Effizienzpotenzial der Betriebsstrategie wird erhöht, sofern eine Abschaltung der Kältemaschinen während der Speicherentladung am Tag zugelassen wird. Die gesamte Kältelast des Netzes muss dann über den Speicher abgefahren werden, was eine entsprechende Regelung der Speicherpumpengruppe voraussetzt. In diesem Fall ist der Wirkungsgrad der Rückkühlwerke bei Nacht gegenüber dem Tag maßgebend für eine Einsparung von Energiekosten. Die maximal mögliche Stillstandszeit der Kältemaschinen leitet sich direkt aus der Größe des Kältespeichers ab. Aufgrund der häufig vorhandenen Bauflächenbeschränkung stellt das Speichervolumen eine individuelle Optimierungsgröße für das betrachtete System dar.

Neben der Lastverschiebung zwischen Tages- und Nachtzeit macht auch die Nutzung der Rückkühlwerke zur Kälteerzeugung den Kältespeicher energetisch interessant. Diese Art der Kältegestehung wird als freie Kühlung bezeichnet und kann bei typischen Prozesskühlanwendungen unterhalb von etwa fünf Grad Celsius Umgebungstemperatur genutzt werden. Die Rückkühlwerke benötigen für diese Aufgabe teilweise deutlich weniger elektrische Energie als Kältemaschinen, da bei letzteren zusätzlich zur Kälteanforderung der Verbraucher auch die elektrische Arbeit des Kältemaschinenverdichters abgeführt werden muss. Freie Kühlung bietet daher hohe Einsparpotenziale, die mit einem Kältespeicher nutzbar gemacht oder weiter ausgeschöpft werden können.

4 Mathematische Modellierung

Die Simulation verschiedener Anlagenkonfigurationen und Betriebsstrategien erfordert eine mathematische Modellierung der Einzelkomponenten. Ziel ist eine Darstellung der Betriebskosten von Einzelkomponenten über den gesamten Lastbereich in Abhängigkeit von den Simulationsrandbedingungen. Bei den Kältemaschinen und den Umwälzpumpen können z. B. die elektrischen Leistungsaufnahmen herangezogen werden. Im Fall der Rückkühlwerke muss neben dem Verbrauch elektrischer Energie eventuell auch die Nutzung von aufbereitetem Wasser betrachtet werden, sofern die Anlagen eine Berieselungsstufe besitzen. Auf Basis der summierten Einzelbetriebskosten können unterschiedliche Betriebsstrategien verglichen und energetisch bewertet werden. Die gewünschten Abhängigkeiten der Kälteversorgungskomponenten können über analytische Ansätze sowie über empirische und physikalische Modelle erhalten werden.

4.1 Kältemaschinen

Die Modellierung der Kältemaschine erfolgt anhand von analytisch gewonnenen Messwerten. Hierfür wird der elektrische Energiebezug der Kältemaschine der zeitlich korrelierten Kälteleistung gegenübergestellt und anschließend der COP bestimmt. Ist der Zusammenhang zwischen Leistungsaufnahme und Kälteleistung der Anlage über große Betriebsbereiche bekannt, so kann eine Funktion aufgestellt werden, welche anhand der thermischen Leistungsanforderung einen erwarteten elektrischen Leistungsbedarf ausgibt. **Bild 4** zeigt die gemessenen Daten einer Schraubenkompressor-Kältemaschine zusammen mit der berechneten Ausgleichsfunktion. Diese Methode liefert für kleine Zeitintervalle im Bereich mehrerer Minuten keine exakte Vorhersage der elektrischen Leistungen. Sie ist aber gut geeignet, um über längere Simulationszeiträume genaue Vorhersagen für den elektrischen Energiebedarf bereitzustellen. Ein großer Vorteil dieser Vorgehensweise ist die schnelle Verfügbarkeit von Messabhängigkeiten, z. B. durch Integration von Wärmemengenzählern.

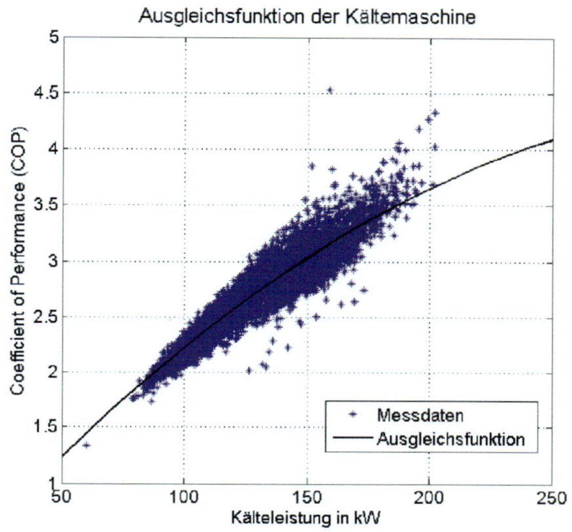

Bild 4 Empirisch bestimmte Abhängigkeit des COPs einer Schraubenkompressions-Kältemaschine von der gelieferten Kälteleistung

Die Genauigkeit der dargestellten Methode hängt von der Ähnlichkeit der zugrundeliegenden Kältelastprofile ab. Durch eine Häufung von großen Kaltwasserspitzenbezügen gegenüber den Messwerten des Modells können Abweichungen im Bereich weniger Prozent der elektrischen Energieaufnahme entstehen. Auch die jahreszeitabhängige Schwankung des Kältebedarfs kann zu Differenzen führen, sofern die Modellfunktion bei anderen klimatischen Bedingungen aufgenommen wurde. Um Fehlern vorzubeugen, sollten die Daten der Ausgleichsfunktion daher aus den gleichen Zeiträumen stammen, welche auch der Simulation zugrunde liegen. Die Abweichung der elektrischen Energieaufnahme zwischen Messwert und Modell desselben Zeitintervalls liegen üblicherweise unterhalb von einem Prozent.

4.2 Thermische Speicher

Den energetischen Nutzen von thermischen Speichern in einem Gesamtsystem zu bestimmen ist nicht trivial. Ein Vergleich der Gesamtkosten zweier unterschiedlicher Betriebsszenarien setzt voraus, dass alle energetisch relevanten Abhängigkeiten im Modell berücksichtigt werden. Zu diesen gehören z. B. thermische Verluste innerhalb des Speicherbehälters und Betriebskosten einzelner Komponenten der Speicherperipherie. Eine Bilanz der gesamten Energiekosten eines abgeschlossenen Simulationszeitraums ermöglicht anschließend die wirtschaftliche Bewertung des Speicherbetriebes. Dagegen lässt die isolierte Darstellung der Speicher-Zyklenzahl meist keine genaue Aussage über den energetischen Nutzen eines thermischen Speichers zu.

4.2.1 Modellierungsansätze

Für die Modellierung thermischer Speicher existieren unterschiedliche Ansätze. Mehrdimensionale Modelle beinhalten die Lösung der Massen-, Impuls- sowie Energieerhaltungsgleichungen und weisen einen erhöhten Aufwand bei der Implementierung und Handhabung auf. Eindimensionale Modelle beruhen dagegen häufig auf der Lösung der Konvektions-Diffusions-Gleichung in Kombination mit einer Energiebilanz über das Kontrollvolumen. Dabei hängt die zeitliche (t) und örtliche (x) Änderung der Temperatur T eines Volumenelementes mit der Strömungsgeschwindigkeit v, der Temperaturleitfähigkeit α, dem Wärmedurchgangskoeffizienten k, der spezifischen Wärmekapazität c_p sowie der Dichte ρ des Wärmeträgermediums, dem Umfang U des Speicherbehälters, der wirksamen Wärmeübertragungsfläche A und der Umgebungstemperatur T_U wie folgt zusammen [6]:

$$\frac{\partial T}{\partial t} + v \cdot \frac{\partial T}{\partial x} = \alpha \frac{\partial^2 T}{\partial x^2} + \frac{k \cdot U}{A \cdot \rho \cdot c_p} \cdot (T_U - T) \qquad (3)$$

Eindimensionale Modelle gelten als einfacher in der Handhabung, erfordern jedoch mehr Annahmen und Anpassungen der Gleichungen an reale Verhältnisse [6]. Eine wichtige Randbedingung ist die Uniformität der Fluidströmung über den gesamten Querschnitt des Speichertanks. Diese Forderung wird gerade im Bereich der Ladeeinrichtungen des Speichers bei realen Anlagen nicht erfüllt und führt zu Abweichungen von den Simulationsergebnissen. Um die Einflüsse komplexer Strömungen im Bereich der Ladeeinrichtungen zu kompensieren, wurden verschiedene Ansätze entwickelt [6, 7, 8]. Durch empirische Anpassung der eindimensionalen Modellgleichungen können somit ähnlich genaue Ergebnisse erzielt werden, wie bei der mehrdimensionalen Formulierung [9]. Dies führt insbesondere bei der Simulation ganzer Energiesysteme über große Zeiträume zu einer deutlichen Reduktion der Rechenzeiten.

4.2.2 Numerische Lösung

Für die im Folgenden dargestellten Simulationen wurde ein Speichermodell nach Ansatz in Gleichung (3) verwendet, bei dem Advektion und thermische Diffusion im Wärmeträgermedium sowie Wärmeverluste über die Speicherwand berücksichtigt werden. Das Modell liefert keine Kompensation für Vermischungseffekte in den Ladezonen bei hohen Ladeleistungen. Die zugrundeliegende Konvektions-Diffusions-Gleichung stellt eine partielle Differentialgleichung zweiter Ordnung dar und wird mit der Finiten-Differenzen-Methode gelöst. Im speziellen Fall wird das Crank-Nicolson-Verfahren eingesetzt, wobei jede Komponente der partiellen Differentialgleichung mit dem zentralen Differenzenverfahren diskretisiert wurde. Der simulierte Kältespeicher wird hierfür virtuell in 110 idealisierte Temperaturzonen zerlegt. Die zeitliche Schrittweite des Speichermodells beträgt ein ganzzahliges Vielfaches der Simulationszeit des Gesamtsystems, um die Genauigkeit des Verfahrens zu gewährleisten [10].

4.3 Übergeordnete Regelstrategie

Im angeführten Beispiel wird ein Zustandsautomat zur Umsetzung der Lastverschiebungsstrategie verwendet, welcher ein ereignisdiskretes System darstellt. Der Systemzustand wird über Eingangsgrößen beeinflusst, die aus den aktuellen Randbedingungen des Anlagenbetriebs (Speicherladezustand, Lastanforderung des Netzes, Ausfall einer Kältemaschine, etc.) und aus einem übergeordneten Fahrplan für die Komponenten generiert werden. Dieser Fahrplan wird im Vorlauf der Simulation erstellt und bietet grundsätzlich die Möglichkeit, Kernladezeiten für den Speicher z. B. anhand von Wettervorhersagen vorzugeben. Die Verwendung eines Zustandsautomaten in der Simulation besitzt den Vorteil, dass die zugrundeliegende Steuer- und Regellogik zunächst modellhaft untersucht und anschließend auch für die Umsetzung auf eine Hardware herangezogen werden kann.

5 Simulationsergebnisse

Die Simulation unterschiedlicher Kälteversorgungsstrategien erfolgt über einen Zeitraum von mindestens einem Tag. Dadurch können Temperaturwechsel sowie Bedarfsschwankungen des Kaltwasserbezuges zwischen Tages- und Nachtzeit ausreichend genau nachgebildet und für den Entwurf einer vorausschauenden Versorgungsstrategie herangezogen werden. Als Grundlage der Simulation dienen reale Daten aus dem Energie-Monitoring-System des Instituts mit einer zeitlichen Auflösung von mindestens einer Minute. Alternativ besteht die Möglichkeit, die Planung des Anlagenbetriebs prognosebasiert anhand von aktuellen Wettervorhersagen und angepassten, historischen Kältelastprofilen des Instituts abzuleiten. **Bild 5** zeigt den historischen Verlauf der Kälteleistung sowie den zeitgleich gemessenen Außentemperaturverlauf des Standorts. Auf der y-Achse im unteren Teil des Diagramms ist der zeitlich korrelierte Ladezustand eines Kältespeichers mit insgesamt 80 m³ Volumen in Prozent aufgetragen. Die Kurve mit dem Titel „*Kältemaschine*" zeigt den Verlauf der abgerufenen Kälteleistung einer simulierten Schraubenkompressor-Kältemaschine ohne Möglichkeit zur Drehzahlregelung. Durch die Speichernutzung kann das Verhältnis von Kälteerzeugung zu aufgenommener elektrischer Energie der Kälteinfrastruktur angehoben werden. Dieser Zusammenhang ist auf den besseren Wirkungsgrad der Rückkühlwerke bei niedriger Außentemperatur in der Nacht zurückzuführen. Die höhere Auslastung der Kältemaschine in den letzten sechs Stunden wird von der Effizienzbetrachtung ausgenommen. Sie leitet sich aus der Beladung des Speichers ab, welcher erst im nachfolgenden Tageszyklus wieder entladen werden kann. Die COP-Werte liegen in diesem Zeitraum oberhalb der realen Messwerte, dürfen aber nicht in den Kostenvergleich einfließen, da die generierten Kälteenergien nicht identisch sind. Zwischen Simulationsbeginn und Stunde 18 liegt der COP des realen Systems bei einem Wert von 1,72. Im Vergleich dazu lässt sich der Gesamt-COP durch gezielte Lastverschiebung mit einem Speicher auf etwa 1,88 erhöhen. Somit können die Kosten für den Betrieb der Kältekomponenten im dargestellten Zeitraum bei einer erzeugten Kälteenergie von 3500 kWh/d unter Annahme eines Strompreises von z.B. 18 ct/kWh$_{el}$ von etwa 366 €/d auf unter 335 €/d gesenkt werden.

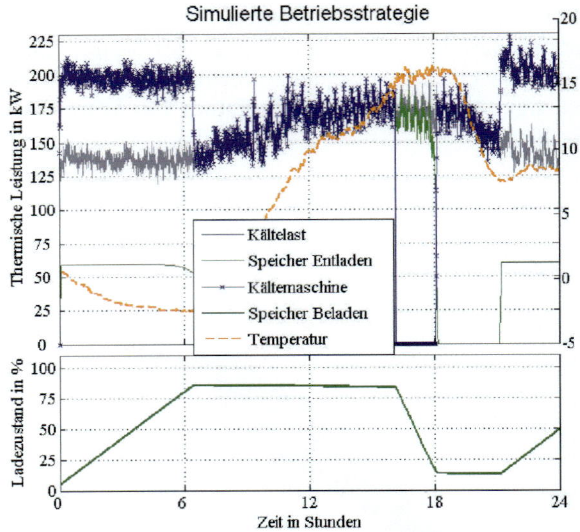

Bild 5 Gemessene Kältelast und optimierter angepasster Verlauf der Kältemaschinenleistung mit zeitlich korreliertem Kältespeicherladezustand

Die Energiekosteneinsparung von etwa 9 % wird im dargestellten Beispiel v. a. durch die simulierte Kältemaschine begrenzt, welche gegenüber ihrer Kühlaufgabe überdimensioniert ist. Durch die nächtliche Beladung des Kältespeichers können in der Simulation maximal 225 kW$_{th}$ abgerufen werden, dies entspricht einer Auslastung der Maschine von unter 65 %. Würde die Kälteversorgung von einer kleineren Maschine realisiert, könnte durch den steileren COP-Verlauf über vergleichbare Kältelastbereiche vermutlich eine höhere Effizienz erreicht werden.

Das dargestellte Einsparpotenzial lässt sich jedoch durch Vergrößerung des Speichervolumens steigern, wodurch eine höhere Auslastung der Kältemaschine in der Nacht ermöglicht wird. Bei doppelter Speichergröße kann der COP der gesamten Kühltechnik somit von 1,72 auf etwa 2,1 angehoben werden. Dies entspricht einer Energiekosteneinsparung von über 18 % bei den oben angenommenen Verbrauchswerten und bedeutet eine tägliche Reduktion der verbrauchsgebundenen Betriebskosten von 66 €/d. Neben der Kältemaschine und der Speichergröße hat auch das Umgebungstemperaturprofil einen großen Einfluss auf die Wirtschaftlichkeit der Lastverschiebungsstrategie. Mit steigender Außentemperatur werden größere Luftdurchsätze der Rückkühlwerke erforderlich, die mit einer Abnahme des Wirkungsgrades einhergehen. Dieser Zusammenhang ist anhand des Gesamt-COP-Verlaufes in **Bild 3** bei 15 °C zu erkennen. Durch den Sprung zu einer größeren Lüfterstufe oberhalb von

190 kW_{th} geforderter Kälteleistung nimmt die Steigung des Gesamt-COP ab. Im Vergleich verschiebt sich dieser Sprung bei einer niedrigeren Außentemperatur von 10 °C zu etwa 240 kW_{th} geforderter Kälteleistung. Der betrachtete Simulationszeitraum in **Bild 5** kann hiervon nicht profitieren, da der gemessene Lastverlauf am Tag nie 200 kW_{th} überschreitet und somit keine höheren Lüfter-Stufen erforderlich sind. Eine Vergleichssimulation bei höheren Tagesaußentemperaturen oder größeren Kälteleistungen würde daher eine weitere Hebung des Effizienzpotenzials aufzeigen.

Die dargestellten Abhängigkeiten gelten nur für Trockenkühler. Bei anderen Rückkühlwerkstypen besitzt die Luftfeuchte eine ebenso große Bedeutung wie die Umgebungstemperatur, da der Wassergehalt der Luft nicht nur die Wärmekapazität, sondern auch die Aufnahme von Wasserdampf beeinflusst. Hybrid- und Nasskühltürme beziehen einen Großteil ihrer Kühlleistung aus der Verdampfung von Wasser, die nur bis zum Erreichen des Sättigungsdampfdruckes in der Luft stattfindet. Da sich die Profile von Temperaturverlauf und relativer Luftfeuchte über den Tag jedoch entgegengesetzt verhalten, sollte der COP-Verlauf solcher Rückkühlwerke z. B. mit Hilfe von Simulationsmodellen prognostiziert werden. Die Zeitpunkte mit höchster Effizienz der Rückkühlwerke können dann der Betriebsstrategie als Basis dienen.

6 Zusammenfassung und Ausblick

Mit thermischen Speichern lässt sich die Effizienz von Kälteversorgungssystemen erhöhen, indem die Kälteerzeugung auf energetisch günstigere Zeitpunkte verlagert wird. Eine große Bedeutung hat dabei die Außentemperatur, welche vor allem Einfluss auf den Wirkungsgrad der Rückkühltechnik ausübt. Die dargestellten Simulationsergebnisse wurden anhand realer Bezugsdaten der Kälteanlagen am IISB erstellt. Hier konnten Energiekosteneinsparungen in der Größenordnung von etwa 9 bis 18 % der verbrauchsgebundenen Betriebskosten durch Lastverschiebung zwischen Tag und Nacht vorausgesagt werden. Das Einsparpotenzial bezieht sich dabei auf die Nutzungsstrategie der Übergangs- und Sommermonate. Eine genaue Aussage über die Amortisationszeit des Speichers erfordert auch ein Nutzungskonzept für die kalte Jahreszeit. Ziel der weiteren Untersuchungen ist daher eine Vorhersage des Effizienzpotenzials im Winter, wenn der Kältespeicher auch der Vorhaltung von Kaltwasser aus freier Kühlung dient. Diese Maßnahme kann nach ersten Abschätzungen höhere Betriebskosteneinsparungen als die dargestellte Lastverschiebung bewirken, steht jedoch nur bei niedrigen Außentemperaturen zur Verfügung.

Eine erste Gegenüberstellung der Betriebs- und Anschaffungskosten des thermischen Speichers indiziert, dass Amortisationszeiten zwischen 3 und 5 Jahren erreicht werden können. Eine Erhöhung der Wirtschaftlichkeit ist vorstellbar, wenn der eingesetzte Speicher auch zur Spitzenlastglättung oder zur Nutzung von Nachtstromtarifen eingesetzt wird. Das angeführte Einsparpotenzial ist weitgehend auch auf andere Standorte in Deutschland übertragbar, da die Effizienz der Kälteversorgung maßgeblich von den klimatischen Randbedingungen abhängt. Für einen optimalen Betrieb der Anlagen sollten gute Kenntnisse der Einzelkomponenten vorliegen, die in einer individuellen Betriebsstrategie Anwendung finden.

Danksagung

Die Forschungen wurden im Rahmen des vom Bayerischen Staatsministerium für Wirtschaft und Medien, Energie und Technologie geförderten Projektes SEEDs durchgeführt.

7 Literatur

[1] Hauer, A.: Wirtschaftlichkeit von Energiespeicher, in: OTTI e.V. (Hrsg.): 2. Anwenderforum Thermische Energiespeicher, S. 87-90 (2013)

[2] VDMA Allgemeine Lufttechnik (Hrsg.): Energieeffiziente Kältetechnik. 1. Aufl. 2010, http://kwt.vdma.org/article/-/articleview/2245536. Zugegriffen: 25. März 2015

[3] Urbaneck, T.: Kältespeicher. Grundlagen, Technik, Anwendung. Oldenbourg, München (2012)

[4] Breidenbach, K.; Der Kälteanlagenbauer. Band 1: Grundkenntnisse. VDE, Berlin Offenbach (2012)

[5] Baehr, H.D., Kabelac, S.: Thermodynamik. Grundlagen und technische Anwendungen. 15. Aufl., Springer Vieweg, Berlin Heidelberg (2012)

[6] Zurigat, Y.H., Ghajar, A.J..: Heat Transfer and Stratification in Sensible Heat Storage Systems. In: Dincer, I., Rosen, M. (Hrsg.): Thermal Energy Storage: Systems and Applications, S.259-302. John Wiley & Sons, LTD, West Sussex (2002)

[7] Oppel, F.J., Ghajar, A.J., Moretti, P.M.: Computer Simulation of Stratified Heat Storage. Appl. Energy 23, 205-224 (1986)

[8] Wildin, M.W., Truman, C.R.: Evaluation of Stratified Chilled-Water Storage Techniques, Volumes 1 and 2. Final Report, The University of New Mexico (1985)

[9] Oliveski, R.C., Krenzinger, A., Vielmo, H.A.: Comparison between Models for the Simulation of Hot Water Storage Tanks. Sol. Energy 75, 121-134 (2003)

[10] Baehr, H.D., Stephan, K.: Wärme- und Stoffübertragung. 6. Aufl., Springer, Berlin-Heidelberg (2008)

Gesamtsystem

Abschätzung des Energieübertragungsbedarfs in Deutschland bei ausschließlicher Einspeisung aus erneuerbaren Energiequellen

N. Erdmann[1], J. Dickert[2], P. Schegner[2]
[1]TU Dresden, Boysen-TUD-Graduiertenkolleg, niels.erdmann@tu-dresden.de
[2]TU Dresden, Institut für Elektrische Energieversorgung und Hochspannungstechnik

Kurzfassung

Bei der Umsetzung der Energiewende stand in den letzten Jahren insbesondere die elektrische Energieversorgung im Fokus. Mit einem Anteil von 27,8 % im Jahr 2014 tragen die erneuerbaren Energien erheblich zur Bereitstellung bei. Im Gegensatz dazu betrug der Anteil an erneuerbaren Energien (EE) beim Wärmeverbrauch nur 9,9 % und im Verkehrssektor 5,4 %. Die Bundesregierung hat mit dem Energiekonzept die Vorgabe gemacht, dass bis 2050 80 % der elektrischen Energie regenerativ bereitgestellt werden soll. Darüber hinaus kann davon ausgegangen werden, dass zukünftig vermehrt elektrische Energie für die Wärmeversorgung mit Wärmepumpen und für die Elektromobilität benötigt wird. Die hier vorgestellte Untersuchung beruht auf Energiebilanzen, und geht im Gegensatz zu den Annahmen der Bundesregierung von einem erhöhten Nettobedarf an elektrischer Energie von 700 TWh/a aus. Dieser soll vollständig durch in Deutschland installierte EE-Anlagen bereitgestellt werden. Es werden zwei Ansätze betrachtet. Zum einen erfolgt im Ansatz A eine lineare Weiterschreibung des bisherigen EE-Zubaus je Region und im Ansatz B ein möglichst verbrauchernaher EE-Zubau. Die fehlenden Energiemengen werden jeweils durch Zubau von Offshore-Windenergieanlagen in Nord- und Ostsee bereitgestellt. Hieraus ergeben sich unterschiedliche Energiemengen, welche innerhalb eines Jahres zwischen den Regionen in Deutschland ausgetauscht werden müssen. Dies ermöglicht Aussagen zum notwendigen Übertragungsbedarf für Energie. Daraus lässt sich ableiten, ob der kontinuierliche Ausbau der elektrischen Netze diese Anforderungen erfüllen kann oder ob durch einen Paradigmenwechsel auf dem Gebiet der Übertragungstechnologie effizientere, realisierbare und tragfähige Lösungen möglich sind.

1 Einleitung

Die hier vorgestellten Untersuchungen gehen über einen prognostizierbaren Zeitraum gewollt weit hinaus. Um die Zukunft der bestehenden Energiesysteme zu hinterfragen werden elementare Annahmen gemacht, welche einen experimentellen Charakter haben. Sie sind jedoch notwendig, um Zukunftsvisionen auszuarbeiten und ermöglichen die Entwicklung neuer Ideen für zukünftige Energieversorgungssysteme. Im ersten Schritt wird Deutschland in 16 Regionen eingeteilt. Die Schwerpunkte der Regionen werden miteinander Verbunden und es resultieren Verbindungskorridore für die der erforderliche Energieübertragungsbedarf bestimmt wird. Dieser ergibt sich aus der Differenz zwischen dem lokalen Bedarf und der Bereitstellung elektrischer Energie. Für die Bereitstellung werden Photovoltaik- (PV), Windenergie- (WEA), Wasser- und Biomasse-Anlagen betrachtet. Der angenommene Bedarf wird aus aktuellen Daten auf 700 kWh/a hochskaliert. Bei der Bereitstellung werden zwei Ansätze untersucht. Das Ergebnis sind Übertragungskorridore mit den dazugehörigen Übertragungsanforderungen.

Zur besseren Lesbarkeit meint im Folgenden der Begriff Energie immer ausschließlich elektrische Energie.

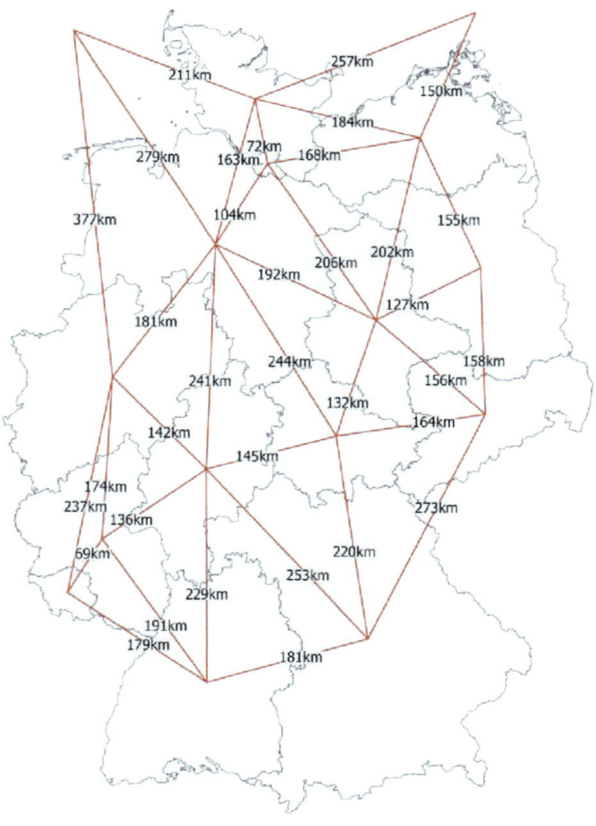

Bild 1 Verbindungskorridore der 16 Regionen in Deutschland

Tabelle 1 Übersicht Energiebedarf, Energiebereitstellung und Energiesaldo nach Regionen, Energiequellen und den Ansätzen A und B (alle Angaben in TWh/a)

Region	Energiebedarf, rd.	Energiebereitstellung								Energiesaldo, rd.	
		Wasser	Biomasse	Photovoltaik		Wind		Gesamt			
				A	B	A	B	A	B	A	B
Baden-Württemberg	81	7,54	6,14	18,71	19,45	3,57	15,32	35,97	48,45	• 45	• 33
Bayern	110	14,68	12,12	40,67	24,10	4,43	34,74	71,90	85,64	• 38	• 24
Berlin-Brandenburg	42	0,10	4,71	4,62	14,10	44,91	37,69	54,33	56,59	12	15
Hamburg	14	0,00	0,33	0,07	2,54	0,47	0,27	0,87	3,14	• 13	• 11
Hessen	60	1,33	1,97	5,37	10,99	5,04	19,19	13,71	33,48	• 46	• 26
Mecklenburg-Vorpommern	8	0,06	3,38	1,50	6,09	17,57	37,82	22,50	47,35	14	39
Niedersachsen-Bremen	81	1,21	11,60	8,53	20,08	70,57	67,16	91,91	100,06	11	19
Nordrhein-Westfalen	165	1,39	6,18	11,52	33,43	27,77	42,82	46,86	83,82	• 118	• 81
Rheinland-Pfalz	36	0,66	1,48	5,35	8,75	11,98	16,06	19,47	26,95	• 17	• 9
Saarland	17	0,21	0,10	1,02	2,49	1,25	1,51	2,58	4,30	• 14	• 13
Sachsen	27	1,65	2,28	3,63	11,80	9,41	26,19	16,95	41,92	• 10	15
Sachsen-Anhalt	24	0,53	3,21	2,98	9,36	35,05	19,09	41,76	32,19	18	8
Schleswig-Holstein	19	0,06	3,88	4,18	6,13	35,22	36,63	43,34	46,70	24	28
Thüringen	16	0,60	2,62	1,94	6,85	7,68	23,18	12,84	33,24	• 3	17
Nordsee	0	0,00	0,00	0,00	0,00	180,00	45,00	180,00	45,00	180	45
Ostsee	0	0,00	0,00	0,00	0,00	45,00	11,25	45,00	11,25	45	11
Deutschland	*700*	*30,00*	*60,00*	*110,09*	*176,15*	*499,91*	*433,91*	*700,00*	*700,00*	*0*	*0*

2 Vorgehen

2.1 Einteilung Deutschlands in Regionen

Deutschland wird in 16 Regionen unterteilt (vgl. **Bild 1**). Berlin und Brandenburg wird zu Berlin-Brandenburg und Niedersachsen und Bremen zu Niedersachsen-Bremen zusammengefasst. Die übrigen Bundesländer entsprechen je einer Region. Zusätzlich werden noch die Flächen von Nordsee und Ostsee hinzugenommen. Jede Region hat einen Energiebedarf und kann Energie durch EE-Anlagen bereitstellen (vgl. **Tabelle 1**).

Eine Erweiterung auf kleinere Einheiten wie z. B. Regierungsbezirke oder Landkreise ist für diese Untersuchung nicht notwendig. Weiterhin unterstützt die grobe Auflösung die Übersichtlichkeit und Nachvollziehbarkeit der Methode und deren Ergebnisse.

2.2 Verbindungskorridore

Um die Energieübertragung zwischen den Regionen zu ermöglichen, werden Verbindungskorridore definiert. Diese direkten Verbindungen zwischen den Regionen symbolisieren ein überregionales Transportnetz. Jede Region besitzt einen zentralen Übergabepunkt, an dem das Transportnetz mit dem unterlagerten Netz verbunden ist. Die Übergabepunkte an Land werden in den jeweiligen Flächenschwerpunkt der Region gelegt. In Nordsee und Ostsee entsprechen die Übergabepunkte den Zentren der aktuell installierten/geplanten Offshore-WEA. Aus den so definierten Punkten wird mittels Delaunay-Triangulation [1] ein Dreiecksnetz erstellt. **Bild 1** zeigt die so gefundenen Verbindungskorridore zwischen den Regionen mit den Längen dieser Direktverbindungen.

2.3 Annahmen zu Bedarf und Erzeugung

2.3.1 Energiebedarf

Es wird davon ausgegangen, dass sich zukünftig zwar der Wirkungsgrad elektrischer Verbraucher weiter verbessert, jedoch die Anzahl der elektrischen Verbraucher z. B. durch Wärmepumpen und Elektromobilität deutlich erhöht. In Summe wird angenommen, dass der Energiebedarf in Deutschland auf 700 TWh/a ansteigt. Dieser erhöhte Energiebedarf wird wie im Referenzjahr 2011 anteilig auf die 16 Übergabepunkte verteilt. Die Fläche innerhalb der schwarzen Kreislinien in **Bild 2** und **Bild 3** entspricht dem Energiebedarf in den Regionen. Die sich ergebenden Bedarfe sind in **Tabelle 1** je Region enthalten.

2.3.2 Energiebereitstellung

Zur Untersuchung der zukünftigen Energiebereitstellung dient der EEG-Statistikbericht[1] 2011 [2] als Ausgangslage. Darauf aufbauend werden hier zwei Ansätze untersucht. In beiden ist der Beitrag aus Geothermie und Deponie-, Klär- und Grubengas vernachlässigbar gering. Weiterhin steigt der Energieanteil durch Biomasse von ca. 28 auf 60 TWh/a und durch Wasser von ca. 5 auf 30 TWh/a. Die verbleibenden 610 TWh/a müssen durch PV und WEA (onshore und offshore) bereitgestellt werden.

[1] Die Quelle beinhaltet ausschließlich EEG-Anlagen und dient hier hauptsächlich als Maß für die räumliche Verteilung der Energieanlagen auf die Regionen. Insbesondere die eingespeiste Energie aller Wasserkraftanlagen (auch nicht-EEG) liegt um ein Vielfaches höher.

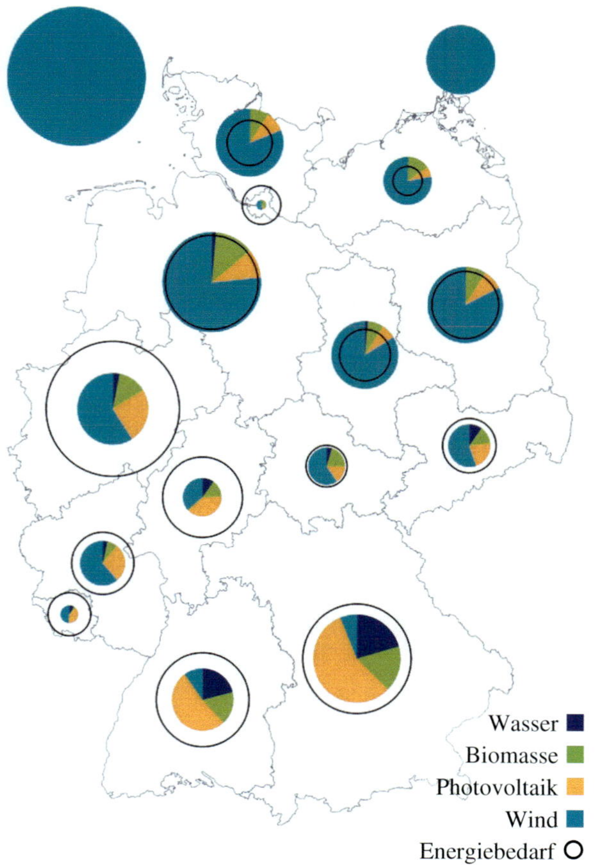

Bild 2 Energiebereitstellung und -bedarf gem. Ansatz A (massiver Zubau Offshore-WEA und linearer Zubau von EE-Anlagen in den Regionen seit 2011)

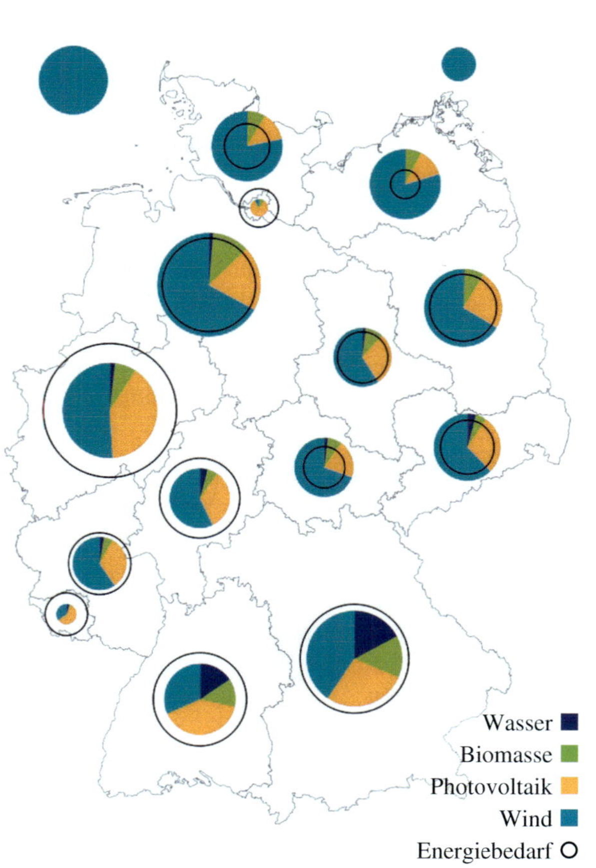

Bild 3 Energiebereitstellung und -bedarf gem. Ansatz B (moderater Zubau Offshore-WEA und Zubau von EE-Anlagen in den Regionen gem. Potentialflächen)

2.3.2.1 Ansatz A

Im Ansatz A wird neben der Weiterschreibung des EE-Zubaus ein massiver Zubau von Offshore-WEA verfolgt. In einer noch unveröffentlichten Studie des VDE [3], gehen die Autoren in Anlehnung an das Energiekonzept der Bundesregierung davon aus, dass Offshore-WEA mit einer Anschlussleistung von 40 GW in der Nordsee und 10 GW in der Ostsee installiert werden. Bei angenommenen Volllaststunden von 4.500 h/a [4] können so 225 TWh/a bereitgestellt werden. Für den verbleibenden Bedarf von 385 TWh/a werden die installierte Leistungen der PV-Anlagen und Onshore-WEA in 2011 linear soweit erhöht, bis der Bedarf gedeckt werden kann. Alle betrachteten Anlagen an Land werden nach der Verteilung in 2011 den Regionen zugeordnet und somit ergibt sich die in **Tabelle 1** in den Spalten A enthaltene Energiebereitstellung. **Bild 2** zeigt die regionale Energiebereitstellung nach Energieträgern und den Energiebedarf.

2.3.2.2 Ansatz B

Im Gegensatz zu Ansatz A wird im Ansatz B das Ziel verfolgt, einen möglichst verbrauchernahen EE-Zubau zu erreichen. Daher werden auch Offshore-WEA nur moderat zugebaut. In der Nordsee werden Offshore-WEA mit einer Anschlussleistung von insgesamt 10 GW und in der Ostsee von 2,5 GW installiert. Dies entspricht unter Berücksichtigung der Volllaststunden einer Energie von 56,25 TWh/a. Für die benötigten PV- und Onshore-WEA wird die potentiell nutzbare Fläche in den Regionen bewertet. Die Flächendaten der einzelnen Regionen werden über CORINE Land Cover 2006 (CLC06) [5] bereitgestellt.

Für die PV-Anlagen werden drei Flächentypen betrachtet: Flächen mit *durchgängig städtischer Prägung* werden zu 10 % mit PV-Anlagen belegt. Beispielsweise beträgt in Hamburg diese Fläche etwa 1 % der gesamten Stadtfläche. In Gebieten mit *nicht durchgängig städtischer Prägung* werden auf 5 % und in *Industrie- und Gewerbeflächen* auf 25 % der Grundfläche PV-Anlagen installiert.
Es wird davon ausgegangen, dass alle PV-Anlagen eine flächenbezogene Leistung[2] von $0{,}1 kW_p/m^2$ und Volllaststunden von 1.000 h/a [6] haben. Unter diesen Annahmen werden in Deutschland 176 TWh/a eingespeist. Die verbleibenden 378 TWh/a müssen durch Onshore-WEA gedeckt werden.

[2] bei einer mittleren horizontaler Jahreseinstrahlung in Deutschland von 1000 W/m² und einem mittleren Wirkungsgrad von 10 %

Aus Winddaten des Deutschen Wetterdienstes (DWD) [7] wird für jede Region eine mittlere Windgeschwindigkeit berechnet. Unter der vereinfachten Annahme, dass eine WEA[3] mit dieser mittleren Windgeschwindigkeit dauerhaft betrieben wird, kann aus der Leistungskennlinie auf eine zu erwartende Jahresenergie pro WEA geschlossen werden. Zur Installation von WEA werden die CLC06-Kategorien *Ackerland* sowie *Wiesen und Weiden* vorgesehen. Die benötigte Energie kann bereitgestellt werden, wenn deutschlandweit auf Ackerland 0,7 WEA/km² sowie auf Wiesen und Weiden 0,2 WEA/km² installiert werden. Die regionale Energiebereitstellung nach Energieträgern und der Energiebedarf ist **Bild 3** zu entnehmen.

2.4 Übertragungskorridore

Übertragungskorridore sind die Verbindungskorridore, über die tatsächlich Energie ausgetauscht wird. In jedem der $m_{\max} = 16$ Übergabepunkte müssen sich Energiebereitstellung und Energieaufnahme mit Energiebedarf und Energieabgabe im Jahresmittel ausgleichen. In der Graphentheorie entspricht dies einem gewichteten Graph mit ungerichteten Kanten.

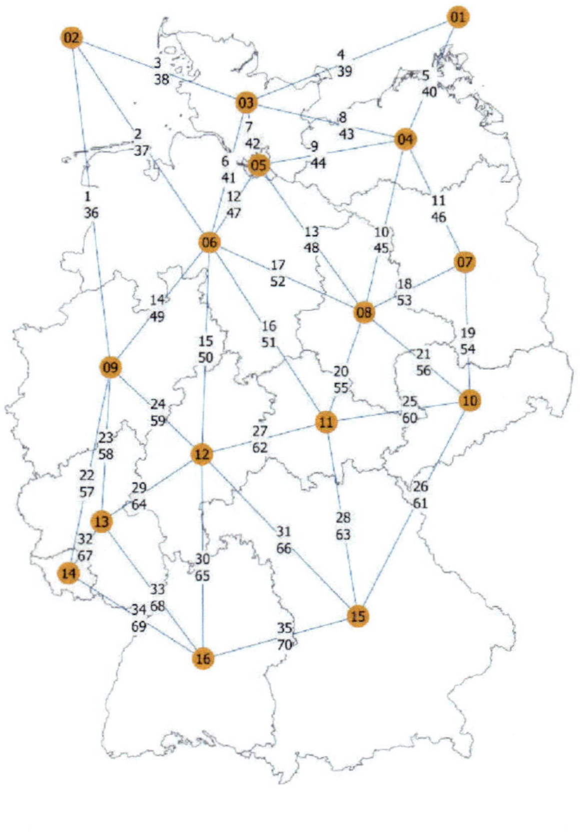

Bild 5 Graphische Darstellung des Optimierungsproblems mit 16 Übergabepunkten ($m = 1 \dots 16$, orange) und 70 unidirektionalen Verbindungskorridoren ($n = 1 \dots 35$ Nord-Süd-Richtung; $n = 36 \dots 70$ Süd-Nord-Richtung)

Daraus ergibt sich ein lineares Problem (LP), welches sich im Allgemeinen darstellen lässt als:

$$\min_{x \in \mathbb{N}} \{\boldsymbol{c}^T \boldsymbol{x} \mid \boldsymbol{A}\boldsymbol{x} = \boldsymbol{b}, \boldsymbol{x} \geq 0\} \qquad (1)$$

Der zu minimierende Zielfunktionswert $F(\boldsymbol{x}) = \boldsymbol{c}^T \boldsymbol{x}$ entspricht dem Skalarprodukt aus den Längen der Verbindungskorridore $\boldsymbol{c}^T$ mit der jeweils übertragenen Energie $\boldsymbol{x}$.

Die bidirektional definierten Verbindungskorridore erlauben die Energieübertragung in beide Richtungen. Folglich muss eine Richtung mit einem negativen Vorzeichen behaftet sein. Da dies unter der Nebenbedingung $\boldsymbol{x} \geq 0$ jedoch nicht zulässig ist, werden die bidirektionalen Verbindungskorridore in je zwei unidirektionale Verbindungskorridore aufgeteilt (siehe **Bild 4** bzw. **Bild 5**).
Die so entstandenen $n_{\max} = 70$ Verbindungskorridore übertragen per Definition Energie mit positivem Vorzeichen. Für $n = 1 \dots 35$ ist die festgelegte Übertragungsrichtung von Nord nach Süd. Für die Übertragung von Süd nach Nord sind $n = 36 \dots 70$ vorgesehen.

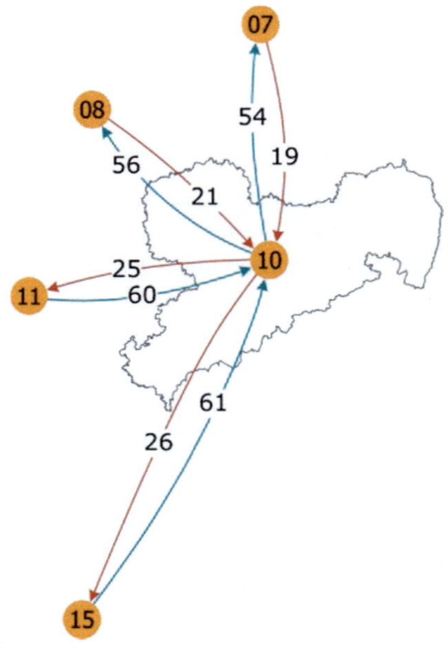

Bild 4 Vergrößerte Darstellung der Region Sachsen (vgl. Bild 5) mit aufgeteilten Verbindungskorridoren für Nord-Süd-Richtung (19, 21, 25, 26) und Süd-Nord-Richtung (54, 56, 60, 61)

Die Nebenbedingungen $\boldsymbol{A}\boldsymbol{x} = \boldsymbol{b}$ gewährleisten ausgeglichene Energiebilanzen. Der rechte-Seite-Vektor $\boldsymbol{b}$ enthält für jeden Übergabepunkt das Energiesaldo (vgl. Tabelle 1). In der dünn besetzten Einflussmatrix $\boldsymbol{A}$ sind die Übergabepunkte (m_{max} Zeilen) mit den Verbindungskorridoren (n_{max} Spalten) verknüpft. Jedes Element $a_{m,n}$ bestimmt, mit welchem Faktor die übertragene Energie x_n in die Energiebilanz des Übergabepunktes m einfließt.

[3] Nennleistung: 3000 kW; Rotordurchmesser: 100 m; Nabenhöhe: 100 m

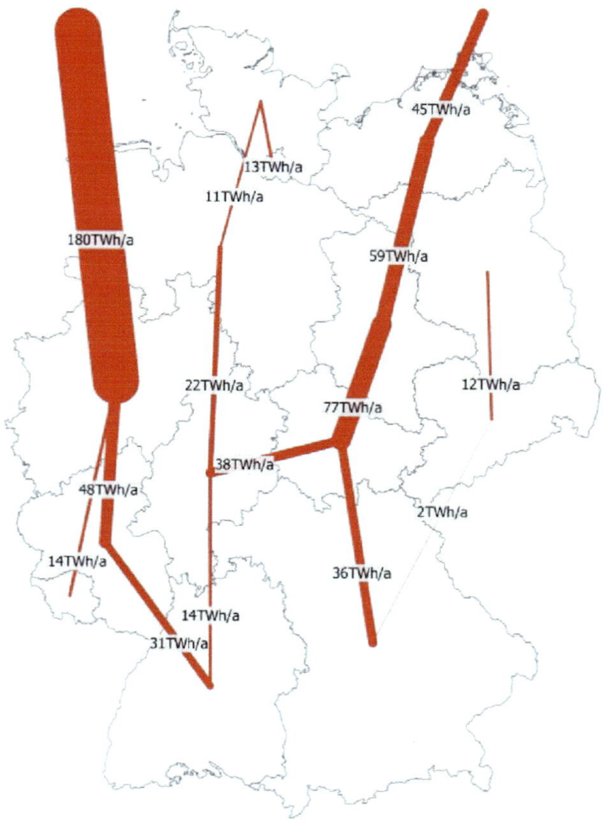

Bild 6 Ansatz A – Übertragungskorridore und die zu übertragende Energie pro Jahr

$$\boldsymbol{A} = \begin{pmatrix} a_{1,1} & \cdots & a_{1,n_{\max}} \\ \vdots & a_{m,n} & \vdots \\ a_{m_{\max},1} & \cdots & a_{m_{\max},n_{\max}} \end{pmatrix} \tag{2}$$

$$\forall\, a_{m,n} \in \{-1, 0, 1\} \tag{3}$$

$m = 1 \ldots 16 \mathrel{\widehat{=}}$ Laufvariable Übergabepunkte
$n = 1 \ldots 70 \mathrel{\widehat{=}}$ Laufvariable Verbindungskorridore
Am Beispiel Sachsen im Ansatz B bedeutet das:

$$m_{\text{Sachsen}} = 10 \tag{4}$$

$$n_{\text{Sachsen}} \in \{19, 21, 25, 26, 54, 56, 60, 61\} \tag{5}$$

Energiesaldo in Sachsen (vgl. **Tabelle 1**):

$$b_m = b_{10} = 15 \text{ TWh} \tag{6}$$

Zugeführte Energie:

$$a_{10,19} = a_{10,21} = a_{10,60} = a_{10,61} = 1 \tag{7}$$

Abgegebene Energie:

$$a_{10,25} = a_{10,26} = a_{10,54} = a_{10,56} = -1 \tag{8}$$

Daraus ergibt sich die Nebenbedingung für Sachsen:

$$x_{19} + x_{21} + x_{60} + x_{61} - x_{25} - x_{26} - x_{54} - x_{56} = b_{10} \tag{9}$$

Nach aufstellen der kompletten Einflussmatrix $\boldsymbol{A}$ mit den Daten für alle Übergabepunkte, lässt sich dieses LP mittels Simplex-Algorithmus [8] lösen.

Das verwendete Optimierungs-Modul aus dem SciPy[4]-Paket [9] findet eine optimale Lösung nach 29 Iterationen. Von den 35 bidirektionalen Verbindungskorridoren werden in beiden Ansätzen lediglich 15 zur Energieübertragung benutzt. Diese Übertragungskorridore und die jeweils übertragene Energie sind in **Bild 6** und in **Bild 7** dargestellt.

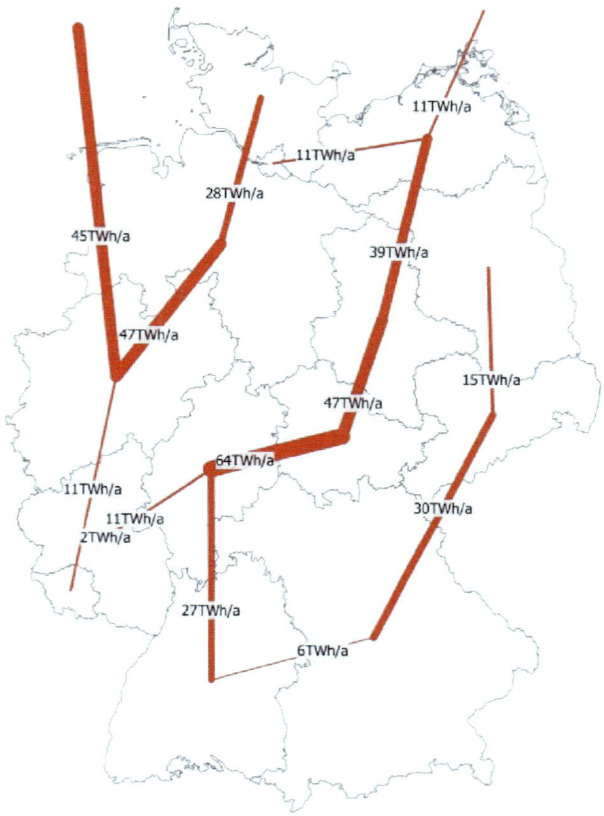

Bild 7 Ansatz B – Übertragungskorridore und die zu übertragende Energie pro Jahr

3 Auswertung

3.1 Energiebedarf und -bereitstellung

Ansatz A und Ansatz B unterscheiden sich nur in der Energiebereitstellung und folglich in der zu übertragenden Energie. Zwischen **Bild 2** und **Bild 3** ist deutlich eine Verschiebung der installierten Leistungen von der Nord- und Ostsee in Richtung Land zu erkennen. In Ansatz A ergeben sich sieben Exportregionen in Norddeutschland. Im Ansatz B wird zusätzlich aus Sachsen und Thüringen Energie exportiert.

Deutliche Unterschiede sind auch in Süddeutschland erkennbar. Während im Ansatz A die Energiebereitstellung in Bayern und Baden-Württemberg durch PV dominiert

[4] SciPy ist eine Python-basierte Open-Source-Softwareumgebung für wissenschaftliche Berechnungen

wird, ist in Ansatz B der Anteil von PV und WEA vergleichbar groß.

Durch den großen Anteil städtisch geprägter Fläche in Nordrhein-Westfalen ergibt sich im Ansatz B eine große Energiebereitstellung durch PV und ein geringerer Importbedarf. Diese und weitere Werte des Energiebedarfs und der Energiebereitstellung sind der **Tabelle 1** zu entnehmen. Um den Energiebedarf zu decken, wird im Ansatz B die stark reduzierte Offshore-Energiebereitstellung kompensiert durch die EE-Anlagen an Land. Im Gegensatz zu Ansatz A gibt es im Ansatz B insgesamt weniger Importregionen in denen zusätzlich auch weniger importiert werden muss.

3.2 Energieübertragung

In beiden Ansätzen muss Energie von Nord nach Süd und von Ost nach West übertragen werden. In Ansatz A muss die komplette Energie aus der Nordsee nach Nordrhein-Westfalen übertragen werden. Der dafür vorgesehene Übertragungskorridor muss eine Übertragungsfähigkeit von 180 TWh/a bei einer Länge 377 km aufweisen. Die größte notwendige Übertragungsfähigkeit im Ansatz B beschränkt sich auf 64 TWh/a und 145 km.

Im Ansatz A müssen pro Jahr 602 TWh, im Ansatz B 394 TWh übertragen werden. In Verbindung mit den Korridorlängen ergibt sich die Kenngröße „Energie-Strecke“. Diese ist ein Maß für die zu erwartenden Investitionen. Sie beträgt im Ansatz A 141.390 TWhkm/a und im Ansatz B 78.936 TWhkm/a. Die „Energie-Strecke“ im Ansatz B beträgt lediglich 55 % der „Energie-Strecke“ im Ansatz A.

4 Zusammenfassung

In dem Beitrag wurde ein Ansatz vorgestellt, mit dessen Hilfe es möglich ist Übertragungskorridore in Abhängigkeit der regional installierten Leistung zu bestimmen. Die Methode ist unabhängig von den betrachteten Energieträgern und Typen der Erzeugungsanlagen. Hierdurch wird es möglich für unterschiedliche Energiebereitstellungskonzepte den zu erwartenden Übertragungsbedarf zu bestimmen.

Es wurden zwei unterschiedliche Ansätze der Bereitstellung elektrischer Energie aus EE-Anlagen untersucht. Der Ansatz A stellt eine Fortschreibung des aktuellen Zubaus an EE-Anlagen dar. Mit dem Ansatz B wird ein stärker regionaler, bzw. zellularer Ansatz verfolgt. Hierbei sollen die erneuerbaren Energien, soweit Potential vorhanden, lokal zur Elektrizitätserzeugung genutzt werden. Das Ergebnis zeigt, dass im ersten Ansatz fast die doppelte „Energie-Strecke“ benötigt wird, dies ist ein Maß für die zu erwartenden Investitionen.

Die Übertragungskorridore entsprechen *nicht* den konkreten Übertragungstrassen. Diese hängen von topographischen, geographischen und weiteren Umweltfaktoren und von der eingesetzten Übertragungstechnologie ab. Die Übertragungskorridore können jedoch Grundlage für die Bewertung verschiedener Übertragungstechnologien sein.

5 Literaturverzeichnis

[1] M. Joswig und T. Theobald, Algorithmische Geometrie: Polyedrische und Algebraische Methoden, Berlin Heidelberg New York: Springer-Verlag, 2008.

[2] Bundesnetzagentur, „EEG-Statistikbericht 2011,“ Bonn, 2013.

[3] Energietechnische Gesellschaft im VDE (ETG), „Der zellulare Ansatz - Grundlagen einer erfolgreichen, regionenübergreifenden Energiewende,“ (Veröffentlichung im III. Quartal 2015), Frankfurt.

[4] Fraunhofer-Institut für Windenergie und Energiesystemtechnik, „Windmonitor,“ [Online]. Available: http://windmonitor.iwes.fraunhofer.de/windmonitor_de/4_Offshore/5_betriebsergebnisse/1_Volllaststunden/. [Zugriff am 04. Mai 2015].

[5] European Environment Agency, „Corine Land Cover 2006, Version 17,“ http://www.eea.europa.eu.

[6] H. Wirth und K. Schneider, „Aktuelle Fakten zur Photovoltaik in Deutschland,“ Fraunhofer ISE, Freiburg, 2015.

[7] Deutscher Wetterdienst, „Windgeschwindigkeiten 1981-2000 100 m,“ ftp://ftp.dwd.de/pub/CDC/grids_germany, Offenbach, 2014.

[8] G. B. Dantzig und A. Jäger, Lineare Programmierung und Erweiterungen, Berlin-Heidelberg-New York: Springer-Verlag, 1966.

[9] The Scipy community, „SciPy.org,“ 18. Janur 2015. [Online]. Available: http://docs.scipy.org/doc/scipy-0.15.1/reference/generated/scipy.optimize.linprog.html. [Zugriff am 26. April 2015].

Analysen zu Markteinsatz und Standortauswahl von Power- & Biomass-to-Gas-Anlagen zur Speicherung von Strom im Erdgasnetz

Dipl.-Ing. Benjamin Schober, Dipl.-Ing. Florian Gutekunst, Univ.-Prof. Dr. techn. Günter Scheffknecht
Institut für Feuerungs- und Kraftwerkstechnik (IFK), Pfaffenwaldring 23, 70569 Stuttgart,
benjamin.schober@ifk.uni-stuttgart.de

Kurzfassung

Das Konzept Power- & Biomass-to-Gas bietet die Möglichkeit, durch den flexiblen Betrieb von Biomassevergasungsanlagen entweder Strom zu erzeugen, oder in Zeiten eines Überangebots elektrischer Energie diese im Erdgasnetz zu speichern. Der Strom wird dabei für eine Wasserelektrolyse genutzt, um zusammen mit Kohlenstoff aus der Biomassevergasung eine Methanisierung zu betreiben. Endprodukt dieses Prozesses ist Synthetic Natural Gas (SNG), das in seiner Zusammensetzung im Wesentlichen natürlichem Erdgas entspricht und in das Erdgasnetz eingespeist werden kann. In diesem Beitrag wird untersucht, inwieweit diese Technologie im künftigen Stromversorgungssystem zum Einsatz kommen kann und welche Standorte sich hinsichtlich der Verfügbarkeit von Biomasse und der Möglichkeit zur Kopplung von Strom- und Gasnetz eignen.

1 Einleitung

Die Europäische Union (EU) hat sich mit ihren 20/20/20-Zielen dazu verpflichtet, neben der Steigerung der Energieeffizienz den Anteil des Bruttoendenergieverbrauchs bis zum Jahr 2020 zu 20 % durch erneuerbare Energien zu decken und den Ausstoß klimaschädlicher Treibhausgase um 20 % gegenüber dem Basisjahr 1990 zu reduzieren [1]. Dies resultierte in länderspezifischen Allokationsplänen, die verbindliche Vorgaben für die einzelnen Mitgliedsstaaten enthalten.

1.1 Herausforderungen der Energiewende

In Deutschland ergibt sich der zu erreichende Anteil an erneuerbaren Energien für den Stromsektor gemäß des zweiten Nationalen Allokationsplans (NAP II) zu 30 %, nach dem 2014 novellierten Erneuerbare-Energien-Gesetz (EEG) soll dieser Anteil im Jahr 2025 sogar 40–45 % betragen [2] [3]. Wie auf **Bild 1** zu erkennen ist, hat sich dieser Anteil seit 1998 von 4,5 % auf 27,8 % im Jahr 2014 bereits versechsfacht und deckt nun nahezu ein Drittel des deutschen Bruttostromverbrauchs.

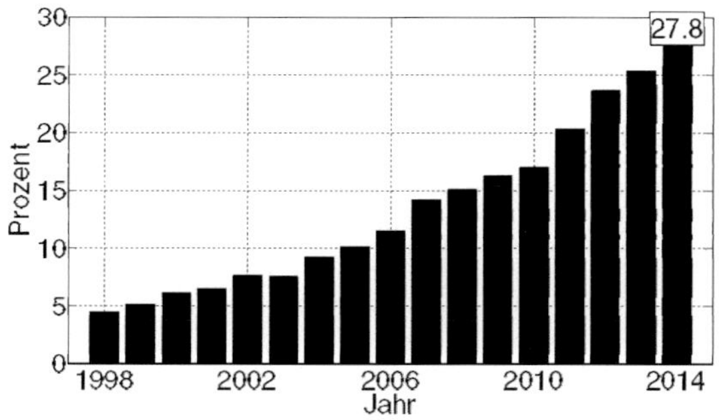

Bild 1 Entwicklung der Anteile der erneuerbaren Energien am Bruttostromverbrauch in Deutschland [4]

Mit der Erhöhung des Anteils der erneuerbaren Energien an der Stromproduktion ergeben sich unter der Annahme einer gleichbleibenden elektrischen Last im Stromversorgungssystem neue Herausforderungen. So sinkt die Residuallast und damit der Anteil der konventionellen Kraftwerke an der Stromerzeugung. **Bild 2** zeigt die geordneten Jahresdauerlinien der prognostizierten Residuallasten für die Jahre 2015, 2020 und 2025 basierend auf den Daten von 2013 und den Ausbauzielen des Erneuerbare-Energien-Gesetzes (EEG) 2014 [3] [5].

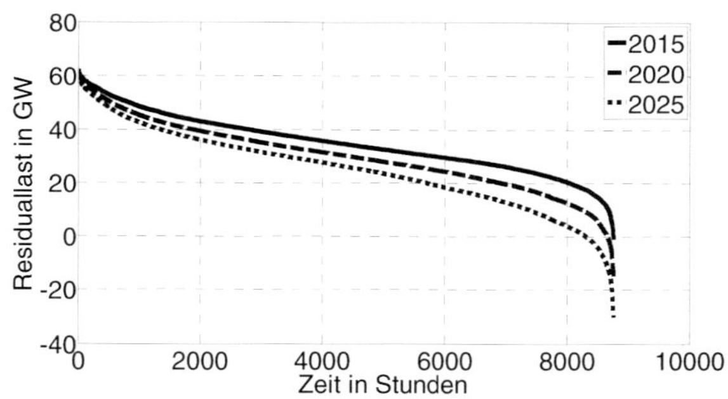

Bild 2 Geordnete Jahresdauerlinien der Residuallast der Jahre 2015, 2020 und 2025

Für die Erbringung von Systemdienstleistungen zur Sicherstellung der Systemstabilität muss zu jedem Zeitpunkt eine Mindesterzeugung elektrischen Stroms durch Synchrongeneratoren garantiert sein. Diese meist in großen konventionellen Kraftwerken verbauten Generatoren sind maßgeblich an der Einhaltung verschiedener Aspekte der Systemstabilität, wie Frequenz- und Spannungshaltung, beteiligt, und stellen über die Trägheit ihrer rotierenden Massen die im transienten Bereich einer Störung wichtige Momentanreserve bereit. Eine Studie der Forschungsgemeinschaft für Elektrische Anlagen und Stromwirtschaft (FGH) geht davon aus, dass eine konventionelle Wirkleistungseinspeisung in Höhe von 13 GW nötig ist um die Systemstabilität jederzeit garantieren zu können [6]. **Tabelle 1** zeigt ausgehend von den prognostizierten Residuallasten aus **Bild 2** die Anzahl der Stunden bzw. den Anteil der Stunden am gesamten Jahr, in denen die Residuallast aufgrund der erhöhten Einspeisung von Strom aus erneuerbaren Energien die von der FGH berechnete notwendigen Mindesterzeugung unterschritte.

Tabelle 1 Anzahl der Stunden mit konventioneller Wirkleistungseinspeisung < 13 GW in den Jahren 2015, 2020 und 2025 (ohne Stromexport)

Betrachtungsjahr	2015	2020	2025
Stunden	181	817	1803
Anteil an Jahresstunden	2,0 %	9,3 %	20,6 %

Um zu vermeiden, dass die notwendige residuale Mindesterzeugung unterschritten wird ohne die Stromerzeugung aus erneuerbaren Energien abzuregeln, rücken Speicher zunehmend in den Fokus. Die derzeit einzige Möglichkeit Strommengen in relevantem Maßstab zu speichern, bieten Pumpspeicherkraftwerke. Aber auch deren Speicherfähigkeit wird selbst bei einem weiteren Zubau von Pumpspeicherkapazitäten nicht ausreichen um die zu den Zeiten aus **Tabelle 1** anfallenden Strommengen speichern zu können [7].

1.2 Das Erdgasnetz als Speicher

Eine vielversprechende Möglichkeit, Strom in großen Mengen auch langfristig speichern zu können, ist die Power-to-Gas-Methode mit nachgeschalteter Methanisierung. Dabei wird dem Netz elektrische Leistung entnommen, um eine Elektrolyse zu betreiben und den erzeugten Wasserstoff im Anschluss der Methanisierung zuzuführen, um Synthetic Natural Gas (SNG) zu erzeugen und ins Erdgasnetz einspeisen zu können. Die Vorteile dieser Methode gegenüber anderen Speichertechnologien liegen in der großen Speicherkapazität und -reichweite der Erdgasinfrastruktur [8], sowie in der unbeschränkten Netzkompatibilität von SNG im Gegensatz zur direkten Einspeisung von Wasserstoff, welcher aufgrund von Verbraucheranforderungen nur bis zu 5 Vol.-% eingespeist werden darf. Zwar wird die Erhöhung dieser Grenze derzeit diskutiert, allerdings wird die Einspeisung von Wasserstoff im Gegensatz zur Einspeisung von SNG begrenzt bleiben [9].

1.3 Die Power- & Biomass-to-Gas-Methode

Power– & Biomass-to-Gas-Anlagen können je nach Stromnetzsituation in zwei Betriebsmodi betrieben werden, wie in **Bild 3** dargestellt.

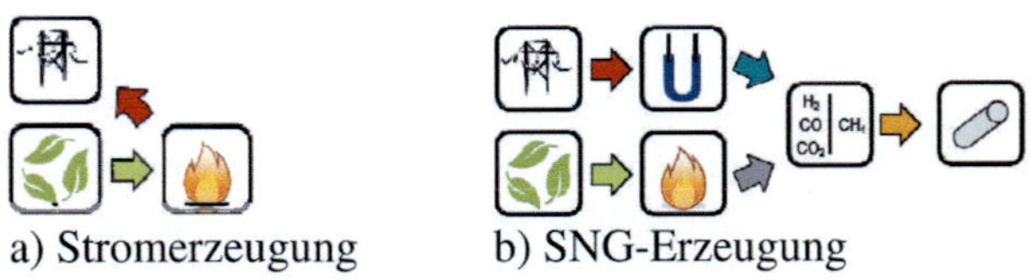

Bild 3 Prinzipielle Betriebsmodi von Biomass- & Power-to-Gas-Anlagen

Bild 3 a) zeigt den Betriebsmodus Stromerzeugung. Dabei wird das aus der Biomassevergasung gewonnene Produktgas über einen Gasmotor mit angeschlossenem Generator verstromt und Elektrizität ins Stromnetz eingespeist. In **Bild 3 b)** ist die Betriebsweise bei der Speicherung von Strom im Erdgasnetz dargestellt. Um SNG erzeugen und einspeisen zu können, bedarf es einer Kohlenstoffquelle, die bei der Power- & Biomass-to-Gas-Methode über die Vergasung von Biomasse bereitgestellt wird. Dabei wird über eine geänderte Prozessführung im Vergleich zum Modus Stromerzeugung neben dem entstehenden Produktgas ein CO_2-reiches Abgas erzeugt, welches direkt in die Methanisierung gegeben und zusammen mit dem Wasserstoff aus der Wasserelektrolyse zu SNG umgesetzt wird. Abgesehen von Revisions- und Wartungsarbeiten können Power- & Biomass-to-Gas-Anlagen durch die flexible Betriebsweise so über die maximal mögliche jährliche Betriebsstundenzahl betrieben werden.

1.4 Prozesstechnologie

Eine Problematik bei der Power-to-Gas-Methode ist die ausreichende Bereitstellung von Kohlenstoff für die Methanisierung zu SNG. Beim Power& Biomass-to-Gas-Prozess wird dieser über ein Biomassevergasungsverfahren gewonnen, für das lediglich bisher ungenutzte Biomassepotenziale wie Waldrestholz, Landschaftspflegematerial und Stroh genutzt werden. Somit ergeben sich keine Konkurrenzen zu bestehenden Bioenergieanlagen. Der Kohlenstoff für die Methanisierung wird dabei von einer dual-zirkulierenden Wirbelschichtvergasung, die nach dem Sorption Enhanced Reforming (SER) Prozess arbeitet, bereitgestellt [10] [11]. Diese erzeugt Synthesegas mit hohen Wasserstoffgehalten und zusätzlich im Modus SNG-Erzeugung neben dem eigentlichen Produktgas ein CO_2 reiches Abgas, wie in **Bild 4** dargestellt.

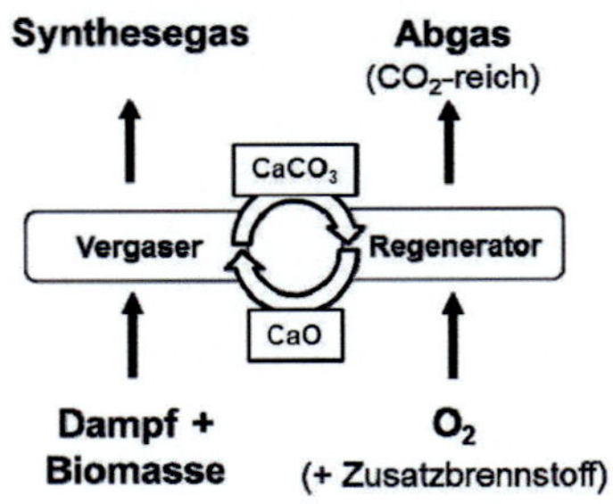

Bild 4 Dual-zirkulierende Wirbelschichtvergasung (SER-Prozess)

Das bei der Vergasung entstehende CO_2 wird durch einen Feststoffstrom, der zwischen Vergaser und Regenerator zirkuliert, gebunden und kann im Regenerator unter Wärmezufuhr wieder entnommen werden.

Dabei wird die Biomassevergasung im SNG-Modus unter Zugabe von Wasserdampf bei Temperaturen von 650–725 °C betrieben und erzeugt ein H_2-reiches Synthesegas. Das zwischen den beiden Reaktoren umlaufende Bettmaterial wird beim Eintritt in den Vergasungsreaktor zu Calciumcarbonat ($CaCO_3$) umgesetzt und in den Regenerator transportiert. Durch die Zugabe von Sauerstoff, der neben Wasserstoff bei der Elektrolyse anfällt, wird das zuvor im Vergaser gebundene CO_2 wieder ausgetrieben und kann ohne weitere Nachbehandlung in die Methanisierung gegeben werden. Das zurückbleibende Calciumoxid (CaO) wird in den Vergaserteil transportiert und rezirkuliert.
Befindet sich die Anlage im Stromerzeugungsmodus, wird durch eine Änderung im Vergaserbetrieb die Synthesegaszusammensetzung so eingestellt, dass damit ein

Gasmotor zur Stromerzeugung betrieben werden kann. Der Regenerator wird dann mit Luft betrieben und die Vergasungstemperaturen auf 800–850 °C angehoben.

2 Ergebnisse der Markteinsatzberechnung und Standortauswahl

Der Nutzen der Flexibilisierung und die Auswirkungen auf das deutsche Stromerzeugungssystem werden durch Marktsimulationen und ein Verfahren zur Standortauswahl untersucht und quantifiziert. Die Ergebnisse dieser Untersuchungen werden im Folgenden dargestellt.

2.1 Schnittpunkte zwischen Strom- und Gasinfrastruktur

Um Strom- und Gasnetzinfrastruktur mittels Power- & Biomass-to-Gas koppeln zu können, muss zunächst die räumliche Nähe der beiden Systeme zueinander untersucht werden. Dazu wurde eine Methodik entwickelt, die ein geografisches Netzmodell des Fernleitungsgasnetzes, die Knotenpunkte des Stromübertragungsnetzes, sowie Daten zum relevanten Biomasseaufkommen in Deutschland enthält. Die Daten zum Gasfernleitungsnetz entstammen der Transparenzplattform des Verbands der Europäischen Fernleitungsnetzbetreiber für Gas (ENTSO-G) [12], sowie öffentlichen Netzkarten einzelner Fernleitungsgasnetzbetreiber [13] [14]. Die Standorte der Stromnetzknotenpunkte liegen dem IFK vor.

Prinzipiell gilt, dass aus technischer Sicht an jeder Stelle des Gasnetzes Erdgas eingespeist und entnommen werden kann. Im Falle des Stromnetzes gilt diese Annahme nicht, so dass für die Berechnung der Schnittpunkte lediglich Standorte von Umspannstationen berücksichtigt werden. Daher wurde die Überprüfung des Abstands so durchgeführt, dass für jeden Knotenpunkt des Stromnetzes der Punkt-zu-Punkt Abstand zu den Gasnetzknoten bzw. der normale Punkt-Gerade Abstand zu den Gasleitungen berechnet wird. Die möglichen Schnittpunkte, die aus einer geforderten Maximaldistanz zwischen den beiden Infrastrukturen von 10 km resultieren, sind in **Bild 5** dargestellt.

Diese Distanz wurde gewählt, da Anlagenbetreiber, die in das Fernleitungsgasnetz einspeisen wollen, bis zu dieser Entfernung zwischen Anlage und gewünschtem Einspeisepunkt nach Gasnetzzugangsverordnung (GasNZV) Anrecht auf finanzielle Beteiligung an den Anschlusskosten durch den Fernleitungsgasnetzbetreiber haben, sofern es sich um erneuerbar erzeugtes SNG handelt [15]. Gasnetzpunkte, die hier ohne Anschluss an das dargestellte Fernleitungsgasnetz dargestellt sind, sind an unterlagerte Netzebenen angeschlossene große Erdgasspeicher und wurden wegen der bereits vorhandenen technischen Einrichtungen ebenfalls in die Standortauswahl einbezogen.

Insgesamt ergeben sich aus diesen Betrachtungen 225 Schnittpunkte, bei denen die gewählte Maximaldistanz zwischen Strom- und Gasnetz nicht überschritten wird.

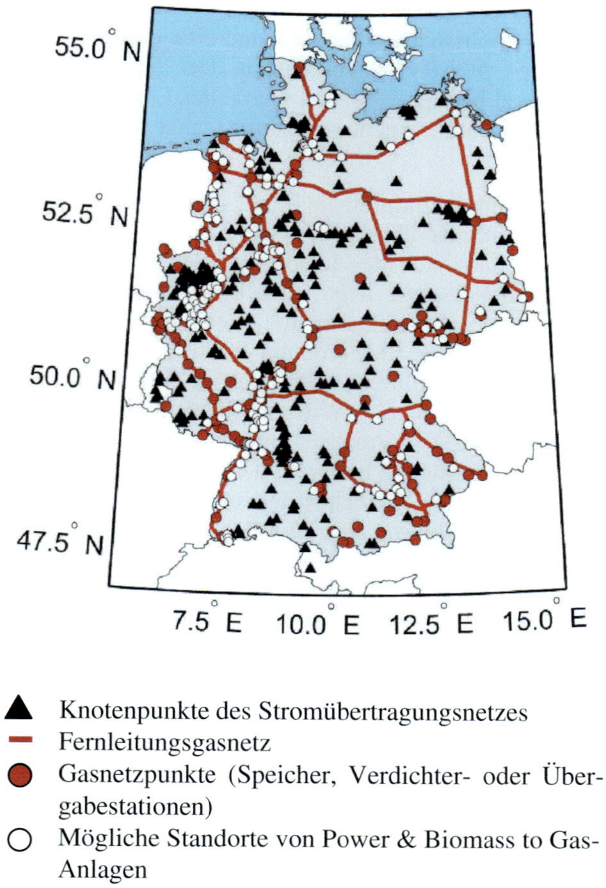

Bild 5 Infrastrukturelle Schnittpunkte bei einer Maximaldistanz von 10 km

2.2 Standortauswahl

Da die Verfügbarkeit von Biomasse für Power- & Biomass-to-Gas-Anlagen ein wesentliches Kriterium zur Auswahl eines Standortes ist, muss diese für die infrastrukturellen Standortmöglichkeiten überprüft werden. Dazu wurden landkreisscharfe Vorkommen der relevanten Biomassearten betrachtet, welche mehreren Studien entnommen und in der Methodik hinterlegt wurden [16] [17] [18].

Für den Biomassebedarf einer Power- & Biomass-to-Gas Anlage wurde folgende Abschätzung verwendet:

$$E_{BM} = \frac{P_{el}}{\eta} \cdot h_{Betr} \tag{1}$$

mit

E_{BM} jährlicher Biomassebedarf
P_{el} elektrische Leistung im Stromerzeugungsmodus
η elektrischer Wirkungsgrad
h_{Betr} jährliche Betriebsstunden der Anlage

Für die folgende Betrachtung wurden die jährlichen Betriebsstunden einer Power- & Biomass-to-Gas-Anlage zu 8000 Betriebsstunden pro Jahr und eine elektrische Leistung von 50 MW_{el} angenommen. Der elektrische Wirkungsgrad beträgt 26 % und wurde der Literatur entnommen und durch eigene Annahmen angepasst [19] [20] [21].

Für alle infrastrukturell in Frage kommenden Standorte wurde zunächst überprüft, wie hoch das Biomassevorkommen innerhalb des Landkreises ist, in welchem der Schnittpunkt selbst liegt. Auf Basis dieses Potenzials wurden die infrastrukturell möglichen Schnittpunkte bewertet und in eine Rangordnung überführt. Dieses Verfahren wird als zielführend betrachtet, da angenommen werden kann, dass die Wirtschaftlichkeit einer Power- & Biomass-to-Gas-Anlage umso höher ist, desto höher die Biomassedichte in ihrem Umfeld und damit desto niedriger der Aktionsradius und somit die Beschaffungs- und Transportkosten der Biomasse sind. Nachdem die verfügbare Biomasse im selben Landkreis dem jeweiligen Standort zugewiesen wurde, wird in einem nächsten Schritt überprüft ob diese den thermischen Bedarf zur Erbringung der elektrischen Leistung der Anlage decken kann. Sollte das Biomasseaufkommen innerhalb des Landkreises des Schnittpunkts nicht ausreichen, wird zusätzlich das Potenzial des Landkreises mit dem nächstgelegenen Flächenschwerpunkt der Anlage zugewiesen. Dieser Ablauf wird so lange durchgeführt, bis der Bedarf aus (1) gedeckt ist. Landkreise, deren Potenziale durch eine Power- & Biomass-to-Gas-Anlage ausgeschöpft wurden, werden für die weitere Auswahl durch andere Schnittpunkte gesperrt. Im letzten Schritt wird der erforderliche Aktionsradius eines Standortes auf Basis der benötigten Fläche seines Biomasseaufkommens berechnet, und zwar so, dass eine Überschneidung der Radien möglichst vermieden wird.Das Ergebnis dieser Berechnung ist in **Bild 6** dargestellt.

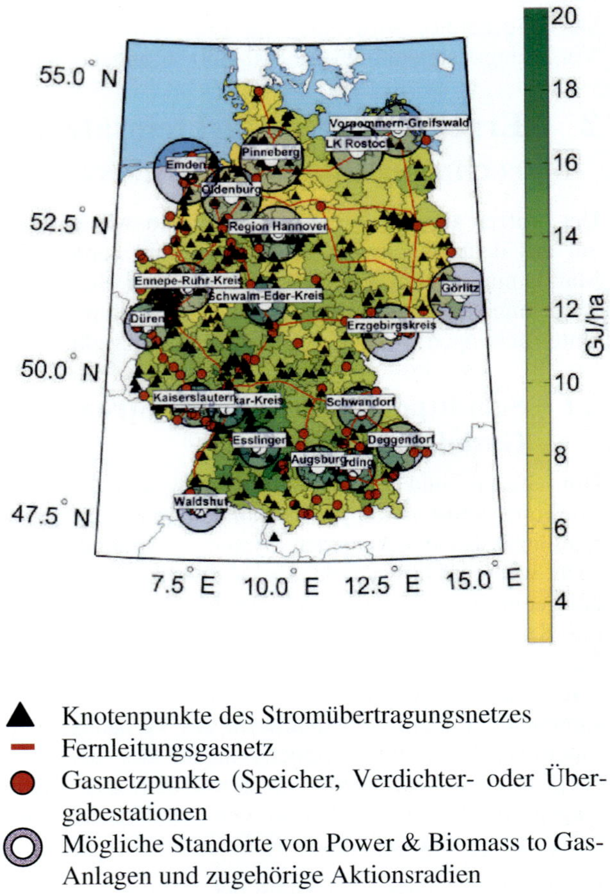

Bild 6 Ergebnis der Standortauswahl für eine Infrastrukturdistanz von 10 km und einer Anlagengröße von 50 MW_{el}

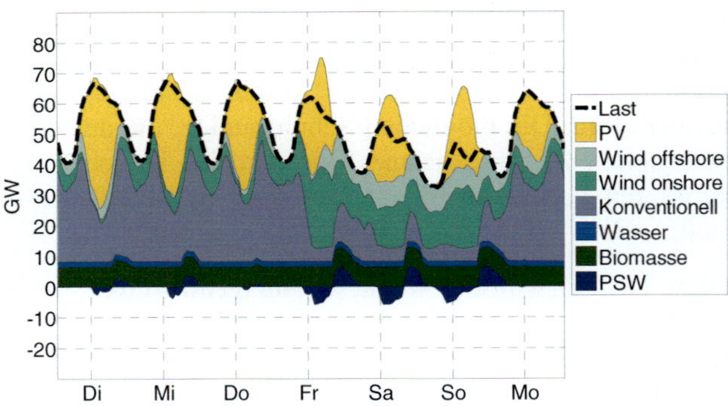

Bild 7 Marktsimulation einer Woche im Juni 2025 ohne Power- & Biomass-to-Gas

In wenigen Fällen tritt eine Überschneidung der Aktionsradien auf, eine Konkurrenz zwischen zwei oder mehreren Standorten wird jedoch aufgrund der Bedingung, dass das Potenzial eines Landkreises nur einmal genutzt werden kann, vermieden. Für das gezeigte Beispiel sind somit von 225 infrastrukturell möglichen Schnittpunkten lediglich 15 für eine Power- & Biomass-to-Gas-Anlage geeignet, nachdem diese hinsichtlich der Biomasseverfügbarkeit überprüft sowie Konkurrenzen ausgeschlossen wurden.

2.3 Marktsimulationen

Um die Auswirkungen großer Stromspeicher auf das Stromversorgungssystem quantifizieren zu können, werden Day-Ahead Marktsimulationen durchgeführt. Die Modellierung stützt sich auf öffentlich zugängliche Kraftwerkslisten, sowie Zeitreihen der Einspeisung von Strom aus erneuerbaren Energien des Referenzjahres 2013 [5] [22] [23]. Die Simulation berücksichtigt dabei keine Austauschleistungen mit dem Ausland und keine netzseitigen Restriktionen. Des Weiteren ist eine Mindesterzeugung konventioneller Anlagen einzuhalten. Das Ergebnis der Simulation für eine Juniwoche des Jahres 2025 ist in **Bild 7** dargestellt.

Der Strombedarf wird dabei von der Stromerzeugung aus erneuerbaren Energien und der sich daraus ergebenden residualen Erzeugung gedeckt, Überdeckungen zur Mittagszeit werden von Pumpspeicherkraftwerken aufgefangen. Ab Freitag reicht die Kapazität der Pumpspeicherkraftwerke nicht aus, um die marktbedingten Stromüberschüsse zu speichern und Abschaltungen erneuerbarer Erzeuger könnten die Folge sein.

Bild 8 zeigt den Verlauf der Einspeisung und Speicherung von Pumpspeicherkraftwerken und Power- & Biomass-to-Gas-Anlagen für dieselbe Woche im Jahr 2025. Dabei wurden die 15 Anlagen aus der Standortanalyse mit einer Leistung von jeweils 50 MW_{el} berücksichtigt und unter unterschiedlichen Marktbedingungen simuliert.

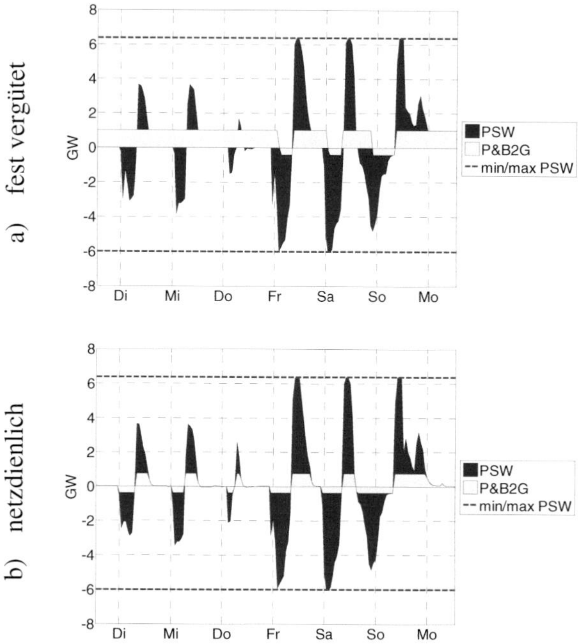

Bild 8 Marktsimulation einer Woche im Juni 2025 mit Power- & Biomass-to-Gas

Die Anlagen werden in **Bild 8a**) als Stromerzeuger im Sinne des EEG mit fester Einspeisevergütung betrachtet, weswegen sie bei Stromnachfrage im Netz dauerhaft einspeisen. Erst wenn die Pumpspeicherkapazitäten nicht ausreichen, wie dies am betrachteten Wochenende der Fall ist, werden die Power- & Biomass-to-Gas-Anlagen systembedingt in den Speicherbetrieb gezwungen. Allerdings entstehen durch dieses Vergütungsmodell unter der Woche Situationen, in denen die beiden Speicher gegeneinander arbeiten. **Bild 8b**) zeigt dieselbe Simulation mit rein netzdienlicher Arbeitsweise. Einziges Kriterium zum Betrieb ist hier die Erfüllung der Leistungsbilanz des Systems, ohne die Betrachtung der Wirtschaftlichkeit der Anlagen. Ein zukünftiges Geschäftsmodell müsste sich zwischen diesen beiden Extrema ansiedeln.

3 Diskussion

Mit 225 möglichen Standorten ist die Bedingung der räumlichen Nähe zwischen Strom- und Gasinfrastruktur als unkritisch zu betrachten. Eine weitere Verfeinerung kann hier noch durch die Einbeziehung weiterer Netzebenen, sowohl auf Strom- als auch auf Gasnetzseite, erzielt werden. Allerdings ist fraglich, ob diese weitere Verfeinerung nach der Überprüfung der Schnittpunkte auf Biomasseverfügbarkeiten und Vermeidung von Konkurrenzen zu signifikant besseren Ergebnissen der Standortauswahl führt, da die Potenziale der Landkreise ein räumlich stark einschränkendes Element darstellen.

Berechnungen mit unterschiedlichen maximal erlaubten Distanzen zwischen Strom- und Gasnetz zeigen, dass diese dennoch zu Verbesserungen in der Anzahl und Auswahl der tatsächlich möglichen Standorte führen. So können bei einer maximal erlaubten Distanz von 5 km lediglich 13 Standorte ermittelt werden, bei einer Distanz von 20 km 21 Standorte.

4 Zusammenfassung und Ausblick

Die Ergebnisse wurde im Rahmen des vom Förderprogramm BWPLUS geförderten Projekts Power&Biomass2Gas (P&B2G) erzielt [24]. Die Resultate der Marktsimulationen zeigen, dass Power- & Biomass-to-Gas eine geeignete Möglichkeit darstellt, über die hohe Speicherkapazität der Erdgasinfrastruktur erneuerbare Energieträger besser in das Stromversorgungssystem zu integrieren. Überdies verdrängt das so erzeugte regenerative SNG fossiles Erdgas und trägt auf diese Weise zusätzlich zur Senkung der CO_2 Emissionen bei.

Die bisherigen Untersuchungen zur Standortauswahl zeigen, dass die ermittelten Standorte günstig für die Integration weiterer erneuerbare-Energien-Anlagen (EE-Anlagen) liegen. Zum einen können Standorte im Nordosten durch die Power-to-Gas-Methode eine Alternative zum Transport häufig anfallender großer Energiemengen von Strom aus Windenergieanlagen (WEA) über das Stromnetz bieten. Die in den Verbrauchszentren im Süden und Westen benötigten elektrischen Energiemengen können dann zum Teil von den dort installierten Power- & Biomass-to-Gas-Anlagen bereitgestellt werden. Zum anderen können Anlagen, die in der Nähe von Leitungen liegen, die besonders häufig von Redispatch-Maßnahmen zur Sicherstellung des (n-1)-Kriteriums betroffen sind, zur Auflösung solcher Engpässe beitragen. Diese Möglichkeit wird im weiteren Projektverlauf untersucht werden.

5 Literatur

[1] Europäische Union: Richtlinie 2009/28/EG des Europäischen Parlaments und des Rates zur Förderung der Nutzung von Energie aus erneuerbaren Quellen und zur Änderung und anschließenden Aufhebung der Richtlinien 2001/77/EG und 2003/30/EG. Amtsblatt der Europäischen Union, 52. Jahrgang, 5. Juni 2009.

[2] Bundesrepublik Deutschland: Nationaler Aktionsplan für erneuerbare Energie gemäß der Richtlinie 2009/28/EG zur Förderung der Nutzung von Energie aus erneuerbaren Quellen. Bundesministerium für Wirtschaft und Energie (BMWi), 2009.

[3] Bundesrepublik Deutschland: Gesetz für den Ausbau erneuerbarer Energien (Erneuerbare-Energien-Gesetz – EEG 2014).

[4] Bundesministerium für Wirtschaft und Energie (BMWi): Zeitreihen zur Entwicklung der erneuerbaren Energien in Deutschland, Stand 02/2014.

[5] Online Transparenzplattformen der Übertragungsnetzbetreiber Tennet, Amprion, Transnet BW, 50 Hertz.

[6] Forschungsgemeinschaft für Elektrische Anlagen und Stromwirtschaft (FGH): Studie zur Ermittlung der technischen Mindesterzeugung des konventionellen Kraftwerksparks zur Gewährleistung der Systemstabilität in den deutschen Übertragungsnetzen bei hoher Einspeisung erneuerbarer Energien. Mannheim, Januar 2012.

[7] Institut für Elektrische Anlagen und Energiewirtschaft, RWTH Aachen: Unterstützung der Energiewende in Deutschland durch einen Pumpspeicherausbau – Potenziale zur Verbesserung der Wirtschaftlichkeit und der Versorgungssicherheit. Aachen, 2014.

[8] M. Specht, M. Sterner et al: Speicherung von Bioenergie und erneuerbarem Strom im Erdgasnetz, FVEE, 2009.

[9] F. Burmeister, J. Senner, J. Brauner, R. Albus: Potenziale der Einspeisung von Wasserstoff im Erdgasnetz – eine saisonale Betrachtung. DVGW, 2012.

[10] A. Pandy, T. Bhaskar, M. Stöcker, R. K. Sukumaran: Recent advances in thermochemical conversion of biomass. Elsevier, 2015.

[11] F. Scala: Fluidized bed technologies for near-zero emission combustion and gasification. Woodhead Publishing, Philadelphia, 2013.

[12] European Network of Transmission System Operators for Gas (ENTSO-G): Transparency Platform. https://transparency.entsog.eu/.

[13] Open Grid Europe: Dashboard. https://www.open-grid-europe.com/cps/rde/xchg/open-grid-europe-internet/hs.xsl/2860.htm.

[14] ONTRAS Gastransport GmbH, ONTRAS Netzkarte. https://www.ontras.com/cms/netz/netzdaten/netzkarte/.

[15] Bundesrepublik Deutschland, Verordnung über den Zugang zu Gasversorgungsnetzen (Gasnetzzugangsverordnung - GasNZV). Stand 2013

[16] Katja Bunzel, Daniela Thrän, Ulrike Seyfert, Vanessa Zeller, Marcel Buchhorn: Forstwirtschaftliche Biomassepotenziale und Reststoffpotenziale in Deutschland. Informationen zur Raumentwicklung, Heft 5/6.2011, S. 297-308.

[17] Ulrike Seyfert, Katja Bunzel, Daniela Thrän, Eric Mauky, Barbara Fritsche, André Schreiber, Jan Liebetrau, Thomas Schmidt, Tobias Ulbricht, Volker Lenz: DBFZ Report Nr. 8: Biomassepotenziale in Deutschland 2008 bis 2020. Deutsches Biomasse-ForschungsZentrum gemeinnützige GmbH (DBFZ), 2011.

[18] L. Leible, S. Kälber, G. Kappler: Systemanalyse zur Gaserzeugung aus Biomasse. Informationen zur Raumentwicklung, Karlsruhe Institute of Technology (KIT), 2011.

[19] Reinhard Rauch: Fluidised bed gasification and synthetic biofuels, the Güssing plant. European Conference on Biorefinery Research, 2006.

[20] Jörgen Held: Gasification – Status and technology. Swedish Gas Centre, 2012.

[21] M. Bolhàr-Nordenkampf, R. Rauch, K. Bosch, C. Aichernig, H. Hofbauer: Biomass CHP Plant Güssing – Using Gasification for Power Generation. 2nd Regional Conference on Energy Technology Towards a Clean Environment, 2003.

[22] Bundesnetzagentur: Kraftwerksliste. Stand Dezember 2013.

[23] Umweltbundesamt: Kraftwerke in Deutschland (ab 100 MW elektrischer Leistung). Stand August 2014.

[24] Power&Biomass2Gas (P&B2G), gefördert von BWPLUS, http://www.pb2g.de.

Techno-ökonomische Bewertung von Lithium-Ionen Batteriespeichern an Windparks

M. Wieland[1, 2], S. Gerhard[2],
[1]Hochschule für Angewandte Wissenschaften Hamburg, 20099 Hamburg, marco.wieland@vattenfall.de
[2]Vattenfall Europe Innovation GmbH, 22297 Hamburg, sebastian.gerhard@vattenfall.de

Kurzfassung

Mit dem zunehmenden Anteil erneuerbarer Energien an der Stromerzeugung in Deutschland steigt der Bedarf an Flexibilität für den Ausgleich von Erzeugung und Verbrauch. Elektrochemische Speicher, insbesondere auf Lithium-Ionen-Basis, eignen sich hervorragend für die Integration erneuerbarer Energien aus z. B. Windparks. Die Kosten für diese Technologie werden durch den zunehmenden Einsatz in der Elektromobilität und den damit verbundenen Skaleneffekt bei der Massenfertigung weiter sinken, sodass ein Einsatz von Lithium-Ionen Batterien für stationäre Anwendungen immer attraktiver wird. Das Ziel dieser Arbeit ist es, den energietechnisch und -wirtschaftlich sinnvollen Betrieb stationärer Lithium-Ionen Batteriespeicher an Windparks zu prüfen. Unter besonderem Fokus steht dabei die Ermittlung von möglichen Anwendungsfällen, die für eine Steigerung der Wirtschaftlichkeit der Windenergieanlagen aus Sicht eines Betreibers sorgen. In einer techno-ökonomischen Bewertung werden diese zu einem Geschäftsmodell zusammengefasst und eine Empfehlung zur technischen Realisierung gegeben. Die Erfassung der wirtschaftlichen Gesamtrentabilität erfolgt als quantitative Abschätzung.

1 Unterstützung der Energiewende mit Flexibilitätsoptionen

Mit der Zustimmung des Bundesrats zur aktuellen EEG-Novelle vom 27.06.2014 soll der Anteil des aus erneuerbaren Energien erzeugten Stroms am Bruttostromverbrauch stetig und kosteneffizient auf mindestens 80 % bis zum Jahr 2050 erhöht werden [1]. Einhergehend sollen die CO_2 Emissionen um die gleiche prozentuale Größe sowie der Primärenergiebedarf gegenüber dem Bezugsjahr 2008 um mindestens 50 % gesenkt werden [2].

Zur Erreichung dieses Ziels sowie der Gewährleistung einer stabilen Energieversorgung, müssen aufgrund der fluktuierenden Einspeisung erneuerbarer Energiequellen mittel- und langfristig vermehrt Flexibilitätsoptionen eigesetzt werden. Der benötigte Bedarf an Strom ist abhängig von den Verbrauchern und ebenfalls volatil. Für die Flexibilitätsoptionen gibt es eine Vielzahl von Ansätzen, welche sich je nach Anwendungszweck unterschiedlich gut eignen. Dargestellt wird diese Problematik in **Abbildung 1**.

Eignung von Batteriespeichern für Windenergieanlagen

Eine Windenergieanlage (WEA) wandelt mithilfe des Rotors kinetische Energie in mechanische und mechanische in elektrische Energie um. Es gibt damit zwei Ansatzpunkte die Energie zu speichern. Für die direkte Speicherung der mechanischen Energie sprechen die geringeren Verluste bei der Energieumwandlung. Allerdings ist die Auswahl an mechanischen Energiespeichern zurzeit sehr gering [3]. Zudem ist der Platz in der Gondel der WEA nur bedingt für größere mechanische Energiespeicher ge

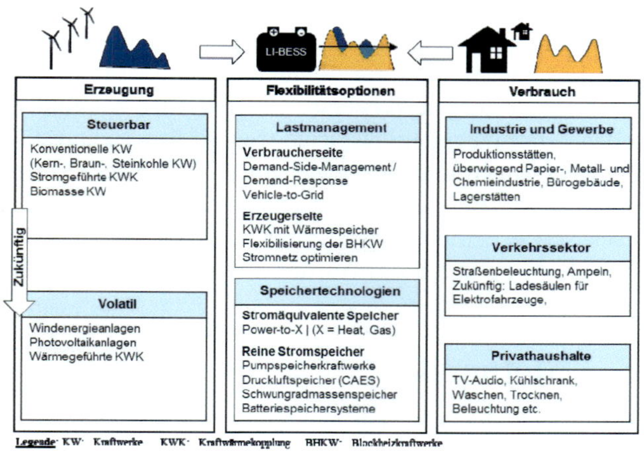

Abbildung 1: Flexibilitätsoptionen für die Energiewende

eignet und der Forschungsaufwand für die benötigte Regelung hoch.

Also wird eine Technologie benötigt, die sich für den zweiten Ansatzpunkt, der Speicherung der elektrischen Energie, eignet. Darüber hinaus wird für den Betrieb an einer WEA eine lange Lebensdauer mit 10 bis 15 Jahren, mehr als 3.000 Vollladezyklen und ein möglichst hoher Lade-, Entladewirkungsgrad pro Zyklus größer 85 % gefordert. Geringe Investitions- und Betriebskosten bei einem hohen Sicherheitsstandard und einfacher Bedienbarkeit (automatische Regelung) sind ebenso relevant für den wirtschaftlichen Betrieb von Speichern [4].

Elektrochemische Speicher insbesondere auf Lithium-Ionen (LI) Basis eignen sich hervorragend für die Integration erneuerbarer Energien aus WEA. Im Gegensatz zu anderen Speicherarten wie Pumpspeicherwerke oder Druckluftspeicher sind sie nicht an geographische Voraussetzungen geknüpft und besitzen eine hohe Energie-

und Leistungsdichte. Außerdem lassen sich Projekte mit einer verhältnismäßig geringen Planungszeit realisieren. Durch hohe Forschungsaktivität wird die Weiterentwicklung dieser Technologien auch in Zukunft gewährleistet sein [5].

Da ein Batteriespeicher zur Speicherung von elektrischer Energie weitere Komponenten wie z. B. Wechselrichter, Schaltschränke etc. benötigt, wird folgend von einem Batterie-Energie-Speicher-System (BESS) gesprochen.

Aktuelle Forschungsprojekte

Aktuell laufen weltweit mehrere Forschungsprojekte zur Untersuchung des Übertragungs-, System- und Marktnutzen von stationären zentralen LI-BESS an WEA bzw. Windparks (WP). In den USA und China existieren seit 2011 drei große Projekte:

- Tehachapi Wind Energy Storage Project (8 MW / 32 MWh) in Kalifornien
- Laurel Mountain Wind Projekt (32 MW) in West Virginia
- Shenzhen Boqing Energy Storage Projekt (10 MW) in Baolong

Laut mehreren Studien führt die Installation eines BESS zur Pufferung und Verstetigung der erneuerbaren Energien bedingt durch die hohen Investitionskosten nur in einem begrenzten Maß zur Reduktion der Gesamtkosten des Stromversorgungssystems. Es wird aber empfohlen, neben dem Erzeugungsausgleich weitere Anwendungsfälle zu betrachten und zu entwickeln [6] [7]. Eine Überlegung, ob ein Betrieb in Deutschland für Betreiber technisch sowie betriebswirtschaftlich von Nutzen ist, muss deshalb zuerst durch das Auffinden von Anwendungsfällen untersucht werden.

2 Methodik

Um den energietechnisch und -wirtschaftlich sinnvollen Betrieb zu identifizieren, wird das Vorgehen in drei Phasen unterteilt.

1. Identifizieren und Entwickeln energietechnischer und -wirtschaftlicher Anwendungsfälle von LI-BESS an WEA mithilfe einer systematischen Literaturrecherche nach aktuellen Stand der Wissenschaft, Technik und Regularien unter Zunahme von Wissensdatenbanken im Internet, Fachbüchern, Publikationen und Expertengesprächen. Besonderes Augenmerk bei der Recherche wird dabei auf die Ergebnisse und Publikationen der Wissensdatenbank des Institute of Electrical and Electronics Engineers (IEE), Science Direct sowie den geförderten Studien des Bundesministerium für Wirtschaft und Energie (BMWi) zum Thema Energiespeicherung gelegt. In einem Experten-Workshop werden die Anwendungsfälle besprochen und nach Wirtschaftlichkeit und Machbarkeit priorisiert. Am Ende von Phase 1 sollen die priorisierten Anwendungsfälle definiert sein und die notwendigen Parameter für eine weitere Untersuchung feststehen.
2. Untersuchung der priorisierten Anwendungsfälle mittels Excel. Dies beinhaltet den Aufbau der Modelle, die Beschaffung der notwendigen Daten und deren Implementierung ins Modell sowie die Durchführung und Auswertung mit einer Datenbasis von 2014. Die Untersuchung erfolgt durch eine statische Modellierung der einzelnen Anwendungsfälle ohne Berücksichtigung der optimalen Betriebsstrategie und Ladezustände des LI-BESS. Im Vordergrund der Untersuchung stehen die quantitative Abschätzung des wirtschaftlichen Potentials und deren Einsatzhäufigkeit in Deutschland. Aktuelle Restriktionen sowie mögliche Kombinationsmöglichkeiten sollten zudem überprüft werden.
3. Die Bewertung der Anwendungsfälle erfolgt anhand der Ergebnisse von Phase zwei. Es werden Aussagen zur technischen Realisierung und der Wirtschaftlichkeit gemacht, sodass in einer zusammenfassenden techno-ökonomischen Gesamtauswertung, der Einsatz von LI-BESS an WEA bewertet werden kann. Zudem werden die sich veränderten Rahmenbedingungen, die mit der Energiewende einhergehen, in einem Ausblick kurz erläutert und eine Empfehlung für die weiteren durchzuführenden Schritte gegeben.

3 Anwendungsfälle für LI-BESS-Speicher an WEA

Im Folgenden werden alle Anwendungsfälle für LI-BESS an WEA aufgelistet, welche über die Literaturrecherche und Expertenbefragungen ermittelt wurden. Die Anwendungsfälle 1 und 2 sind direkt der WEA zugeordnet, 3 bis 9 sind Anwendungsfälle für den netzdienlichen Einsatz und 10 bis 12 können der Vermarktung und sonstigen Bereichen zugeordnet werden. Die Einordnung der Anwendungsfälle erfolgt nach dem Nutzen für Betreiber von WEA in Deutschland und wurde im Rahmen eines Expertenworkshops erstellt. Dargestellt werden die Ergebnisse anhand einer Portfolio-Analyse in **Abbildung 2**. Die Anwendungsfälle werden hinsichtlich des wirtschaftlichen Potentials und der Machtbarkeit bzw. dem Stand der Forschung analysiert.

<u>Systemnutzen</u>

1) Stromerzeugung erhöhen
   - Leistungsgradienten beim An- und Abfahren erhöhen
   - Leistungsbeschränkungen aufgrund mangelnder Netzkapazitäten nach §13.2 EnWG zwischenspeichern
2) Strombezug aus dem Netz vermeiden

<u>Übertragungsnetznutzen</u>

3) Asymmetrische Regelleistung bereitstellen
4) Momentanreserve bereitstellen
5) Netzausbau reduzieren
6) Blindleistung kompensieren
7) Flicker Balancing
8) Schwarzstartfähigkeit ermöglichen
9) Autarke Versorgung (Inselbetrieb) ermöglichen

Marktnutzen

10) Prognosefehler ausgleichen
11) Direktvermarktung verbessern
12) Synergieeffekte und weitere Nutzen
    - Wiederverwendung zu 2nd-Life BESS
    - Verhinderung von lokalen Überlastungen durch zunehmende Elektromobilität
    - Industriedienstleistungen bereitstellen bzw. Kopplung mit industriellen Verbrauchern

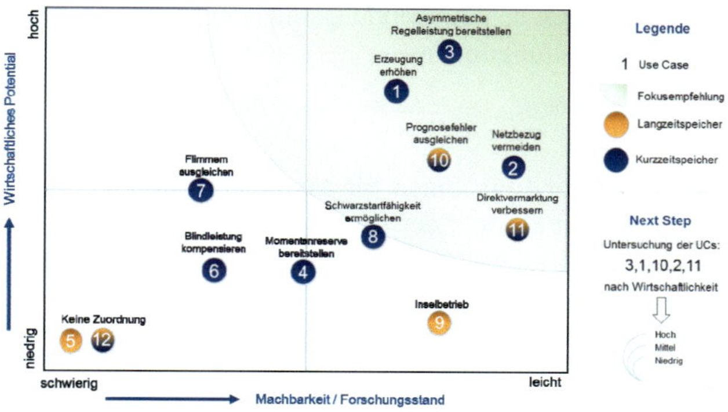

Abbildung 2: Einordnung der Anwendungsfälle, Ergebnisse aus Expertenworkshop vom 18.11.2014

Es lassen sich zur Bewertung der Anwendungsfälle folgende Aussagen treffen:

- Anwendungsfälle für den unmittelbaren Einsatz von LI-BESS an WEA sind nur im begrenzten Maß vorhanden und lassen sich in zwei übergeordnete Themen – Erzeugung erhöhen und Netzbezug vermeiden – zusammenfassen.
- Eine Betrachtung von LI-BESS für den netzdienlichen Einsatz ist aufgrund der derzeitigen Vergütungsmodelle und regulatorischen Rahmenbedingungen nur für den Regelleistungsmarkt und den Ausgleich von Prognosefehlern interessant. Zukünftige bilaterale Geschäfte mit dem Übertragungsnetzbetreiber (ÜNB) und Verteilnetzbetreiber (VNB) sowie ein Eingreifen der Politik können dies aber ändern.
- Die Verbesserung der Direktvermarktung des Stroms könnte aufgrund der in 2012 eingeführten Marktprämie aus derzeitiger Sicht interessant sein. Zudem sind nach der Novellierung des EEG-2014, Betreiber verpflichtet ihre Neuanlagen ab 2017 direkt zu vermarkten.
- Die Anwendungsfälle, die ein LI-BESS als Langzeitspeicher benötigen, sind wegen der hohen Investitionskosten derzeit und auch in näherer Zukunft nicht interessant.

Auf Basis dieser Erkenntnisse wird in dieser Arbeit eine Untersuchung folgender priorisierter Anwendungsfälle hinsichtlich ihres wirtschaftlichen Potentials durchgeführt:

1. Asymmetrische Regelleistung bereitstellen
2. Direktvermarktung verbessern
3. Erzeugung erhöhen
4. Netzbezug vermeiden

Die anderen Anwendungsfälle können zu einem technologischen Mehrwert im elektrischen Stromnetz beitragen, jedoch sind sie für BESS-Betreiber aus betriebswirtschaftlicher Sicht derzeit nicht attraktiv. Unabhängig von der Erfüllung der priorisierten Anwendungen, besteht auch zukünftig die Notwendigkeit zur Bereitstellung von Systemdienstleistungen für die Versorgungssicherheit, da eine Mehrzahl dieser zurzeit durch konventionelle Kraftwerke gedeckt wird. Eine Änderung der rechtlichen, wirtschaftlichen Rahmenbedingungen ist zu erwarten.

Da die priorisierten Anwendungsfälle Erzeugung erhöhen (1), Netzbezug vermeiden (2) sowie Direktvermarktung verbessern (13) nach eigenen näheren Analysen im Verhältnis zur Bereitstellung von asymmetrischer Regelleistung ein geringeres wirtschaftliches Potential besitzen, wird im Weiteren nur letzteres näher erläutert.

Asymmetrische Primärregelleistung bereitstellen

Regelleistung ist eine Reserveleistung, die bei unvorhergesehenen Ereignissen mithilfe von Frequenzhaltung für eine Stabilisierung des Netzes sorgt. Diese ist positiv bei Energiebereitstellung oder negativ bei Energieaufnahme und wird unterteilt nach Primär-, Sekundär-, und Tertiärregelleistung (PRL, SRL, TRL). Hinsichtlich der unterschiedlichen Vergütung dieser Märkte, sollte eine Erbringung auf dem PRL-Markt bevorzugt erfolgen. Im Verbund mit einer WEA kann asymmetrische PRL (mehr negative als positive PRL) angeboten werden.

a) Positive PRL wird ausschließlich durch die Leistung des LI-BESS erbracht.
b) Negative PRL kann über die Leistungsfähigkeit des BESS hinaus durch die Einsenkung der Energieproduktion der WEA erbracht werden.
c) Das Anbieten von asymmetrischer PRL ist heute nur in einem Anlagenpool bspw. in Verbindung mit Kohlekraftwerken möglich, die den positiven Anteil, der über die Leistungsfähigkeit des BESS hinaus geht übernehmen.

Der Bedarf an PRL wird nach der DENA-Netzstudie II bis 2020 konstant bleiben [8]. Die Teilnahme von WEA am PRL-Markt ist nach heutigem Technologiestand nur sehr eingeschränkt möglich. Die Teilnahme am PRL- und SRL-Markt erfordert garantierte Leistungsbereitstellungen über die Dauer einer Woche (PRL) oder eines Monats (SRL). Diese sind durch die Ungenauigkeit der Windprognosen für den Zeitraum nicht möglich. Bei neueren WEA mit Pitch- oder Stallregelung besteht aber die Möglichkeit sich an den täglichen Ausschreibungen für den TRL-Markt zu beteiligen. Positive TRL wird dabei durch eine gedrosselte Fahrweise der Anlage bereitgestellt, was jedoch nur selten im Jahr – bei starker Einspeisung aus erneuerbaren Energien, niedriger Last und einem hohen Lastgradienten im Netz – wirtschaftlich ist [8]. Negative

Regelleistung wird durch eine lastabhängige Absenkung der WEA erreicht und kann zukünftig wohl auch durch Wind- und Photovoltaikanlagen angeboten werden [9].

In **Abbildung 3** wird die Bereitstellung der PRL in Abhängigkeit ihrer Komponenten dargestellt. Die Kombination aus LI-BESS und WEA liefert in erster Linie asymmetrische – mehr negative – PRL, was auf die Einsenkungsmöglichkeit der WEA zurückzuführen ist. Durch den Zusammenschluss von Anlagen kann im „Anlagenverbund" aber weitere positive PRL aus z. B. einem Kraftwerk vorgehalten werden [10]. Die Erfüllung der Anforderungen von LI-BESS zur Erbringung der verschiedenen Regelleistungen ist bereits durch mehrere Studien bewiesen [11].

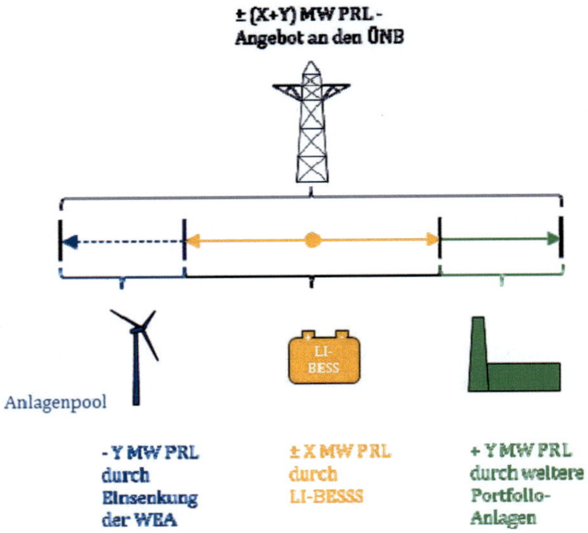

Abbildung 3: Prinzip Asymmetrische PRL bereitstellen

Es erfolgt daher ein Hinweis auf die möglich auftretenden Probleme. Bei der Auswahl des Netzes spielt es generell keine Rolle, in welchem Übertragungsnetzgebiet und auf welcher Spannungsebene der Anschluss erfolgt. Allerdings wird in tendenziell überlasteten Verteilnetzen eine Zustimmung durch den VNB eher abgelehnt. Zur Erfüllung der Erbringungsdauer von einer Woche ist zu berücksichtigen, dass die Anbieter ihre Anlagen zu einem Anlagenverbund zusammenschalten können, dessen Zuordnung zu Beginn einer jeden Viertelstunde geändert werden kann [12] [13]. Damit wäre bei einem Einsatz von LI-BESS im Verbund mit weiteren Portfolio-Anlagen auch das Zeitkriterium erfüllt.

Das Hauptziel ist es, mithilfe eines LI-BESS-WEA-Systems den Erlös des WEA-Anlagevermögens zu steigern. Die Mehreinnahmen durch die Teilnahme an den Regelleistungsmärkten müssen mit der verminderten Erzeugung durch Einsenkung der WEA, gelieferten Regelleistung und Wirkungsgradverlusten sowie den evtl. höheren Kosten durch eine Zwischenspeicherung in dem LI-BESS verglichen werden.

Nach §19 EEG-2014 besteht der Anspruch auf EEG-Förderung auch dann, wenn der Strom vor der Einspeisung in das Netz zwischengespeichert wurde. Die EEG-Förderung für den aus WEA erzeugten und im BESS zwischengespeicherten Strom bleibt also erhalten. Die zusätzlichen Kosten entstehen derzeit nur durch die zusätzlichen Gebühren und Umlagen, die je nach Netzanschluss und Anwendung unterschiedlich hoch ausfallen. Angemerkt sei auch, dass nach § 39 EEG-2014 nur Anlagen am Regelenergiemarkt teilnehmen dürfen, die ihren Strom direkt vermarkten und nicht die fixe Einspeisevergütung erhalten.

Für den Erlös des Anwendungsfalls ergeben sich damit folgende Zielfunktionen:

$$\text{Erlös}_{\text{UC3}} = \text{Erlös}_{\text{Asym. PRL}} - \text{Aufwand}_{\text{EEG}} - Aufwand_{Sonstiges} \quad (1)$$

Die Höhe der Einnahmen für asymmetrische PRL ergibt sich aus dem Produkt der bereitgestellten Leistung [MW] mit dem Leistungspreis [€/MW] über die Zeit intergiert. Für die zur Verfügung gestellte Leistung durch Einsenkung der WEA erhält der Betreiber nur die Hälfte der Regelleistungsvergütung abzüglich einer Vermittlungsgebühr für einen positiven Regelleistungserbringer.

$$Erlös_{Asym.\ PRL} = \int_{t=0}^{t_b} \begin{bmatrix} P_{LI-BESS}(t) * X_{PRL-LP}(t) + \\ P_{WEA}(t) * \left(\frac{X_{PRL-LP}(t)}{2} - X_G\right) \end{bmatrix} dt \quad (2)$$

$P_{LI-BESS}$: Angebot der bereitgestellten Leistung +/- des LI-BESS
P_{WEA}: Angebot der eingesenkten Leistung der WEA
X_{PRL-LP}: Leistungspreis PRL
X_G: Vermittlungsgebühren für einen positiven PRL-Erbringer
t_b: Zeit/ Dauer der bereitgestellte Leistung

Der Wert, der dem WEA Betreiber zu vergüten ist, ergibt sich aus der Differenz der tatsächlich erbrachten und erhaltenen – also positiven und negativen – PRL sowie dem Erzeugungsverlust durch Einsenkung der WEA, multipliziert mit der jeweiligen EEG-Vergütung. Zudem muss der Wirkungsgradverlust der elektrochemischen Energiewandlung von ca. 8 % (AC-DC-Wandlung) für die zwischengespeicherte Energie mitberücksichtigt werden.

$$\text{Aufwand}_{\text{EEG}} = (\text{E}_{\text{LI-BESS}} + \text{E}_{\text{WEA}})_{\text{L}} * \text{X}_{\text{EEG}} \quad (3)$$

$$\text{E}_{\text{LI-BESS}_\text{L}} = \frac{1}{\eta_{\text{LI-BESS}}} * \begin{pmatrix} \int_{t=0}^{t_{pos.\ PRL}} \text{P}_{\text{pos.PRL}}(t)\,dt \\ - \int_{t=0}^{t_{neg.\ PRL}} \text{P}_{\text{neg. PRL}}(t)\,dt \end{pmatrix} \quad (4)$$

$$\text{E}_{\text{WEA}_\text{L}} = \int_{t=0}^{t_{\text{WEA}}} \text{P}_{\text{WEA}_\text{E}}(t)\ dt \quad (5)$$

X_{EEG}: EEG Vergütung für WEA
$E_{LI-BESS}$: Erbrachte Regelenergie LI-BESS
E_{WEA}: Erbrachte Regelenergie WEA
$\Pi_{LI-BESS}$: Wirkungsgrad des LI-BESS bei Zwischenspeicherung
$P_{pos.\ PRL}$: Lieferung positiver PRL
$P_{neg.\ PRL}$: Lieferung negativer PRL
t_{WEA}: Zeitraum der Einsenkung der WEA

Der sonstige Aufwand beinhaltet z. B. die Kosten für eine Präqualifikation oder zusätzliche Messstellen.

Eine genaue wirtschaftliche Analyse des Erlöses für diesen Anwendungsfall ist schwierig, da neben den vielen Freiheitsgraden, eine optimale Betriebsstrategie für die WEA und dem LI-BESS in Abhängigkeit der verschiedenen Märkte erfolgen muss. Zum Beispiel ist der Abruf der PRL von mehr als 90 % der angebotenen Leistung auf wenige Stunden im Jahr begrenzt (1,75 Stunden im Jahr 2013) und eine technische Einheit zur Erbringung von PRL muss im Schnitt nur ca. 10 % der angebotenen PRL tatsächlich erbringen [14].

4 Techno-ökonomische Bewertung

Empfehlung zur technischen Realisierung

Im Folgenden wird die technische Realisierung eines LI-BESS an einem Windpark (WP) zur Erfüllung des Anwendungsfalls asymmetrische PRL bereitstellen vorgeschlagen.

In Deutschland speisen WEA ihre elektrische Energie zu 95 % in ein Stromnetz ein [3]. Die Betrachtung für einen Anschluss an ein autarkes, netzunabhängiges System kann also wegen der geringen Anwendungsgebiete in Deutschland vernachlässigt werden. Ebenso verpflichtend ist der Anschluss des LI-BESS an das Übertragungs- oder Verteilnetz zur Erbringung von Regelleistung.

Da die Gebiete, bei denen technisch nutzbare Windgeschwindigkeiten existieren, regional begrenzt sind, wird angenommen, dass auch in Zukunft eher zentral aufgestellte WP statt Einzelanlagen errichtet werden. Die Fluktuationen der Einspeiseleistung sind im Verhältnis zur angeschlossenen Leistung bei einem WP deutlich geringer als bei einer einzelnen WEA. Die abgegebene Leistung variiert von Aufstellungsort und Ausrichtung der Rotorblätter und ist nicht bei allen WEA eines WPs gleich. Der Anschluss eines zentralen LI-BESS an einem WP ist demnach besser als dezentrale LI-BESS für jede einzelne WEA in Hinblick auf die benötigte Kapazität des LI-BESS [15].

Für einen großflächigen Einsatz in Deutschland sollten LI-BESS an Onshore-WPs angeschlossen werden, da die Anlagenzahl in diesem Segment deutlich höher liegt und weitere Anwendungsfälle nicht nur auf küstennahe Gebiete beschränkt sind. Eine zentrale, erzeugungsnahe Speicherung gewährleistet zudem geringere Stromverluste beim Transport.

Die EEG-Anfangsvergütung entspricht bei einem Offshore-WP derzeit in etwa das Doppelte oder Dreifache von dem eines Onshore-WP. Für die Bereitstellung von Regelleistung durch Einsenkung des WP, bei der die erbrachte Regelleistung nicht gesondert vergütet wird, ist eine Offshore-Anlage weniger attraktiv.

Das Alter und die damit verbundene EEG-Förderung des anzuschließenden WP spielen eine Rolle bei der WP-Auswahl. Zur Bereitstellung von Regelleistung sind Onshore-Anlagen attraktiver, die älter als 5 Jahre sind, da sie statt der hohen Anfangsvergütung nur noch die geringere Grundvergütung erhalten, die in etwa der Hälfte entspricht. Hinzu kommt, dass WEA eine theoretische Laufzeit von 20 Jahren und LI-BESS eine kalendarische Laufzeit von ca. 15 Jahren besitzen. Bei einem Anschluss des LI-BESS nach 5 Jahren, würde die theoretische Restlaufzeit der Anlagenteile gleich sein.

Die Dimensionierung des LI-BESS richtet sich nach der installierten Nennleistung des WP sowie des durchzuführenden Anwendungsfalls. Da der primäre Anwendungsfall zurzeit in der Bereitstellung von asymmetrischer PRL liegt und die bereitgestellte Leistung nur für eine kurze Dauer, bis zur Ablösung von SRL (ca. 15 min) benötigt wird, empfiehlt es sich beim BESS wegen der relativ hohen Investitionskosten pro kWh, nur eine geringe Kapazität zu installieren bzw. eine hohe C-Rate zu verwenden.

Nachteilig sind die derzeitigen Netzentgelte und Gebühren und Umlagen, jedoch gilt es abzuwarten, was die Gesetzgebung für ein neues Strommarktdesign beschließt bzw. einführt. Eine Einsparung der Kosten könnte evtl. auch durch einen Anschluss an das Arealnetz des WP erreicht werden. Die Nachweispflicht der Strombezüge und Lieferungen aus dem WP und dem öffentlichen Netz können mit dem Anschluss von zwei oder mehreren Stromzählern realisiert werden.

Es ist zu prüfen, inwieweit die gleichzeitige Bedienung der Anwendungsfälle möglich ist. Bei einer Bereitstellung von PRL muss ein gewisser Ladezustand im LI-BESS gehalten werden, um positive oder negative Regelleistung zur Verfügung zu stellen. Dies vermindert die verfügbare Energiemenge für weitere Anwendungsfälle stark.

Wirtschaftliche Gesamtrentabilität

Eine Ermittlung der wirtschaftlichen Gesamtrentabilität ist aufgrund der sich veränderten Rahmenbedingungen z. B. durch ein novelliertes Strommarktdesign in den nächsten Jahren nicht allgemeingültig und muss in geraumer Zeit überprüft werden.

Abbildung 4 zeigt die jährlichen Erlöse sowie den jährlichen Cashflow des Investitionsvorhabens eines LI-BESS an einem 12 MW WP zur Bedienung des Anwendungsfalls „asymmetrische PRL bereitstellen". Es kann ein Kapitalwert (NPV) von knapp 300 T€ innerhalb einer Nutzungszeit von 15 Jahren bei der Verwendung eines kalkulatorischen Zinssatzes von 9 % erwirtschaftet werden. Der interne Zinsfuß (IRR) beträgt 11 %.

Entscheidend für die Einhaltung dieser Werte sind die Einnahmen am PRL-Markt. In die Gewinnberechnung fließen Vattenfall-interne Vorhersagen für die Preisentwicklung an der Strombörse sowie dem PRL-Markt ein. Bei einem Wegfall dieses Anwendungsfalls besteht ein hohes Risikopotential für ein Verlustgeschäft, woraus sich schließen lässt, dass einzig der Einsatz eines LI-BESS in Verbund mit einem WP zur Bereitstellung von PRL derzeitig einen wirtschaftlich sinnvollen Betrieb zulässt.

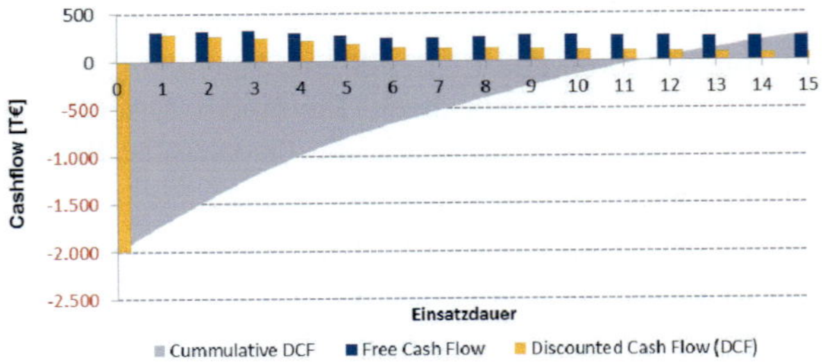

Abbildung 4: Cashflow, Beispiel an einem 12 MW WP

5 Zusammenfassung und Fazit

Das Ziel dieser Arbeit war es den energietechnisch und -wirtschaftlich sinnvollen Betrieb stationärer LI-BESS an Windparks zu prüfen. Wie sich herausgestellt hat, kann ein LI-BESS in Kombination mit einer WEA eine Vielzahl von Anwendungsfällen im Stromsektor bedienen. Diese befinden sich vor allem in den Bereichen des kurzfristigen Erzeugungsausgleichs sowie dem Erhalt der Versorgungssicherheit des elektrischen Stromnetzes. Jedoch rechtfertigen die meisten Anwendungsfälle hinsichtlich der Wirtschaftlichkeit für Betreiber keinen Betrieb.

Der Hauptanwendungsfall für LI-BESS an WEA aus betriebswirtschaftlicher Sicht liegt in der Bereitstellung von asymmetrischer PRL. Dabei ist die Höhe des Erlöses maßgeblich von den hohen Vergütungen zur Bereitstellung der PRL abhängig. Der Bedarf bleibt nach Aussage der DENA-Netzstudie II auch zukünftig stabil. WEA erfüllen die Präqualifikationsbedingungen für die garantierte Leistungsbereitstellung wegen der Ungenauigkeit der Windprognosen nicht von allein. Die Kombination mit einem LI-BESS kann diesen Umstand jedoch ändern und macht damit WEA auch für Systemdienstleistungen zwecks Versorgungssicherheit attraktiv.

6 Literatur

[1] DEUTSCHER BUNDESTAG: Gesetz zur grundlegenden Reform des Erneuerbare-Energien-Gesetzes und zur Änderung weiterer Bestimmungen des Energiewirtschaftsrechts (in Kraft getr. am 24.07.14) (24.07.14))

[2] MAUBACH, Klaus-Dieter: *Energiewende : Wege zu einer bezahlbaren Energieversorgung*. 2. Aufl. Wiesbaden : Springer VS, 2014 (SpringerLink : Bücher)

[3] HAU, ERICH: *Windkraftanlagen : Grundlagen, Technik, Einsatz, Wirtschaftlichkeit*. 5. Aufl // 5., neu bearbeitete Aufl. Berlin, Heidelberg : Springer; Springer Vieweg, 2012 // 2014

[4] HUANG, XIAOBIN ; JIANG, BEIBEI: Research on Lithium battery energy storage system in wind power. In: *2011 International Conference on Electrical and Control Engineering (ICECE)*, S. 1200–1203

[5] KORTHAUER, Reiner: *Handbuch Lithium-Ionen-Batterien*. Berlin, Heidelberg : Imprint: Springer Vieweg, 2013 (SpringerLink : Bücher)

[6] DANIEL FÜRSTENWERTH, LARS WALDMANN (Mitarb.): Agora Energiewende - *Stromspeicher in der Energiewende : Untersuchung zum Bedarf an neuen Stromspeichern in Deutschland für den Erzeugungsausgleich, Systemdienstleistungen und im Verteilnetz*. Deutschland, 10.09.2014

[7] KORPAAS, Magnus ; HOLEN, Arne T. ; HILDRUM, Ragne: *Operation and sizing of energy storage for wind power plants in a market system*. In: *International Journal of Electrical Power & Energy Systems* 25 (2003), Nr. 8, S. 599–606

[8] STEPHAN KOHLER, ANNEGRET-CL. AGRICOLA, HANNES SEIDL (Mitarb.): dena-Netzstudie II – Integration erneuerbarer Energien in die deutsche Stromversorgung im Zeitraum 2015-2020 mit Ausblick auf 2025. - Publikationen - dena.de. November 2010

[9] BUNDESMINISTERIUM FÜR WIRTSCHAFT UND ENERGIE ; *Ein Strommarkt für die Energiewende (Grünbuch)*. Berlin, Oktober 2014

[10] SEBASTIAN GERHARD: *Asymmetrische Regelleistung - Expertenworkshop - Gerhard*

[11] BUNDESMINISTERIUM FÜR WIRTSCHAFT UND ENERGIE ; ÖFFENTLICHKEITSARBEIT, Referat: *Studie: Eignung von Speichertechnologien zum Erhalt der Systemsicherheit.*

[12] BUNDESNETZAGENTUR FÜR ELEKTRIZITÄT, GAS, TELEKOMMUNIKATION, POST UND EISENBAHNEN,: *Festlegung zu Verfahren zur Ausschreibung von Regelenergie in Gestalt der Primärregelung, insbesondere zu Mindestangebotsgrößen* (in Kraft getr. am 12. 4. 2011) (2011-04-12).

[13] CONSENTEC GMBH: *Beschreibung von Regelleistungskonzepten und Regelleistungsmarkt.*

[14] SEBASTIAN GERHARD ; FELIX HALFMANN: *Entwurf einer Betriebsstrategie für Batteriespeicher zur Teilnahme am Primärregelleistungsmarkt*, NEIS 2014, Hamburg, DOI: 10.13140/2.1.1534.8006

[15] JANNATI, M. ; HOSSEINIAN, S. H. ; VAHIDI, B. ; LI, Guo-Jie: *A survey on energy storage resources configurations in order to propose an optimum configuration for smoothing fluctuations of future large wind power plants*. In: *Renewable and Sustainable Energy Reviews* 29 (2014), S. 158–172

Aktuelle Ergebnisse der wissenschaftlichen Begleitforschung zum KfW-Förderprogramm „Erneuerbare Energien – Speicher“

Kai-Philipp Kairies[1,2,3], David Haberschusz[1,2,3], Dirk Magnor[1,2,3], Dirk Uwe Sauer[1,2,3]
Elektrochemische Energiewandlung und Speichersystemtechnik
[1]Institut für Stromrichtertechnik und Elektrische Antriebe (ISEA), RWTH Aachen University
[2]Institute for Power Generation and Storage Systems (PGS), E.ON ERC, RWTH Aachen University
[3]Jülich Aachen Research Alliance, JARA Energy

Kurzfassung

Dezentrale Solarstromspeicher zur Erhöhung des lokalen Eigenverbrauchs stehen seit einigen Jahren vermehrt im Fokus von Forschung, Industrie und Verbraucherinteresse. Im Zuge von kontinuierlich sinkenden garantierten Einspeisevergütungen für Strom aus Photovoltaikanlagen (PV-Anlagen) und steigenden Strombezugspreisen für private Endkunden wird der lokale Eigenverbrauch von Solarstrom zunehmend attraktiver. Da Solarstromspeicher heute noch einen jungen Markt mit verhältnismäßig hohen Endverbraucherpreisen und geringen Betriebserfahrungen darstellen hat die Bundesregierung zusammen mit der KfW-Bankengruppe ein Förderprogramm aufgelegt. Um die Wirkung des Förderprogramms zu bewerten und ein besseres Verständnis der Technologie zu ermöglichen wurde zusätzlich ein wissenschaftliches Monitoringprogramm etabliert, in dem sowohl Markt- als auch Betriebsdaten von PV-Speichern durch die RWTH Aachen erfasst und analysiert werden. Das vorliegende Paper gibt einen Überblick über die KfW-Förderung sowie das wissenschaftliche Monitoringprogramm und präsentiert aktuelle Analysen zur Marktentwicklung. Zusätzlich werden erste Ergebnisse der hochauflösenden Vermessung privat betriebener PV-Speichern dargestellt: Am Beispiel eines energieintensiven Haushalts im Raum Aachen werden für den Monat Juli Auswertungen der Eigenverbrauchsquote und des Autarkiegrades durchgeführt, zusätzlich werden das Netzverhalten des Gesamtsystems sowie die belastungsabhängige Temperaturentwicklung der Batterie anhand der Messdaten eines Sommertags exemplarisch diskutiert.

1 Das KfW-Förderprogramm Erneuerbare Energien ‚Speicher‘

Die Bundesregierung fördert seit dem 01. Mai 2013 gemäß den „Richtlinien zur Förderung von stationären und dezentralen Batteriespeichersystemen zur Nutzung in Verbindung mit Photovoltaikanlagen“ die Anschaffung von dezentralen Solarstromspeichern. Ziel der Maßnahme ist die Förderung von PV-Speichersystemen, die eine Netzdienlichkeit aufweisen und den Eigenverbrauch erhöhen. Als volkswirtschaftlicher Gegenwert für die Förderung müssen geförderte PV-Speicher so betrieben werden, dass die Problematik der Netzbelastung zu Zeiten hoher Einspeisung reduziert wird.

1.1 Art der Förderung

Die Förderung von dezentralen Solarstromspeichern ist als KfW-Programm („Erneuerbare Energien – Speicher“, Programmnummer 275) ausgestaltet. Sie erfolgt für eine individuelle Anlage durch einen zinsgünstigen Kredit der KfW-Bank sowie einen durch das Bundesministerium für Wirtschaft und Energie (BMWi) finanzierten Tilgungszuschuss in Höhe von maximal 30 % der förderfähigen Investitionskosten des Batteriespeichersystems. Die Möglichkeit einer Förderung steht grundsätzlich Privatpersonen, Freiberuflern, Landwirten, gemeinnützigen Antragstellern und Unternehmen offen. Der Tilgungszuschuss reduziert dabei die Kreditschuld des Fördernehmers und verkürzt somit die Laufzeit des Kredites, eine Auszahlung des Tilgungszuschusses ist nicht vorgesehen.

1.2 Technische Rahmenbedingungen der Förderung

Um einen netzdienlichen Betrieb der Speicher sowie eine nachhaltige Entwicklung der Technologie anzureizen ist die Förderung an weitgehende technische Randbedingungen gebunden. Die aus technischer Sicht wichtigsten Forderungen sind im Folgenden zusammenfassend dargestellt:

- Die maximale Nennleistung der PV-Anlage, die mit dem Batteriespeichersystem verbunden wird, darf 30 kW nicht überschreiten.

- Die maximale Leistungsabgabe der PV-Anlage am Netzanschlusspunkt ist durch geeignete Maßnahmen auf 60 % der Nennleistung der Photovoltaikanlage zu begrenzen (siehe **Bild 1**). Die Verpflichtung zur Leistungsbegrenzung besteht dauerhaft für die gesamte Lebensdauer der Photovoltaikanlage, mindestens aber 20 Jahre, und erstreckt sich damit auch auf einen eventuellen Weiterbetrieb der Photovoltaikanlage

- Alle Betreiber von geförderten PV-Speichersystemen verpflichten sich zu einer Teilnahme an einem wissenschaftlichen Monitoringprogramm. Als Teil des Monitoringprogramms werden unter anderem Leistungs- und Betriebsdaten der installierten Systeme über das Webportal www.speichermonitoring.de abgefragt und anonymisiert zu wissenschaftlichen Auswertungen herangezogen.

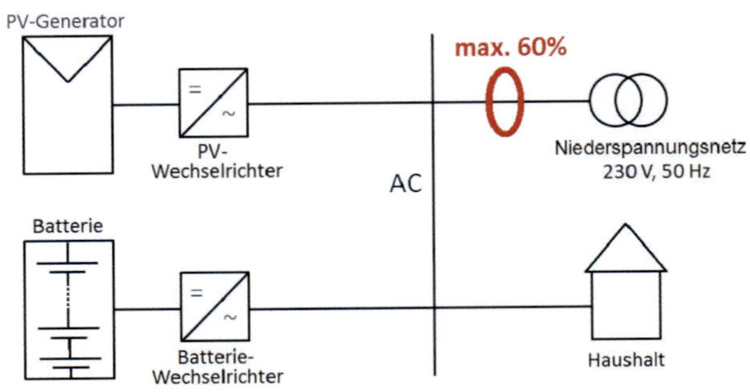

Bild 1 Schematische Darstellung der Reduzierung der maximalen Einspeiseleistung am Beispiel eines AC-gekoppelten Speichersystems

- Für die Batterien des Batteriespeichersystems muss eine Zeitwertersatzgarantie für einen Zeitraum von 7 Jahren vorliegen. Hierbei wird bei Defekt der Batterien der Zeitwert der Batterien ersetzt. Der Zeitwert berechnet sich anhand einer über den Zeitraum von 7 Jahren linear angenommenen jährlichen Abschreibung. Die Zeitwertersatzgarantie ist vom Händler/Hersteller dem Käufer des Batteriesystems gegen-über zu garantieren oder über eine geeignete Versicherungsleistung, deren Kosten der Händler/Hersteller trägt, zu gewährleisten.

2 Das wissenschaftliche Mess- und Evaluierungsprogramm Solarstromspeicher

Dezentrale Solarstromspeicher die den lokalen Eigenverbrauch erhöhen und eine Netzdienlichkeit aufweisen bieten das Potenzial, langfristig einen positiven Einfluss auf die erfolgreiche Umsetzung der deutschen Energiewende zu entwickeln. Neben dem in der Bevölkerung verbreiteten Ideal einer weitgehenden Selbstversorgung mit erneuerbaren Energien sowie zukünftigen monetären Vorteilen eines erhöhten Eigenverbrauchs haben zahlreiche Studien anhand von Systemmodellen einen positiven Einfluss dezentraler PV-Speicher auf die Verteilnetze gezeigt (unter anderem [1–5]). Der tatsächliche Einfluss einer signifikanten Anzahl von dezentralen Batteriesystemen in der Verteilnetzebene kann heute jedoch nur abgeschätzt werden. Um ein umfassendes Verständnis dieser Effekte unter realistischen Bedingungen zu gewinnen und um den Einfluss des Förderprogramms auf die Marktentwicklung von Solarstromspeichern zu evaluieren, wird das Förderprogramm von Beginn an durch ein wissenschaftliches Monitoringprogramm begleitet. Das wissenschaftliche Monitoring wird durch das Bundesministerium für Wirtschaft und Energie (BMWi) gefördert und vom Institut für Stromrichtertechnik und elektrische Antriebe (ISEA) der RWTH Aachen durchgeführt. Das wissenschaftliche Monitoringprogramm gliedert sich in drei Abschnitte:

Im Rahmen des ***Basis-Monitorings*** werden die Marktdaten aller geförderten Speichersysteme in Deutschland erfasst und kontinuierlich ausgewertet. Auf einem dazu eingerichteten Webportal tragen die Betreiber geförderter PV-Speichern die Stammdaten Ihrer Speichersysteme ein um einen Anspruch auf Ihren Tilgungszuschuss zu erhalten. Zu den erfassten Stammdaten gehören unter anderem: Datum der Installation, Nennleistung der PV-Anlage, Hersteller des Speichersystems, installierte und nutzbare Energiekapazität des Batteriespeichers, verwendete Batterietechnologie, Systemtopologie, Kaufpreis des Speichersystems, etc. Ausgesuchte Ergebnisse der Marktanalyse werden in Abschnitt 3 präsentiert.

Im Rahmen des ***Standard-Monitorings*** werden anhand von monatlich erfassten Zählerständen und Logfiles von über 1.000 Speicherbetreibern Aussagen über durchschnittliche Eigenverbrauchsraten und Autarkiegrade in Abhängigkeit der PV-Anlagen- und Speichergröße generiert. Weiterhin können die volkswirtschaftlichen Effekte eines vermehrten Eigenverbrauchs abgeschätzt werden.

Im Rahmen des ***Intensiv-Monitorings*** werden durch die RWTH Aachen hochauflösende Messsysteme (1s-Intervall) an 20 privat betriebenen PV-Speichern installiert. Die Auswertung dieser Daten erlaubt weitreichende Aussagen über die unter realistischen Umständen erreichten Wirkungsgrade sowie die Batterie- und Energiemanagementsysteme der betrachteten Speicher. Erste qualitative Ergebnisse dieser Messungen werden in Abschnitt 4 präsentiert.

3 Marktsituation von Solarstromspeichern in Deutschland

Der Markt für PV-Speichern ist seit der Einführung des KfW-Förderprogramms im Mai 2013 kontinuierlich angewachsen. Bis zum 31. März 2015 wurden in Deutschland ca. 17.000 dezentrale Batteriespeicher zur Erhöhung des lokalen Eigenverbrauchs an die Niederspannungsnetze angeschlossen. Etwa 55 % dieser Speicher wurden durch das KfW-Förderprogramm „Erneuerbare Energien – Speicher" gefördert und bilden somit die Datengrundlage des wissenschaftlichen Monitoringprogramms.

3.1 Systemmerkmale von geförderten PV-Speichern

In **Bild 2** sind die wesentlichen Systemmerkmale (*Batterietechnologie, Systemtopologie, Installationstyp*) der beim Speichermonitoring registrierten Batteriespeicher zusammenfassend dargestellt.

Es zeigt sich, dass der Anteil von Blei-Säure Batterien und Lithium-Ionen Batterien bezogen auf ihre Anzahl und nutzbare Kapazität heute nahezu ausgeglichen ist. Der deutliche Überhang von Blei-Säure Speichern bezogen auf ihre installierte Kapazität ist dabei durch die typische Dimensionierung von Batteriespeichern zurückzuführen: Während Lithium-Ionen Batterien in stationären Anwen-

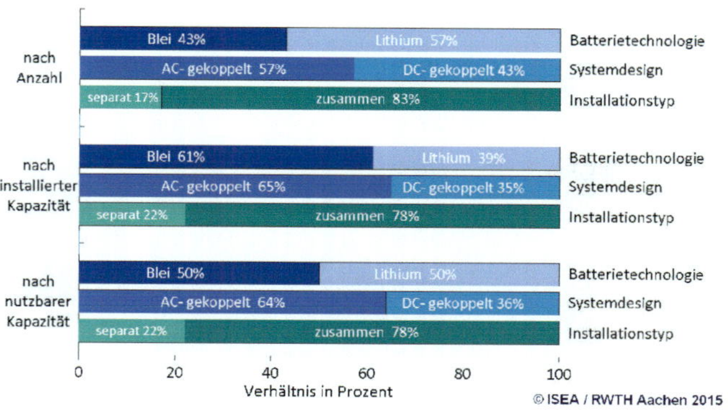

Bild 2 Systemmerkmale geförderter Solarstromspeicher

dungen üblicherweise zwischen 80 und 100 % ihrer installierten Batteriekapazität nutzen, verwenden Bleibatterien typischerweise zur Optimierung ihrer Lebensdauer nur 50–60 % ihrer installierten Kapazität Zur Nutzbarmachung der gleichen Speicherkapazität werden Bleibatterien daher im Allgemeinen überdimensioniert. Diese beobachtete typische Dimensionierung von Batteriespeichern ist nicht auf den PV-Speichermarkt beschränkt, sondern deckt sich mit weithin bekannten Forschungsergebnissen zur Batteriealterung und Lebensdaueroptimierung stationärer Batteriespeichersysteme, wie unter anderem in [6–8] für Blei-Säure Batterien und in [9,10] für Lithium-Ionen Batterien dargestellt ist.

Auch die Betrachtung des Systemdesigns zeigt einen tendenziell ausgewogenen Markt: AC-gekoppelte Systeme, die an die Wechselspannungsseite des Haushaltes angeschlossen werden und somit leicht nachgerüstet werden können, und DC-gekoppelte Speichersysteme, bei denen sich PV-Anlage und Speichersystem einen Wechselrichter teilen, sind im Markt etwa gleich stark vertreten. Der höhere Anteil in den kapazitätsbezogenen Kategorien zeigt, dass AC-Systeme dabei tendenziell größer ausgelegt werden als DC-Systeme. Es zeigt sich außerdem, dass der überwiegende Anteil an dezentralen Batteriespeichersystemen zusammen mit einer neu angeschafften PV-Anlage installiert wurde. Dies lässt sich direkt auf die Randbedingungen der Förderung zurückführen, welche eine Nachrüstung nur für PV-Anlagen zulässt, die nach dem 31.12.2012 installiert wurden (siehe auch Abschnitt 1.2). Der Markt für potenzielle Nachrüstungen ist somit im Rahmen des Förderprogramms klar begrenzt.

3.2 Entwicklung des Endkundenpreises von dezentralen Solarstromspeichern

In **Bild 3** ist die Entwicklung der Endkundenpreise von PV-Speichern seit Beginn des KfW-Förderprogramms im Mai 2013, aufgeteilt nach der verwendeten Batterietechnologie, dargestellt. Aufgetragen ist dabei jeweils der Endkundenpreis (inkl. MwSt) einer nutzbaren Kilowattstunde Speicherkapazität. Weitere Randbedingungen der Darstellung sind im Folgenden kurz zusammengefasst:

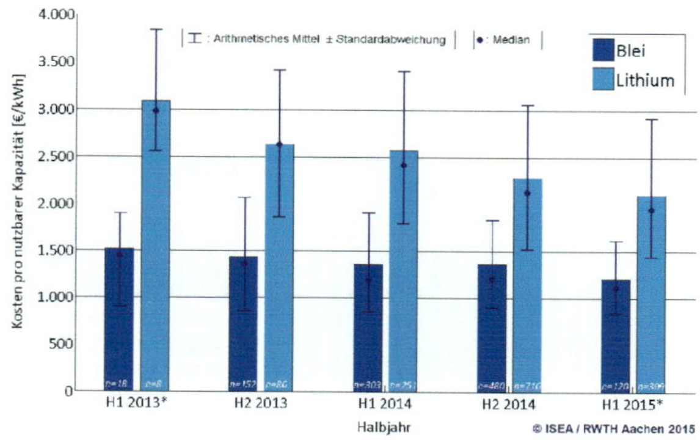

Bild 3 Entwicklung des durchschnittlichen Endkundenpreises dezentraler Solarstromspeicher pro Kilowattstunde nutzbarer Speicherkapazität (inkl. MwSt.)

- In die Berechnung geht das gesamte Speichersystem (*Batterien, Leistungselektronik, Schütze, Sensoren, etc*) ein.
- PV-Module und Installationskosten sind nicht Teil der Darstellung.
- Nachgerüstete Systeme und Komplettinstallationen werden gemischt betrachtet.
- DC- und AC-Systeme werden gemischt betrachtet. Bei AC-gekoppelten Speichersystemen geht der PV-Wechselrichter nicht in die Berechnung des Preises mit ein.

Bezogen auf den Zeitraum Halbjahr-1-2014–Halbjahr-1-2015 zeigt sich, dass die durchschnittlichen Endverbraucherpreise für Speichersysteme mit Bleibatterien derzeit jährlich um ca. 11 % und die durchschnittlichen Endverbraucherpreise für Speichersysteme mit Lithium-Ionen Batterien jährlich um ca. 18 % fallen. Dies spiegelt die hohen Kostensenkungspotenziale insbesondere in der Zellherstellung von Lithium-Ionen Batterien wider und entspricht der typischen Lernkurve neuer Technologien, laut derer eine Verdopplung der Produktion eine Kostensenkung von ca. 10–20 % folgen. Diese Entwicklung macht PV-Speichersysteme langfristig wirtschaftlich attraktiver.

4 Hochauflösende Vermessung dezentraler Solarstromspeicher in Feldversuchen

Im Rahmen des Intensiv-Monitorings werden 20 privat betriebene dezentrale PV-Speichersysteme durch ein an der RWTH Aachen University entwickeltes Messsystem hochauflösend vermessen. Dabei werden neben allen relevanten Spannungen, Strömen, Leistungs- und Energieflüssen an Haushalt, Speicher und Erzeugungsanlage (jeweils dreiphasig erfasst) auch die lokale Solareinstrahlung sowie die Temperatur von Batterie und Aufstellraum des Speichers aufgenommen. In den folgenden vier Abschnitten sind einige der seit dem zweiten Quartal 2015 gewonnenen Messdaten exemplarisch aufbereitet. Die Analyse der Messdaten des vollständigen Jahresverlaufs wird in

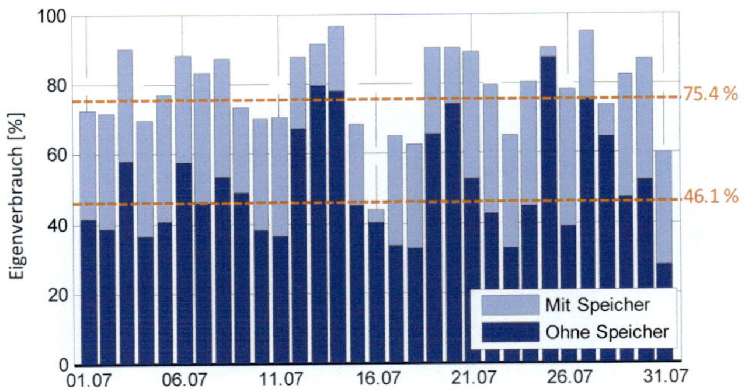

Bild 4 Gemessene Eigenverbrauchserhöhung durch Solarstromspeicher an einem energieintensiven Privathaushalt im Raum Aachen.

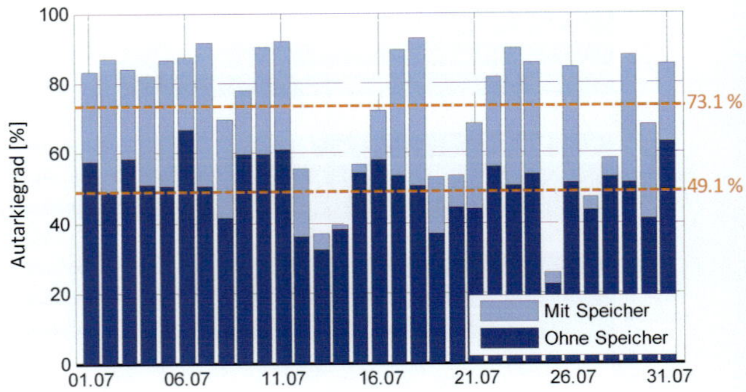

Bild 5 Gemessener Autarkiegrad durch Solarstromspeicher an einem energieintensiven Privathaushalt im Raum Aachen.

Zukunft weitreichende Aussagen über die tatsächlichen Wirkungsgrade und Alterungsmechanismen von dezentralen Speichern sowie deren Netznutzen ermöglichen. Die präsentierten Daten beziehen sich jeweils auf einen energieintensiven 2-Personen Haushalt im Raum Aachen mit einem angegeben Stromverbrauch von 8.000 kWh pro Jahr. 2014 wurde dort eine 6,5 kWp PV-Anlage zusammen mit einem Blei-Säure Solarstromspeicher mit einer nutzbaren Speicherkapazität von 8 kWh installiert. Die Installation des hochauflösenden Messsystems erfolgte im April 2015.

4.1 Erhöhung des Eigenverbrauchs durch PV-Speicher

Das primäre Ziel privater Speicherbetreiber ist in der Regel zunächst die Erhöhung der selbstverbrauchten elektrischen Energie aus der PV-Anlage. Die Eigenverbrauchsquote (EVQ) beschreibt dabei den Anteil an selbstgenutzter Solarenergie im Verhältnis zur insgesamt erzeugten Strommenge der PV-Anlage (siehe Formel 4-1).

$$EVQ = 1 - \frac{Netzeinspeisung}{PV_{Erzeugung}} \qquad (4\text{-}1)$$

In **Bild 4** ist exemplarisch die Auswirkung des vermessenen Speichersystems auf den Eigenverbrauch des beschriebenen Haushaltes im Juli 2015 dargestellt. Die hellblauen Balken geben dabei jeweils die täglich erzielte Eigenverbrauchsquote an. Die obere rote gestrichelte Linie stellt die für den gesamten Monat ermittelte durchschnittliche Eigenverbrauchsquote dar. An Tagen mit diesigem oder verregnetem Wetter, an denen die PV-Anlage nur geringe Strommengen produziert, konnte so bis zu 90 % des Solarertrages im Haushalt genutzt werden. Im Durchschnitt verbleibt etwas über Dreiviertel der Solarenergie im Haushalt und nur ein Viertel wird dementsprechend in das öffentliche Stromnetz eingespeist.

Auf Basis der erfassten Messdaten ist es zudem möglich zu berechnen, wie hoch die Eigenverbrauchsquote ohne Speichersystem gewesen wäre. Diese ist in **Bild 4** durch die dunkelblauen Balken dargestellt. Im Monatsschnitt hätte der Haushalt ohne Speichersystem eine Eigenverbrauchsquote von 46,1 % erzielt (untere rot gestrichelte Linie). Bei einer Eigenverbrauchsquote von 75,4 % mit Speichersystem bedeutet dies, dass im Juli 2015 durch den installierten Solarstromspeicher eine Erhöhung des Eigenverbrauchs um 29,3 Prozentpunkte erreicht werden konnte.

4.2 Erhöhung der Autarkiequote durch PV-Speicher

Der Autarkiegrad (ATG) eines Haushaltes beschreibt, wie unabhängig der Speicherbetreiber vom öffentlichen Stromnetz ist (siehe Formel 4-2).

$$ATG = 1 - \frac{Netzbezug}{Stromverbrauch} \qquad (4\text{-}2)$$

Ein Autarkiegrad von 100 % bedeutet somit, dass der Speicherbetreiber bilanziell keine zusätzliche Energie aus dem öffentlichen Stromnetz beziehen muss. In **Bild 5** ist analog zu **Bild 4** der Autarkiegrad des betrachteten Haushaltes für den Monat Juli 2015 dargestellt. Die hellblauen Balken zeigen wiederum die Messergebnisse unter Berücksichtigung des installierten Speichersystems während die dunkelblauen Balken eine Berechnung des Autarkiegrads „ohne Speichersystem" darstellen.

Durch das installierte Speichersystem erreicht der Betreiber im Monat Juli an einzelnen Tagen Autarkiegrade von bis zu 90 %. Eine vollständige Versorgung des Haushaltes durch selbsterzeugten Solarstrom wurde dabei vor Allem durch die begrenzte Leistungskapazität des Speichersystems verhindert: Um Leistungsspitzen oberhalb von 2 kW bedienen zu können, musste zusätzlich auf Netzstrom zurückgegriffen werden. Insgesamt konnte somit für den Monat Juli ein durchschnittlicher Autarkiegrad von 73,1 % erzielt werden. **Bild 5** zeigt jedoch auch die starken Schwankungen des täglichen Autarkiegrades von mehr als fünfzig Prozentpunkten, abhängig vom jeweiligen Tageslastprofil sowie der aktuellen PV-Erzeugung - hinzu kommen jahreszeitliche Änderungen der solaren

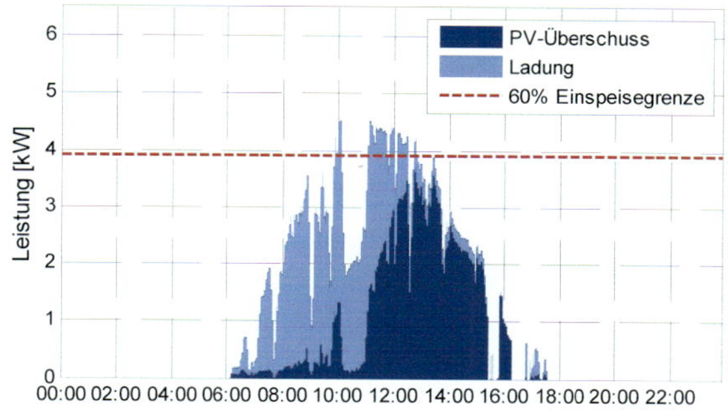

Bild 6 Gemessene Netzeinspeisung an einem Haushalt mit Solaranlage und Batteriespeichersystem

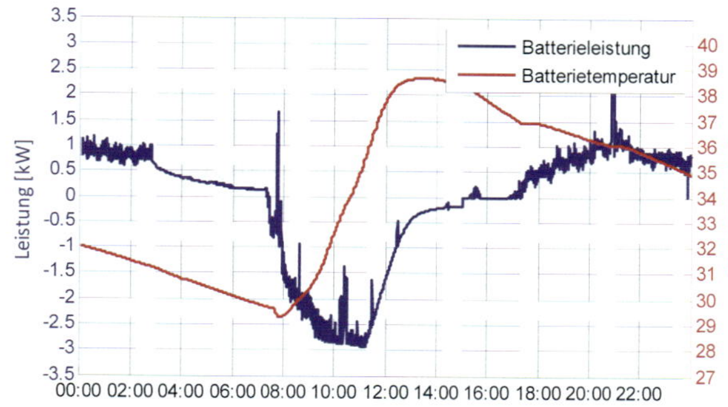

Bild 7 Gemessener Verlauf der Batterietemperatur in Abhängigkeit der Lade- und Entladeleistung

Einstrahlung und eine Erhöhung des Strombedarfs in den Wintermonaten. Eine sichere Versorgung ausschließlich durch eine hauseigene Photovoltaikanlage und einen Batteriespeicher ist somit für die allermeisten Haushalte im Allgemeinen nicht möglich. Um die im Jahresverlauf auftretenden Lasten verlässlich zu decken wird stets ein Netzanschluss oder eine zusätzliche Erzeugungsquelle (zum Beispiel Blockheizkraftwerke) benötigt.

4.3 Einfluss von PV-Speichern auf die Netzeinspeisung

Aus Sicht des Versorgungsnetzes ist es von besonderem Interesse, wie sich Speichersysteme am Netzanschlusspunkt auf die Einspeiseleistung der zugeordneten PV-Anlagen auswirken. Systembasierte Simulationen haben gezeigt, dass durch ein „Abschneiden“ der Erzeugungsspitzen in der Mittagszeit die Aufnahmefähigkeit der Niederspannungsnetze für dezentrale Erzeugungsanlagen um ein Vielfaches erhöht werden kann [1–5]. Um einen ersten Eindruck der speichergestützten Netzentlastung unter realistischen Randbedingungen zu gewinnen, wird in **Bild 6** die gemessene Einspeise- bzw. Bezugsleistung am Netzanschlusspunkt des Haushalts (durchgezogene blaue Linie) für einen Tag im Juli 2015 exemplarisch dargestellt. Die rot gestrichelte Linie entspricht der im Rahmen der KfW-Förderung vorgesehenen maximalen Einspeiseleistung von 60 % der PV-Nennleistung von 6,5 kWp. Die blau gestrichelte Linie entspricht der Netzaustauschleistung die in das öffentliche Stromnetz eingespeist worden wäre, wenn im Haushalt kein Batteriespeichersystem betrieben worden wäre. Die Fläche zwischen beiden Kurven entspricht somit der Energie, die durch das Speichersystem aufgenommen und zwischengespeichert wurde.

Bild 6 verdeutlicht, dass das Energiemanagement des betrachteten Speichersystems eine reine Maximierung des Eigenverbrauchs betreibt: Das Speichersystem beginnt bereits mit dem ersten Überschuss an PV-Leistung am Morgen mit der Ladung der Batterie und speist über den Tag hinweg nach Möglichkeit die gesamte Residualleistung in die Batterie. Dieses Ladeverhalten garantiert eine maximale Ausnutzung des installierten Speichersystems und ist nicht auf Wetterprognosen oder eine Kommunikationsfähigkeit angewiesen (siehe auch [11, 12]). Sollte die Batterie allerdings bereits im Verlauf des Vormittags vollständig geladen werden, kann zur Zeit der maximalen Einspeisung keine weitere Solarenergie aufgenommen werden; im Rahmen der Einspeisebegrenzung müsste dann der Anteil der Einspeiseleistung, der die 60-Prozent-Grenze überschreitet, abgeregelt werden. Aufgrund der im Verhältnis zur Nennleistung der PV-Anlage großen Speicherkapazität von 8 kWh hat das Speichersystem allerdings im Verlauf des betrachteten Tages noch ausreichend Kapazität frei, um auch die auftretenden Mittagsspitzen aufzufangen. Insgesamt trägt das betrachtete Speichersystem somit auch ohne intelligente Betriebsstrategie effektiv zu einer netzdienlichen Minderung der PV-Erzeugungsspitzen bei. Zukünftige Auswertungen der Netzeinspeisung im gesamten Jahresverlaufs werden hier auch quantitative Analysen ermöglichen, in denen der Netznutzen von dezentralen Speichern mit alternativen Möglichkeiten der Netzstabilisierung (zum Beispiel Netzausbau, vergütete Abregelung, regelbare Ortsnetztransformatoren, …) verglichen werden kann.

4.4 Verlauf der Batterietemperatur im Be-trieb

Bei der Wirtschaftlichkeitsbetrachtung von dezentralen Speichersystemen zur Eigenverbrauchserhöhung stellt die erreichbare Lebensdauer der Batterien einen wesentlichen Faktor zur Berechnung der Lebenszykluskosten dar. Die kalendarische Alterung von Batteriespeichern wird dabei insbesondere durch ihre lokalen Betriebstemperaturen bedingt. Um den Einfluss der Temperatur auf die Leistungsfähigkeit und Lebensdauer der betrachteten Speichersysteme zu quantifizieren werden im Rahmen des Intensiv-Monitorings neben den elektrischen Messdaten des Gesamtsystems auch mittels eines analogen PT100 Temperatursensors durchgängig die Batterietemperaturen erfasst. In **Bild 7** ist der Verlauf der Batterietemperatur des betrachteten Blei-Säure Speichers in Abhängigkeit der Lade- bzw. Endladeleistung des Speichersystems für einen Tag im Juli 2015 dargestellt.

Entsprechend der in Abschnitt 4.3 dargelegten Betriebsstrategie beginnt der Blei-Säurespeicher bereits mit den ersten PV-Überschüssen des Tages gegen 08:00 Uhr morgens mit der Ladung; die Batterietemperatur steigt infolge der auftretenden Ladeleistung von bis zu 3 kW innerhalb von 6 Stunden um etwa 9 K auf über 38,5 °C an. Ab 14:00 Uhr beginnt das Speichersystem mit der Entladung – die dadurch geringer belastete Batterie kühlt langsam ab, um gegen Morgen wieder unterhalb einer Temperatur von 30 °C zu gelangen. Der hier beobachtete thermische Zyklus von etwa 9 K hat langfristig Einfluss auf die zu erwartende Lebensdauer der Batterie: Eine Erhöhung der durchschnittlichen Batterietemperatur um 10 K wird mit einer Verdopplung der kalendarischen Alterung des elektrochemischen Systems, bei Blei-Säure Batterien insbesondere durch Korrosionseffekte und Verbleiung, assoziiert [13–14]. Die daraus resultierende Minderung der Lebensdauer erhöht die spezifischen Durchsatzkosten einer gespeicherten Kilowattstunde Solarstrom und verringert somit für den Betreiber die Wirtschaftlichkeit des Speichersystems. Zur Minderung der beobachteten Problematik sind ein kühlerer Aufstellungsraum des Speichers sowie ein angepasstes Energiemanagementsystem, das die Batterien über den Tag verteilt mit geringeren Leistungen lädt, vorstellbar. Die andauernden hochauflösenden Messungen werden zusammen mit geplanten Kapazitätstests in Zukunft die Auswirkung der Betriebsstrategie und der erhöhten Batterietemperaturen auf die Lebensdauer des Speichersystems unter realistischen Bedingungen besser verständlich machen. Hieraus können konkrete Handlungsempfehlungen zur Anpassung der Technologie entwickelt werden.

5 Zusammenfassung und Ausblick

Das öffentlich geförderte Forschungsprojekt „Wissenschaftliches Mess- und Evaluierungsprogramm Solarstromspeicher" stellt die Begleitforschung zum Förderprogramm „Erneuerbare Energien – Speicher" der KfW-Bankengruppe und des Bundesministerium für Wirtschaft und Energie dar. Im Rahmen dieses Forschungsprojektes wird ein umfassendes Bild der aktuellen Entwicklung des deutschen Marktes für dezentrale Solarstromspeicher erstellt. Neben der Entwicklung der Marktdaten werden anhand von hochauflösenden Messsystemen in privaten Haushalten die Effekte auf die Haushalte und das Niederspannungsnetz sowie volkswirtschaftliche Effekte untersucht.

Die Ergebnisse der ersten eineinhalb Jahre des Speichermonitorings wurden Mitte 2015 in einem umfangreichen öffentlichen Jahresbericht zusammengestellt, der unter www.speichermonitoring.de frei verfügbar ist.

5.1 Danksagung

Das vorliegende Beitrag entstand im Rahmen des Forschungsvorhabens „Wissenschaftliches Mess- und Evaluierungsprogramm Solarstromspeicher" (WMEP PV-Speicher), das durch das Bundesministerium für Wirtschaft und Energie gefördert wird.

6 Literatur

[1] Kerber, G.: Aufnahmefähigkeit von Niederspannungsverteilnetzen für Strom aus Photovoltaikanlagen. Dissertation. Fakultät für Elektrotechnik und Informationstechnik, Technische Universität München, München (2011)

[2] Wieland, T., Otto, F., L. Fickert, T. K. Schuster: Analyse, Bewertung und Steigerung möglicher Einspeisekapazität dezentraler Energieerzeugungsanlagen in der Verteilnetzebene. 8. Internationale Energiewirtschaftstagung an der TU Wien, Wien (2013)

[3] Degner, T. et al: Integration von Photovol-taikanlagen in Verteilungsnetze. Kasseler Symposium Energie-Systemtechnik, Kassel 2011

[4] Predki, H.: System- und Marktintegration von Photovoltaik-Anlagen durch dezentrale Stromspeicher? – Eine Analyse der technischen Potentiale und rechtlichen Rahmenbedingungen. Leuphana Schriftenreihe Nachhaltigkeit & Recht, Lüneburg (2013)

[5] Weniger, J. Bergner, J., Tjaden, T., Quaschning, V.: Dezentrale Solarstromspeicher für die Energiewende. Hochschule für Technik und Wirtschaft HTW Berlin, Berlin (2015)

[6] Schiffer, J. et al: Model prediction for ranking lead-acid batteries according to expected lifetime in renewable energy systems and autonomous power-supply systems. J Power Sources 2007;168; 66-78

[7] P. Ruetschi: Aging mechanisms and service life of lead-acid batteries. J Power Sources 2004; 127; 33-44

[8] DIN 60896-1, Stationary lead-acid batteries, Part 11: Vented types, General requirements and methods of test (IEC 60896-11:2002); German version EN 60896-11:2003

[9] Käbitz, S. et al: Cycle and calendar life study of a graphite |LiNi1/3Mn1/3Co1/3O2 Li-ion high energy system. Part A: Full cell characterization. J Power Sources 2013; 239; 572-583

[10] Ecker, M. et al: Calendar and cycle life study of Li(NiMnCo)O2-based 18650 lithium-ion batteries. J Power Sources 2014; 248; 839-851

[11] Kairies, K.-P- et al: Wissenschaftliches Mess und Evaluierungsprogramm Solarstromspeicher – Jahresbericht 2015. RWTH Aachen University, Aachen, 2015

[12] Moshövel, J. et al: Analysis of the maximal possible grid relief from PV-peak-power impacts by using storage systems for increased self-consumption. In: Applied Energy 137 (2015), S.567–575

[13] A. Jossen, W. Weydanz: Moderne Akkumulatoren richtig einsetzen. 1. Auflage, Reichardt Verlag, 2006

[14] Brauchbarkeitsdauer – Betrachtungen bei stationären Bleibatterien. ZVEI Merkblatt Nr. 19 Ausgabe Januar 2013. Zentralverband Elektrotechnik- und Elektroindustrie e.V., Fachverband Batterien

Posterbeiträge

Energieforschungsverbund Hamburg

- Kompetenzen der Energieforschung vernetzen -

www.energieforschungsverbund.hamburg

Hintergrund

Die Umstellung auf eine nachhaltige Energieversorgung gehört zu den größten Herausforderungen der nächsten Jahre. Mit Unterstützung des Senats der Freien und Hansestadt Hamburg wurde der *Energieforschungsverbund Hamburg* (EFH) als Zusammenschluss der fünf großen Hamburger Hochschulen gegründet, um die unterschiedlichen Kompetenzen im Bereich Energieforschung untereinander zu vernetzen und somit einen wissenschaftlichen Mehrwert zu generieren.

Die Realisierung der zukünftigen Energieversorgung wird in erster Linie durch vier Aspekte getragen, die im Rahmen des *Energieforschungsverbunds Hamburg* wissenschaftlich begleitet werden sollen:

- **FLEXIBLE ENERGIEERZEUGUNG,**
- **ENERGIEÜBERTRAGUNG UND –VERTEILUNG,**
- **ANGEBOTSGERECHTER VERBRAUCH,**
- **SPEICHERUNG VON ENERGIE.**

Diese beinhalten neben ingenieurwissenschaftlich/technischen und naturwissenschaftlichen Fragestellungen auch rechtliche und ökonomische Gesichtspunkte.

Foto: www.mediaserver.hamburg.de/ Christian Spahrbier

Unsere Aufgabe

Foto: Dr. Frank Hoffmann

Die Hamburger Hochschulen weisen besonderes und komplementär verteiltes Forschungspotential in den Bereichen

- **WINDENERGIE / SOLARENERGIE**
- **BIOMASSE / BIOENERGIE**
- **REGENERATIVE ENERGIE AUS DEM MEER**
- **NETZINTEGRATION / SMART GRIDS**
- **ENERGIESPEICHER**
- **ENERGIEEFFIZIENTE STADT**

auf und decken somit das Gebiet der Energieforschung nahezu vollständig ab. Durch den *Energieforschungsverbund Hamburg* werden verschiedene Forschungsideen – schwerpunktmäßig diese Themenfelder betreffend – aus Hochschulen und Wirtschaft zusammengeführt und zu potentiellen neuen Fachprojekten ausgearbeitet. Somit können gezielt neue und umfassende Verbund-Förderanträge hervorgebracht werden, die wesentlich zur Energiewende beitragen. Dazu kooperieren wir bei unserer Arbeit auch eng mit dem Cluster Erneuerbare Energien Hamburg.
Als Knotenpunkt zwischen den Hochschulen stellen wir einen Anlaufpunkt für Wissenschaftler/innen und Unternehmer/innen dar, die an einem gemeinsamen Verbundprojekt rund um das Thema Energie interessiert sind.

Bei Interesse oder Fragen sprechen Sie uns gerne an!

Energieforschungsverbund Hamburg - EFH

Universität Hamburg
DER FORSCHUNG | DER LEHRE | DER BILDUNG

Universität Hamburg:

Prof. Dr. Michael Fröba
Tel.: 040 42838 3100
Dr. Elisabeth Ziemann
Tel.: 040 42838 3103
E-Mail:
unihh@energieforschungsverbung.hamburg

Hochschule für Angewandte Wissenschaften Hamburg
Hamburg University of Applied Sciences

Hochschule für Angewandte Wissenschaften:

Prof. Dr. Werner Beba
Tel.: 040 42875 6937
Dr.-Ing. Hans Schäfers
Tel.: 040 42875 9895
E-Mail:
haw@energieforschungsverbung.hamburg

TUHH
Technische Universität Hamburg-Harburg

Technische Universität Hamburg-Harburg:

Prof. Dr.-Ing. Martin Kaltschmitt
Tel.: 040 42878 3008
Dipl.-Ing. Hannes Wagner *Tel.: 040 42878 4549*
E-Mail:
tuhh@energieforschungsverbung.hamburg

Helmut Schmidt Universität

Helmut Schmidt Universität:

Prof. Dr.-Ing. Detlef Schulz
Tel.: 040 6541 2757
Dipl.-Ing. Thomas Weiß
Tel.: 040 6541 2163
E-Mail:
hsu@energieforschungsverbund.hamburg

HCU | HafenCity Universität Hamburg

HafenCity Universität Hamburg:

Prof. Peter Braun
Tel.: 040 42827 4571
Dr. Stefan Krümmel
Tel.: 040 42827 5303
E-Mail:
hcu@energieforschungsverbund.hamburg

Ausbau und Erhöhung der Energieeffizienz des Energiesystems der Leuphana Universität Lüneburg

Nikolai Strodel[1*]; Oliver Opel[1], Karl F. Werner[1], Wolfgang Ruck[1]

[1]Leuphana Universität Lüneburg, Nachhaltigkeitsforschung Energie, Scharnhorststraße 1, 21335 Lüneburg
*strodel@leuphana.de , +494131 / 677-2087

Nachhaltigkeitsforschung Energie

Das Energiesystem der Leuphana Universität Lüneburg und das neue Zentralgebäude

Der Lüneburger Universitäts-Campus im Gebiet Scharnhorstraße und Bockelsberg wird sowohl städtebaulich als auch energetisch weiterentwickelt [1]. Das Areal wird über ein Nahwärmesystem mit BHKW-Abwärme versorgt. Ergänzend zu der Umstellung auf Biomethanbefeuerung der BHKW's im Jahr 2013 wird das Energiesystem zunehmend auf erneuerbare Energien umgestellt und exergetisch optimiert. Im Jahr 2014 wurden 99 % der elektrischen Energie und 47 % der abgenommenen Wärmeenergie aus erneuerbaren Energieträgern bereitgestellt. Photovoltaik-Module mit einer Fläche von mehr als 3500 m² und einer installierten Leistung von 650 kWp decken 20 % des jährlichen Strombedarfs ab. Die Modul-Ausrichtungen nach Ost und West gewährleisten dem Campusleben entsprechend hohe Strahlungseinträge von früh morgens bis abends. Die Eigenstromquote innerhalb des eigenen Campus-Niedrigspannungsnetzes beträgt planungsgemäß etwa 90 %.

Funktionaler Bestandteil wird das Niedrigexergiesystem des neuen Zentralbaus (siehe Abbildung 1). Das Konzept sieht eine stufenweise Wärmenutzung auf verschiedenen Temperaturniveaus vor.

Abbildung 1: Neuer Zentralbau der Universität mit Kältebedarfsreduktion durch selbstverschattende Fassade und intelligentes Glas

Durch eine der Netz- und Bedarfsstruktur angepasste flexible Wärmeversorgung auf Mittel- und Niedertemperaturniveau, wird die Rücklaufauskühlung des Campussystems unterstützt und es können Temperaturen bis 25°C genutzt werden.

Der Endenergiebedarf wird bei einem für öffentliche Gebäude hervorragenden spezifischen Wert von unter 100 kWh/m²a liegen. Das Gebäude wird Anfang 2017 in das Nahwärmesystem integriert und in Betrieb genommen. Ein neuartiges Monitoringkonzept sowie ein innovatives Gebäude-Nutzer-Kommunikationssystem („ambient intelligence" und „sustainable ambient computing") sind gegenwärtig in der Entwicklung [2].

Für die Betriebsanalysen werden Betriebszustände (z.B. Nutzung, Komfort, Stand-by), Energieflüsse im Gebäude und Daten der Gebäudeautomation (z.B. Systemtemperaturen, Betriebszustände, Ventilstellungen etc.) in hoher zeitlicher Auflösung erfasst. Die Daten werden ausgewertet, mit den zuvor festgelegten Planungszielen, sofern vorhanden, verglichen und analysiert.

Flexibilisierung des Energiesystems durch Integration von (Wärme-)Speicherkonzepten

Das Nahwärmenetz besteht aus den Abnehmern Wohngebiet Bockelsberg, Campus und Zentralgebäude (ab 2017). Versorgt werden die Abnehmer neben den zwei Biomethan-BHKW's zusätzlich durch zwei Spitzenlastkessel. Der jährliche Gesamtwärmebedarf beträgt derzeit etwa 10-12 GWh. Zentraler Bestandteil der Energieeffizienzsteigerung ist die Integration neuer Speicherkonzepte zur Flexibilisierung des Nahwärmenetzes und zur Erhöhung des KWK-Anteils. Bislang sind Wärmeerzeuger und Wärmeabnehmer ausschließlich über Heißwasser-Kurzzeitspeicher gepuffert.

Derzeit wird an einem Konzept gearbeitet, welches zusätzlich eine Mittel- und Langzeitspeicherung der Wärmeenergie vorsieht.

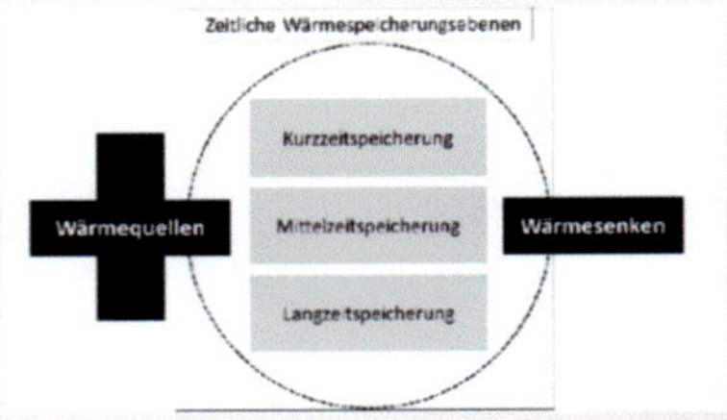

Abbildung 2: geplante zeitliche Wärmespeicherungsebenen zur Entkopplung von Wärmequellen und Wärmesenken

Mittelzeitspeicherung durch Thermochemische Wärmespeicherung:

Die thermochemische Wärmespeicherung nutzt die Reaktionsenthalpie reversibler chemischer Reaktionen. Dabei verläuft eine Reaktionsrichtung exotherm unter Wärmeabgabe (Entlademodus) und eine Reaktionsrichtung endotherm unter Wärmeaufnahme (Laden). Das Arbeitsmedium kann beispielsweise Wasserdampf sein (vgl. Abbildung 3). Vorteile dieser Technologie ergeben sich aus hohen spezifischen Wärmekapazitäten, verlustarmer Wärmespeicherung und Anwendung in breiten Temperaturspektren. An der Leuphana wurden in den letzten Jahren über 100 Salze zur thermochemischen Wärmespeicherung im Nieder- und Hochtemperaturbereich untersucht [3,4].

Vielversprechendste Materialien:

Niedertemperaturbereich bis 60°C	⇨	Mischung aus $CaCl_2/MgCl_2$
Mittel- bzw. Hochtemperaturbereich bis über 400°C	⇨	$Mg(OH)_2$ bzw. $Ca(OH)_2$

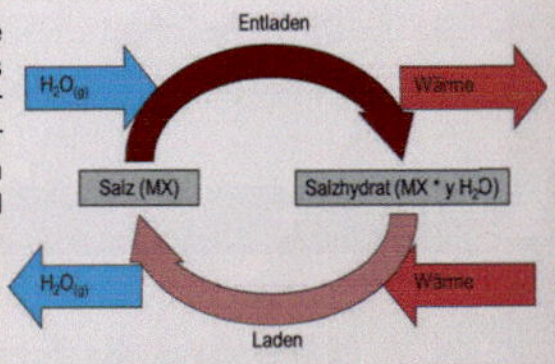

Abbildung 3: Prinzip der Wärmespeicherung durch reversible chemische Reaktionen

Langzeitspeicherung durch die Nutzung eines Aquifers in 500m Tiefe:

Aquiferspeicher nutzt natürliche Grundwasser führende Schichten, sog. Aquifere, um Wärme ein- bzw. auszuspeichern. Dabei fungieren die Brunnen den Jahreszeiten entsprechend abwechselnd als Förder- bzw. Schluckbrunnen. In Deutschland eignen sich etwa 70 % des Untergrunds zur Nutzung als Aquiferspeicher [5]. Obgleich dieses Potentials bestehen in Deutschland derzeit nur drei in Betrieb befindliche Speichersysteme [6].

In einer knapp dreijährigen Machbarkeitsstudie für den Universitätsstandort Lüneburg konnten gute hydro-geologische Randbedingungen nachgewiesen werden. Da das Abnehmersystem (Wärmesenken) im Verhältnis zur Kapazität des Speichers jedoch ungünstig ist [7], werden Erweiterungs- und Integrationsmöglichkeiten auf Gesamtsystemebene untersucht. In Abbildung 4 sind unter der Prämisse eines Aquiferwärmespeichers fünf Ansätze zur Energieeffizienzerhöhung gekennzeichnet, die in einer Gesamtsystemsimulation mit TRNSYS untersucht und in einem Variantenvergleich energetisch und wirtschaftlich bewertet werden.

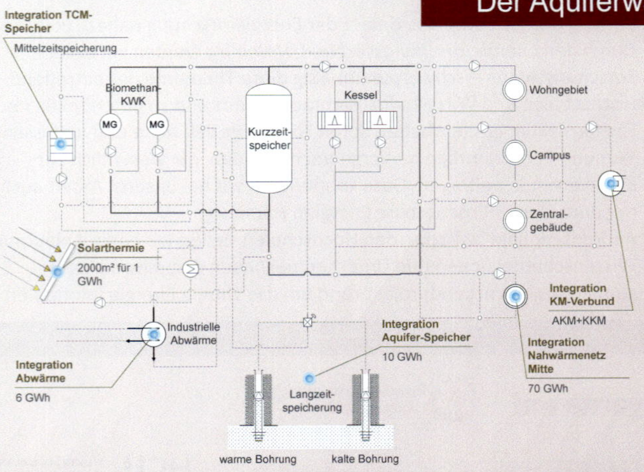

Abbildung 4: vereinfachtes Energiesystem mit Optionen zur Energieeffizienzsteigerung und verbesserten Einbindung des Aquifers

Der Aquiferwärmespeicher als Basis- und Schlüsseltechnologie

Eigenschaften des untersuchten Nutzhorizontes

- Teufe (m. u. Gel.): 450-485
- Lithologie: Fein-, Mittelsand, kalkig
- Schichtwassertemperatur: 25 °C
- Kapazität > 10 GWh
- Max. Einspeichertemperatur: 90 °C

→ Direkte Entladung möglich

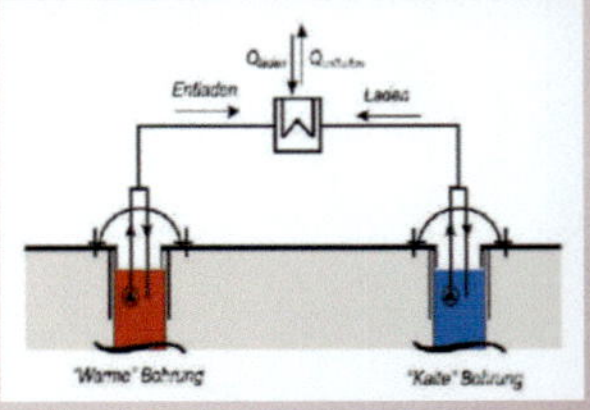

Abbildung 5: Darstellung der Funktionsweise einer Bohrungsdublette für einen Aquiferspeicher nach [8]

Ist-Zustand des Energiesystems

- **100 % der elektrischen Energie aus erneuerbaren Energieträgern**
- **75 % der Wärmeenergie aus erneuerbaren Energieträgern (Campus)**
- **95 % der in den PV-Modulen generierten Energie wird selbst genutzt**
- **20 % der verbrauchten elektrischen Energie wird selbst erzeugt (Campus)**

Literatur und Referenzen

[1] Strodel N, Werner K, Opel O, Ruck W. (2015): Integrales Energiekonzept für eine klimaneutrale Universität. Projekt "klimaneutraler Campus". In: Ostbayerisches Technologie-Transfer-Institut e. V.: 4. Fachforum Thermische Energiespeicher, Bd. 4. Neumarkt in der Oberpfalz, S. 89–91.
[2] Tribel A, Geffken J, Opel O. *Fostering Energy Effi-ciency in Buildings – The implementation of social media patterns as symbols in Building Management Systems' Graphical User Interfaces using Peirce's semeiosis as a communication concept.* Oldenburg: BIS-Verlag.
[3] N'Tsoukpoe KE, Schmidt T, Rammelberg HU, Watts BA, Ruck WK. A systematic multi-step screening of numerous salt hydrates for low temperature thermo-chemical energy storage. *Applied Energy* 2014;124:1–16, doi:10.1016/j.apenergy.2014.02.053.
[4] Rammelberg HU, Schmidt T, Ruck W. Hydration and dehydration of salt hydrates and hydroxides for ther-mal energy storage - kinetics and energy release. *En-ergy Procedia* 2012;30:362–9, doi:10.1016/j.egypro.2012.11.043.
[5] Seibt P. KF. *Aquifer Thermal Energy Storage-Pro-jects Implemented in Germany.* New Jersey, USA; 2006.
[6] Informationsdienst B. Statusseminar Forschung für Energieoptimiertes Bauen.
[7] Wolfgramm M, Lenz G, Bartels J, Diaz S, Hoffmann F, Kabus F, Mertens S, Kabus B. *Machbarkeitsstudie zur Bewertung der Möglichkeit der Speicherung von Abwärme einer Kraft-Wärme-Kopplung im Erdboden im Rahmen der klimaneutralen Campusentwicklung der Universität Lüneburg.* Neubrandenburg; 2014.

Betrieb und Regelung von Brennstoffzellen-systemen zur hocheffizienten Energieversorgung

PROFESSUR FÜR REGELUNGSTECHNIK

Dr.-Ing. Martin Schultze, Dipl.-Ing. Christian Hähnel und Univ.-Prof. Dr.-Ing. Joachim Horn

HELMUT SCHMIDT UNIVERSITÄT
Universität der Bundeswehr Hamburg

Dem Polymer-Elektrolyt-Membran (PEM)-Brennstoffzellensystem werden Wasserstoff H_2 und Sauerstoff O_2 aus komprimierter Luft (79% Stickstoff N_2, 21% O_2) zugeführt. Die PEM-Brennstoffzelle wandelt die chemische Energie von H_2 elektrochemisch in elektrische Energie und Wärme. Die Reaktion entzieht der Luft den Sauerstoff, bildet Wasser H_2O und liefert feuchte, sauerstoffarme Abluft. Ein Brennstoffzellensystem ermöglicht eine schadstoffarme und effiziente elektrische Energieversorgung.

Eine zentrale Komponente ist der PEM-Brennstoffzellenstack, der aus einzelnen Brennstoffzellen zusammengesetzt ist. Eine Gasverteilungsstruktur sorgt für eine gleichmäßige Gasversorgung von Anode und Kathode, an der die H_2-Oxidation und die O_2-Reduktion stattfinden. Die Membran trennt die Reaktionsräume von Anode und Kathode voneinander. Sie ist für Protonen und Wasser durchlässig aber nicht elektrisch leitfähig. Der Elektronenfluss erfolgt über den externen elektrischen Kreis.

Brennstoffzellenprüfstand der Professur für Regelungstechnik auf dem Gelände der Helmut-Schmidt-Universität

Brennstoffzellenprüfstand mit PEM-Brennstoffzellenstack, DC/DC-Hochsetzsteller und elektrischer Last

Die Professur für Regelungstechnik entwickelt Regelungs- und Betriebsstrategien für PEM-Brennstoffzellensysteme. Zur Anpassung und Verifikation der Simulationsmodelle für die modellbasierte Entwicklung als auch zur Integration und den Test der Strategien steht ein Brennstoffzellenprüfstand mit dem 4,4kW PEM-Brennstoffzellenstack PM200 von Proton Motor und einer regelbaren elektrischen Last zur Verfügung. Der Prüfstand umfasst weiterhin einen DC/DC-Hochsetzsteller von Zahn Electronics Inc. für einen Spannungsbereich von 48-60V am Ausgang mit einem nominellen Wirkungsgrad von 96%.

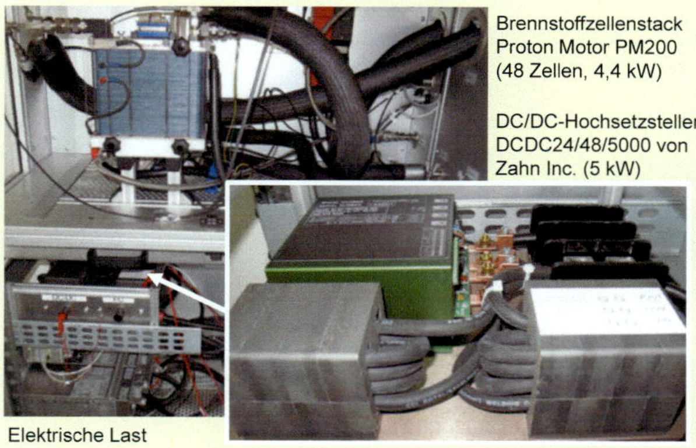

PEM-Brennstoffzellenstack mit DC/DC-Hochsetzsteller (Wirkungsgrad 96% bei 5 kW el. Leistung, Schaltfrequenz 31 kHz) und regelbarer Last

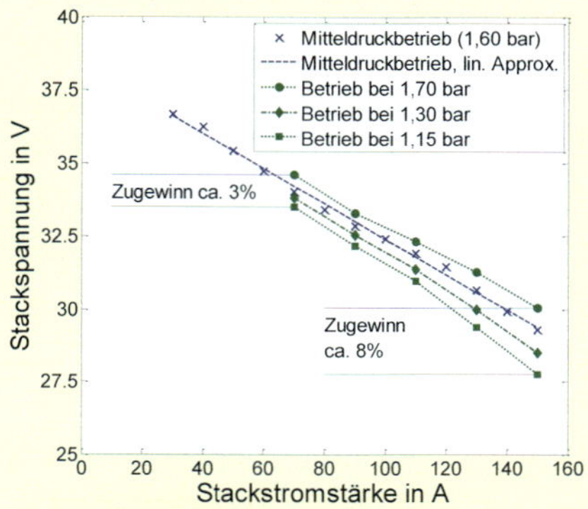

Abhängigkeit der Polarisationskurven des PEM-Brennstoffzellenstacks vom Kathodendruck

Forschungsthemen der Professur für Regelungstechnik zu PEM-Brennstoffzellen

Modellprädiktive Regelung der Luft- und Brennstoffversorgung

Entwicklung Modellprädiktiver Regelungen für den Kathoden- und Anodendruck zur Erlangung eines effizienten Betriebspunkts des Brennstoffzellenstacks mit der Bedingung einer geringstmöglichen mechanischen Belastung der Polymermembran.

Bereitstellung sauerstoffarmer Abluft

Entwicklung von Regelungs- und Betriebsstrategien zur Bereitstellung sauerstoffarmer Abluft mit einem geringen Sauerstoffanteil z.B. zur Anwendung in Flugzeugen zur Inertisierung von Treibstofftanks und Frachträumen.

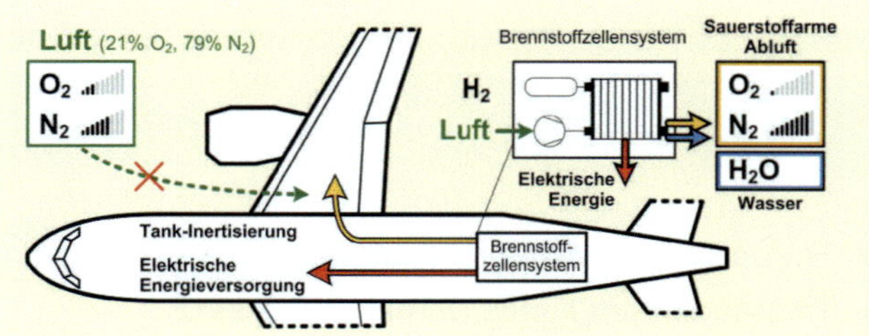

Lernfähige Betriebsstrategien

Abweichungen zwischen Brennstoffzellensystem und dem Modell (z.B. durch Alterung) können durch lernfähige Strategien kompensiert werden. Zum An- und Abfahren des Brennstoffzellenstacks werden lernfähige Betriebsstrategien entwickelt.

Hardware-in-the-Loop Simulation

Kopplung des Brennstoffzellensystems mit der Simulation eines Anwendungsszenarios (hier: Spannungswandler, elektrisches Netz, Speicher und Verbraucher) und Entwicklung von Betriebs- und Regelungsstrategien für das Gesamtsystem.

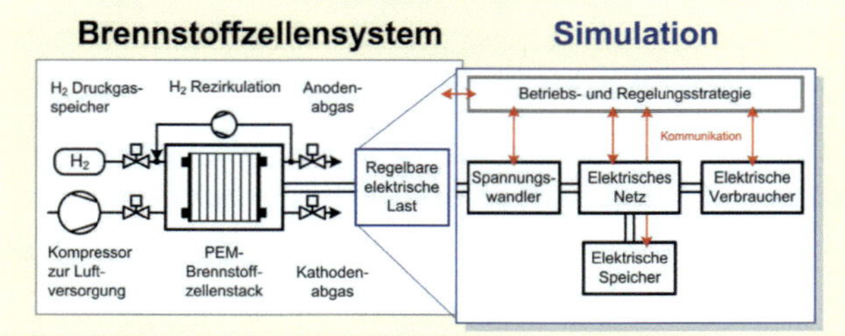

Ausblick

Erweiterung um elektrische Speicher

Das bestehende Brennstoffzellensystem wird um einen elektrischen Speicher (z.B. Akkumulator oder Kondensator) zum Anschluss an den DC/DC-Hochsetzsteller erweitert. Ziel ist der Entwurf und die Integration von Betriebs- und Regelungsstrategien des Gesamtsystems.

Erweiterung der Hardware-in-the-Loop Simulation

Die Hardware-in-the-Loop Simulation des Brennstoffzellensystems wird um weitere Anwendungsszenarien (z.B. Energieversorgung, Automobil) ergänzt.

Helmut Schmidt Universität, Professur für Automatisierungstechnik
Centrum Wiskunde & Informatica, Intelligent Systems

Bottom-Up Demand Response durch freiwillige Lastverschiebung auf Endkundenbasis

M.S.,Dipl.-Ing. Tobias Linnenberg[1], Dr. Michael Kaisers[2], Prof. Dr.-Ing. Alexander Fay[1]

1. Motivation

Kommende Herausforderungen der Energiewende

- Wachsender Anteil erneuerbarer Energien
  - Entkopplung von Erzeugung und Verbrauch
- 30% elektrischer Verbraucher in Privathaushalten
  - 60% davon flexible Lasten
- Privathaushalte müssen an Demandside Management teilnehmen
  - Finanzielle Anreize fehlen
- Elektrizität muss Commodity-Status ablegen
  - Soziale Integration des Nutzers
  - Added Value Services etablieren

2. OEEX-Konzept

Traditionelle Grenzen überkommen:

- Endkunden direkten Handel miteinander ermöglichen
  - Über Grenzen der EVUs hinweg
- EVUs behalten Kundenstamm
  - Kunden können Vertrag behalten
- OEEX Plattform dient der Vernetzung

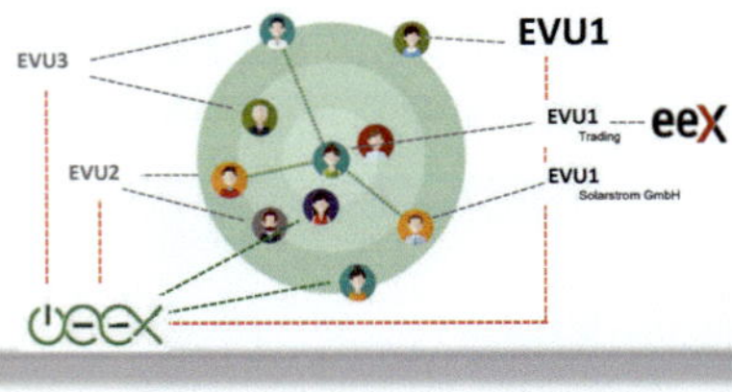

3. Beziehungen etablieren – P2P Handel ermöglichen - Enabler

App:

- Nutzerkommunikation mittels App
- Darstellung der umliegenden EEG-Anlagen
- Soziale Profile der Anlagenbetreiber hinterlegt
- Einbindung sozialer Netzwerke wie Facebook & co.
- Verständliche Statistiken zu CO_2 Einsparungen
- Plug & Play Hardware

Smart Plug:

- Messen & Schalten
- Mittels WiFi mit OEEX Cloud verbunden

OEEX Cloud:

- Future Internet-Ware basiert
- Verbindet Nutzer & Maschinen
- Optimiert Strombezug & rechnet ab

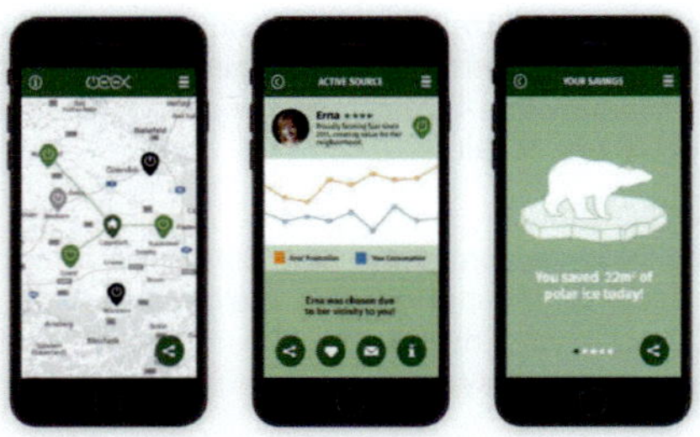

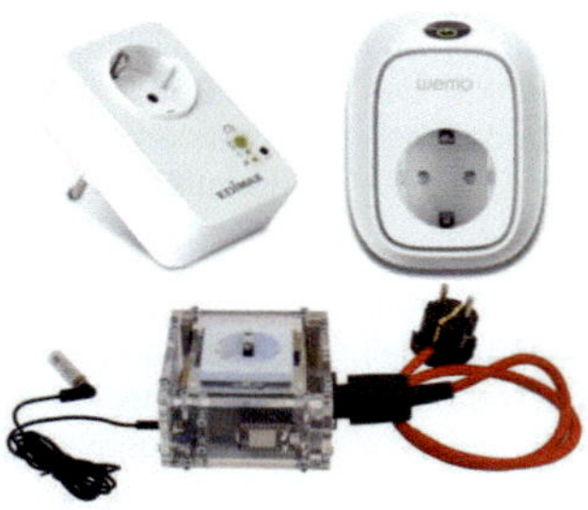

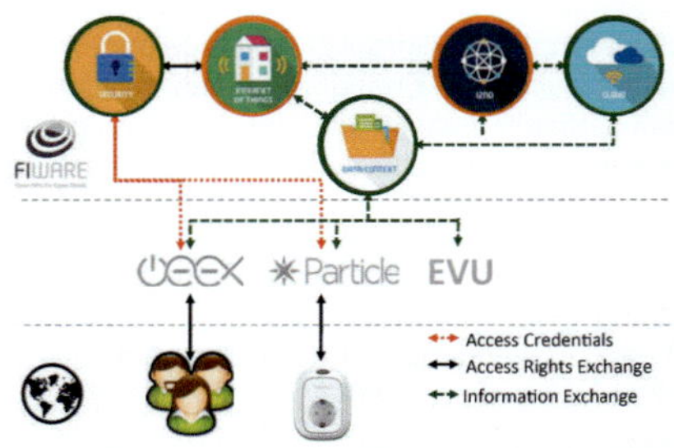

4. Anwendungsfälle

Vorteile für Energieversorgungsunternehmen:

- Demandside Management für EVUs und Netzbetreiber
- Portfolio Optimierung für EVUs
- Transparenz der Stromherkunft für Endkunden
- Transparenz des Kundenverhaltens für EVUs

Relevante Veröffentlichungen

Linnenberg, T., Fay, A., Kaisers, M.: Bottom-Up Demand Response by Following Local Energy Generation Voluntarily.
In: Proceedings of Twenty-Ninth AAAI Conference on Artifical Intelligence 2015

Linnenberg, T., Kaisers, M., Fay, A.: Open Energy Exchange – A new way of social energy networking.
In: Proceedings of Smarter Europe 2015

Visit us on: www.oeex.org

[1]Institut für Automatisierungstechnik
Helmut-Schmidt-Universität
Holstenhofweg 85, 22043 Hamburg
E-Mail: {tobias.linnenberg, alexander.fay}@hsu-hh.de

[2]Centrum Wiskunde & Informatika
Sciencepark 123, 1098 XG Amsterdam, Niederlande
E-Mail: m.kaisers@cwi.nl

Entwicklung eines Messgerätes zur Bestimmung der zeit- und frequenzabhängigen Netzimpedanz auf der Mittelspannungsebene

HELMUT SCHMIDT UNIVERSITÄT
Universität der Bundeswehr Hamburg

Hauke Langkowski, Michael Jordan, Trung Do Thanh, Detlef Schulz

Helmut-Schmidt-Universität, Holstenhofweg 85, 22043 Hamburg, Tel. +49 40 6541-2905, Fax +49 40 6541-3083
Hauke.Langkowski@hsu-hh.de, Michael.Jordan@hsu-hh.de, Trung.Dothanh@hsu-hh.de, Detlef.Schulz@hsu-hh.de

Kurzfassung

Der stetig steigende Einsatz von erneuerbaren Energien mit leistungselektronischer Netzkopplung hat zunehmend Einfluss auf die Spannungsqualität. Die von diesen Anlagen in das Netz eingespeisten Oberschwingungsströme führen in Abhängigkeit von der Netzimpedanz zu unerwünschten Oberschwingungsspannungen. Die Netzimpedanz ist deshalb für die Beurteilung der entstehenden Netzrückwirkungen von großem Interesse. Das neuartige Messgerät erlaubt die Bestimmung der zeit- und frequenzabhängigen Netzimpedanz auf der Mittelspannungsebene. Es kann zur Bewertung der möglichen Einspeiseleistung verschiedener Netzanschlussknoten und zur verbesserten Filterauslegung herangezogen werden.

Messverfahren

- Netzanregung durch Takten eines Hochleistungswiderstands mit einem IGBT-Schalter (14 x 4,5 kV Module)
- Messung von Strom und Spannung im Leerlauf- und Lastfall
- Anwendung der FFT zur Bestimmung des Frequenzverlaufs

$$\underline{Z}_{\mathrm{N}} = R_{\mathrm{N}} + j\omega L_{\mathrm{N}} = \frac{\Delta \underline{U}}{-\Delta \underline{I}} = \frac{\underline{U}_1 - \underline{U}_2}{\underline{I}_2 - \underline{I}_1} = \frac{\underline{U}_1 - \underline{U}_2}{\underline{I}_2}$$

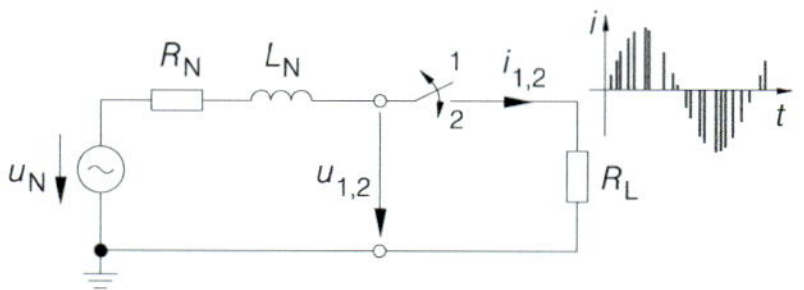

Einphasige Bestimmung der Netzimpedanz

- Bestimmung der dreiphasigen Netzimpedanz auf der Mittelspannungsebene durch asynchrones Schalten (zyklisches Verschalten und Takten von je zwei Außenleitern durch eine B6-Thyristorbrücke)

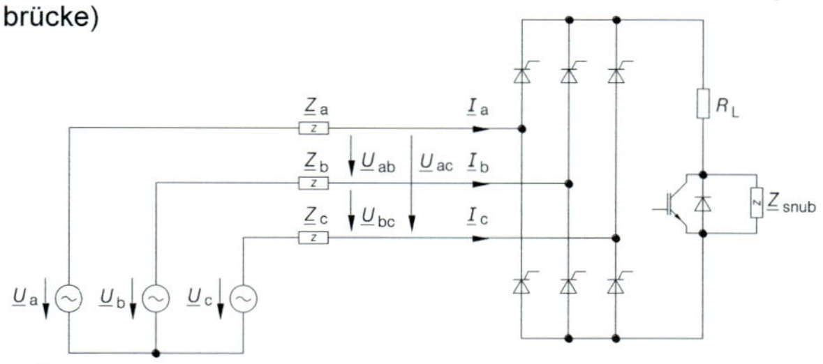

Bestimmung der Netzimpedanz in einem Dreileitersystem

$$\begin{pmatrix} \underline{Z}_a \\ \underline{Z}_b \\ \underline{Z}_c \end{pmatrix} = \frac{1}{2} \begin{pmatrix} \frac{1}{\underline{I}_{ab,2}} & -\frac{1}{\underline{I}_{bc,2}} & \frac{1}{\underline{I}_{ac,2}} \\ \frac{1}{\underline{I}_{ab,2}} & \frac{1}{\underline{I}_{bc,2}} & -\frac{1}{\underline{I}_{ac,2}} \\ -\frac{1}{\underline{I}_{ab,2}} & \frac{1}{\underline{I}_{bc,2}} & \frac{1}{\underline{I}_{ac,2}} \end{pmatrix} \cdot \begin{pmatrix} \Delta \underline{U}_{ab} \\ \Delta \underline{U}_{bc} \\ \Delta \underline{U}_{ac} \end{pmatrix}$$

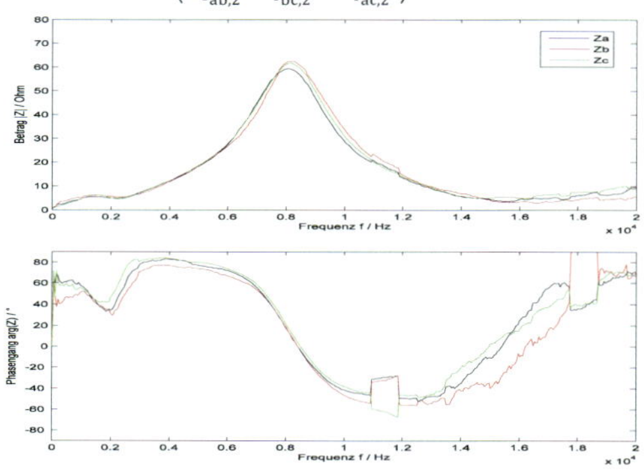

Betrag und Phase der Netzimpedanzen gemessen am 10 kV Anschluss der HSU

Messcontainer

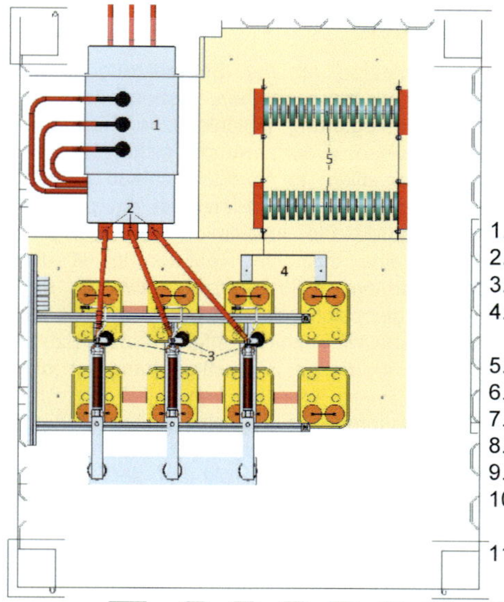

1. Leistungsschalter
2. Stromsensoren
3. Spannungssensoren
4. Verbindungsschienen
5. Widerstände
6. Leistungselektronik
7. Sicherungsgestell
8. Messtechnik
9. Kabelzuführung
10. Überspannungsableiter
11. Klimaanlage

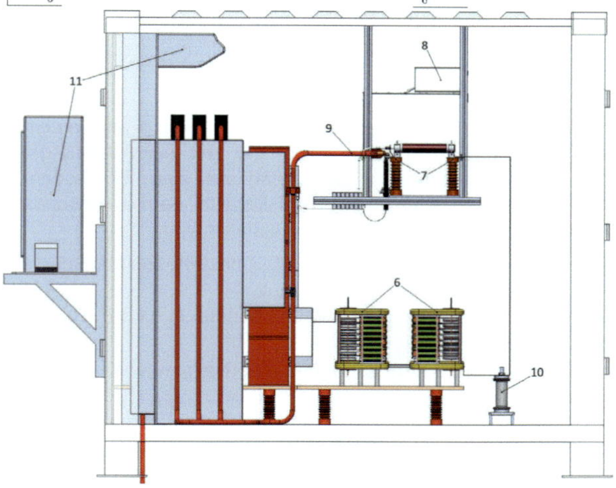

Aufbau des mobilen Messcontainers

- Typgeprüfter Messcontainer ausgelegt für 20 kV
- Leistungsschalter zum Netzanschluss
- Hochgenaue Strom-/Spannungswandler zur Auswertung der Netzanregungen

Patentanmeldungen:

- Vorrichtung und Verfahren zur Messung der Netzimpedanz auf der Mittelspannungsebene, DE 10 2006 050 766 A1 Offenlegungsschrift 30.04.2008, Deutsche Patentanmeldung 2006
- Vorrichtung zur Messung der Netzimpedanz des elektrischen Versorgungsnetzes, DE 10 2010 018 996.0, Deutsche Patentanmeldung 2010

Danksagung

Dieses Projekt „Entwicklung eines Messgerätes zur Bestimmung der zeit- und frequenzabhängigen Netzimpedanz auf der Mittelspannungsebene" wird unter dem Förderkennzeichen 0325049 vom Bundesministerium für Umwelt, Naturschutz und Reaktorsicherheit gefördert.

Entwicklung eines Messgerätes zur Bestimmung der frequenzabhängigen Netzimpedanz auf der Hochspannungsebene bis 110 kV zur Bewertung der Verfügbarkeit von Netzkapazitäten als Systemgröße zur Dimensionierung von Energiespeichern

HELMUT SCHMIDT UNIVERSITÄT
Universität der Bundeswehr Hamburg

Hauke Langkowski, Michael Jordan, Trung Do Thanh, Detlef Schulz

Helmut-Schmidt-Universität, Holstenhofweg 85, 22043 Hamburg, Tel. +49 40 6541-2905, Fax +49 40 6541-3083
Hauke.Langkowski@hsu-hh.de, Michael.Jordan@hsu-hh.de, Trung.Dothanh@hsu-hh.de, Detlef.Schulz@hsu-hh.de

Kurzfassung

Der Anteil erneuerbarer Energien an der Stromerzeugung in Deutschland wird in Zukunft weiter zunehmen, insbesondere im Rahmen der von der Bundesregierung beschlossenen Energiewende. Um das Gelingen dieser Energiewende sicherzustellen müssen große Windparks vor allem in Norddeutschland in das Energieversorgungsnetz integriert werden. Hierzu ist ein parallel stattfindender Netzausbau sowie die Installation von Energiespeichern erforderlich. Für eine effiziente Realisierung ist die genaue Vermessung des bestehenden Stromnetzes notwendig, um die Standorte für nötige Energiespeicher festzulegen und diese zu dimensionieren sowie um Ausbaumaßnahmen effektiv umzusetzen. Die systemrelevante Größe in diesem Zusammenhang ist die Netzimpedanz.

Im Rahmen dieses Forschungsprojektes wird eine transportfähige Anlage entwickelt, die es erlaubt die zeit- und frequenzabhängige Netzimpedanz von Netzanschlusspunkten direkt auf der 110-kV-Hochspannungsebene zu bestimmen. Mit den objektiven Messergebnissen kann die Bewertung von Netzrückwirkungen und die Bestimmung der Anschlusskapazität vorgenommen werden. Auf der Grundlage der Vermessungen von mehreren Netzanschlusspunkten kann im Anschluss die Dimensionierung von benötigten Energiespeichern erfolgen.

Messverfahren

Das Messverfahren beruht auf einer Netzanregung mit einem ohmschen Hochleistungslastwiderstand, der mit einer leistungselektronischen Anregeschaltung bestehend aus einer B6-Diodenbrücke mit einem nachgeschalteten IGBT-Schalter dem elektrischen Energieversorgungsnetz pulsförmig zugeschaltet wird.

- Frequenzabhängige Netzimpedanz $\underline{Z}_N(\omega)$ besteht aus ohmschen Anteil $R_N(\omega)$ und induktiven Anteil $L_N(\omega)$
- Messung von Strom und Spannung im Leerlauf- ($\underline{I}_1$ und $\underline{U}_1$) und Lastfall ($\underline{I}_2$ und $\underline{U}_2$)
- Anwendung der FFT zur Bestimmung des Frequenzverlaufs

$$\underline{Z}_N(\omega) = R_N(\omega) + j\omega L_N(\omega) = \frac{\underline{U}_1(\omega) - \underline{U}_2(\omega)}{\underline{I}_2(\omega) - \underline{I}_1(\omega)} = \frac{\Delta\underline{U}(\omega)}{-\Delta\underline{I}(\omega)}$$

Auf der Hochspannungsebene sind im Allgemeinen Dreileiternetze mit isoliert betriebenen Sternpunkten installiert. Ziel ist die Messung der Frequenzverläufe der Impedanzen der Außenleiter. Aus mehreren Messungen in zeitlich vorgegebenen Abständen können unterschiedliche Lastzustände und verschiedene Netzverschaltungen erfasst werden.

- Bestimmung der Netzimpedanzen der einzelnen Außenleiter auf der Hochspannungsebene durch asynchrones Schalten (zyklisches Verschalten und Takten von je zwei Außenleitern)
- Leistungsschalter kann Phasenselektion für Messprinzip umsetzen

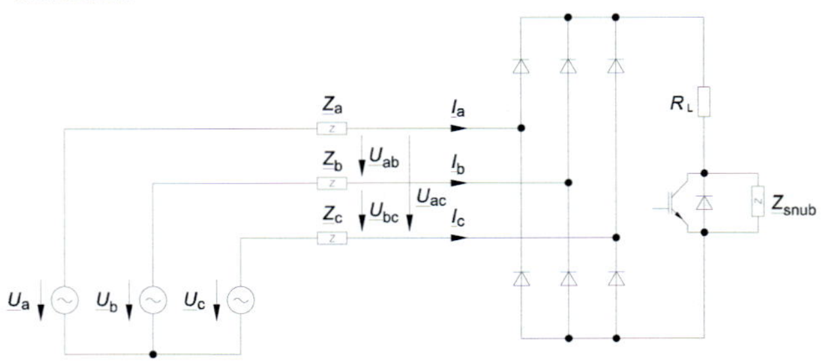

Bestimmung der Netzimpedanzen auf der Hochspannungsebene

Messcontainer

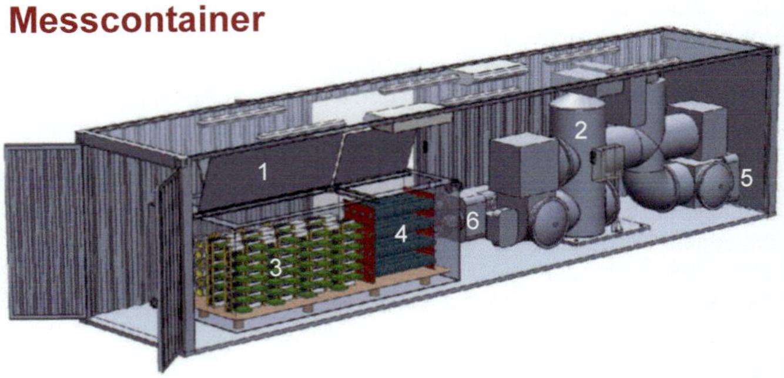

Aufbau des mobilen Messcontainers (Abmessungen: 12,2 x 2,5 x 2,7 m)

1	Tank mit Ester als Isoliermedium	**4**	Hochleistungslastwiderstände
2	110-kV-Leistungsschalter	**5**	Netzanschluss
3	Leistungselektronik	**6**	Hochspannungskabel

Die beiden Hauptkomponenten innerhalb des mobilen Messcontainers stellen der gasisolierte Leistungsschalter sowie der mit Ester gefüllte Spezialtank dar. Aufgrund der hohen Spannung ist ein luftisolierter Aufbau nicht innerhalb eines kompakten Messcontainer realisierbar.

- Leistungsschalter mit SF6 zur Netzzuschaltung; Auslegung der Druckentlastung des Leistungsschalters auf Kurzschlusswerte von $I_k = 31{,}5\ kA$ bei $T_k = 1\ s$
- Ester als Isolationsmedium im Spezialtank mit im Vergleich zu Öl besseren Umwelt- und Sicherheitseigenschaften
- Verbindung der Hauptkomponenten über ein Hochspannungskabel mit Hochspannungssteckern
- Bemessungsspannung der Leistungselektronik: 240 kV
- Bemessungsstrom der Leistungselektronik: 200 A
- Anregeleistung von bis zu 22 MW
- 6 Stacks à 12 Dioden in Reihenschaltung bei jedem Halbbrückenmodul (insgesamt 432 im Messcontainer)
- 15 Stacks à 6 IGBTs in Reihenschaltung mit Spannungsversorgung aus den Netzspannungen und LWL-Ansteuerung; Active-Clamping und Metalloxid-Varistoren zum Überspannungsschutz
- Stromwandlern zu Mess- und Schutzzwecken, induktiver und hochgenauer breitbandiger RC-Teiler zur Spannungsmessung
- Klimaanlage zur Wärmeabfuhr im Messbetrieb und –zufuhr bei geringen Außentemperaturen
- Mess- und Ansteuerelektronik

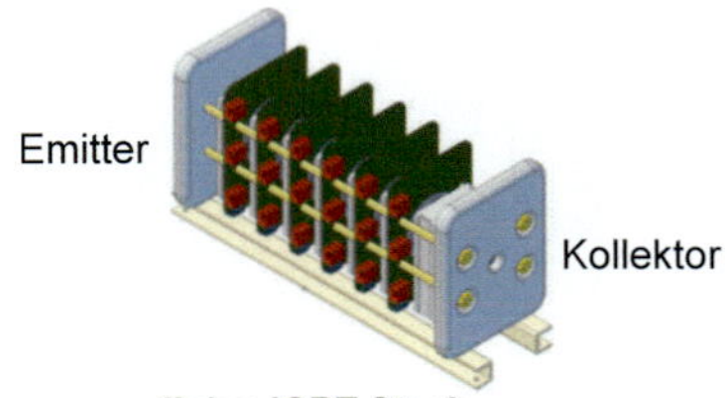

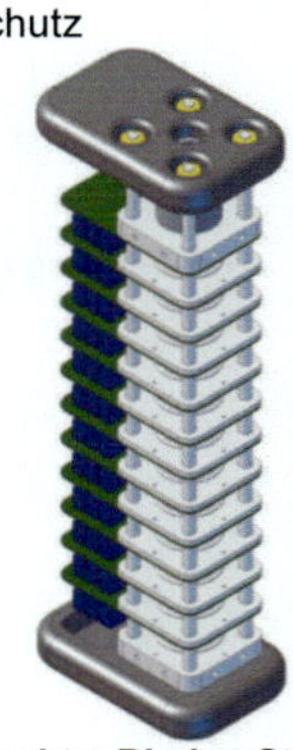

links: IGBT-Stack **rechts: Dioden-Stack**

Danksagung

Dieses Projekt „Entwicklung eines Messgerätes zur Bestimmung der frequenzabhängigen Netzimpedanz auf der Hochspannungsebene bis 110 kV zur Bewertung der Verfügbarkeit von Netzkapazitäten als Systemgröße zur Dimensionierung von Energiespeichern “ wird unter dem Förderkennzeichen 0325562 vom *Bundesministerium für Wirtschaft und Energie* gefördert. Die beiden Projektpartner Stromnetz Hamburg GmbH und Astrol Electronic AG leisten nicht nur durch eine enge Zusammenarbeit und technische Unterstützung, sondern auch durch finanzielle Eigenmittel einen entscheidenden Beitrag zum Gelingen des Forschungsprojektes.

Using Already Existing Artificial Structures for Energy Storage in Areas with High Shares of Renewable Energies

Thomas Weiss, Detlef Schulz

Helmut-Schmidt-University, Holstenhofweg 85, 22043 Hamburg, Tel. +49 40 6541-2163, Fax +49 40 6541-3083
Thomas.Weiss@hsu-hh.de, Detlef.Schulz@hsu-hh.de

HELMUT SCHMIDT UNIVERSITÄT
Universität der Bundeswehr Hamburg

Abstract

Energy Storage is one of the key technologies when facing a future power supply based on renewable energies. Especially in rural areas a strong penetration of fluctuating energy from renewables can stress the transmission and distribution system. Storage Technologies can improve the grid stability by taking the surplus of energy produced by e.g. wind or solar power and provide it at times of shortages. The amplitude of the power that has to be transmitted can be lowered as well. Especially already existing artificial structures like former open cast mining structures and German waterways with its locks and canal lifts can be used for a cost effective and decentralized way of storing energy.

Approaches

Pumped Hydro Energy Storage (PHES) is today the only technology that is concretely available and in a relevant scale. Due to limited natural sites for the installation of PHES other approaches have to be considered.

Long term storage

Technology	Actual costs in€ct/kWh	Future costs in €ct/kWh
Hydrogen	23	9 (in >10 years)
CAES	38	22 (in >10 years)
PHES	3 to 11	Not considered

Short term storage

Technology	Actual costs in €ct/kWh	Future costs in €ct/kWh
Hydrogen	24	11 (in >10 years)
CAES	6	3 (in >10 years)
PHES	2,5 to 5,5	Not considered

PHES in Open Cast Mining Structures

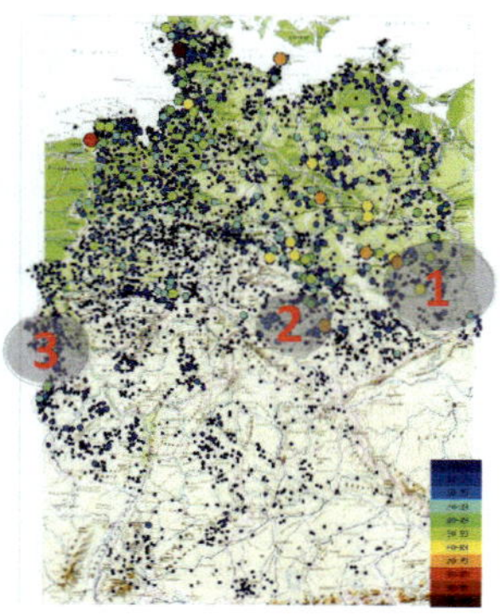

Installed wind power (Source: ISET, IWET, 2010) and areas of open cast mining structures in Germany marked with numbers 1 to 3
1: Lusatian mining district
2: Central German mining district
3: Rhine mining district

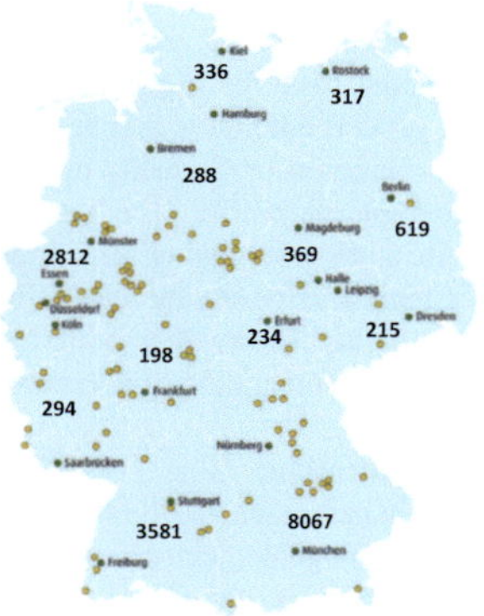

German chalk stone mining areas (source BV Kalk, May 2011) and installed PV power in bold numbers (source SolarServer, December 2011)

Central German and Lusatian mining district have a total of 68 mining lakes with

- A total surface of 26,518 ha
- A total volume of 4,511 M m^3

Reckoning a tenth of the lakes with a mean fall height of 80m, this incorporates a storage capacity of **79,5 GWh**

Using the Federal Waterways as Energy Storage

The canal system in the north of Germany has a total length of 1300 km and includes 45 locks and 3 canal lifts. Each of the locks or canal lifts is equipped with pumps to allow navigation upwards as well as downwards. For downward navigation the water is just released in the following canal or river. The surpluses of water due to rainfall or natural inflow are released directly in a lower part of the system as well. Therefore it seems likely to modify the pumps or install turbines to use this water flow.

Overview of the canal system in the north of Germany

Lock/ canal lift	Height difference in m	Boundaries in cm	Installed pumps in m^3/s	Storable Energy in kWh
Scharnebeck	38	20	23.00	50,097
Uelzen	23	20	28.00	54,401
Anderten	14.7	20	9.60	34,769
Suelfeld	9	20	24.00	21,287
Rothensee	16	40	17.50	63771

Future investigation with EnERgioN project: Creating a virtual power plant in the north of Germany with the canal system as an energy storage facility.

Acknolegdment

The authors gratefully acknowledge the contributions of the EU founded project stoRE (www.store-project.eu), Contract-Nr.: IEE/10/222. stoRE aims at creating the right regulatory and market conditions for an increased development of energy storage infrastructure to the extent necessary for the accommodation of the planned renewable energy installations to the electricity grid.

Department of Electrical Engineering
Electrical Power Systems
Prof. Dr.-Ing. habil. Detlef Schulz
http://evhs.hsu-hh.de

Potential of new pumped-storage power plants in open cast mining structures and comparison of the storage capability with electro-mobility

HELMUT SCHMIDT UNIVERSITÄT
Universität der Bundeswehr Hamburg

Thomas Weiß, Detlef Schulz

Helmut Schmidt University, Holstenhofweg 85, 22043 Hamburg, Tel. +49 40 6541-2163, Fax +49 40 6541-3083
Thomas.Weiss@hsu-hh.de, Detlef.Schulz@hsu-hh.de

Abstract

In this contribution the use of open cast mining structures as pumped-storage power plants (PSPP) is introduced. After coal production has been finished quarries with big volumes still remain in open cast mining areas. In some of these mining lakes PSPP could be installed. This new application offers a few technical, organizational and economical advantages. The potential of PSPP for some selected open cast mining areas is outlined and compared with the storage capacity of battery-driven electric vehicles.

Problem description

- Increasing power from renewable energy plants requires future energy storage technology
- Central storage units like compressed-air storage plants (CASP) and PSPP are preferred since they are simple to handle in controlling the grid
- The realization potential of both technology in Germany is limited
  - CASP need the highly competitive used air-tight caverns
  - Due to geographical profile, costly approval procedures, acceptance problems and high demands for natural protection new PSPP are hardly to realise

Pumped storage power plants in open cast mining structures

- Using open cast mining structures for installing PSPP could solve the problems above
- Coal layer quarries with depths between 120 m and 300 m remain after coal production

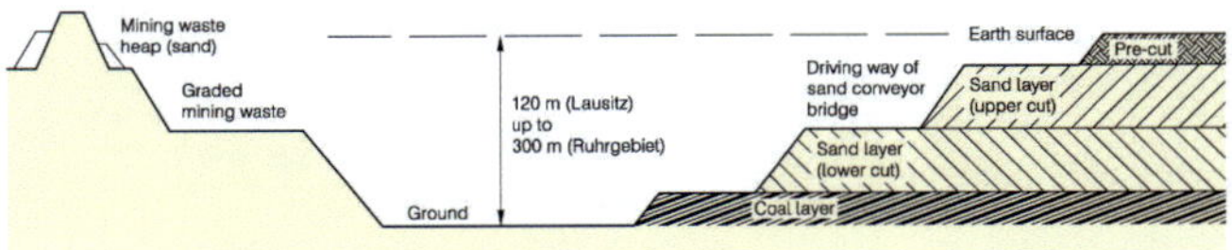

- One side of the coal-mine is natural ground and the other side consists of mining waste
- Due to settle and flow processes the mining waste side is geological instable
- The natural ground should be used for the installation of components
- Natural ground side is characterized by the pre-cut and the two cut levels of the conveyor bridge
- The coal-mine is used as the lower storage lake
- The upper storage lake and a dam must be built

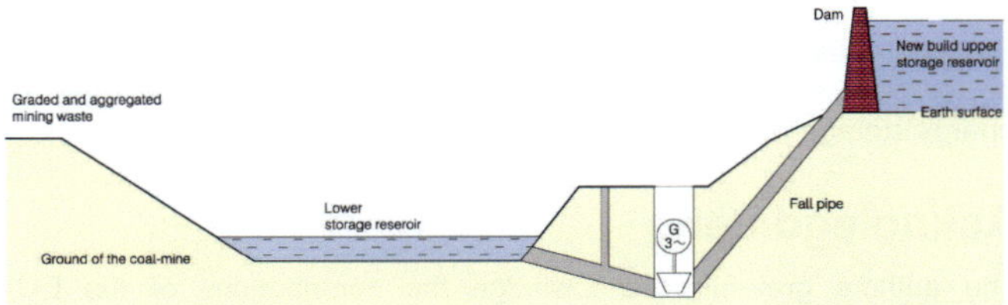

- The fall pipe and the machine house with the motor/generator and pump/turbine units are installed between upper and lower lake.
- The main advantage of this setup is the existing lower storage lake
- Bisection of the costs is expected compared to conventional PSPP.
- The installation could be a part of the recultivation-process
- Higher environmental acceptability
- Supporting the local population in mainly underdeveloped regions due to long term local business tax

Technical potential

Investigation were focused on the open cast mining areas Lausitzer Revier (federal state of Brandenburg) and Mitteldeutsches Revier (between federal state of Saxony and Saxony-Anhalt)

- Wind energy share per capita 37 % for Saxony-Anhalt and 30 % for Brandenburg in 2008
- Only 10 % of the coal-mining quarries in these areas could provide a storage capacity of nearly 80 GWh in case of a supposed average fall height of 80 m
- Compared to the installed wind energy power in Germany of 24 GW in 2009, the full power of all wind turbines could be stored over more than three hours.

Comparison storage capacity of electro-mobility with future needs

Example for the future storage capacity of electric cars in Berlin:

- Assuming that 75 % of the second car portfolio of 200,000 cars deliver power to the grid.
- With a maximum of 5 kW over 30 minutes per car the energy amount results to 375 MWh.

➢ Compared to the installed onshore and expected offshore wind power in the future of 44 to 52 GW, the storage capacity of 1000 cities with the size of Berlin with a complete portfolio of an electrical second car portfolio would be needed.

Peak load example:

- The yearly peak load in Germany is about of 77.8 GW
- Assume that 50 % of the electrical power is delivered by renewable sources until 2030 and 10 % of the yearly peak load have to be stored over one week, the needed energy amount E is:

$$E_{10\,\%} = 77.8\,\text{GW} \cdot 0.1 \cdot 7 \cdot 24\,\text{h} = 1307\,\text{GWh}$$

➢ The energy storage of electro-mobility is in the short-time range and is not suitable for the bridging of time periods over some days

Conclusion

Comparing the technical properties of suitable technologies, pumped-storage power plants have the highest potential because of its total efficiency of 81%. A general opinion is that the capacity of pumped-storage power plants is fully exhausted in Germany due to the complicate permission rules and the geographical profile. The proposed installation of new water storage plants in open cast mining structures could overcome this limitation and deliver high energy storage capabilities in the future. A comparison with electro-mobility shows that both technologies should be applied for an optimised grid integration of decentralised power generation.

Patent Application:
Building, in particular structural design of buildings for pumped storage power plants in geologically unstable formations for energy recovery of abandoned quarries of coal and ore mining, DE 10 2008 002 801 A1, published application 17.09.2009, German Patent Application 2008

Fakultät für Elektrotechnik
Elektrische Energiesysteme
Prof. Dr.-Ing. habil. Detlef Schulz
http://evhs.hsu-hh.de

Forschungsschwerpunkt Nachhaltige Energieversorgung

Demonstratoren

Demonstrator zur interdisziplinären Bewertung von virtuellen Kraftwerken

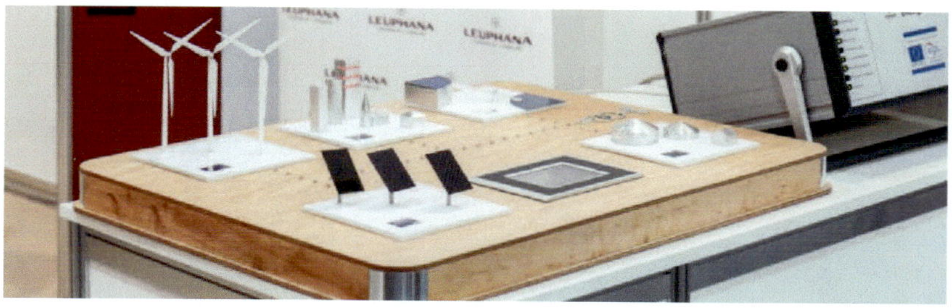

Das Kompetenztandem EnERgioN (Erneuerbare Energien in der Region Nord) ging der Frage nach, wie sich regionaler Strom aus erneuerbaren Energien besser in der Region speichern und verteilen lässt. Dafür prüfte das interdisziplinäre Forschungsteam Geschäftsansätze zu virtuellen Kraftwerken, die kleine, dezentrale und häufig von Privatpersonen betriebene Stromerzeuger zusammenschalten sollten. Die virtuellen Kraftwerke verknüpften Versorgungsunternehmen, Netzbetreiber und Konsumenten. Sie koordinierten Erzeugung, Speicherung und Verbrauch von Energie. Für die anschauliche Darstellung der Daten wurde ein Demonstrator entwickelt, der aus einer Simulationssoftware sowie aus Hardwarekomponenten zur Visualisierung der Ergebnisse bestand. Das Demonstrationsobjekt ermöglicht sowohl Fachpublikum als auch Laien einen anschaulichen Zugang zur Funktionalität eines virtuellen Kraftwerks.

Das System besteht aus 5 modularen Erzeugungs-, Speicher- und Verbrauchseinheiten, die über ein Kommunikationsnetzwerk miteinander verbunden sind. Das Modul, dass die Abbildung der zu untersuchenden Verbrauchseinheiten repräsentiert wird in Form einer abstrahierten Stadt dargestellt (siehe Abbildung 1). In ähnlicher Form dienen weitere abstrahierte Modelle der anschaulichen Darstellung einzelner Module des virtuellen Kraftwerks. Die Funktion als Testsystem wird durch eine eigenständige Stromversorgung sowie einen getrennten Signalübertragungspfad gewährleistet.

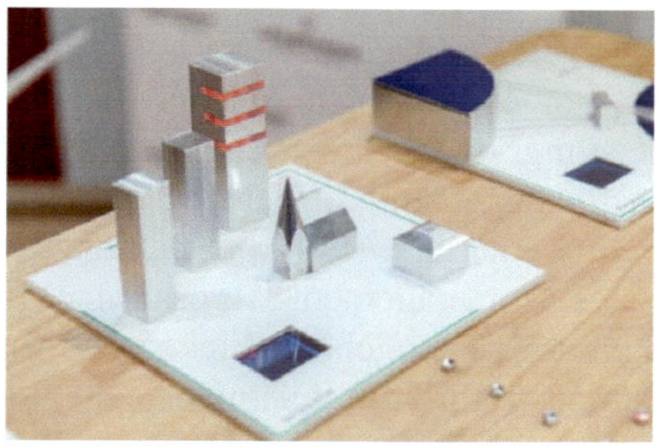

Abbildung 1: Moduls zur Darstellung der Verbraucher mittels einer virtuellen Stadt

Abbildung 2: Operation Terminal zur zentralen Steuerung des virtuellen Kraftwerks

Das zentrale Steuerungs- und Überwachungssystem wurde ergänzend mit analogen sowie digitalen Steuerungsmodulen ausgestattet. Als weitere Steuerungseinheit wurde ein Operation Terminal integriert. Dieses ist in Abbildung *2* dargestellt. Es kann durch manuelle Bedienung an die jeweiligen Anforderungen individuell angepasst werden. Gleichzeitig ermöglicht es dem Benutzer, eine Rückmeldung über die vorgenommenen Handlungen und Veränderungen in Echtzeit zu bekommen. Die Steuerungen sind über Ethernet miteinander verbunden und führen einen ständigen Datenabgleich durch.

Innovative elektrische Flugzeugkabinennetze für eine optimierte Bordnetzversorgung

Demonstrators für Lastmanagement-Systeme

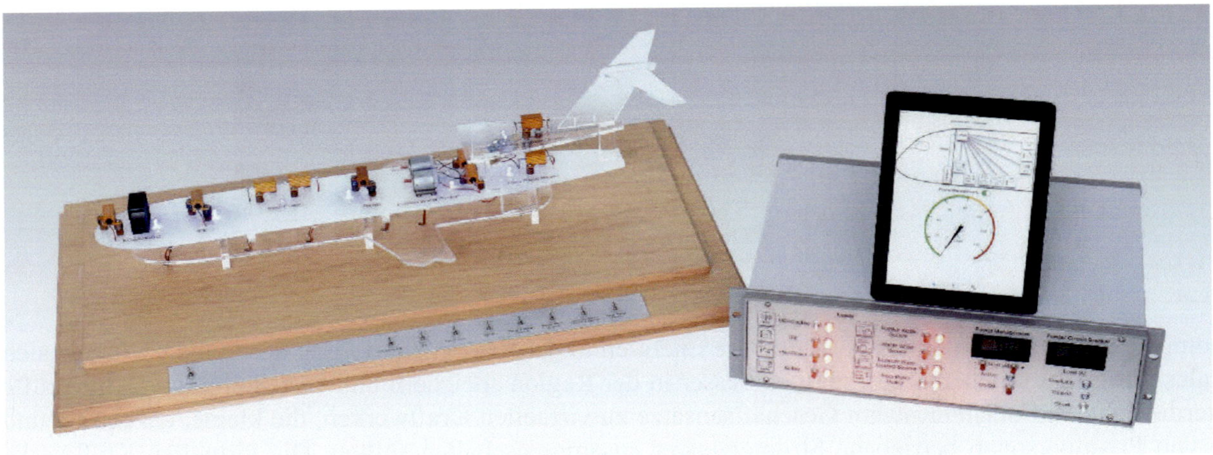

Ziel der untersuchten Aufgabenstellung ist eine Optimierung der Kabinenenergieversorgung in modernen Verkehrsflugzeugen. Dabei liegen die Schwerpunkte auf dem Einsatz hoher Gleichspannungen zur Versorgung der Kabinenlasten und auf der Entwicklung intelligenter Lastmanagement-Systeme und Algorithmen. Ein Lastmanagement verkleinert die Auslegungsgrundlage des betreffenden Energieversorgungssystems. Dadurch wird eine Gewichtseinsparung bei den Erzeugern und den Leitungen (Feedern) möglich.

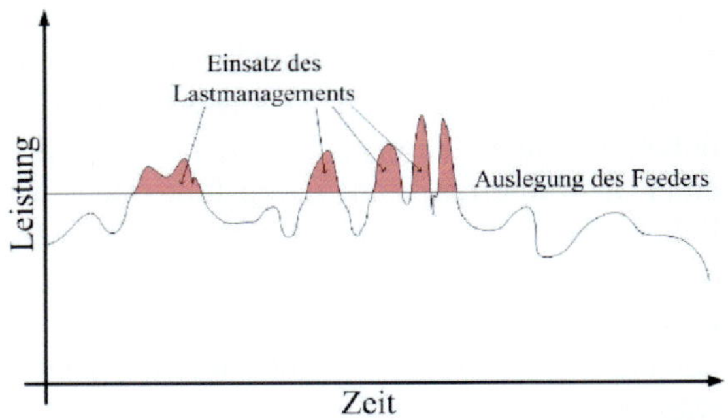

Die Abbildung zeigt den Einsatz eines Lastmanagements. Die rot eingefärbten Flächen liegen über der zulässigen Leistung und werden abgeregelt. Wichtig ist, dass ein Lastmanagement „unsichtbar" arbeitet, also vom Passagier nicht bemerkt wird. Zur Demonstration und Evaluation von Lastmanagement-Algorithmen für Flugzeuge entwickelte das Fachgebiet „Elektrische Energiesysteme" einen Demonstrator für Lastmanagement-Systeme. Dieser ist vollständig in einen 19-Zoll-Einschub integriert und beinhaltet die Spannungsversorgung, die Schalteinrichtungen, die Messwandler sowie Steuer- und Regelungshardware. Die Regelalgorithmen sind auf einer industriellen speicherprogrammierbaren Steuerung (SPS) implementiert. Eine MATLAB/SIMULINK-Softwareanbindung der SPS ermöglicht die Entwicklung und Implementierung neuer Algorithmen.

Im Zusammenhang mit der Demonstrator-Entwicklung werden eigene Applikationen für Tablet-PCs erstellt. Mit diesen Applikationen können Demonstratoren gesteuert und überwacht werden. Für das Projekt wurde eine eigene Schnittstelle zwischen dem Hardware-in-the-Loop-System und einem iPad entwickelt. Mithilfe dieser Applikationen und einem iPad wird eine einfache Steuerung und Präsentation verschiedener Funktionen der Demonstratoren ermöglicht.

Innovative elektrische Flugzeugkabinennetze für eine optimierte Bordnetzversorgung

Demonstrator moderne elektrische Kabinenarchitektur

In diesem Teilprojekt werden innovative Architekturkonzepte zur Technologiereife entwickelt. Dabei werden unterschiedliche elektrische Bordnetzarchitekturen anhand konventioneller Flugzeugkabinennetze untersucht. Die Abbildung zeigt den schematischen Aufbau im Kabinen- und Frachtbereich eines traditionellen Bordenetzes (links) und eines Architekturkonzeptes mit höherer Gleichspannung (HVDC) (rechts).

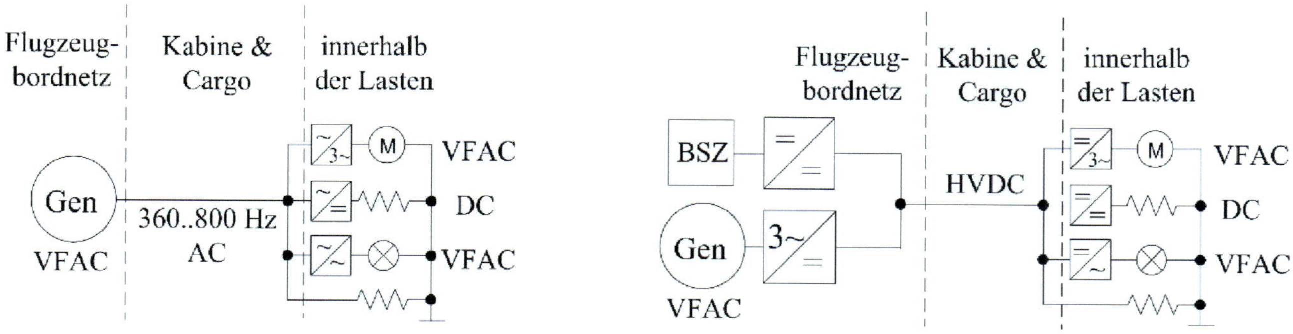

Ein Forschungsschwerpunkt stellt die Gewichtsoptimierung der Verkabelung in der Kabine sowie im Frachtraum dar. Darüber hinaus werden Szenarien im Zusammenhang mit der Einführung von Faserverbundstoffen (CFK) und Hybrid-Werkstoffen analysiert.

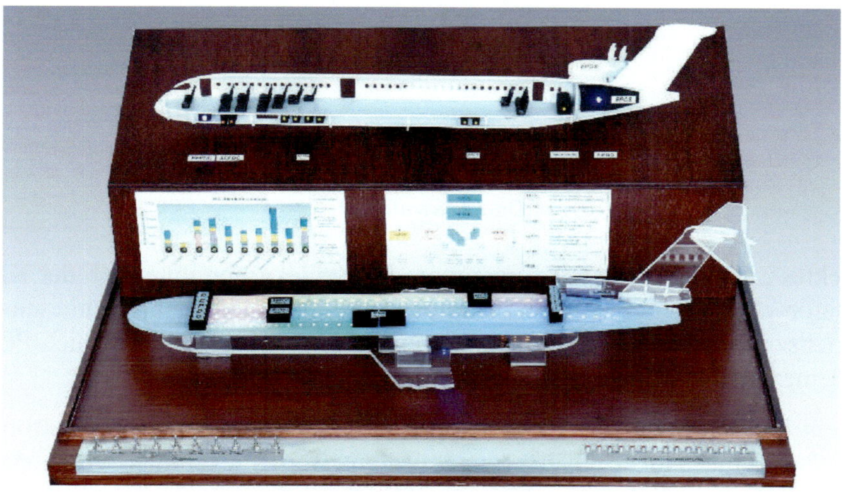

Der Demonstrator zeigt ein optimiertes Flugzeugkabinennetz eines Mittelstreckenflugzeugs sowie die Leistungsanforderung der einzelnen Flugphasen zwischen Start und Ziel der Reiseroute. Die Hauptverteilung des Energiever-sorgungsnetzes ist im PEPDC (Primary Power Distribution Center) gebündelt. Über Power-Feeder werden Unterverteiler (SPDBs – Secondary Power Distribution Boxes) mit Spannung versorgt. Die Lasten wiederum sind an den Unterverteilern angeschlossen und werden dort abgesichert. Dabei ergeben sich folgende Optimierungsansätze:

- Finden einer gewichtsoptimierten Verteilerstruktur
- Erhöhung der Versorgungsspannung
- Optimale Positionierung der Unterverteiler im Flugzeug
- Einsatz von gewichtsoptimierten Kabelbündeln
- Gewichtsoptimierte Verteilung des erlaubten Spannungsabfalls auf Power- und Lastverkabelung
- Schutzkleinspannungskonvertierung innerhalb der SPDB

Demonstrator: Autarke Notstromversorgung der Bevölkerung Unterhalb der KRITIS-Schwelle

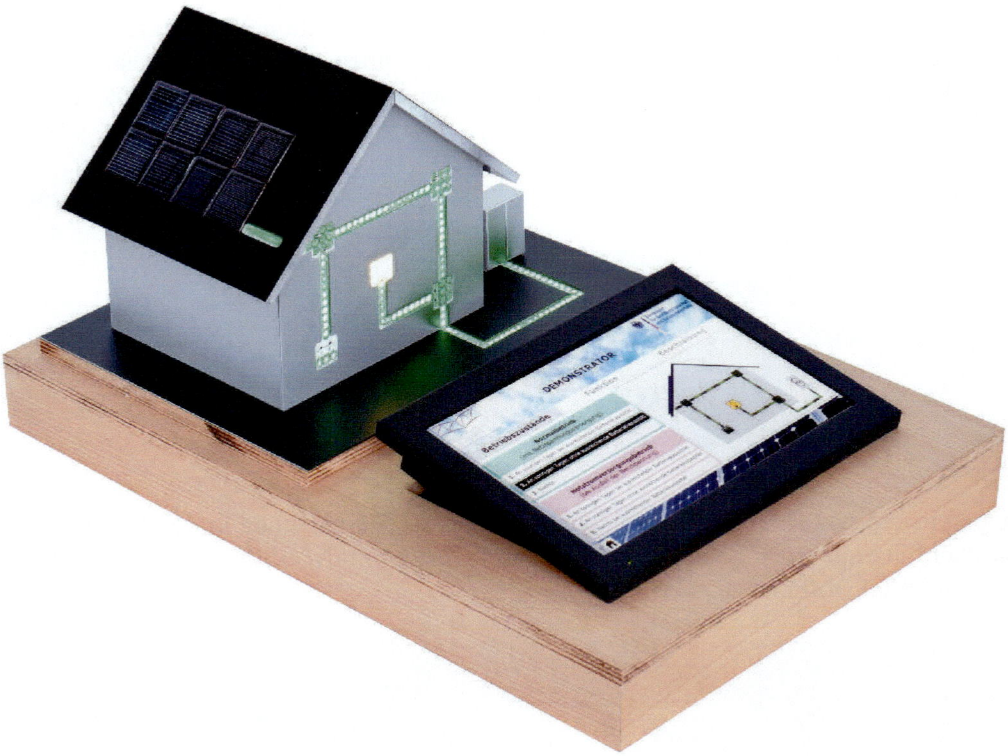

Demonstrator

Im Rahmen des Projektes „Autarke Notstromversorgung der Bevölkerung unterhalb der KRITIS-Schwelle“ wurde ein Demonstrator entworfen. Der Demonstrator besteht aus einem abstrahierten Hausmodell zur Darstellung der Betriebszustände der Notstromversorgung, einem eingebauten Touch-PC zur Bedienung sowie der darauf implementierten Software.

Der Demonstrator dient der anschaulichen Darstellung der Funktionsweise eines Notstromversorgungssystems. Die Betriebszustände eines Notstromversorgungssystems werden exemplarisch an einem Einfamilienhaus gezeigt. In ähnlicher Weise können auch Mehrfamilienhäuser und sogar ganze Siedlungen mit den hier vorgestellten Notstromversorgungstypen ausgestattet werden.

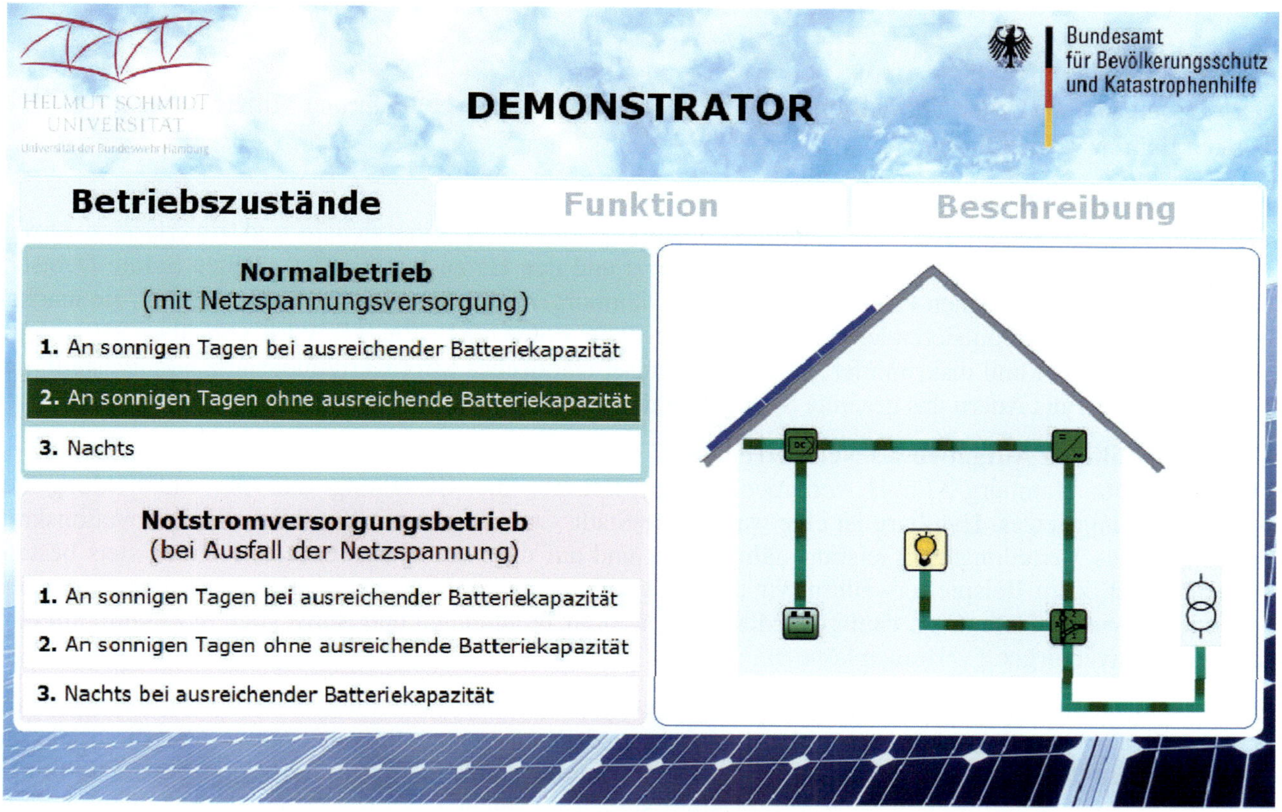

Beispiel eines Betriebszustands

Projektergebnisse

Die Studie hat gezeigt, dass neben Photovoltaik-Inselanlagen und Energiespeichersystemen zur Realisierung von Notstromversorgungssystemen auch weitere Technologien einsetzbar sind wie z. B. Batteriesysteme, Brennstoffzellensysteme, Windkraftanlagen, Blockheizkraftwerke, Dieselgeneratoren oder manuell betriebene Generatoren. All diese Technologien weisen jedoch spezifische Vor- und Nachteile bezüglich der betrachteten Bewertungskriterien (Stromausfalldauer, Preis, Leistung, Umwelteinfluss, Handhabung und Genehmigungsaufwand) auf. Nichtsdestotrotz lassen sich ihre Potentiale durch optimale Einsatzbedingungen (Verbraucher/Erzeuger-Abhängigkeit oder szenariobedingte Erzeugungstechnologie) erhöhen.

Exkursion am 10.09.2015 zur Stromnetz Hamburg GmbH: Unser Verteilungsnetz für Hamburg

Gero Boomgaarden (Leiter Netzbetrieb), Claus-Peter Siemens (Leiter Netzführung Mittelspannung), Torsten Behrens (Netzführung)

Versorgungsqualität mit Tradition

Seit mehr als 120 Jahren steht die Stromnetz Hamburg GmbH für eine sichere und zuverlässige Stromversorgung und damit für die Menschen, die Industrie und den Hafen in einer der größten Städte Deutschlands. Die rund 1,8 Millionen Hamburgerinnen und Hamburger liegen uns besonders am Herzen, sie machen unsere Stadt zu einer pulsierenden Metropole. Die dafür notwendige Energie wird von über 350 Stromanbietern angeboten und diskriminierungsfrei über unser städtisches Stromverteilungsnetz bereitgestellt, das wie ein Geflecht aus Adern das gesamte Stadtgebiet durchzieht.

Unsere vielfältigen Aufgaben als Netzbetreiberin

Die Stromnetz Hamburg GmbH verantwortet den Betrieb, die Gestaltung und die Entwicklung des Stromverteilungsnetzes. Hamburg ist eine wachsende Stadt – wir sorgen mit Weitblick und Innovationskraft dafür, dass das Verteilungsnetz leistungsfähig bleibt und auf die Bedürfnisse unserer Kunden stets bestens vorbereitet ist. Zum Beispiel erweitern wir unser Verteilungsnetz in dem neu geplanten und erschlossenen Stadtviertel wie die HafenCity, damit die Menschen in ihren Wohnungen und Büros auch in Zukunft auf eine sichere Stromversorgung vertrauen können.

Sollte es doch einmal zu einer Störung kommen, beseitigen wir diese schnell und effektiv: Unsere hohen Ansprüche an die Versorgungsqualität schlagen sich in niedrigen Stromausfallzeiten unter dem Bundesdurchschnitt nieder. Darüber hinaus ermöglichen wir schnell und sicher den Anschluss und den diskriminierungsfreien Zugang zu unserem Netz. Stromzähler stellen wir nicht nur zur Verfügung, wir kümmern uns auch um den eichrechtlichen einwandfreien Betrieb und die Ablesung.

Im Interesse der Stadt und ihrer Bürger

Wir verstehen uns im wahrsten Sinne des Wortes als Dienstleister, der seine Leistungen in den Dienst der Stadt stellt und im Interesse seiner Bürgerinnen und Bürger handelt – kompetent, wirtschaftlich und effizient. Auch der Schutz von Umwelt und Klima steht bei all unseren Aktivitäten im Fokus.

Zentrale des Verteilungsnetzes

Die Netzführung sorgt als Schaltzentrale des Hamburger Verteilungsnetzes rund um die Uhr für einen sicheren Betriebszustand. Von hier aus steuert und überwacht ein topgeschultes Team rund um die Uhr sämtliche Anlagen. Angefangen, von den Übergabepunkten des Übertragungsnetzes von 50Hertz Transmission GmbH in der Hochspannung bis zur Mittelspannung.

Hohe Versorgungsqualität

Kommt es dennoch einmal zu einer Störungen, koordiniert die Netzführung die Aktivitäten, um die betroffenen Kunden von der 0,4- bis zur 110-kV-Ebene schnellstmöglich wieder zu versorgen. Die Hamburger Haushalte waren, im Vergleich zum Bundesdurchschnitt von 15,3 Minuten, lediglich rund 11,7 Minuten ohne Strom.

Laufender Betrieb und Aufgaben im Hintergrund

Umspannwerke, Schaltanlagen, Netzstationen, Verteilerschränke und nicht zuletzt tausende Kilometer Leitungen und Kabel – redundant aufgebaut. Wer einmal den Netzplan Hamburgs in der Übersicht gesehen hat, ahnt, welche komplexen Aufgaben sich daraus für die Netzführung ergeben. Der laufende Betrieb mit der Steuerung der Anlagen ist der Kernbereich. Zusätzlich steht hinter der operativen Netzführung eine Vielzahl von Aufgaben „im Hintergrund". Hier wird unter anderem ständig an technischen und organisatorischen Möglichkeiten gearbeitet, um die Netz- und damit die Versorgungsqualität noch weiter zu verbessern.

Stromnetz Hamburg in Zahlen

Kabel (unterirdisch) 25.500 km

Freileitungen 1.344 km

Verkabelungsgrad 95 %

Anzahl Kunden bzw. Entnahmestellen/Zähler 1.330.000

Entnommene Jahresarbeit 12,6 TWh

Anzahl Umspannwerke (110 kV/10 kV) 53

Anzahl Kunden- und Netzstationen 7.500

Anzahl der in Hamburg aktiven Stromlieferanten > 350

Zukunft und Entwicklung

Neben der eigenen Modernisierung des Verteilungsnetzes und vielen Erweiterungen zum Smart Grid, gibt es viele Projekte, an denen Stromnetz Hamburg beteiligt ist und diese aktiv unterstützt, wie zum Beispiel Partner der Energiewende, Ladeinfrastruktur für Elektromobilität, intelligente Messsysteme. Mehr Informationen entnehmen Sie unserer Webseite http://www.stromnetz.hamburg/.

Unter www.energieportal-hamburg.de finden Sie viele Informationen rund um den aktuellen Stromverbrauch und der Erzeugung aus regenerativer Energie in Hamburg.

Autorenverzeichnis

MIX
Papier aus verantwortungsvollen Quellen
Paper from responsible sources
FSC® C105338

FSC
www.fsc.org

If you have any concerns about our products,
you can contact us on
ProductSafety@springernature.com

In case Publisher is established outside the EU,
the EU authorized representative is:
Springer Nature Customer Service Center GmbH
Europaplatz 3, 69115 Heidelberg, Germany

Printed by Libri Plureos GmbH
in Hamburg, Germany